CW01149628

ISBN 978-1-5281-2659-5
PIBN 10897724

FB &c Ltd, Dalton House, 60 Windsor Avenue, London, SW19 2RR.
Company number 08720141. Registered in England and Wales.

For support please visit www.forgottenbooks.com

Professor James
Columbia College New York
from Colonel Sabine

MAGNETICAL AND METEOROLOGICAL OBSERVATIONS

AT

LAKE ATHABASCA AND FORT SIMPSON,

AND AT

FORT CONFIDENCE.

MAGNETICAL AND METEOROLOGICAL

OBSERVATIONS

AT

LAKE ATHABASCA AND FORT SIMPSON,

By CAPTAIN J. H. LEFROY,

ROYAL ARTILLERY;

AND AT

FORT CONFIDENCE,

IN GREAT BEAR LAKE,

By SIR JOHN RICHARDSON, C.B., M.D.

PRINTED BY ORDER OF HER MAJESTY'S GOVERNMENT.

LONDON:

PUBLISHED FOR HER MAJESTY'S STATIONERY OFFICE,

BY

LONGMAN, BROWN, GREEN, AND LONGMANS.

1855.

CONTENTS OF THIS VOLUME.

PREFACE

BY COLONEL SABINE, R.A.

THE observations of Admiral Löwenorn, in 1786, at Reikiavik in Iceland, confirmed by Lottin in 1836, and those made by myself in 1823 at Fairhaven in Spitzbergen, also confirmed by the observations of the "Commission du Nord" at Magdalena Bay in Spitzbergen, in 1839, showed that, in the high magnetic latitudes of the northern hemisphere, the horary variation of the magnetic declination is subject to wide differences in respect of the turning hours, and the direction of the movement at the same hours of local time, from the phænomena which in the middle latitudes of the same hemisphere are found to prevail generally, and, with very slight modifications, in all meridians. The progress which, since the results of the magnetic observatories established in the last few years have been known and discussed, has been made towards the physical explanation of many of the magnetic phænomena, renders it desirable that facts which at first sight, and to minds accustomed to the comparative regularity of the diurnal variation elsewhere, have somewhat the aspect of anomalies, should be more extensively investigated and better understood. The differences which they present from the ordinary march of the phænomena are far too considerable and too consistent to be ascribed to accident: they are obviously specialities; and the particular laws which govern them will no doubt ultimately be found to be consistent with, and to form, in fact, a part of, the general laws by which the diurnal variation in all parts of the globe shall be comprehended.

But the parts of the globe where such observations can be made are little frequented, and are difficult of access; and the observations cannot be effectively made without considerable sacrifices of personal convenience. The Magnetic Survey of the British Possessions in North America—undertaken by Her Majesty's Government at the recommendation of the Royal Society, and executed by Captain Lefroy, of the Royal Artillery—and the expedition in search of Sir John Franklin and his companions, under the direction of Sir John Richardson, afforded opportunities which the zeal and public spirit of those gentlemen did not suffer to pass unimproved.

The instruments with which the observations were made were supplied from the establishment under my direction at Woolwich; and on the completion of the services, the observations were transmitted to me. On application to the Treasury, a sanction was obtained for their publication in the present form. The observations of Captain Lefroy, both magnetical and meteorological, have been arranged and discussed by himself, as have the meteorological observations of Sir John Richardson by himself; but on learning from Sir John Richardson, soon after his return, that his professional duties at Haslar would prevent him from undertaking the examination and reduction of his magnetical observations, they were placed in the hands of Captain Younghusband, of the Royal Artillery, then my assistant at Woolwich, by whom that portion of the volume has been prepared. The proof-sheets of the whole have been read and compared with the original manuscripts by the non-commissioned officers of the Royal Artillery permitted by the Master General of the Ordnance to be employed in my office for purposes of a similar nature.

EDWARD SABINE.

Woolwich, December 14th, 1854.

MAGNETICAL AND METEOROLOGICAL OBSERVATIONS

AT

LAKE ATHABASCA AND FORT SIMPSON,

Territory of the Hudson's Bay Company.

INTRODUCTION

BY CAPTAIN J. H. LEFROY, *Royal Artillery.*

THE stations of magnetical observations established in North America in the year 1840, namely, Philadelphia, Washington, and Cambridge near Boston, in the United States; Toronto in Canada; and Sitka in Russian America; might all, with the exception of the last, be comprised in a circle of little over 200 miles radius; nor were any means at that time provided for attaining a knowledge of the absolute or relative values of the magnetical elements, or of their regular and irregular changes, in the northern parts of the British possessions; a region of peculiar interest, as comprising both the focus of maximum magnetic intensity in the northern hemisphere, and the point or pole of vertical dip. It was the principal purpose of the magnetic survey of British North America, authorized by the Government in 1841, and in part executed in the years 1843 and 1844, to supply the former deficiency; but with a view also to the latter, I was provided, in addition to other instruments, with a complete set of transportable magnetometers, of the construction of Dr. Weber, as improved by Captain Riddell; and it was arranged with the authorities of the Hudson's Bay Company, that the excursion of the first summer should terminate at Moose Factory on Hudson's Bay, where it was left optional with me to pass the whole of the ensuing winter, or to return in the course of it to Canada. The employment of these instruments in the magnetical term days, and in the observation of disturbances, was in either case the special duty of the time to be so employed. On arriving at the Red River settlement, in June 1843, I found various difficulties in the way of an execution of this part of my instructions, and was led to believe that their object would be better attained by wintering at some more northern station. As Colonel Sabine, foreseeing the difficulty of precisely defining the details of a task which involved many contingencies, had kindly left me considerable discretionary latitude to be guided by circumstances, I decided on giving up the journey to Moose Factory, for that time, and selected in its place Fort Chipewyan on Lake Athabasca; not only the most northerly station which could be conveniently reached in the season, but one also whose resources would make an unexpected addition of eight persons, to the number of its occupants, a matter of no inconvenience. I reached this post with my assistant, Corporal William Henry, Royal Artillery, since Adjutant of Pensioners, on the 23d September 1843. Observations were here made every hour of the 24^h from the 16th October 1843 to the 29th February following; together with very numerous extra observations on magnetic disturbances. On the 3d March 1844 I started, in company with the same assistant, and a party of four or five servants of the Company, for Fort Simpson on Mackenzie's River; we were provided with three trainaux, each drawn by three dogs, for the conveyance of the instruments and provisions; and a cariole, to which a team of four dogs was allotted, was very kindly provided by Mr. Colin Campbell for my own use, if required. The distance, which is about 350 geographical miles in a straight line, but considerably more by the course of the Slave and Mackenzie rivers, which is the route travelled, was accomplished in twenty-one days, including one day of detention at Great Slave Lake; and without other hardship or inconvenience than that occasioned by the severity of the cold, which ranged on several occasions between 30° and 40° below zero of Fahrenheit.

Fort Chipewyan is situated in latitude 58° 43′ N., longitude 7^h 35′ 15″ W. from Greenwich, and is distant about 1,700 geographical miles from Toronto. By the exertions of Mr. Campbell,—to whose kindness, as well as to that of Mr. Lewis, the chief factor resident at Fort Simpson, and to Sir George Simpson, the Governor of the Hudson's Bay Company, I have to acknowledge the greatest obligations,—a small detached log building was erected, 18 × 13 feet in dimensions, especially for my use as an observatory; it was begun on the 27th September, and finished on the 13th October. No iron was used in the construction, it was furnished with an open fire place, and received light from three parchment windows, each having a small panel of glass, and so disposed as to throw light on the scales of the instruments, the arrangement of which is shown in the annexed diagram.

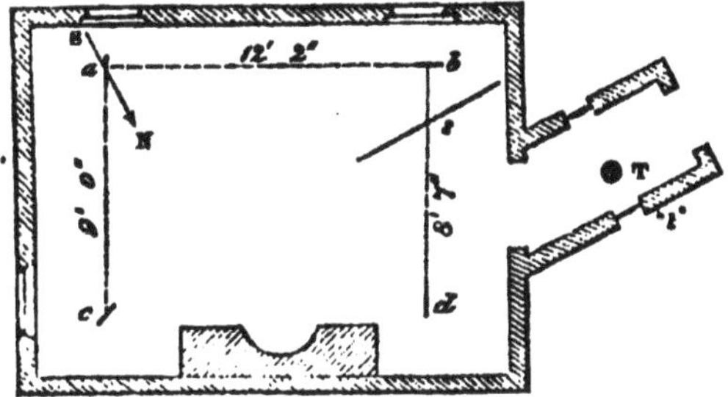

a Portable Declination Magnetometer.
b Portable Bifilar.
c Portable Induction Inclinometer.
d Second or spare Declinometer.
T Portable transit instrument.
t Thermometers.
s Screen.

The Bifilar was screened from the direct action of the fire by a leather curtain, the Inclinometer was screened by the projection of the chimney; the whole were mounted on firm wooden pillars disconnected from the floor. The internal temperature ranged from +61°·0 on 19th October, to −1°·2 on the 22d January; we have even the mean for 24^{h} as high as 52°·8 on 18th October, and as low as 15°·2 on the 8th January. The extremes of cold usually occurred on Monday morning, the room not being occupied on the Sunday.

The system of relief adopted to carry out a series of hourly observations with only one assistant, was this: A observed from 8 P.M. to midnight, and on retiring aroused B, who observed from 1 to 5 A.M.; he in turn retiring, again aroused A, who resumed the observations at 6 A.M., and so on for four hours alternately. It will not be found that the omissions are numerous, the fatigue of this system, maintained for so many months, being considered. I have much pleasure here in acknowledging the assistance rendered by Mr. T. Dyke Boucher, the junior resident of the fort, upon several occasions. I have before acknowledged the zeal and spirit with which Corporal Henry devoted himself to his laborious duties throughout the magnetic survey.

The building given up to my use at Fort Simpson as an observatory and dwelling-room, was also a detached wooden building on the north side of the principal house; which has since been removed to a point a little further back from the river. It was close to the then north-west angle of the inclosure. Care was taken to keep out of it, while occupied as a sitting-room and bed-room, all guns, axes, and utensils of iron. The annexed diagram represents the arrangement of the instruments:—

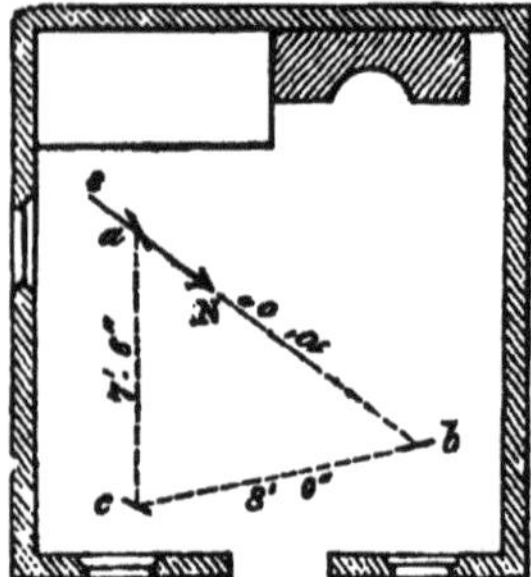

a Declinometer.
b Bifilar.
c Induction Inclinometer.

The Declination Magnet produced an effect of −0·3 div. on the scale reading of the Bifilar, and of −1·8 div. on that of the Inclinometer. The Bifilar Magnet produced an effect of +1·7 div. on the scale reading of the Inclinometer, but no sensible effect on that of the Declinometer; the effect of the Inclinometer Magnet was +0·6 div. on the Bifilar, and of −0·5 div. on the Declinometer.

Fort Simpson is situated in latitude 61° 51′·7 N., longitude 8^h 5′ 40″ W. from Greenwich, and is about 1,800 geographical miles distant from Toronto; its distance from the Russian Observatory at Sitka is about 460 miles, that of Fort Chipewyan from the same point being 780 miles.

I have endeavoured in the following pages to pursue the comparison of the phenomena observed, as far as the data admitted, through the registers of all the Magnetical Observatories in North America; reducing the results to a common unit, by means of the scale co-efficients given in the respective publications of the Observatories. As the observations at Toronto have been published since the completion of these reductions, it is necessary to observe that the scale co-efficient of the Bifilar at Toronto, here employed,—namely, $k = \cdot 0001057$ X,—was determined by an extensive series of experiments of Deflection made in 1848, in conformity with a circular of Instructions addressed at that time to Directors of Magnetical Observatories.

It has been found necessary to omit the detail of a part of the observations on Magnetical Disturbances and Term Days, for want of space.

J. H. Lefroy.

Woolwich, July 1854.

INDEX

TO THE

OBSERVATIONS AT LAKE ATHABASCA AND FORT SIMPSON.

MAGNETICAL DECLINATION.

ERRATA.

The reader is requested to make the following corrections :—

Page 18. Near the foot, *for* fourteen-inch, *read* fifteen-inch.

„ 19. Heading of Table X., *for* 14-inch, *read* 15-inch.

„ 65. Near the foot *for* $\frac{\Delta R}{R}$ *read* $\frac{\Delta \phi}{\phi}$.

„ 75. Near the top, *after* evident, *insert* that.

„ 75. Near the foot, *for* substracted, *read* subtracted.

„ 79. Line three from foot, *for* does, *read* it does.

„ 80. Heading, *insert* VIII., *after the word* Table.

„ 106. Table L., heading, *for* extremes of each instrument, &c., *read* comparison of selected days at three stations, with reference to the degree of disturbance which prevailed.

„ 140. Above the centre, *after* or Fort Simpson, *add* with two exceptions, and bracket [October 24] and [December 8].

„ 140. Below the centre, *for* all coinciding, *read* two of them coinciding, and bracket [October 19].

„ 142. Table LXV., midnight, January No. of A., *for* 7, *read* 6; at the foot, *for* 51, *read* 50; in column of total, *for* 23, *read* 22. See below, p. 169. Again, 17^h February No. of A., *for* 2, *read* 3; in column of total, *for* 12, *read* 13.

„ 148. First line, *for* Table LXVII., *read* Table LXVI.

„ 152. Before the Table, *dele* together with the values of those quantities.

„ 153. Near the top, *for* but five instances, *read* but four instances; and *dele* November 27.

„ 167. Foot-note, *for* -3°, *read* $-37^\circ\cdot 8$.

„ 169. At 18^d 20^h, the entry of Aurora at 20^h 45^m belongs to January 17^d.

„ 178. February 29^d 0^h, after the entry, *add*, *idem* 1^h.

„ 298. Second line, *for* second, *read* first.

MAGNETICAL OBSERVATIONS

AT

LAKE ATHABASCA AND FORT-SIMPSON.

ADJUSTMENTS, ABSTRACTS, AND COMMENTS.

SECTION I.

DECLINATION.

MAGNETICAL OBSERVATIONS

AT

LAKE ATHABASCA AND FORT-SIMPSON.

ADJUSTMENTS, ABSTRACTS, AND COMMENTS.

SECTION I.

DECLINATION.

ιation,
h Ameri
' and

MAGNETICAL OBSERVATIONS.

SECTION I.

MAGNETIC DECLINATION.

Declinometer, 12th October 1843.—The adjustment of this instrument consists in levelling the base, and turning the arm which carries the Telescope in azimuth, until the central division of the scale coincides with its line of collimation. This being done, the value of the ratio $\frac{H.}{F.}$, for the coefficient of torsion, was found to be $\frac{1}{1452}$, whence one division of the scale $= a \ (1 + \frac{H.}{F.}) = 1'\cdot00069$. The Magnet was 3 inches in length, and suspended by a single thread of silk. The effect of the massive copper box in which it was suspended such, that the Magnet was generally at rest, and underwent derable changes without vibration. Increasing numbers on the denote an easterly movement of the north end of the bar.

Absolute Declination.—The following observation was made with the Collimator Magnet, c. 9. October 16th 1843, to determine a zero value of the Declinometer scale. The portable Theodolite was levelled, and made to coincide approximately with the magnetic axis of the Collimator, then directed to the sun, and the transit of both limbs observed; after reading off the verniers, it was again directed to the magnetic axis of the Collimator, and a series of simultaneous readings of the scale and of the Declinometer were taken. The sun was too low at the conclusion to allow the Theodolite to be referred to it again.

Mean Scale reading of the Collimator, 76·72, corresponding to 409·20 on the scale of the Declinometer. Point of scale on magnetic axis, 82'·06, each division is equal to 2'·51, showing a deviation of the Telescope of 13'·13 to the West.

Mean reading of Verniers - - -	249° 29'·83
Deviation of Telescope to the West -	13 ·13
Reading of magnetic axis - - -	249 42 ·96
Reading of Sun's centre - - -	99 9 ·83
Magnetic azimuth of Sun's centre at 3^h 52^m 49^s App. T. - - - -	150 33 ·13
The Sun's true azimuth at 3^h 52^m 49^s, App. T.	122 2 ·31
Variation East, corresponding to 409'·2 on scale - - - - -	28 30 ·82

The absolute values corresponding to the mean scale reading for each fortnight, will be found at p. 13, Table IX. The mean of the whole is 420·93, and the corresponding absolute Declination 28° 42'·6.

Diurnal Variation of the Declination.

Before proceeding to examine the mean diurnal curves for the four and a half months of observation at Lake Athabasca, it will be useful to obtain a general idea of the magnitude of the changes which the Declination is liable at this station and at Fort Sim The following Table has been drawn out with this view, showi difference between the highest and lowest hourly readings between the highest and lowest readings, observed in each Götti day; the latter shows the actual range of the element, the forme requisite for comparison with other stations. The Table may b referred to, also, for the dates of disturbances.

Table I.

Daily Range of the Declination.

Date.	In the hourly Series.		Observed.		Range.	
	Highest.	Lowest.	Highest.	Lowest.	Hourly.	Total.
1843:					′	′
October 1	—	—	—	—	—	—
″ 2	—	—	—	—	—	—
″ 3	—	—	—	—	—	—
″ 4	—	—	—	—	—	—
″ 5	—	—	—	—	—	—
″ 6	—	—	—	—	—	—
″ 7	—	—	—	—	—	—
″ 8	—	—	—	—	—	—
″ 9	—	—	—	—	—	—
″ 10	—	—	—	—	—	—
″ 11	—	—	—	—	—	—
″ 12	—	—	—	—	—	—
″ 13	—	—	—	—	—	—
″ 14	—	—	—	—	—	—
″ 15	—	—	—	—	—	—
″ 16	491·6	405·1	507·9	352·4	86·5	155·5
″ 17	448·3	392·3	466·0	356·2	56·0	109·8
″ 18	424·4	403·0	432·0	369·6	21·4	62·4
″ 19	436·0	382·0	436·0	382·0	54·0	54·0
″ 20	430·0	406·0	—	—	24·0	—
″ 21	426·6	403·0	—	—	23·6	—
″ 22	S.	—	—	—	—	—
″ 23	422·0	411·0	—	—	11·0	—
″ 24	436·0	408·2	456·2	408·2	27·8	48·0
″ 25	479·8	406·0	523·4	396·0	73·8	127·4
″ 26	434·6	388·0	437·0	388·0	46·6	49·0
″ 27	450·5	408·0	450·5	404·0	42·5	46·5
″ 28	425·0	391·0	458·0	391·0	34·0	67·0
″ 29	S.	—	—	—	—	—
″ 30	490·0	387·5	490·0	386·0	102·5	104·0
″ 31	457·8	406·0	470·0	406·0	51·8	64·0
November 1	423·6	412·0	—	—	11·6	—
″ 2	423·5	375·0	428·4	326·0	48·5	102·4
″ 3	425·0	408·6	460·2	359·4	16·4	100·8
″ 4	432·0	404·2	—	—	27·8	—
″ 5	S.	—	—	—	—	—
″ 6	464·0	410·0	464·0	404·0	54·0	60·0
″ 7	422·0	400·2	—	—	21·8	—
″ 8	475·4	407·8	475·4	404·0	67·6	71·4
″ 9	422·8	415·2	430·0	415·2	7·6	14·8
″ 10	423·8	408·2	429·8	408·2	15·6	21·6
″ 11	420·8	404·0	—	—	16·8	—
″ 12	S.	—	—	—	—	—

TABLE I.—*continued.*

Date.	In the hourly Series.		Observed.		Range.	
	Highest.	Lowest.	Highest.	Lowest.	Hourly.	Total.
1843:					′	′
November 13	438·0	402·6	438·0	402·6	35·4	35·4
„ 14	426·0	407·0	432·6	349·0	19·0	83·6
„ 15	433·2	470·2	—	—	23·0	—
„ 16	440·0	408·0	440·0	403·0	32·0	37·0
„ 17	422·0	412·0	—	—	10·0	—
„ 18	421·1	405·0	—	—	16·1	—
„ 19	S.	—	—	—	—	—
„ 20	433·0	413·4	—	—	19·6	—
„ 21	422·0	411·8	—	—	10·2	—
„ 22	436·5	411·8	—	—	24·7	—
„ 23	422·0	412·5	—	—	9·5	—
„ 24	436·2	408·2	444·0	408·2	28·0	36·8
„ 25	420·2	412·8	422·0	412·8	7·4	9·2
„ 26	S.	—	—	—	—	—
„ 27	423·0	414·0	—	—	9·0	—
„ 28	430·0	408·0	—	—	22·4	—
„ 29	425·5	401·6	—	—	23·9	—
„ 30	434·0	409·0	434·0	404·0	25·0	30·0
December 1	430·8	390·0	453·2	324·0	40·8	129·2
„ 2	450·1	414·2	484·0	414·2	35·9	69·8
„ 3	S.	—	—	—	—	—
„ 4	422·0	412·8	—	—	9·2	—
„ 5	427·0	405·0	439·0	394·4	22·0	45·0
„ 6	436·5	411·8	450·0	411·8	24·7	38·2
„ 7	421·5	413·9	—	—	7·6	—
„ 8	464·2	396·0	464·2	396·0	68·2	68·2
„ 9	424·0	406·4	—	—	17·6	—
„ 10	S.	—	—	—	—	—
„ 11	427·0	405·0	—	—	22·0	—
„ 12	428·4	410·4	—	—	28·0	—
„ 13	428·2	416·2	—	—	12·0	—
„ 14	434·0	422·6	—	—	11·4	—
„ 15	431·0	416·4	—	—	14·6	—
„ 16	428·2	415·0	—	—	13·2	—
„ 17	S.	—	—	—	—	—
„ 18	429·5	413·4	—	—	16·1	—
„ 19	451·5	416·8	452·0	416·8	34·7	35·2
„ 20	438·2	408·0	438·2	408·0	30·2	30·2
„ 21	426·0	410·0	426·3	407·8	16·0	18·5
„ 22	425·6	410·6	—	—	15·0	—
„ 23	422·8	414·0	—	—	8·8	—
„ 24	S.	—	—	—	—	—
„ 25	Christmas Day.		—	—	—	—
„ 26	429·5	416·0	—	—	13·5	—
„ 27	447·0	406·8	454·0	406·8	40·2	47·2

TABLE I.—*continued.*

Date.	In the hourly Series.		Observed.		Range.	
	Highest.	Lowest.	Highest.	Lowest.	Hourly.	Total.
					′	′
1843:						
December 28	477·8	400·4	477·8	400·4	77·4	77·4
" 29	438·5	412·0	438·5	368·0	26·5	70·5
" 30	433·4	411·8	—	—	21·6	—
" 31	S.	—	—	—	—	—
1844:						
January 1	—	—	—	—	—	—
" 2	443·0	414·4	—	—	28·6	—
" 3	429·5	419·0	—	—	10·5	—
" 4	460·0	384·0	470·0	384·0	76·0	86·0
" 5	497·2	408·4	500·0	408·4	88·8	91·6
" 6	452·0	392·0	458·0	392·0	60·0	66·0
" 7	S.	—	—	—	—	—
" 8	436·4	408·0	447·8	408·0	28·4	39·8
" 9	436·5	416·0	—	—	20·5	—
" 10	443·8	405·0	—	—	38·8	—
" 11	436·0	416·0	—	—	20·0	—
" 12	450·8	417·2	—	—	33·6	—
" 13	434·0	420·2	—	—	13·8	—
" 14	S.	—	—	—	—	—
" 15	434·0	419·0	—	—	15·0	—
" 16	433·0	418·9	—	—	14·1	—
" 17	438·0	412·2	438·0	412·2	25·8	25·8
" 18	440·0	420·0	—	—	20·0	—
" 19	448·0	423·0	454·0	423·0	25·0	31·0
" 20	434·0	414·8	—	—	19·2	—
" 21	S.	—	—	—	—	—
" 22	444·2	406·0	—	—	38·2	—
" 23	436·5	420·0	—	—	16·5	—
" 24	480·4	413·8	515·0	379·2	66·6	135·8
" 25	551·0	419·0	551·0	416·0	132·0	136·0
" 26	443·0	414·0	—	—	29·0	—
" 27	436·0	418·2	445·0	418·2	17·8	26·8
" 28	S.	—	—	—	—	—
" 29	434·4	410·4	—	—	24·0	—
" 30	432·0	413·0	—	—	19·0	—
" 31	432·2	394·4	—	—	37·8	—
February 1	476·0	411·4	486·7	341·0	64·6	145·7
" 2	439·0	373·0	439·0	373·0	66·0	66·0
" 3	428·0	396·0	—	—	32·0	—
" 4	S.	—	—	—	—	—
" 5	449·6	399·1	504·4	348·0	50·5	156·4
" 6	437·8	409·0	442·6	398·6	28·8	44·0
" 7	439·0	410·2	—	—	28·8	—
" 8	450·6	408·6	459·0	408·6	42·0	50·4
" 9	428·6	417·9	—	—	10·7	—
" 10	451·5	417·0	—	—		

nearly. There were 116 days of observation at Lake between 16th October 1843 and 29th February 1844, ıng 23d December for their middle period. There were 46 days f hourly observation at Fort Simpson, between 1st April and 25th May, having 27th April for their mean period. By classifying the ranges according to magnitude, we have the following results:—

TABLE II.

Daily Change of Declination.	Lake Athabasca.		Fort Simpson.	
	Hourly Observations.	Hourly and extra.	Hourly Observations.	Hourly and extra.
Less than 10′ - -	7	6	0	0
10′— 15′ - -	16	17	0	0
15′— 20′ - -	19	13	1	1
20′— 25′ - -	21	20	2	2
25′— 30′ - -	16	12	5	4
30′— 35′ - -	8	6	0	0
35′— 40′ - -	6	10	6	5
40′— 45′ - -	4	2	7	4
45′— 50′ - -	2	4	2	2
50′— 55′ - -	4	2	4	3
55′— 60′ - -	1	0	3	3
60′— 65′ - -	2	3	1	3
65′— 70′ - -	4	5	1	3
70′— 75′ - -	1	2	2	2
75′— 80′ - -	2	1	2	1
80′— 90′ - -	2	2	1	0
90′—100′ - -	0	1	0	2
100′—110′ - -	1	4	0	0
110′—120′ - -	0	0	4	0
2°—3° - -	1	7	5	6
3°—4° 10′ -	0	0	0	4
Above 7° - -	0	0	0	1

The greatest range in any one day during the winter at Lake Athabasca was 2° 35′, on the 16th October 1843; and the greatest during the spring at Fort Simpson was 7° 27′, on the 16th April 1844. Upon the last occasion, however, the actual difference of scale reading observed was 8° 10′, the westerly extreme falling on April 16^d 19^h 50^m, and the easterly on April 17^d 1^h 24^m; this is believed to be the greatest range hitherto recorded. During the same season ... ution at the three permanent Observatories in America was

...	21	
,,	22	401
,,	23	422·0
,,	27	447·0

TABLE III.

—	16th Oct. 1843—29th Feb. 1844.			April—May 1844.		
	Philadelphia.	Toronto.	Sitka.	Philadelphia.	Toronto.	Sitka.
Less than 10′ -	103	94	81	23	8	6
10′—15′ -	13	19	29	22	32	31
15′—20′ -	0	1	12	6	5	18
20′—25′ -	0	2	4	1	3	3
25′—30′ -	1	1	3	0	1	0
More than 30′	0	1	3	1	2	3
	117	118	132	53	51	61

The means of all the daily ranges during the above periods by the regular hourly observations, that is to say, the square roots of the mean of their squares are:—(1.) For the winter, 7′′77 at Philadelphia, 9′′56 at Toronto, 11′′64 at Sitka, and 33′′8 at Lake Athabasca: (2.) For the spring, 12′′32 at Philadelphia, 15′′17 at Toronto, 18′′40 at Sitka, and 75′′6 at Fort Simpson. For the several months again, we have the means as follows:—

TABLE IV.

	Philadelphia.	Toronto.	Sitka.	L. Athabasca. / Fort Simpson.
	′	′	′	′
1843. October (the whole) -	8·74	10·44	12·63	—
„ (16th to 31st)	8·08	9·70	12·50	53·31
November - -	7·15	8·44	9·80	27·32
December - -	7·34	8·07	10·39	30·30
1844. January - -	6·72	8·38	10·44	30·85
February - -	8·64	12·25	14·33	32·03
March - -	13·47	16·94	21·02	—
April - -	12·90	15·62	21·19	85·95
May - -	11·72	14·73	15·19	57·10
June - -	12·14	13·75	15·35	—
July - -	12·85	14·27	15·65	—
August - -	15·01	15·95	22·93	—
September - -	14·50	19·69	18·25	—

The observations were made at the same moment of time, at all the stations; the difference in the number of days arises from the Sunday being a day of observation at the Russian stations. There is no marked preponderance of number under any one daily range at Lake Athabasca between 13′ and 29′, and these two values include half the days of observation.

rs, then, that during the winter under comparison, the ent of the Declination Magnet, observed hourly, exceeded 15′, e proportion of fifteen days to each hundred at Philadelphia, eventeen days to each hundred at Toronto, and exactly the same at Sitka, but on eighty days of each hundred at Lake Athabasca. During April and May they exceeded 30′ in the proportion of about two to a hundred at Philadelphia, four to a hundred at Toronto, five to a hundred at Sitka, but of eighty-five to a hundred at Fort Simpson, showing an increase in the liability to disturbance at these stations, which it appears difficult to attribute to the merely negative influence of a diminished directive power in the magnet.

The next Table contains the hourly means of all the observations during the winter period, with the exception of six days which are omitted at Lake Athabasca as incomplete, namely, October 20th, November 3d and 4th, January 2d, 9th, and 27th. Each Value at this station is, therefore, the mean of 110 observations at the same hour. Since the principal novelty of these means consists in their maxima being found at a period of the 24^h which is not marked by a similar inflexion at any other station, they are, in the same Table, compared with the means for the corresponding periods at the other three American stations.

Table V.

Mean Diurnal Curves of Declination at all the American Stations, for the Period included between the 1*st or* 16*th October* 1843 *and the* 29*th February* 1844; *together with the Difference of each hourly Value from the Mean of the whole; expressed in arc.*

Local Mean Time.	Athabasca.		Sitka.*		Toronto.		Philadelphia.	
	Scale.	Diff.	Scale.	Diff.	Scale.	Diff.	Scale.	Diff.
		′		′		′		′
Midn.	419·07	−1·86	429·48	+0·92	126·51	+0·36	547·52	+0·38
13	419·39	−1·54	429·42	+0·89	126·25	+0·17	547·12	+0·19
14	421·83	+0·90	430·08	+1·26	126·02	+0·01	546·98	+0·13
15	422·54	+1·61	430·06	+1·25	126·50	+0·35	547·46	+0·35
16†	426·76	+5·83	430·20	+1·32	126·91	+0·65	547·66	+0·54
17	429·20	+8·23	430·54	+1·51	127·09	+0·78	548·00	+0·[illegible]
18	426·81	+5·84	430·74	+1·63	127·30	+0·93	548·82	+[illegible]
19	426·49	+5·56	431·08	+2·81	128·14	+1·54	550·98	+[illegible]
20	424·83	+3·90	430·90	+1·71	128·74	+1·97	551·38	+[illegible]
21	424·69	+3·76	429·50	+0·93	128·71	+1·95	550·82	+1[illegible]
22	422·26	+1·33	426·82	−0·55	126·98	+0·70	547·30	+0[illegible]
23	417·95	−2·98	424·92	−1·60	124·32	−1·22	543·82	−1·[illegible]

* In every instance in which the observations at Sitka are referred to, the mean fo[r] month of November 1843, as given in the *Annuaire Magnetique*, &c., has been correcte[d] adding 23·2 div. to each scale reading of Declination from $1^d\ 0^h$ to $14^d\ 10^h$ Gott., being difference between the means for the 24^h before and after the last-named hour. T[his] difference is permanent, and appears due to some accidental cause, although no explanati[on] of it is given.

† 0^h Göttingen mean time at Lake Athabasca.

TABLE V.—*continued.*

Local Mean Time.	Athabasca. Scale.	Athabasca. Diff.	Sitka.* Scale.	Sitka.* Diff.	Toronto. Scale.	Toronto. Diff.	Philadelphia. Scale.	Philadelphia. Diff.
		′		′		′		′
Noon	417·00	−3·93	423·14	−2·59	122·27	−2·70	540·92	−2·62
1	416·04	−4·89	422·82	−2·77	121·39	−3·33	540·14	−2·97
2	415·72	−5·21	422·52	−2·93	121·83	−3·02	541·02	−2·57
3	416·74	−4·19	424·10	−1·95	123·00	−2·17	542·50	−1·90
4	417·70	−3·23	425·22	−1·44	124·17	−1·32	543·90	−1·27
5	418·05	−2·88	426·58	−0·68	125·23	−0·56	545·42	−0·57
6	419·13	−1·80	427·44	−0·20	126·32	+0·22	546·62	−0·03
7	419·61	−1·32	426·64	−0·65	126·73	+0·52	547·58	+0·40
8	419·75	−1·18	428·44	+0·29	127·34	+0·96	548·56	+0·84
9	420·01	−0·92	428·52	+0·39	127·87	+1·34	549·10	+1·10
10	420·70	−0·23	428·86	+0·58	127·88	+1·35	548·94	+1·02
11	420·90	−0·03	429·44	+0·90	126·87	+0·62	548·14	+0·65
	420·97		427·81		126·01		546·69	

The observations were taken 5ᵐ before the hours named at Lake Athabasca, 28ᵐ after the hours named at Sitka, 3ᵐ after at Toronto, and 19ᵐ after at Philadelphia.

It appears that the mean diurnal changes of Declination at Lake Athabasca follow the same law as at all the other stations, so far as relates to the principal minimum or westerly extreme of the 24^h, which occurs at 2 P.M.; from this hour the Declination continues to increase until 11 P.M.; it shows a westerly tendency at midnight and 1 A.M., after which it increases again, at first slowly, but between 3^h and 4^h A.M. with rapidity, until it attains its maximum or easterly extreme between 4^h and 7^h A.M., after which it begins a westerly course, conducting to the minimum at 2 P.M. This occurrence of a strongly marked maximum at the earlier hours of the morning not been observed at any other station.*

It has been shown by Colonel Sabine, from the observations Toronto, that no continuance of observation will give a stric

* A second Declinometer, having a Magnet of only two inches in length, was obse from the 16th December to the 29th February. The means by 58 complete days observation in this period are given below:

TABLE VI.

M.T.	Midnt.	13.	14.	15.	16.	17.	18.	19.	20.	21.	22.	23.	
Scale -	228·89	228·53	232·09	233·28	236·46	239·29	235·93	235·77	234·67	234·98	231·05	226·77	
	′	′	′	′	′	′	′	′	′	′	′	′	
Difference	−1·46	−1·82	+1·74	+2·93	+6·11	+8·94	+5·58	+5·42	+4·32	+4·63	+0·70	−3·58	
	Noon.	**1.**	**2.**	**3.**	**4.**	**5.**	**6.**	**7.**	**8.**	**9.**	**10.**	**11.**	**Mean.**
Scale -	226·07	224·57	223·63	225·59	226·81	227·28	228·32	229·40	229·68	229·39	229·24	230·80	230·35
	′	′	′	′		′	′	′	′	′	′	′	
Difference	−3·28	+5·78	−6·72	−4·76	−3·54	−3·07	+2·03	−0·95	−0·67	−0·96	−1·11	+0·45	

normal curve, or one wholly free from the effects of disturbance, since the disturbing causes are not entirely irregular in their action, but have a preponderating influence in one direction. It will be shown below that this remark applies equally at the stations under consideration, consequently the foregoing mean cannot be regarded as a true representation of the normal curve at Lake Athabasca. This can only be obtained by some selection of undisturbed days, and the next Table has been formed as an approximation to it. It contains the mean by all those days on which no extra observations were made, assuming that circumstance to be a proof of the absence of any decided disturbance. They amount to 46, and the mean of the same 46 days at Toronto, which have been formed into a similar abstract, furnishes a direct comparison of the mean diurnal movement, uninfluenced by disturbance, or nearly so, at these two stations.

TABLE VII.

Local Mean Time.	L. Athabasca.		Toronto.		Local Mean Time.	L. Athabasca.		Toronto.	
	Scale.	Diff.	Scale.	Diff.		Scale.	Diff.	Scale.	Diff.
		′		′			′		′
Midn.	420·85	−0·91	416·36	+0·06	Noon	418·68	−3·08	412·87	−2·46
13	421·68	−0·08	416·34	+0·04	1	417·21	−4·55	411·97	−3·11
14	421·37	−0·39	416·27	−0·01	2	417·81	−3·95	412·24	−2·92
15	422·92	+1·16	416·77	+0·35	3	418·44	−3·32	413·65	−1·90
16	424·91	+3·15	417·55	+0·92	4	419·39	−2·37	414·85	−1·03
17	425·57	+3·81	417·81	+1·10	5	419·90	−1·86	415·89	−0·28
18	425·31	+3·55	417·48	+0·87	6	420·30	−1·46	416·58	+0·22
19	425·57	+3·81	418·05	+1·28	7	428·88	−0·88	416·82	+0·39
[illegible]	425·86	+3·60	418·72	+1·76	8	420·93	−0·83	417·38	+0·79
[illegible]	425·40	+3·64	418·60	+1·67	9	420·98	−0·78	417·60	+0·95
[illegible]	424·04	+2·35	417·11	+0·60	10	422·25	+0·49	418·06	+1·28
[illegible]3	420·65	−1·11	414·85	−1·03	11	421·91	+0·15	417·11	+0·60
						421·76		416·29	

It appears that a partial rejection of the more disturbed days at Lake Athabasca has the effect of throwing back the hour of greatest westerly deviation to 1 P.M., but occasions very little change, and none of a systematic character, in the mid-day or afternoon branches of the curve. The principal effect is shown in the reduction of the daily variation between midnight and 10 A.M. The maximum at 5 A.M. disappears, and in its place we have a nearly uniform value prevailing from 5^h to 9^h A.M. constituting the easterly extreme of the 24^h, but materially less in amount than the corresponding value before the rejection, thus proving the unusual maximum in question to be the effect of disturbance. The inferior maximum at 11 P.M. is thrown back to 10 P.M., but the succeeding minimum is scarcely affected. The mean curve given by the corresponding 46 days at

Toronto differs so little from that of the whole period, that when drawn on a scale of 10′ to one inch they can scarcely be distinguished.

Proceeding now to the observations at Fort Simpson, we find the same peculiarity in a more marked degree, as may be expected from the greater magnitude of the daily ranges observed at this station. The incomplete days here are not omitted, as they form rather too large a proportion of the whole to be passed over. There were 46 days of observation, of which number twelve are imperfect. The omissions occur as follows, at

0^h Gött.	or	15^h of Table VIII.		5
1	„	16	„	6
2	„	17	„	2
3	„	1	„	1

consequently, the two first alone can be materially affected, and at twenty of the twenty-four hours the means are strictly comparable. In one of the cases at each of the above-named hours an observation taken *late* has been employed.

Table VIII.

Mean Diurnal Curves of Declination at all the American Stations for April and May 1844, *together with the Difference of each hourly Value from the Mean of the whole; expressed in arc.*

Local Mean Time.	Fort Simpson.		Sitka.		Toronto.		Philadelphia.	
	Scale.	Diff.	Scale.	Diff.	Scale.	Diff.	Scale.	Diff.
		′		′		′		′
Midn.	385·31	− 5·24	432·80	+1·91	126·17	+1·13	552·95	+0·91
13	387·98	− 2·60	433·15	+2·10	126·34	+1·26	553·10	+0·98
14	394·54	+ 3·96	433·05	+2·04	126·40	+1·30	552·70	+0·79
15*	399·05	+ 8·47	433·35	+2·22	126·67	+1·49	552·90	+0·88
16	402·64	+12·06	432·80	+1·91	126·89	+1·65	553·35	+1·09
17	410·47	+19·89	435·25	+3·27	127·67	+2·21	555·10	+1·88
18	410·90	+20·32	436·40	+3·91	129·19	+3·31	557·50	+2·97
19	412·49	+21·91	438·15	+4·88	129·70	+3·68	558·45	+3·40
20	408·98	+18·40	439·35	+5·55	130·40	+4·18	558·50	+3·42
21	401·86	+11·28	437·95	+4·77	129·21	+3·33	556·00	+2·29
22	395·22	+ 4·64	433·45	+2·27	125·45	+0·61	551·30	+0·16
23	388·72	− 1·86	429·05	−0·17	121·60	−2·16	546·65	−1·95
Noon	385·02	− 5·56	424·15	−2·89	118·45	−4·44	542·55	−3·81
1	382·31	− 8·27	421·45	−4·38	117·05	−5·45	540·35	−4·57
2	380·03	−10·55	418·85	−5·83	116·88	−5·57	541·25	−4·40
3	380·01	−10·57	418·50	−6·02	118·14	−4·66	542·70	−3·29
4	380·70	− 9·88	419·30	−5·57	120·07	−3·27	546·05	−2·22
5	379·35	−11·23	420·80	−4·74	122·16	−1·76	548·85	−0·95
6	379·61	−10·97	424·55	−2·66	123·19	−1·02	549·40	−0·70
7	382·63	− 7·95	426·65	−1·50	124·33	−0·19	551·70	+0·34
8	380·23	−10·35	427·30	−1·14	124·42	−0·13	551·10	+0·07
9	381·58	− 9·00	428·05	−0·72	125·76	+0·84	552·45	+0·68
10	381·13	− 9·45	429·60	+0·14	127·33	+1·97	554·15	+1·45
11	383·07	− 7·51	430·40	+0·58	126·77	+1·56	553·65	+1·23
Means -	390·58		429·35		124·60		550·95	

* 0^h of Göttingen mean time at Fort Simpson.

The observations at Fort Simpson were made 15^{m} after the hour named, the rest as before stated.

It will be seen that a remarkable difference exists in the amount of the daily movement at Fort Simpson and Lake Athabasca, although the diurnal law is nearly the same. It is difficult to attribute this altogether to change of locality, which in this case involves but a slight change of magnetical position*; we are therefore led to connect it with the advance of the season. On referring to the Table, it will be seen that the westerly extreme at 2 P.M., which is so well marked at Toronto and the lower stations, is here prolonged for nearly four hours, during which space the magnet does not sensibly deviate from it. After 7 P.M. it returns to the eastward, but so gradually as not to attain its mean, or zero value, before 2 A.M. instead of 8 P.M., as at Toronto. Then commences a sudden and rapid increase, leading to a maximum at 7 A.M., after which a steady and rapid westerly movement takes place, which attains its limit, as already stated, at or about 2 P.M., and amounts to the very large mean quantity of 32′. The sun was above the horizon 15^{h} 44^{m} on the middle date at this station, and 5^{h} 56^{m} on the middle date at Lake Athabasca, which correspond to 13^{h} 48^{m} and 8^{h} 44^{m} on the same dates at Toronto respectively. Notwithstanding this great difference in the length of the day in December and April, it will be observed that the magnet in its regular westerly movement from 7 A.M. to 2 P.M. cuts the line of mean at or near the same hour in both periods, as it does also on its return at night towards its easterly extreme. When referred to the westerly limit, or lowest mean value, as zero, instead of to the mean of the whole twenty-four hours, the diurnal changes at Lake Athabasca for the afternoon branch do not exceed, but, upon the whole, rather fall short of those at Toronto, and at Fort Simpson they are less than those at Toronto for ten hours following that epoch, namely from 3^{h} to 12^{h} inclusive. It is the nocturnal branch at both stations which occasions the great apparent excess of their diurnal changes above the changes at Toronto.

The persistency of the magnet at its westerly extreme, in April and May, is strongly contrasted with a similar persistency at its easterly extreme, which appears to be one of the principal characteristics of the curve at Lake Athabasca in the winter. We see by Table IV. that when free from the effect of disturbances, the easterly maximum is maintained from 5^{h} to 9^{h} A.M., which suggests the inference that the effect of the advance of the season on the

* The Horizontal Force is 2·02 at Lake Athabasca, and 1·95 at Fort Simpson; the inclination is 81° 37′·6 and 81° 52′·3, the total force 13·92 and 13·84, at the same stations respectively.

daily curve is analagous to that of the sun's diurnal progress, the increased power of the sun being attended by a determination of the north end of the magnet to the west, its diminished power by a determination of the same end to the east.

Comparing together the means of the four American stations for the two periods under discussion, we find that besides the different amount of the daily changes shown, they have peculiarities of a systematic character, which is apparent more particularly in those of the spring months, but not so decidedly in either group as in those of the bifilar. Referring first to the curves for April and May, it appears that the magnet attains its daily limit to the westward at nearly the same hour at Toronto and Philadelphia, namely, 2 P.M. or thereabouts, and then commences an immediate return towards its mean position; it reaches the same limit about half-past three at Sitka, and its return is more gradual. It would appear from the general character of the curve at Fort Simpson that its turning hour is there as late as 5 or 6 P.M., and the return is effected still more slowly. The hour of attaining the easterly limit does not appear to vary much. The number of hours at which the magnet is to the eastward of its mean position is, however, considerably less at the more northern than at the southern stations, which results from the easterly tendencies during the night having the effect of raising the mean scale reading, and throwing the readings in the latter hours of the afternoon and evening relatively to the westward. The numbers for the winter group are sixteen at Philadelphia, seventeen at Toronto, fourteen at Sitka, and nine at Lake Athabasca; for the spring group, sixteen, fourteen, thirteen, and nine respectively.

The mean scale readings at both stations show an increase of Easterly Declination, which was slight in the winter months, and very rapid in the spring. Taking the means for each fortnight, and referring them to absolute values, we have the following series. The approximate means for the same periods at Sitka, are inserted for comparison.

TABLE IX.

1 scale division = 0'·553.	Sitka.	Lake Athabasca.
		° ′
1843. Oct. 16 to Oct. 21	- - -	414·24=28 35·8 E.
Oct. 22 to Nov. 4	- - -	415·84=28 37·4
Nov. 5 to Nov. 19	- - -	416·61=28 38·2
Nov. 20 to Dec. 3	427·94	417·58=28 39·2
Dec. 4 to Dec. 17	423·02	419·28=28 40·9
Dec. 18 to Dec. 31	427·74	421·26=28 42·9
1844. Jan. 1 to Jan. 15	427·54	425·71=28 47·3
Jan. 16 to Jan. 29	424·25	427·32=28 48·9

TABLE IX.—*continued.*

	Sitka.	Lake Athabasca.
		° ′
1844. Jan. 30 to Feb. 12 - -	420·69	423·32=28 44·9 E.
Feb. 13 to Feb. 28 - -	423·47	424·63=28 46·2
Feb. 26 to Mar. 10 - -	426·95	Fort Simpson.
Mar. 11 to Mar. 24 - -	427·84	° ′
Mar. 25 to April 7 - -	431·93	358·14=36 53·1 E.*
April 8 to April 21 - -	429·44	385·92=37 26·9
April 22 to May 5 - -	431·58	400·33=37 41·3
May 6 to May 19 - -	433·17	418·21=37 59·2
May 20 to June 2 - -	431·69	

The regular character of the progression makes it difficult to attribute the easterly movement, from October 16th to the end of January, to any accidental or instrumental cause, still more so the much more rapid movement in April and May; the latter is the more difficult to account for, as the effect of Disturbances at Fort Simpson in those months appears to have been less exclusively easterly than during the previous period at Lake Athabasca. The means at Sitka, which are derived for the present comparison from the three equidistant observations of 0^h, 8^h, and 16^h, Gött., show no similar tendency from November to January, and but a slight one from February to May. We have, however, a similar result from the observations of Sir John Franklin at Great Bear Lake in 1826, which give the mean Declination in February 38° 41′·6, and in March 38° 50′·2 E. As no observations were made by Sir J. Franklin at those hours of the night which are now shown to include, in these latitudes, the extreme easterly movements, the difference in this case is probably within the truth. The effect of Disturbances being greater in March than in February, would have increased the relative easterly value of the former month.

In all the foregoing tables, the movements of a three inch-magnet at one station have been regarded as comparable with those of a fourteen-inch magnet at the other, without consideration of the difference of their size and weight. It appears, however, from a comparison of the mean diurnal curves, deduced from observations of two magnets of these dimensions respectively at Toronto, that there is a slight and apparently specific difference between them. The diurnal variation by the small bar is greatest from 8 P.M. to 5 A.M., and least from 6 A.M. to 8 P.M., with an exception at 1 P.M. only; that is to say, the Declination by the larger magnet is slightly more

* The regular division is continued at Sitka; at Fort Simpson the fortnights are taken from April 1st to 14th, and so on.

easterly during the day, and less easterly during the night, than the Declination by the smaller bar, and this peculiarity presents itself in every month, but the amount is not sufficient to affect any of the foregoing conclusions. A difference of a similar nature is observable between the mean diurnal curves by two bifilar magnets of corresponding dimensions.

TABLE X.

Comparison of the Diurnal Variation of Declination from the Register of a large and small Declinometer at Toronto in 1847.

Mean Time.	14-Inch Magnet.				3-Inch Magnet.			
	June.	July.	Aug.	Mean.	June.	July.	Aug.	Mean.
h	′	′	′	′	′	′	′	′
16	+1·53	+0·40	+2·27	+1·40	+1·53	+0·51	+2·40	+1·48
17	+2·73	+1·60	+3·68	+2·64	+2·80	+1·83	+4·04	+2·89
18	+5·25	+6·04	+6·05	+5·78	+5·05	+5·77	+6·08	+6·63
19	+5·98	+7·17	+8·47	+7·21	+5·73	+6·67	+7·95	+6·78
20	+6·16	+6·93	+8·15	+7·08	+5·88	+6·74	+7·75	+6·79
21	+4·49	+4·99	+5·92	+5·13	+4·34	+4·61	+5·55	+4·83
22	+1·77	+1·01	+0·89	+1·49	+1·62	+1·00	+0·68	+4·10
23	−1·89	−2·01	−3·12	−2·34	−1·89	−2·80	−3·15	−2·61
Noon	−4·87	−5·61	−7·43	−5·97	−5·05	−5·64	−7·36	−6·02
1	−6·31	−6·78	−9·20	−7·43	−6·39	−6·34	−8·95	−7·23
2	−5·95	−5·93	−8·23	−6·70	−6·08	−6·15	−8·28	−6·84
3	−5·02	−4·76	−6·05	−5·28	−5·19	−4·97	−6·50	−5·55
4	−3·43	−3·03	−3·84	−3·43	−3·65	−3·20	−4·06	−3·64
5	−1·88	−0·97	−2·03	−1·63	−2·05	−1·02	−2·21	−1·76
6	−1·02	−0·29	−0·20	−0·50	−1·10	−0·36	−0·33	−0·60
7	−0·11	−0·74	−0·48	−0·43	−0·15	−0·73	−0·42	−0·43
8	+0·14	−0·27	−0·32	−0·15	+0·43	−0·08	−0·18	+0·06
9	−0·17	−0·59	+1·35	+0·20	+0·12	−0·30	+1·71	+0·51
10	+0·74	+1·36	+0·52	+0·87	+1·08	+1·75	+0·75	+1·19
11	+1·00	+1·10	+0·29	+0·80	+1·20	+1·39	+0·68	+1·09
Midn.	+1·25	+0·82	+0·53	+0·87	+1·49	+1·07	+0·73	+1·10
13	−0·07	−0·05	+1·24	+0·37	+0·29	+0·10	+1·45	+0·61
14	−0·26	−0·13	+0·76	+0·12	−0·05	+0·03	+0·86	+0·28
15	−0·07	+0·54	+0·68	+0·38	+0·05	+0·59	+0·81	+0·48

SECTION II.

BIFILAR MAGNETOMETER.

The instrument employed was a small transportable bifilar, provided with a hollow 3·0-inch magnet, weighing 207 grains. The reading telescope was carried by an arm connected with the base of the instrument, so that the copper box and suspension apparatus turned with the telescope. The azimuth circle read by verniers to single minutes, the torsion circle to 5′·0. The suspension silk was about 9 inches long. The arc value of each division of the scale attached to the telescope was 2′·0.

Adjustment, 13th October 1843.—(1.) The reading telescope was brought to the magnetic meridian by attaching the magnet to an unifilar suspension and bringing the central division of the scale (210) to the wire. The azimuth circle read 302° 55′, declination 428, whence 302° 45′ corresponds to the scale reading 418·0 found at the next step.

(2.) The magnet was next attached to a bifilar suspension of waxed silk, the north end pointing to the north, and the torsion circle turned until the same division was on the wire, showing the plane of detorsion to coincide with the meridian. At the first trial the force of torsion was too great, the interval of the threads was in consequence reduced; this, however, altered the position of the plane of detorsion, and it appears that in adjusting it afresh to the meridian, the telescope was inadvertently moved 34′ to the west, the next reading of the horizontal circles being 302° 11′ instead of 302° 45′. The plane of detorsion, therefore, instead of being in the meridian, was in such a position as to deflect the magnet 34′ to the westward, or rather, correcting for a small change of declination, 33′·2 to the westward. If (u) and (v) be the angles at which the magnetic force and the force of torsion act on the magnet, we have in every position of equilibrium $\frac{F}{G} = \frac{\sin v}{\sin u}$, and in the present case $\frac{F}{G} = 0·86$, and $u = 33'·2$, hence $v = 28'·4$, and the angle $u + v = 61'·6$ being the angle by which the plane of detorsion differed from the meridian to the westward. The reading of the torsion circle was 58° 50′, whence the true position of the plane of detorsion was 58° 50′ + 1° 1′′6 = 59° 51′′6, which is the datum employed. Declination 418·0 scale reading of bifilar 209·6 (for 210).

(3.) The telescope was now turned 90° in azimuth, from the assumed meridian, or 90° 33′′2 from the true one; the division of the scale at right angles to it was therefore 226·6 instead of 210, and the torsion circle being turned until the scale read 208, the magnet was carried 37′′2 beyond the position intended, namely, that at right angles with the magnetic meridian. The torsion circle read 0° 15′ Declination 416·5.

In the equation $\frac{F}{G} = \frac{\sin v}{\sin u}$ we have therefore $u = 90° 37''2$ instead of 90°, which somewhat diminishes the sensibility of the adjustment, but is otherwise unimportant. $v = 59° 51''6 - 0° 15' = 59° 36'6$, also $a = 2''0$, whence the value of one division of the scale in parts of the Horizontal Force was

$$k = a \cot v = {}^{\cdot}0003432.$$

Increasing numbers denote increasing Horizontal Force.

It appears that the scale readings decreased regularly until January, showing the north end of the magnet to be carried from the meridian to the south of west; the readings increase again in February, and this circumstance, coupled with that of a contrary change in the inclinometer scale readings, makes it appear that in part, at least, the effect was due to a real increase of Inclination and decrease of Horizontal Force in the winter months, having respectively their maximum and minimum in the latter portion of the month of January. Mixed up with this periodical change there is probably also a loss of magnetic moment in the bifilar bar, and an increase of permanent magnetism in the soft iron bar of the inclinometer. The fortnightly mean scale readings of both instruments are collected in the following table. Dr. Lloyd having shown that any three equi-distant observations give, very nearly, the true mean of the twenty-four hours, that principle has been applied in the present and all the similar tables, to obtain true mean scale readings for days on which one or more observations were omitted, by the simple proceeding of omitting, also, the corresponding readings of the imperfect triplet or triplets, and taking a mean of the remainder or perfect triplets.

Table XI.

Fortnightly Mean Scale Readings of the Bifilar and Inclinometer at Lake Athabasca.

Date.	Period.	Bifilar.				Inclinometer corrected.
		Observed.	Temp.	Corrected.	Angle *u*.	
			°		° ′	
1843	Oct. 15 to Oct. 22 -	247·32	51·07	254·39	89 04	—
„	Oct. 23 to Nov. 5 -	232·01	43·68	234·60	89 44	—
„	Nov. 6 to Nov. 19 -	210·78	43·63	213·31	90 27	120·12
„	Nov. 20 to Dec. 3. -	183·93	41·64	185·08	91 23	139·55
„	Dec. 4 to Dec. 17 -	174·17	41·33	175·10	91 43	161·80
„	Dec. 18 to Dec. 31 -	171·41	37·19	169·46	91 54	169·80
1844	Jan. 1 to Jan. 15 -	152·72	30·90	146·32	92 41	194·65
„	Jan. 16 to Jan. 29 -	140·67	25·51	130·29	93 13	209·74
„	Jan. 30 to Feb. 12 -	144·94	41·76	146·17	92 41	195·83
„	Feb. 13 to Feb. 27 -	174·20	44·87	177·63	91 38	197·14
					91 34	

The first fortnight has only half weight.

The coefficient of Temperature of the bifilar magnet was ascertained after its return to Toronto, in January 1845. A series of deflections on different days gave the following results:—

January 8, 1845, $q = 0·0003128$
10 „ ·0003009
11 ·0001784
14 „ ·0001941
17 „ ·0001818

The mean of the whole is 0·0002336. The mode of proceeding was the same in each case. The temperature of the magnet was raised at once from the lowest to the highest point, being about 40° Fahr. and 90° respectively, and three sets of readings taken at each, with an interval of 5^m between them, the bar being previously allowed 15^m to acquire the temperature of the surrounding water; the Declination and Horizontal Force were also observed at the same time. As the two first of the above values differ materially from the rest, the experiment was repeated in 1848 under similar arrangements, except that the change of temperature at each alternation was from 40° to 60° and no higher temperature was employed, the intervals were also reduced to 4 minutes between each set, and 10 minutes between each alternation:

March 4, $q = 0·0002472$
6 ·0002138
7 ·0002874

The mean of these is 0·0002495, not materially different from the former value. Lastly, the mean of both sets is,

$$q = 0·0002396$$

And adopting this value, we have the ratio $\frac{q}{k}$, or the change in scale divisions for a change of temperature of 1° Fahr. = 0·70. The whole of the readings in the abstracts have been reduced to the uniform temperature of 40° with an approximate coefficient $\frac{q}{k} =$ 0·66, which occasions, however, an error of only 2·1 divisions at the extreme temperature recorded, a quantity so small, as compared with the extent of other changes, that it has not been thought necessary to correct the work. The more accurate coefficient has been employed below in deducing the mean diurnal curve, and in correcting the observations on term day, and disturbances.

HORIZONTAL FORCE.

In Absolute Measure. – The observations of the Absolute Horizontal Force made at Lake Athabasca and Fort Simpson have been published in detail by Colonel Sabine (Contributions to Terrestial Magnetism, No. VII.), but it may be convenient to repeat the particulars here. Six deflecting magnets, varying in length from 3·6 inches to 2·0 inches, were employed at both stations; two of these, Nos. 30 and 31, of three inches in length, were considered the standard bars, and the others employed for verification. The whole of the magnets were suspended for vibration by a silk fibre attached directly to the bar, without the addition of any stirrup; bars 30 and 31 were also vibrated in a stirrup, the weight of which was 322 grains, thus giving a second and independent value of the term m X. The amount of inertia of each suspension was found by vibrating the magnet with and without the addition of carefully turned brass rings, according to the method recommended by Dr. Lamont. The following are the mean values:—

	Bar 30.	Bar 31.
Log. $\pi^2 k$ for vibration without the stirrup	1·33408	1·33952
Log. $\pi^2 k$ for vibration in the stirrup	1·50167	1·50521

The observation at Lake Athabasca was commenced on the 13th October, shortly after completing the adjustment of the bifilar already described; the experiments of deflection with four bars were completed on that day, and with the other two on the following day. The bars were vibrated without the stirrup on the 14th, and with it on the 20th, each experiment being connected with scale readings of the bifilar. The observation was repeated in March 1844.

TABLE XII.

Determination of Absolute Horizontal Intensity with Bars 30 *and* 31 *at Lake Athabasca. Length of Suspended Magnet* 2·45 *inches; length of Deflecting Magnet* 3·0 *inches.*

DEFLECTION.						VIBRATION.								
Date.	Bar.	Dist.	Observed Angle.	Temp.	Corrected Angle.	Date.	Observed Time.	Temp.	Corrected Time.	m	X	Means.	General Mean.	
1843:		Foot.	° ′	°	° ′		s	°	s					
October 13 -	30	1·0257	22 28·2	38·0	22 26·8	October 14 -	5·0354	38·0	5·0514	0·418	2·022	2·024	2·026	2·022
" -	30	1·3257	10 13·0	38·0	10 6·8	" -	- -	- -	- -	0·417	2·028			
October 13 -	31	1·0257	20 31·0	36·6	20 27·7	October 14 -	5·2409	37·0	5·2816	0·385	2·034	2·030		
" -	31	1·3257	9 23·7	36·4	9 24·8	" -	- -	- -	- -	0·387	2·026			
October 13 -	30	1·0257	22 28·2	38·0	22 26·8	October 20 -	6·0943	34·0	6·1381	0·418	2·018	2·027		
" -	30	1·3257	10 13·0	38·0	10 6·8	" -	- -	- -	- -	0·416	2·024			
October 13 -	31	1·0257	20 31·0	36·6	20 26·7	October 20 -	6·3767	34·0	6·4017	0·385	2·031	2·027		
" -	31	1·3257	9 23·7	36·4	9 24·8	" -	- -	- -	- -	0·386	2·023			
1844:														
March 1 -	30	1·0257	22 34·8	41·8	22 32·1	March 2 -	5·0258	34·5	5·0328	0·421	2·025	2·025	2·018	
" -	30	1·3257	10 15·8	43·8	10 15·0	" -	- -	- -	- -	0·421	2·025			
March 1 -	31	1·0257	20 14·9	45·4	20 13·6	March 2 -	5·3643	31·8	5·3740	0·376	2·009	2·008		
" -	31	1·3257	9 13·5	45·4	9 12·6	" -	- -	- -	- -	0·376	2·007			
March 1 -	30	1·0257	22 34·8	41·8	22 32·1	March 2 -	6·1011	29·0	6·1234	0·419	2·019	2·018		
" -	30	1·3257	10 15·8	43·8	10 15·0	" -	- -	- -	- -	0·419	2·018			
March 1 -	31	1·0257	20 14·9	45·4	20 13·6	March 2 -	6·4381	29·0	6·4728	0·378	2·019	2·020		
" -	31	1·3257	9 13·5	45·7	9 12·6	" -	- -	- -	- -	0·378	2·021			

The mean value by the other four bars in October was

$$X = 2{\cdot}030$$

which agrees nearly with the mean deduced from the observations of Bars 30 and 31; but it has not been employed, as three of them were corrected with an assumed coefficient of temperature, and the value of the moment of inertia employed has not been so well determined as that for the standard magnets.

The value of the Absolute Horizontal Force was found to be

$$X = 2{\cdot}040$$

by the standard bars in July 1844, being an increase of 0·0088X, or $\frac{1}{113}$ of its whole amount, over the value found in October and March, which may probably be attributed to a real increase of that element in the summer months.

Since the publication of the preceding observations by Colonel Sabine, Captain Younghusband has succeeded in obtaining a satisfactory determination of the value of the ratio $\frac{\Delta m}{m}$ for Bar 30, the previous results having been obtained without any correction for the change in the induced magnetism of the bars under the different circumstances of vibration and deflection. Five experiments made at Woolwich in May 1847 gave the following results:—

$$\frac{\Delta m_0}{m_0} = 0{\cdot}00042$$
$$0{\cdot}00048$$
$$0{\cdot}00037$$
$$0{\cdot}00042$$
$$0{\cdot}00056$$

mean of the whole, 0·00045. Assuming the magnetic moment of the bar itself at the period of these experiments to have been nearly the same as in June 1846, when the absolute intensity was determined last with it at Woolwich, we have the means of reducing this value of the correction to the value applicable at the northern station.

$$\frac{\Delta m}{m} = \frac{\Delta m_0}{m_0} \quad \frac{A_0}{A}$$

where A_0 = the ratio $\frac{m}{X}$ at Woolwich, and A the same at the northern station. $A_0 = \frac{0{\cdot}372}{3{\cdot}728}$ and $A = \frac{0{\cdot}419}{2{\cdot}022}$ (at Lake Athabasca), whence $\frac{\Delta m}{m} = {\cdot}000218$ On applying this correction to the times of vibration and the angles of deflection, by the formula given in the *Supplement to Magnetical Instructions, &c.*, 1846, the difference in the resulting value of X proves to be insignificant; it is less than 0·0003, and does not affect the above values, which are not carried beyond the third decimal, the effect is negative.

The foregoing value of the Horizontal Force is not referable to any one division on the bifilar scale. The readings of the instrument on the 14th and 20th October, taken in connexion with the experiments of vibration, differed considerably from the readings on the 13th, which were taken with the experiments of deflection. The mean on the 13th was 176·0 at 64°·6, on the 14th 255·5 at 57°·4, and on the 20th 254·4 at 50°·4; there is therefore a change of about 74 divisions between the 13th and 14th of October. The greatest difference between two successive daily means in the subsequent series is 27·4 div., which makes it probable that the change between the 13th and 14th October was partly instrumental, the magnet not at once taking up its position of adjustment. As the change of reading on the first day of regular observation amounted to 223·8 div., the amount of this difference did not appear at the time to be so great as to call for re-adjustment, but in calculating the observation of absolute intensity the observations on the 13th and 14th were reduced separately to the mean reading on those days respectively, and those on the 20th to the mean of the first week of regular observations, which commences on 16th October. The instrument had been dismounted before the observation of March, but the regular change evinced by the fortnightly means would have rendered the zero division of the scale in October, had it been determined, inapplicable in succeeding months.

Changes of the Horizontal Force and Inclination.—These elements, especially the last named, like the declination (p. 1.) were found to be liable to daily change far exceeding in magnitude anything commonly observed at stations in lower magnetic latitudes. The daily extremes of both are brought together in the next table. The section relating to the inclinometer may be referred to for the grounds on which the value of the scale has been assigned. The ranges of inclination here given are considered subject to an uncertainty not exceeding one tenth of their amount.

Table XIII.

Date Gött. Day.	BIFILAR. $k = \cdot 0003412$.						INCLINOMETER.						
	In the Hourly Series.		Observed.		Range of Hor. Force.		In the Hourly Series.		Observed.		Approx. Range of Inclination.		Scale.
	Highest.	Lowest.	Highest.	Lowest.	Hourly.	Observed.	Highest.	Lowest.	Highest.	Lowest.	Hourly.	Observed.	Coeff.
1843.											′	′	′
Oct. 16	300·6	77·6	300·6	77·0	·0761	·0763	325·9	86·9	341·6	86·9	43·1	45·9	0·1805
17	292·6	115·3	292·6	68·0	·0605	·0766	287·7	115·8	362·8	97·1	30·0	47·9	—
T { 18	311·0	231·7	314·1	231·7	·0271	·0278	152·6	83·6	153·6	74·8	12·4	14·2	—
T { 19	280·2	143·8	322·7	132·4	·0465	·0649	269·3	91·4	281·2	50·6	31·9	41·5	—
20	274·1	225·6	—	—	·0165	—	265·7	232·0	—	—	5·9	—	0·1747
21	277·5	260·3	—	—	·0059	—	245·1	185·0	—	—	10·5	—	—
22	S.	—	—	—	—	—	S.	—	—	—	—	—	—
23	276·9	226·1	—	—	·0173	—	269·5	241·9	—	—	4·8	—	—
24	246·8	153·6	246·8	144·9	·0318	·0348	323·2	236·1	349·7	236·1	15·2	19·8	—
25	240·2	164·4	240·2	0·9	·0259	·0816	332·2	248·3	556·2	248·3	14·6	53·8	—
26	240·8	111·8	240·8	111·8	·0440	·0440	391·5	240·6	391·5	240·6	26·0	26·0	—
27	231·5	63·4	231·5	52·5	·0573	·0611	414·7	249·3	449·8	249·3	28·9	34·9	—
28	272·2	160·2	—	—	·0382	—	307·7	252·0	—	—	9·7	—	—
29	S.	—	—	—	—	—	S.	—	—	—	—	—	—
30	288·5	156·4	288·5	83·6	·0451	·0699	390·2	258·9	492·7	258·9	20·9	40·8	—
31	279·7	178·3	293·9	155·1	·0346	·0480	341·0	255·6	378·0	255·6	14·9	21·3	—
Nov. 1	268·7	219·4	—	—	·0168	—	133·4	81·6	—	—	8·8	—	0·1708
2	266·3	117·9	278·3	81·2	·0506	·0673	255·3	81·1	286·5	51·7	29·7	40·1	—
3	238·9	129·9	—	—	·0031	—	222·0	99·9	—	—	20·8	—	—
*4	248·6	193·0	—	—	·0190	—	117·1	99·8	—	—	2·9	—	—
5	S.	—	—	—	—	—	S.	—	—	—	—	—	—
6	267·1	156·8	267·1	156·8	·0137	·0137	223·4	102·6	223·4	100·5	20·6	20·9	—
7	246·9	227·1	—	—	·0067	—	125·9	101·2	—	—	4·2	—	—
8	249·0	133·1	249·0	122·1	·0394	·0433	217·2	70·1	232·7	70·1	25·1	27·8	—
9	240·9	156·7	240·9	127·8	·0287	·0386	191·5	91·8	224·3	91·8	17·0	22·6	—
10	223·6	169·8	223·6	122·7	·0183	·0344	168·2	109·2	221·5	109·2	10·1	19·1	—
11	233·8	185·5	—	—	·0165	—	144·1	93·3	—	—	8·7	—	—
12	S.	—	—	—	—	—	S.	—	—	—	—	—	—
13	239·6	109·6	239·6	105·6	·0444	·0457	243·9	100·5	243·9	100·5	24·5	24·5	—
14	216·1	196·2	231·4	196·2	·0068	·0120	134·5	102·9	134·5	86·0	5·4	8·4	—
15	215·7	176·6	—	—	·0133	—	151·1	109·5	—	—	7·1	—	—
16	237·5	195·9	237·5	195·9	·0142	·0142	134·1	84·2	134·1	83·2	8·5	8·8	—
17	221·3	195·5	—	—	·0088	—	126·9	115·8	—	—	1·9	—	—

* Imperfect day.

Table XIII.—*continued.*

Date Gött. Day.	BIFILAR.	$k = \cdot 0003412$.					INCLINOMETER.						
	In the Hourly Series.		Observed.		Range of Hor. Force.		In the Hourly Series.		Observed.		Approx. Range of Inclination.		Scale.
	Highest.	Lowest.	Highest.	Lowest.	Hourly.	Observed.	Highest.	Lowest.	Highest.	Lowest.	Hourly.	Observed.	Coeff.
1848.											′	′	′
Nov. 18	203·6	187·2	—	—	·0054	—	131·5	115·8	—	—	2·7	—	0·1706
19	S.	—	—	—	—	—	S.	—	—	—	—	—	—
20	216·3	176·0	—	—	·0137	—	145·6	117·0	—	—	4·9	—	—
21	209·6	190·3	—	—	·0066	—	142·7	121·9	—	—	3·5	—	—
22	203·8	178·1	—	—	·0088	—	147·9	126·1	—	—	3·9	—	—
23	200·2	183·1	—	—	·0058	—	137·6	124·1	—	—	2·3	—	—
24	192·1	105·6	192·1	105·6	·0295	·0295	235·5	118·8	235·5	118·8	19·9	19·9	—
25	197·4	182·7	—	—	·0050	—	142·0	129·4	—	—	2·2	—	—
26	S.	—	—	—	—	—	S.	—	—	—	—	—	—
27	193·0	181·0	—	—	·0041	—	151·0	133·6	—	—	3·6	—	—
28	190·6	175·9	—	—	·0043	—	153·6	131·4	—	—	4·0	—	—
29	196·6	141·6	196·6	136·0	·0188	·0207	182·2	131·7	195·8	131·7	9·5	10·9	—
30	184·0	165·8	—	—	·0062	—	152·4	134·7	—	—	3·0	—	—
Dec. 1	192·6	90·4	192·6	57·7	·0349	·0400	279·6	128·8	313·2	128·8	25·8	31·5	—
2	185·0	63·8	185·0	41·7	·0414	·0490	271·7	134·1	328·9	134·1	23·5	33·3	—
3	S.	—	—	—	—	—	S.	—	—	—	—	—	—
4	185·9	167·4	—	—	·0063	—	159·0	146·4	—	—	2·2	—	—
5	182·7	96·6	182·7	73·9	·0294	·0371	237·9	143·5	287·1	143·5	16·1	24·5	—
6	220·0	126·5	220·0	116·5	·0319	·0353	205 1	117·9	218·9	117·9	14·9	15·5	—
7	186·6	174·6	—	—	·0041	—	161·2	146·6	—	—	2·5	—	—
8	220·4	161·5	220·4	161·5	·0201	·0201	194·0	117·1	194·0	117·1	13·1	13·1	—
9	205·5	151·8	—	—	·0183	—	185·5	129·2	—	—	9·6	—	—
10	S.	—	—	—	—	—	S.	—	—	—	—	—	—
11	195·4	163·0	—	—	·0110	—	185·2	132·4	—	—	9·0	—	—
12	199·4	165·2	—	—	·0117	—	177·6	142·9	—	—	5·9	—	—
13	183·4	149·2	—	—	·0117	—	191·0	136·7	—	—	5·8	—	—
14	171·3	143·0	—	—	·0096	—	188·5	165·8	—	—	3·9	—	—
15	178·0	152·4	—	—	·0087	—	183·6	160·9	—	—	3·9	—	—
16	171·6	161·1	—	—	·0036	—	181·2	166·3	—	—	2·5	—	—
17	S.	—	—	—	—	—	S.	—	—	—	—	—	—
18	179·0	161·2	—	—	·0061	—	179·3	153·1	—	—	4·5	—	—
19	175·8	101·2	175·8	41·8	·0254	·0457	255·8	159·6	357·8	159·6	16·4	33·9	—
T { 20	183·1	154·8	181·8	139·9	·0096	·0144	189·1	157·7	204·2	157·7	5·4	7·9	—
T { 21	183·3	163·0	183·3	163·0	·0060	·0060	181·0	154·2	181·0	134·2	4·6	4·6	—
22	182·3	161·7	—	—	·0070	—	176·7	160·4	—	—	2·8	—	—

TABLE XIII.—*continued.*

Date Gött. Day.	BIFILAR. $k = {}^{\cdot}0003412$.						INCLINOMETER.						
	In the Hourly Series.		Observed.		Range of Hor. Force.		In the Hourly Series.		Observed.		Approx. Range of Inclination.		Scale.
	Highest.	Lowest.	Highest.	Lowest.	Hourly.	Observed.	Highest.	Lowest.	Highest.	Lowest.	Hourly.	Observed.	Coeff.
1843.											′	′	′
Dec. 23	175·7	162·6	—	—	·0045	—	175 6	162·5	—	—	2·2	—	0·1708
24	S.	—	—	—	—	—	S.	—	—	—	—	—	—
25	Christmas Day.			—	—	—	Christmas Day.			—	—	—	—
26	187·3	77·3	187·3	52·8	·0375	·0456	299·2	156·5	328·9	156·5	24·4	29·4	—
27	203·5	123·0	212·5	123·0	·0275	·0305	223·2	146·6	229·2	129·7	13·6	17·0	—
28	200·4	112·6	200·4	112·6	·0299	·0299	229·5	143·7	235·7	141·2	14·7	16·1	—
29	183·4	73·8	183·4	40·7	·0375	·0488	276·8	163·2	340·5	163·2	19·4	30·1	—
30	170·4	148·5	—	—	·0075	—	182·1	158·4	—	—	4·0	—	—
31	S.	—	—	—	—	—	S.	—	—	—	—	—	—
1844.													
Jan. 1	—	—	—	—	—	—	—	—	—	—	—	—	—
2	181·9	80·8	—	—	·0345	—	281·6	149·1	—	—	22·6	—	—
3	160·9	141·7	—	—	·0065	—	198·4	175·0	—	—	4·0	—	—
4	224·9	−37·5	239·0	−51·9	·0895	·0991	472·9	110·3	477·0	85·5	61·9	66·9	—
5	182·7	−49·0	182·7	−49·0	·0790	·0790	380·3	152·0	380·3	152·0	39·0	39·0	—
6	161·0	29·8	161·0	29·8	·0447	·0447	328·4	164·7	328·4	164·7	27·9	27·9	—
7	S.	—	—	—	—	—	S.	—	—	—	—	—	—
8	154·6	121·4	170·4	87·9	·0113	·0281	221·2	175·8	261·0	158·9	7·7	17·4	—
9	157·1	125·7	—	—	·0107	—	222·5	159·5	—	—	10·8	—	—
10	157·1	126·9	—	—	·0103	—	238·1	179·6	—	—	10·0	—	—
11	153·3	112·4	—	—	·0139	—	236·9	184·6	—	—	8·9	—	—
12	163·6	141·6	—	—	·0075	—	213·5	138·3	—	—	12·9	—	—
13	154·2	138·3	—	—	·0054	—	208·3	187·1	—	—	3·6	—	—
14	S.	—	—	—	—	—	S.	—	—	—	—	—	—
15	159·2	138·2	—	—	·0072	—	215·1	193·8	—	—	3·6	—	—
16	157·0	144·8	—	—	·0042	—	208·5	190·4	—	—	3·1	—	—
17	162·0	123·5	162·0	98·9	·0131	·0215	233·5	180·5	252·6	180·5	9·1	12·3	—
18	145·5	123·5	—	—	·0075	—	220·2	195·8	—	—	4·2	—	—
19	160·3	104·5	160·3	104·5	·0190	·0197	262·8	199·6	262·8	199·6	10·8	10·8	—
20	150·5	110·0	—	—	·0138	—	216·8	196·8	—	—	3·4	—	—
21	S.	—	—	—	—	—	S.	—	—	—	—	—	—
22	157·1	102·3	—	—	·0187	—	261·3	173·6	—	—	15·0	—	—
23	138·4	120·8	—	—	·0060	—	212·5	194·6	—	—	3·1	—	—
{24	147·3	35·6	153·0	−37·6	·0381	·0650	381·2	195·6	401·9	130·2	31·7	46·4	—
{25	145·1	24·5	150·3	14·7	·0411	·0463	392·9	192·0	404·4	177·9	34·3	38·7	—

TABLE XIII.—*continued.*

Date Gött. Day.	BIFILAR. $k=\cdot0003412$.						INCLINOMETER.						
	In the Hourly Series.		Observed.		Range of Hor. Force.		In the Hourly Series.		Observed.		Approx. Range of Inclination.		Scale.
	Highest.	Lowest.	Highest.	Lowest.	Hourly.	Observed.	Highest.	Lowest.	Highest.	Lowest.	Hourly.	Observed.	Coeff.
1844.											′	′	′
Jan. 26	134·4	82·3	134·4	74·0	·0178	·0206	254·3	196·5	259·0	196·5	9·9	10·7	0·1708
27	135·5	111·3	—	—	·0082	—	219·2	193·2	—	—	4·4	—	—
28	S.	—	—	—	—	—	S.	—	—	—	—	—	—
29	145·5	119·1	—	—	·0090	—	238·5	189·1	—	—	8·4	—	—
30	159·4	117·7	—	—	·0142	—	229·7	178·4	—	—	9·2	—	—
31	159·5	112·2	—	—	·0161	—	219·4	167·7	—	—	8·8	—	—
Feb. 1	185·5	− 3·4	185·5	− 9·7	·0645	·0666	353·8	115·1	390·4	115·1	40·7	47·0	—
2	204·8	67·0	226·9	66·2	·0470	·0548	276·7	143·3	289·7	109·1	22·8	30·7	—
3	173·7	118·2	—	—	·0189	—	224·8	160·5	—	—	11·0	—	—
4	S.	—	—	—	—	—	S.	—	—	—	—	—	—
5	214·8	−18·0	219·3	−71·0	·0794	·0952	494·8	121·8	564·0	83·1	63·7	82·1	—
6	164·2	67·4	164·2	67·4	·0330	·0330	285·5	171·2	285·5	154·1	19·5	22·4	—
7	183·7	122·6	—	—	·0208	—	231·9	158·9	—	—	12·5	—	—
8	175·7	71·5	175·7	71·5	·0355	·0355	294·6	169·3	294·6	169·3	21·4	21·4	—
9	172·9	131·3	—	—	·0142	—	193·1	124·8	—	—	11·7	—	—
10	179·4	108·1	—	—	·0243	—	246·3	172·3	—	—	12·6	—	—
11	S.	—	—	—	—	—	S.	—	—	—	—	—	—
12	158·4	96·0	—	—	·0213	—	259·5	179·5	—	—	13·7	—	—
13	159·3	134·5	—	—	·0085	—	204·7	184·9	—	—	3·4	—	—
14	172·6	144·7	—	—	·0095	—	210·5	167·5	—	—	7·3	—	—
15	183·6	126·3	—	—	·0195	—	224·1	188·6	—	—	6·1	—	—
16	200·2	157·9	200·2	150·3	·0144	·0170	218·0	174·7	225·6	174·7	7·4	8·7	—
17	204·6	164·6	—	—	·0136	—	215·7	182·3	—	—	5·7	—	—
18	S.	—	—	—	—	—	S.	—	—	—	—	—	—
19	190·7	175·8	—	—	·0051	—	205·5	191·4	—	—	2·4	—	—
20	197·0	172·6	—	—	·0083	—	215·3	189·2	—	—	4·5	—	—
21	196·1	158·2	—	—	·0129	—	232·6	185·1	—	—	8·1	—	—
22	202·2	184·4	—	—	·0061	—	209·5	186·2	—	—	4·0	—	—
23	205·0	189·1	—	—	·0054	—	199·0	178·1	—	—	1·9	—	—
24	188·6	175·1	—	—	·0046	—	203·1	191·2	—	—	2·0	—	—
25	S.	—	—	—	—	—	S.	—	—	—	—	—	—
26	202·9	106·3	202·9	94·8	·0330	·0368	277·9	175·8	296·3	175·8	17·4	20·6	—
27	217·7	163·6	—	—	·0184	—	232·2	186·7	—	—	7·8	—	—
28	—	—	—	—	—	—	304·0	144·8	—	—	27·2	—	—
29	—	—	—	—	—	—	346·1	170·9	—	—	29·9	—	—

Table XIII.—*continued.*

Date Gött. Day.	BIFILAR. k=·0003412.						INCLINOMETER.						
	In the Hourly Series.		Observed.		Range of Hor. Force.		In the Hourly Series.		Observed.		Range of Inclination.		Scale.
	Highest.	Lowest.	Highest.	Lowest.	Hourly.	Observed.	Highest.	Lowest.	Highest.	Lowest.	Hourly.	Observed.	Coeff.
1844.											′	′	′
April 1	266·5	113·4	—	—	·0490	—	394·1	61·2	—	—	43·0	—	0·1293
2	284·7	79·3	284·7	<79·0	·0335	·065	467·7	58·0	>743·2	58·0	53·0	88·6	—
3	278·6	115·9	278·6	66·7	·0518	·0674	368·2	83·4	484·5	83·4	36·8	51·9	—
4	300·8	124·7	—	—	·0560	—	368·1	69·7	—	—	38·6	—	—
5	Good Friday.		—	—	—	—	Good Friday.		—	—	—	—	—
6	298·0	228·8	304·2	228·8	·0220	·0240	167·1	63·0	167·1	51·5	13·5	15·0	—
7	S.	—	—	—	—	—	S.	—	—	—	—	—	—
8	262·8	241·0	—	—	·0069	—	194·7	129·3	—	—	8·5	—	—
9	296·4	187·8	296·4	168·8	·0345	·0406	264·0	124·0	291·8	124·0	18·1	21·7	—
10	251·7	18·0	251·7	>0·0	>·0744	>·080	272·2	96·8	467·1	96·8	22·7	47·8	—
11	244·8	188·9	—	—	·0177	—	189·8	112·2	—	—	10·0	—	—
12	238·2	210·4	—	—	·0088	—	143·4	118·8	—	—	3·2	—	—
13	287·6	214·3	—	—	·0233	—	135·1	102·6	—	—	4·2	—	—
14	S.	—	—	—	—	—	S.	—	—	—	—	—	—
15	291·9	127·4	291·9	91·5	·0522	·0637	383·3	100·7	453·8	100·7	36·5	44·5	—
16	377·6	137·8	378·1	9·9	·0763	·1170	355·2	−11·0	631·0	−29·5	47·3	85·3	—
17	343·9	−60·8	343·9	−20·0	·1284	>·173	837·9	20·9	837·9	20·9	105·6	105·6	—
18	283·4	255·1	—	—	·0077	—	153·1	104·0	—	—	6·3	—	—
19	272·5	161·5	272·5	127·5	·0031	·0410	319·6	124·6	368·7	124·6	25·2	31·9	—
20	271·5	191·1	—	—	·0227	—	255·8	121·2	—	—	17·4	—	—
21	S.	—	—	—	—	—	S.	—	—	—	—	—	—
22	277·4	221·1	—	—	·0159	—	194·6	114·3	—	—	10·4	—	—
23	282·4	178·3	—	—	·0294	—	304·8	127·3	—	—	22·9	—	—
T { 24	304·2	264·9	316·2	171·3	·0111	·0408	173·6	109·4	332·4	87·3	8·3	31·7	—
T { 25	316·9	86·6	316·9	60·8	·0651	·0725	526·6	88·2	586·0	87·1	56·7	64·7	—
26	297·1	26·8	297·1	−46·7	·0764	·0973	645·9	108·2	781·5	108·2	69·5	100·0	—
27	309·2	137·8	309·2	133·2	·0486	·0499	377·4	85·2	415·2	85·2	37·8	42·8	—
28	S.	—	—	—	—	—	S.	—	—	—	—	—	—
29	285·1	137·0	285·1	114·3	·0419	·0484	423·3	126·1	455·5	126·1	38·4	42·5	—
30	326·4	80·0	335·0	40·0	·0697	·0835	469·1	80·7	562·7	80·7	52·8	59·7	—
May 1	289·0	219·5	—	—	·0197	—	256·9	129·4	—	—	16·8	—	—
2	336·4	207·8	336·4	207·8	·0362	·0362	Imperfect.		—	—	—	—	—
3	271·2	43·0	271·2	39·0	·0645	·0656	614·1	204·0	627·1	204·0	63·6	96·5	0·1540
4	286·7	215·1	—	—	·0203	—	263·0	184·0	—	—	11·2	—	—.

TABLE XIII.—*continued.*

Date Gött. Day.	BIFILAR. k = ·0003412.						INCLINOMETER.						
	In the Hourly Series.		Observed.		Range of Hor. Force.		In the Hourly Series.		Observed.		Range of Inclination.		Scale.
	Highest.	Lowest.	Highest.	Lowest.	Hourly.	Observed.	Highest.	Lowest.	Highest.	Lowest.	Hourly.	Observed.	Coeff.
1844.											′	′	′
May *5	S.	—	258·5	149·7	—	·0308	S.	—	385·5	202·1	—	23·6	—
6	280·7	229·6	—	—	·0144	—	282·4	117·0	—	—	25·6	—	—
7	300·3	167·9	300·3	167·9	·0373	·0373	407·2	165·4	407·2	165·4	37·5	37·5	—
8	306·7	176·1	306·7	176·1	·0368	·0368	400·8	160·2	400·8	160·2	37·3	37·3	—
9	276·1	162·1	—	—	·0322	—	422·6	221·4	—	—	31·2	—	—
10	285·7	192·0	—	—	·0262	—	281·6	228·2	—	—	8·3	—	—
11	274·1	238·4	—	—	·0101	—	291·8	222·1	—	—	10·8	—	—
12	S.	—	—	—	—	—	S.	—	—	—	—	—	—
13	325·6	242·1	325·6	242·1	·0236	·0236	282·0	134·0	280·2	134·0	23·0	23·6	—
14	325·7	223·4	225·7	214·3	·0289	·0316	311·0	138·4	338·0	138·4	26·8	30·9	—
15	291·8	236·9	—	—	·0156	—	270·0	198·0	—	—	11·0	—	—
16	267·5	212·1	267·5	212·1	·0137	·0157	324·8	221·1	324·8	221·1	16·1	16·1	—
17	263·9	241·9	—	—	·0062	—	258·7	227·4	—	—	4·9	—	—
18	283·3	242·4	—	—	·0115	—	239·6	218·0	—	—	6·4	—	—
19	S.	—	—	—	—	—	S.	—	—	—	—	—	—
20	273·3	238·7	—	—	·0098	—	286·7	236·0	—	—	7·9	—	—
21	288·0	224·0	—	—	·0181	—	300·2	216·7	—	—	13·9	—	—
22	308·9	112·8	344·3	28·3	·0553	·0904	519·2	161·6	692·3	121·9	55·4	88·3	—
23	272·7	98·9	272·7	98·9	·0492	·0520	519·1	243·8	545·9	243·8	42·7	46·8	—
T { 24	285·3	100 3	285·3	140·5	·0354	·0410	440·0	211·0	468·8	184·4	35·5	44·1	—
T { 25	—	—	—	214·9	Imperfect.		—	—	—	—	—	—	—

* Extra observations were taken from $5^d\ 21^h$ to $5^d\ 23^h$.

It appears from the foregoing table, that the greatest change of Horizontal Force observed in any Göttingen day in the winter, was ·0991X, on the 4th January 1844, the change of Inclination observed being 1° 6′·9, according to the approximate scale value employed. The greatest change of Horizontal Force observed in any Göttingen day of the two spring months was not less than 0·173 X, on the 17th April, or one sixth of the whole amount of that element, and was accompanied by a change of inclination of 1° 45′·7; the movements of both instruments upon this occasion went beyond the limits of their scales, and could only be valued approximately, by holding up some object and afterwards measuring its distance from the zero of the scale. The mean inclination at Lake Athabasca was

81° 37′·7, and at Fort Simpson 81° 52′·0; a change of ± 1′ of these elements would therefore produce a change of ∓ ·001996 X at the former, and of ∓ ·002037 X at the latter station; in round numbers ± 1′·0 of inclination corresponds to ∓ ·002 of horizontal force at both stations; classifying the daily ranges upon this scale, we have the following results:

TABLE XIV.

Range of Horizontal Force.	Lake Athabasca.		Fort Simpson.		Range of Inclination Approximate.	Lake Athabasca.		Fort Simpson.	
	Hourly Observations.	Including Disturbances.	Hourly Observations.	Including Disturbances.		Hourly Observations.	Including Disturbances.	Hourly Observations.	Including Disturbances.
	Days.	Days.	Days.	Days.		Days.	Days.	Days.	Days.
Less than ·010 X	40	38	6	5	Less than 5′	37	37	3	3
·010 to ·020 X	33	30	10	9	5′ to 10′	27	23	7	6
·020 - ·030 X	12	11	6	5	10′ - 15′	18	15	6	6
·030 - ·040 X	13	11	6[*]	4[*]	15′ - 20′	8	7	4	3
·040 - ·050 X	8	11	4	7	20′ - 25′	9	10	3	3
·050 - ·060 X	2	1	6	4	25′ - 30′	9[†]	6	3	1
·060 - ·070 X	2	6	4	5	30′ - 35′	3	6	1	4
·070 - ·080 X	3	4	3	1	35′ - 40′	1	2	8	3
·080 - ·090 X	1	0	0	2	40′ - 45′	2	3	2	5
·090 - ·100 X	0	2	0	2	45′ - 50′	0	4	1	2
More than ·100	0	0	1	2	Above 50′ -	2	3	7	9
	114	114	46	46		116	116	45	45

[*] A day is here included which is wanting in the Inclinometer series, namely, May 2.
[†] Two days are included which are wanting in the Bifilar series, namely, Feb. 28 and 29, 1844.

It may be remarked in reference to this table, that were the total force to undergo no changes, we should expect to find an exact coincidence between the number of days giving certain ranges of the horizontal force, and their equivalents in terms of the inclination, unless there existed an error in one of the co-efficients; the great changes of the former element being always positive, and those of the horizontal force negative, we have an indication, in the excess under the higher ranges of inclination, which is apparent above, that the tendency in disturbances is to an increase of total force.

The mean range of horizontal force during the winter months at Lake Athabasca, that is to say, the square root of the mean of the squares of the differences between the highest and lowest scale reading, included in the hourly observations of each day, is ·0286 X, and that of the inclination 17′·0 (minutes). The corresponding means for the two spring months at Fort Simpson are ·0421 X, and

47″·5 of inclination. The other American stations give the following values for the mean range of horizontal force found in the same way, and for the same periods. For the winter months, 16th October 1843 to 29th February 1844, Philadelphia ·00149 X, Toronto ·00240 X, and 3″·25 of inclination*, Sitka ·00377 X. For the two spring months, April and May 1844, Philadelphia ·00157 X, Toronto ·00357 X, Sitka ·00429 X. For the several months again, we have the mean ranges in the following table, which, like that of the declination, I have extended to include twelve months at the permanent stations.

TABLE XV.

Month.	Philadelphia.		Toronto.				Sitka.		Lake Athabasca. — Fort Simpson.			
			Bifilar.		Inclinom.				Bifilar.		Inclinom.	
	Scale Divs.	$\frac{\Delta X}{X}$	Scale.	$\frac{\Delta X}{X}$	Scale.	$\Delta\theta$	Scale. Divs.	$\frac{\Delta X}{X}$	Scale.	$\frac{\Delta X}{X}$	Scale.	$\Delta\theta$
1843.												
October -	45·7	·00183	—	—	—	—	35·4	·00428	—	—	—	—
16th to 31st -	36·9	·00148	22·9	·00242	5·02	4′·23	31·3	·00388	118·3	·0402	127·0	22′·5
November -	30·2	·00121	21·7	·00229	3·46	2·92	28·0	·00347	62·5	·0213	75·2	12·8
December -	32·6	·00130	20·5	·00217	3·25	2·75	26·4	·00327	63·1	·0215	73·1	12·5
1844.												
January -	32·7	·00131	22·7	·00240	3·57	3·01	29·9	·00383	84·8	·0289	112·3	19·2
February -	34·1	·00136	26·3	·00278	—	—	[illegible]·9	·00472	89·8	·0306	105·0	17·9
March - -	43·8	·00175	31·5	·00333	—	—	46·3	·00592	—	—	—	—
April - -	38·9	·00156	35·4	·00268	—	—	38·9	·00498	166·8	·0507	305·2	39·4
May - -	39·9	·00159	32·6	·00344	—	—	30·2	·00386	113·1	·0324	189·2	29·3
June - -	31·7	·00127	28·5	·00301	—	—	24·8	·00317	—	—	—	—
July - -	28·1	·00112	34·4	·00363	—	—	29·0	·00359	—	—	—	—
August - -	43·8	·00175	39·6	·00418	—	—	31·5	·00386	—	—	—	—
September -	49·7	·00159	36·4	·00385	—	—	42·0	·00520	—	—	—	—

The scale co-efficients used above were 1 division = − ·00040 X at Philadelphia, (magnetic and met. observations at Girard College, p. 1819), = ·0001236 at Sitka for the observations of 1843, and ·000128 for those of 1844, = ·000105 for those at Toronto; the observations at the two former stations were not reduced to a uniform temperature, but from the irregularity in the hours of occurrence of the greatest and least values, the effects of inequalities of

* By observations with an instrument of one bar from 16th October 1843 to 10th February 1844, 1 division = 0′·820.

temperature must be in a great measure neutralized in the final result for each month.

Diurnal Variation of the Horizontal Force.—The following Tables contain the mean of the scale readings of the Bifilar as observed, and the same reduced to the uniform temperature of 40°. As the observatory was artificially warmed, the mean daily range of internal temperature is small, although the occasional fluctuations were very considerable; the uncorrected curve differs in consequence comparatively little from the corrected one.

Table XVI.

Monthly means of the Bifilar readings, uncorrected for temperature, and with omission of incomplete days, namely, October 20th, November 4th, January 2d and 9th, February 28th and 29th.

Civil time -	h. m. 15 55	h. m. 16 55	h. m. 17 55	h. m. 18 55	h. m. 19 55	h. m. 20 55	h. m. 21 55	h. m. 22 55	Noon.	h. m. 0 55	h. m. 1 55	h. m. 2 55
Gött. time -	0	1	2	3	4	5	6	7	8	9	10	11
1843. October -	208·53	205·01	228·63	242·06	247·14	245·39	250·56	250·71	248·07	247·77	256·34	255·96
November -	196·44	199·06	201·96	204·29	203·14	205·99	202·89	206·02	202·99	204·19	204·88	206·84
December -	164·53	166·06	167·85	165·66	172·86	174·36	172·30	170·44	172·19	172·90	176·94	177·58
1844. January -	133·34	135·53	142·10	145·34	147·70	147·57	145·77	145·82	146·80	144·85	145·77	148·94
February -	140·36	147·70	155·69	157·62	156·64	153·50	155·77	157·57	160·80	159·22	164·57	165·48
Mean -	164·53	167·66	174·63	177·35	179·64	179·74	179·24	179·27	180·24	179·88	183·29	184·55

Civil time -	h. m. 3 55	h. m. 4 55	h. m. 5 55	h. m. 6 55	h. m. 7 55	h. m. 8 55	h. m. 9 55	h. m. 10 55	Mid.	h. m. 12 55	h. m. 13 55	h. m. 14 55	
Gött. time -	Mid.	13	14	15	16	17	18	19	20	21	22	23	Mean.
1843. October -	254·04	255·19	253·23	253·45	249·84	251·36	252·82	237·92	217·41	233·20	219·89	236·31	241·60
November -	207·86	209·00	209·90	210·33	210·78	208·43	208·66	207·49	208·46	195·42	196·56	192·02	203·98
December -	177·78	178·46	177·94	176·76	177·50	177·96	177·51	178·49	174·73	168·97	157·17	157·21	172·29
1844. January -	148·76	149·73	149·86	151·47	152·86	150·04	153·58	154·70	145·12	137·82	140·60	134·07	145·21
February -	167·57	169·24	170·43	170·64	169·81	170·48	172·79	173·14	168·36	160·13	155·61	153·92	161·34
Mean -	185·16	186·27	186·41	186·42	186·63	185·86	187·26	185·80	178·51	173·92	169·39	165·37	179·47

Table XVII.

Mean temperature of the Bifilar magnet.

Civil time -	h. m. 15 55	h. m. 16 55	h. m. 17 55	h. m. 18 55	h. m. 19 55	h. m. 20 55	h. m. 21 55	h. m. 22 55	Noon	h. m. 0 55	h. m. 1 55	h. m. 2 55
Gött. time -	0	1	2	3	4	5	6	7	8	9	10	11
1843.	°	°	°	°	°	°	°	°	°	°	°	°
October -	43·98	44·33	44·67	44·55	44·77	44·30	44·06	44·25	45·75	46·63	49·50	47·01
November -	43·31	42·82	43·62	43·74	43·29	42·52	42·61	43·00	43·54	43·75	43·89	43·91
December -	40·32	39·96	39·72	39·96	39·96	39·30	39·20	40·72	41·69	41·62	41·14	40·96
1844.												
January -	28·42	27·85	27·25	28·10	28·32	28·06	27·4[illegible]	28·47	29·67	31·38	30·87	30·81
February -	42·02	42·10	42·07	42·33	41·80	41·39	40·87	42·13	43·30	44·36	44·62	45·10
Mean - -	39·20	38·93	38·96	39·28	39·13	38·60	38·34	39·28	40·31	41·27	41·24	41·00

Civil time -	h. m. 3 55	h. m. 4 55	h. m. 5 55	h. m. 6 55	h. m. 7 55	h. m. 8 55	h. m. 9 55	h. m. 10 55	Mid.	h. m. 12 55	h. m. 13 55	h. m. 14 55	Mean.
Gött. time -	Mid.	13	14	15	16	17	18	19	20	21	22	23	
1843.	°	°	°	°	°	°	°	°	°	°	°	°	°
October -	46·12	45·43	45·26	45·49	45·46	45·43	46·39	46·43	46·81	44·82	44·34	44·83	45·50
November -	43·66	43·66	43·46	42·62	42·26	42·35	43·28	43·68	43·74	43·21	43·48	43·11	43·27
December -	41·19	41·06	40·43	39·82	39·77	40·50	41·08	40·34	39·68	39·17	39·71	40·20	40·32
1844.													
January -	30·91	31·22	31·05	30·45	29·73	29·06	28·95	29·39	29·72	27·62	28·13	28·56	29·23
February -	44·68	44·39	44·33	44·28	44·29	44·27	44·03	44·13	44·32	43·50	43·04	42·55	43·33
Mean - -	40·82	40·72	40·46	40·02	39·76	39·80	40·18	40·22	40·24	39·11	39·26	39·28	39·81

The general mean for each hour in the above Tables has been obtained by dividing the sum of all the observations by the total number, which is 110.

In the next Table, the difference from 40° of each mean temperature in Table XVII has been multiplied by the co-efficient $\frac{q}{k} = 0\cdot702$, anp applied to the values in Table XVI.

Table XVIII.

Mean Bifilar readings reduced to a uniform temperature of 40°.

Civil time	h. m. 15 55	h. m. 16 55	h. m. 17 55	h. m. 18 55	h. m. 19 55	h. m. 20 55	h. m. 21 55	h. m. 22 55	Noon.	h. m. 0 55	h. m. 1 55	h. m. 2 55	
Gött. time	0	1	2	3	4	5	6	7	8	9	10	11	
1843. October -	(−) 206·80	208·08	231·94	245·29	250·51	246·44	233·44	253·73	252·15	258·90	(+) 263·28	260·94	
November	198·79	201·06	204·58	208·94	205·47	207·68	204·74	205·15	205·50	206·85	207·63	209·61	
December	164·74	166·03	167·65	165·63	172·85	173·80	171·56	170·95	173·59	174·06	177·75	178·28	
1844. January -	124·92	126·90	133·05	136·89	139·41	139·09	136·86	137·63	139·47	138·73	139·29	141·72	
February -	(−) 141·79	149·19	157·16	159·27	158·02	154·49	156·39	159·08	163·14	162·31	167·85	169·66	
Means -	163·97	166·84	173·90	176·85	179·03	178·76	178·06	178·77	180·46	180·77	184·16	185·23	
Civil time	h. m. 3 55	h. m. 4 55	h. m. 5 55	h. m. 6 55	h. m. 7 55	h. m. 8 55	h. m. 9 55	h. m. 10 55	Mid.	h. m. 12 55	h. m. 13 55	h. m. 14 55	Mean.
Gött. time	Mid.	13	14	15	16	17	18	19	20	21	22	23	
1843. October -	258·38	239·04	256·88	257·35	255·72	255·23	257·88	242·48	222·26	236·62	222·97	229·88	244·98
November	210·46	211·60	(+) 212·46	212·28	212·88	210·10	210·97	210·10	206·11	197·10	199·18	(−) 194·28	206·80
December	178·63	(+) 179·21	178·25	176·63	177·84	178·32	178·28	178·73	174·50	168·38	(−) 156·97	157·07	172·52
1844. January -	142·31	143·30	143·51	144·60	145·50	142·28	145·74	(+) 147·17	137·82	129·03	133·18	(−) 115·96	137·66
February -	170·89	172·36	173·50	173·66	172·85	173·46	175·65	(+) 176·07	171·42	162·61	157·77	155·88	168·70
Mean -	185·73	186·77	186·73	186·43	186·46	185·74	187·41	185·93	178·66	173·30	169·67	164·87	179·33

It is remarkable, that the above means have a decided feature in common, which is not found in the corresponding ones at Toronto, or at any other American station,—they all exhibit a minimum of Horizontal Force at or near 3 A.M. By omitting all days on which extra observations for Disturbance were taken, as in Table VII., the lowest value of the 24^h is still at 3 A.M., but the amount of the daily change is most materially reduced, proving this feature to be, in great measure, due to the effect of disturbances, which has already been shown to be the case with the extreme of Declination at the same hour. The following Table exhibits, side by side, the mean diurnal curve of Horizontal Force at all the American stations, for the period included in the observations under discussion, to which is added the mean by the 46 days selected as free from disturbance.

TABLE XIX.

Comparison of the mean diurnal curve of Horizontal Force at all the American stations for the period included between October 1843 *and February* 1844.

Local mean time.	Philadelphia.		Toronto.		Sitka.		Lake Athabasca.			
							The whole period.		Forty-six selected days.	
	Scale.	$\frac{\Delta X}{X}$	Scale.	$\frac{\Delta X}{X}$	Scale.	$\frac{\Delta X}{X}$	Scale.	$\frac{\Delta X}{X}$	Scale.	$\frac{\Delta X}{X}$
Mid.	161·74	−·000071	493·15	−·000098	509·08	+·000194	178·68	−·00022	173·90	+·00124
1 A.M.	163·72	+·000008	493·22	−·000178	506·24	−·000159	173·30	−·00206	169·21	−·00038
2	163·98	+·000019	493·61	−·000066	506·06	−·000183	169·07	−·00350	167·03	−·00116
3	165·32	+·000072	494·02	−·000023	505·44	−·000259	164·87	−·00513	166·33	−·00137
4	167·42	+·000156	495·10	+·000091	504·74	−·000347	163·97	−·00522	166·90	−·00117
5	167·79	+·000171	496·11	+·000191	505·02	−·000312	166·84	−·00424	167·07	−·00112
6	168·12	+·000184	496·66	+·000258	503·98	−·000443	173·90	−·00183	167·43	−·00099
7	167·08	+·000134	496·23	+·000212	503·52	−·000500	176·85	−·00083	167·67	−·00091
8	163·22	−·000011	494·02	−·000023	504·66	−·000357	179·08	−·00010	168·41	−·00066
9	161·24	−·000100	492·22	−·000218	504·62	−·000362	178·76	−·00019	167·94	−·00082
10	158·56	−·000207	490·39	−·000407	505·26	−·000282	178·08	−·00042	166·51	−·00131
11	156·94	−·000272	488·78	−·000577	504·90	−·000327	178·77	−·00019	166·69	−·00124
Noon.	158·00	−·000229	489·11	−·000542	505·52	−·000249	180·46	+·00039	167·85	−·00095
1 P.M.	161·32	−·000096	491·04	−·000338	507·06	−·000058	180·77	+·00049	168·59	−·00060
2	164·60	+·000035	494·30	+·000006	507·80	+·000037	184·16	+·00165	171·37	+·00035
3	166·70	+·000119	496·53	+·000242	508·92	·000177	185·25	+·00202	172·31	+·00067
4	167·40	+·000147	498·03	+·000401	510·60	+·000387	185·73	+·00218	173·88	+·00122
5	167·42	+·000148	498·12	+·000410	511·14	+·000455	186·77	+·00254	173·73	+·00116
6	165·22	+·000060	497·12	+·000304	512·28	+·000587	186·73	+·00252	173·96	+·00123
7	163·92	+·000008	496·16	+·000203	511·84	+·000543	186·43	+·00242	173·00	+·00091
8	163·80	+·000003	495·84	+·000179	511·10	+·000450	186·46	+·00243	173·34	+·00102
9	163·32	−·000016	495·00	+·000080	510·18	+·000335	185·74	+·00219	173·45	+·00106
10	162·06	−·000067	493·80	−·000046	510·14	+·000330	187·41	+·00276	175·49	+·00176
11	161·72	−·000080	493·28	−·000101	510·12	+·000327	185·95	+·00226	176·13	+·00197
	163·73	—	494·24	—	507·51	—	179·33	—	170·34	—

The observations were taken 19[m] after the hour named at Philadelphia, 3[m] after at Toronto, 28[m] after at Sitka, and 5[m] after at Lake Athabasca.

Fort Simpson.—The Bifilar magnetometer at Fort Simpson received an accidental shock on the 10th April, which rendered it necessary to readjust it*, this was done on the 13th, a correction being applied to the intermediate readings. We have therefore two series, the first of only eleven days, of which one is incomplete, the second of thirty-five days, nine of which, however, want one observation or more. A separate mean for each will be found in the abstract. The whole forty-six days have also been combined in a general mean, without omission of any one, for the reasons already stated in reference to the Declination Observations, page 15.

Table XX.

Mean scale reading and temperature of the Bifilar Magnet at Fort Simpson in April and May 1844, *to which are added the mean for the same two months at all the American stations, and the difference of each reading from the mean of the whole, in terms of the Horizontal Force. The scale readings at Philadelphia are the complement to* 1100 *of the actual readings.*

Fort Simpson.						Local mean time.	Philadelphia.		Toronto.		Sitka.	
Gött. Hour.	Local mean time.	Observed. Scale.	Observed. Temp.	Means red. to Temp. 60°.	$\frac{\Delta X}{X}$		Scale.	$\frac{\Delta X}{X}$	Scale.	$\frac{\Delta X}{X}$	Scale.	$\frac{\Delta X}{X}$
	h. m.		°									
21	12 15	232·11	64·7	235·85	−·00190	12	106·70	+·00004	504·04	−·00020	515·20	−·00016
22	13 15	227·25	64·5	230·83	−·00336	13	107·65	+·00008	503·87	−·00022	514·20	−·00028
23	14 15	228·26	64·0	231·48	−·00317	14	108·90	+·00013	503·92	−·00021	513·70	−·00035
Noon	15 15	205·47	64·3	208·94	−·00973	15	110·55	+·00019	503·40	−·00027	515·30	−·00014
1	16 15	202·20	63·8	205·27	−·01080	16	112·60	+·00028	506·02	+·00001	515·15	−·00016
2	17 15	211·70	63·3	214·36	−·00865	17	112·80	+·00028	506·44	+·00005	514·45	−·00025
3	18 15	217·23	63·2	219·77	−·00658	18	110·45	+·00019	503·67	−·00024	514·40	−·00026
4	19 15	224·56	63·0	226·98	−·00448	19	109·70	+·00016	504·08	−·00019	514·65	−·00023
5	20 15	230·04	62·8	232·30	−·00293	20	104·85	−·00003	502·87	−·00032	514·45	−·00025
6	21 15	238·54	60·5	238·95	−·00099	21	97·35	−·00033	499·06	−·00073	514·25	−·00028
7	22 15	244·67	61·7	246·05	+·00107	22	94·80	−·00044	497·05	−·00094	512·50	−·00050
8	23 15	247·70	62·5	249·68	+·00213	23	94·40	−·00045	498·30	−·00081	512·10	−·00054
9	0 15	244·99	62·8	247·27	+·00143	Noon	94·50	−·00045	501·27	−·00049	510·55	−·00075
10	1 15	244·82	63·2	247·36	+·00145	1	99·25	−·00026	506·52	+·00006	511·40	−·00064
11	2 15	250·13	63·6	253·04	+·00310	2	106·35	+·00003	511·22	+·00056	512·90	−·00045
Mid.	3 15	256·33	63·2	258·92	+·00482	3	106·95	+·00005	514·53	+·00091	516·85	+·00005
13	4 15	256·54	63·8	259·56	+·00500	4	109·90	+·00017	514·55	+·00091	520·45	+·00051
14	5 15	258·89	64·3	262·30	+·00580	5	110·20	+·00018	516·19	+·00108	521·70	+·00067
15	6 15	259·30	64·4	262·84	+·00596	6	107·35	+·00007	512·04	+·00065	522·00	+·00071
16	7 15	258·02	64·5	261·64	+·00561	7	106·30	+·00002	508·94	+·00032	521·75	+·00068
17	8 15	257·65	64·6	261·33	+·00552	8	106·50	+·00003	506·82	+·00009	522·20	+·00074
18	9 15	253·68	65·0	257·66	+·00445	9	105·65	−·00000	505·71	−·00002	522·35	+·00076
19	10 15	252·12	65·1	256·21	+·00401	10	106·85	+·00005	506·70	+·00008	522·35	+·00076
20	11 15	243·94	65·3	248·21	+·00170	11	106·05	+·00001	505·04	−·00009	519·50	+·00039
	—	—	—	242·37	—		105·69	—	505·93	—	516·43	—

See remark at Table XIX. as to the difference of the actual observation from the hours given.

* The details of the adjustments will be found in a future section.

Mean diurnal curve of Horizontal Force.—The mean diurnal curve of the Horizontal Force, as given by observation, and influenced by disturbances, appears to consist at Lake Athabasca of a single progression, having its minimum at 4 A.M. and its maximum at 10 P.M., and agreeing in neither respect with the diurnal changes of this element at Toronto. The latter have at the same season two maxima, namely, at 3 P.M. and 6 A.M., and two minima, at 11 A.M. and 1 A.M. respectively.

Upon closer examination, it is evident that a second progression is superadded to the first at Lake Athabasca, which produces a subordinate maximum at 8 A.M., and a minimum at 10 A.M. Lastly, by omitting days most influenced by disturbance, and confining our attention to the mean given by 46 days, which were in a great measure free from it, we obtain evidence of the two diurnal maxima and two minima, as at Toronto, but accompanied by a third and more considerable maximum at 11 P.M., of which there is no trace at the latter station. The first of these maxima occurs at 8 A.M., and appears to correspond to that which occurs two hours earlier at Toronto; the second occurs at 4 or 5 P.M., and corresponds to the principal daily maximum at Toronto; the third is caused by the Horizontal Force retaining its high value after the hour just named at Athabasca, and even exhibiting an increase of it at 10 and 11 P.M. immediately before its great diurnal decline, whereas at Toronto it uniformly and steadily declines from 5 P.M. to 1 A.M.

Comparing together the values at the American stations as a group, from Table XIX., as laid down, plate 2, we find that the mean diurnal curve of the horizontal component at the two most southern stations, Philadelphia and Toronto, which are about 300 miles distant from each other, are similar in their hours of maxima and minima, but differ considerably in the value of their ordinates for the same hours, those at Toronto being much the larger, especially about the time of the morning minimum at 11 A.M., and of the principal maximum at 5 P.M.; they both present an increasing force at the hours at which it is decreasing to its lowest value at Lake Athabasca. At Sitka, which geographically is not far distant from the last-named station, while magnetically it belongs to the same group as the former, we have a curve of intermediate character; the great decrease of the Horizontal Force from 11 P.M. to 4 A.M., which occurs at Lake Athabasca, and is there followed by an equally rapid return towards mean values until 8 A.M., is, it is true, wanting, but we have a continuous slightly decreasing value, from 11 P.M. to 7 A.M., being the same period in which it is increasing at Toronto and Philadelphia. The curves in fact exhibit a striking progression of character, both in respect to the minimum and maximum of force;

we see the latter increasing rapidly in amount, and tending more and more towards an advanced period of the afternoon, as we proceed to the north, each culminating point falling above and in advance of that of the curve belonging to the stations to the southward from the lowest, which is that of Philadelphia, to the greatest, which belongs to Lake Athabasca. Again, we find the maximum at 6 A.M., which at Philadelphia exceeds that at 5 P.M. in amount, at Toronto is considerably less than the latter; at Sitka it cannot be distinguished with certainty upon the observations of one winter; and at Lake Athabasca we find in its place the very low values already pointed out, or if the small relative maximum of 8 A.M. be identified with it, it exists only as an inconsiderable undulation upon a much larger movement, determined probably by other causes.

At Fort Simpson the mean curve does not differ in general character from that at Athabasca, but is enormously increased in amplitude; the extreme deviations, both positive and negative, are doubled in amount, and there are other proofs of the influence of the advance of the season, the subordinate maximum just referred to being reduced to a still smaller relative amount, and shown three hours later; we find also no trace of the increase of the element preceding its great nocturnal decline, which was remarked in every one of the winter months; it declines slightly from 6 P.M. to 10 P.M., and then the great movement commences. At this station, as at Lake Athabasca, the mean curve by the induction inclinometer follows all the inflexions of that of the Horizontal Force, and gives a satisfactory confirmation of the accuracy with which they are represented.

A comparison of all the American stations for the two spring months, confirms the previous remark as to the systematic character of their differences, but shows also the curious fact that the *relative* change from winter to spring was less at Sitka than at either of the other stations, which is also apparent in the declination.

Induction Inclinometer.—The instrument employed for measurement of the changes of Inclination was the Unifilar, with which the absolute determinations of the Horizontal Force were made. The arm opposite to the one which carried the reading telescope and scale was provided with a socket, at the distance of 5 inches from the suspended magnet, for the reception of a single soft iron bar of 12 inches in length. The length of the suspended magnet was 2·5 inches; the arc value of the scale was 1'·0.

This instrument was one of the first of the kind that were made, and the first employed in any of the colonial observatories. I believe, also, that the present observations with it are the first that have been published at large, and as the Induction Inclinometer is less known than any other of the magnetical instruments referred to in this account, and has been less generally employed than its merits appear to deserve*, it will be proper to state at some length the grounds for that degree of confidence in the results which has led to their being included in the present volume.

The principle of the instrument may be stated from the explanations of Dr. Lloyd, as follows †:—If a soft iron bar, perfectly devoid of magnetic polarity, be held in a vertical position, it immediately becomes a temporary magnet under the inducing action of the earth's magnetic force, the lower extremity becoming a north pole, and the upper a south pole; accordingly, if a freely suspended magnet, whose dimensions are small in comparison with those of the bar, be situated near, and in a plane passing through one of these poles, it will be deflected from the magnetic meridian. The deflecting force is the induced force of the bar, which is a function of the vertical component (Y) of the earth's magnetic force and of the temperature, but depends also upon the quantity and distribution of the magnetism in the bar, and its distance from the suspended magnet. In practice it may also contain a term depended upon the *permanent* magnetism of the bar, which is seldom wholly evanescent. The tendency of this force is to turn the magnet; it is resisted by the horizontal component (X) of the same force; under the opposing influence of these two forces the bar assumes a position of equilibrium at a certain angle (u) from the magnetic meridian. This position serves to determine the ratio which subsists between them, and from the changes which it undergoes, may be, in like manner, inferred the changes of this ratio, and therefore those of the magnetic inclination.

The moment of free magnetism of the suspended magnet being denoted by m, let mU be the moment of the force exerted on it by the iron bar, U being, as already stated, a function of the vertical component and of the temperature; then, since mX sin u is the

* The advantages of the Instrument are these: Its construction is not attended with any of the mechanical difficulties which have led to the failure of the Balance Magnetometers. The changes of inclination being given directly, the deduction of those of the total force is much facilitated. It can be employed with increased advantage where the Balance Magnetometer, which is its only substitute, becomes nearly useless from its limited range of scale, and its unsteadiness in disturbances. It is easily adjusted, and not liable to get out of adjustment. It is observed with the same facility as a Declinometer, and its coefficient can be verified as often as we please without interrupting the series of observations; this last circumstance was not known at the period of the present observations, which was prior to the suggestion by Dr. Lamont of the method of deflection, to be referred to presently.

† Proceedings of the Royal Irish Academy, 1842 and 1850; also Letter to Colonel Sabine, dated 12th October 1848, printed and circulated for the information of the Directors of the British Colonial Observatories.

moment of the opposing force of the horizontal component X exerted at the angle u, the equation of equilibrium is

$$U = X \sin u$$

now let the two components of the earth's force undergo any small changes ΔX and ΔY, and let $V\Delta Y$ be the change of U, then Δu, denoting the corresponding change of the angle u, in parts of radius

$$V\Delta Y = X \cos u \Delta u + \Delta X \sin u$$

whence, dividing by the equation $Y = X \tan \theta$, in which θ denotes the magnetic inclination,

$$V \tan \theta \frac{\Delta Y}{X} = \cos u \Delta u + \sin u \frac{\Delta X}{X}$$

or if $p = V^{-1} \cot \theta$ we have

$$\frac{\Delta Y}{Y} = p \left(\cos u \Delta u + \sin u \frac{\Delta X}{X}\right.$$

assuming that the induced magnetism of the iron bar is proportional to the inducing force, the co-efficient p may be found by inverting the bar and observing the angle of deflection in the direct and inverted positions; denoting these angles by u and u', it is shown that

$$p = \frac{2}{\sin u + \sin u'}$$

whence

$$\frac{\Delta Y}{Y} = \frac{\cos u}{\sin S \cos D} \Delta u + \frac{\sin u}{\sin S \cos D} \quad \frac{\Delta X}{X}.$$

where $S = \frac{1}{2}(u + u')$ and $D = \frac{1}{2}(u - u'.)$
also since

$$\Delta \theta = \sin \theta \cos \theta \left\{ \frac{\Delta Y}{Y} - \frac{\Delta X}{X} \right\}$$

by substitution

$$\Delta \theta = \frac{\sin 2\theta \cos u}{2 \sin S \cos D} \left\{ \Delta u + \frac{\cos S \sin D}{\cos u \sin 1'} \cdot \frac{\Delta X}{X}. \right\}$$

The angle u in this formula being the deviation of the suspended magnet from the position which it would assume under the action of the earth alone, its changes Δu are the differences between the observed changes of position, measured from a fixed line, and the corresponding changes of declination. The effect of temperature upon the iron bar may be corrected by substituting $(\Delta u + a \Delta t)$ for Δu, Δt being the actual change of temperature, and a the change of angle in parts of radius, corresponding to a change of 1°. Dr. Lloyd states, that the effect of an increase of temperature upon a soft iron bar, in all his experiments, has been an *increase* of its induced magnetism, being the reverse of its effect upon the permanent magnetism of an artificial magnet. The same effect was observed in the case of the present instrument, and in that of the observatory instrument with two bars at Toronto, but the amount was very small in both, as was also found by him.

Since the date of the observations under discussion, Dr. Lamont has shown that the assumption, that the induced magnetism of the bar is proportional to the inducing force, is not strictly in accordance with the fact, and has proposed a method of determining the scale co-efficient of the instrument, "which is independent of all hypothesis, and necessarily includes all the circumstances upon which the quantity sought depends." The principle of his method consists in altering the induced force artificially, by a small but known amount, and observing the change of angle produced thereby, and this is effected by placing a magnet at a considerable distance* above or below the suspended magnet, their centres being in the same vertical line, and observing the scale readings with this magnet, first vertical, in which position it exerts no direct action upon the suspended magnet, but only on the iron bar, and next horizontal, and at right angles to the suspended magnet, in which position it should exercise no action on the iron bar, but only on the suspended magnet. It will be shown below that at the distances of deflection which it is necessary to employ in practice, the assumption that in its horizontal position the magnet exercises no effect on the induced magnetism of the iron bar is not quite in accordance with the fact, but the effect produced can be eliminated very nearly, by reversals. Now if n be the angle of deflection with the magnet vertical, n' with the magnet horizontal, a the distance of the magnets from centre to centre, e the length of a line connecting the centre of the fixed magnet with the centre of the iron bar, and ϕ the angle which that line forms with the vertical, it is shown by Dr. Lloyd, that

$$\frac{n}{n'}=\frac{a^3}{e^3}\,(1+\cos^2\theta)\ \mathrm{V}$$

By substituting the value of V thus found, in the formula above,

$$\frac{\Delta \mathrm{Y}}{\mathrm{Y}}=\mathrm{V}^{-1}\cot\theta\left(\cos u\Delta u+\sin u\frac{\Delta \mathrm{X}}{\mathrm{X}}\right)$$

a new and more accurate expression is obtained for the changes of the Vertical Force; also since

$$\Delta\theta=\sin\theta\cos\theta\left(\frac{\Delta \mathrm{Y}}{\mathrm{Y}}-\frac{\Delta \mathrm{X}}{\mathrm{X}}\right)$$

By substituting the last expression for $\frac{\Delta \mathrm{Y}}{\mathrm{Y}}$, we have

$$\Delta\theta=\mathrm{V}^{-1}\cos^2\theta\cos u\Delta u+f.\ \frac{\Delta \mathrm{X}}{\mathrm{X}}$$

* Seven or eight times the length of the deflecting magnet is the distance recommended, but it does not appear to be sufficient.

where $f = V^{-1} \cos^2\theta \sin u - \sin\theta \cos\theta^*$; or if we put

$$\sin \phi = V^{-1} \cos \theta \cos u,$$

then $f = 2 \cos \theta . \cos \frac{1}{2} (\phi + \theta) \sin \frac{1}{2} (\phi - \theta)$

The instrument made use of at Lake Athabasca has been subsequently sent to the East Indies, and I have had no opportunity of putting in practice both these methods of determining the scale co-efficient with it, for the purpose of ascertaining in what manner the value obtained by the original method must be modified, to agree with that resulting from the experiments of deflection. Both methods have been tried, however, with this view, with two different instruments at Toronto, and with other instruments elsewhere; the results appear to warrant the conclusion, that the ratio between the values thus obtained is not only constant for the same instrument, but so nearly the same for all instruments of similar construction, and furnished with iron bars of similar quality, that we may obtain a pretty good approximation to the true scale value, when, as in the present case, it cannot be directly determined, by multiplying the value given by the formula of the instructions by the mean ratio deduced from all the experiments.

The following Table contains the particulars of a series of experiments of deflection made with an Induction Inclinometer with one iron bar, at Toronto; this instrument is precisely similar to the one used in the northern observations, except that the magnet suspended is 3·0 instead of 2·6 inches in length; it was made and sent to America at the same time.

* The foregoing explanation is, in substance, derived entirely from the Papers of Dr. Lloyd referred to in a previous note, and is given as much as possible in his own words.

TABLE XXI.

Experiments to determine the scale co-efficient of a one-bar Inclinometer by the method of deflections, under various adjustments. In this instrument $b = 4{\cdot}96$ inches.*

	Date.	Acting End of Iron Bar a	Distances. Inches.			$\cos^2\phi$	Deflection.		Deduced Scale Value.	Co-efficient by Adjustment.	Ratio.
			a	e	h		$2u$	$2u'$			
	1851.								′	′	
I.	June 6 -	South Pole -	30·50	25·16	5·83	·961	11·58	59·08	1·097	0·851	1·290
	" -	" -	33·75	28·36	5·83	·969	7·61	42·82	1·152	—	1·353
II.	June 9 -	North Pole -	26·62	32·06	5·09	·974	6·78	80·81	0·829	0·854	0·970
	" -	" -	30·87	36·30	5·09	·981	4·81	52·57	0·822	—	0·962
III.	June 11 -	" -	26·80	32·20	5·02	·976	6·44	80·62	0·878	0·838	1·046
IV.	June 12 -	South Pole -	30·55	25·40	5·63	·965	10·40	50·05	1·192	0·845	1·411
V.	June 13 -	South Pole -	33·29	28·34	5·39	·969	7·98	45·35	1·126	0·849	1·326
VI.	June 13 -	South Pole -	33·29	28·34	5·39	·969	8·35	46·37	1·094	0·843	1·296
VII.	" -	North Pole -	33·29	38·91	5·39	·984	4·33	42·71	0·756	0·843	0·896
	" -	" -	30·35	36·09	5·39	·981	5·27	55·34	0·769	—	0·910
VIII.	June 16 -	South Pole -	33·03	27·64	5·39	·969	7·93	46·15	1·151	0·847	1·359
	" -	" -	29·36	23·97	5·39	·959	12·44	66·26	1·112	—	1·312
IX.	" -	North Pole -	33·03	38·42	5·39	·984	4·43	42·66	0·730	0·847	0·862
	" -	" -	29·36	34·75	5·39	·980	5·33	60·84	0·816	—	0·963

In the foregoing experiments, the deflecting magnet employed was the one used in the determinations of Absolute Horizontal Force, its length 3·66 inches. It will be observed that all the co-efficients obtained when the acting end of the bar was a south pole, or the bar was upwards, are greater than those given by the formula of the instructions, and all those obtained when the acting end was a north pole, or the bar downwards, with one exception, are less. It was proved, by reversing the deflecting magnet in the vertical position when the iron bar was away, that it has no effect on the suspended magnet in

* The particulars of the previous adjustments are as follows: n = number of reversals of the iron bar to obtain mean values of S and D; $\theta = 75°\ 19'$.

		n	u	S	D	
			° ′	° ′	° ′	
I.	June 5 - -	8	16 33·1	16 2·2	30·9	
II.	June 9 - -	8	16 3·8	16 0·7	3·1	Acting end of bar changed.
III.	June 11 - -	7	16 29·8	16 18·0	12·0	
IV.	June 12 - -	7	15 52·2	16 7·1	14·8	
V.	June 13 - -	×	14 52·0	16 12·0	6·1	
VI.	June 13 - -	5	16 18·0	16 12·0	6·1	
VII.	June 13 - -	×	16 5·0	16 12·0	6·0	
VIII.	June 16 - -	7	16 6·3	16 8·7	0 2·3	
IX.	June 16 - -	×	16 8·2	16 8·7	0 2·3	

this position; but it would appear that the difference in question may result, in part at least, from an effect on the induced magnetism of the iron bar when the magnet is horizontal, which renders the angle n' greater in each case when the acting end is a south pole, for the same value of a, than when it is a north pole. According to the theory, the angle of deflection when the magnet is horizontal should be the same for the same distance, whatever the position of the iron bar, which is supposed not to be affected by it.

As it appeared desirable to establish this point, and it might be suggested that the effect was in consequence of not taking a sufficient distance of deflection, although in some of the foregoing experiments it was between nine and ten times the length of the magnet, a second series was made, with a deflector of 7·5 inches, which allowed a considerably greater distance to be used. The particulars are contained in the next Table.

Table XXII.

*Experiments of deflection continued, deflector of 7·5 inches.**

	Date.	Acting End of Iron Bar a	Distances. Inches.			$\cos^2\varphi$	Deflection.		Deduced Scale Value.	Coefficient by Adjustment.	Ratio.
			a	e	h		$2n$	$2n'$			
	1851.								′		
X.	June 23 -	North Pole -	55·09	60·67	5·38	0·993	5·68	55·17	0·898	0·829	1·084
	„ -	„ -	58·83	64·43	5·38	0·994	5·50	43·43	0·935	„	1·129
XI.	June 24 -	South Pole -	55·09	49·95	5·39	0·990	9·43	59·48	1·029	0·818	1·258
	„ -	„ -	58·83	53·66	5·39	0·991	7·04	47·84	1·089	„	1·332
XII.	June 25 -	North Pole -	55·09	60·67	5·33	0·992	5·96	56·07	0·856	0·805	1·064
	„ -	„ -	58·83	64·43	5·33	0·994	5·08	44·76	0·818	„	1·016
XIII.	„ -	South Pole -	55·09	49·95	5·38	0·990	10·21	58·04	0·942	0·821	1·147
	„ -	„ -	58·83	53·66	5·38	0·991	8·03	46·45	0·996	„	1·213

It appears that, notwithstanding the increased distance of deflection, we have the same result as before. The ratio which the experimental scale value bears to the theoretical one evidently depends upon the nature of the acting pole, or rather upon the position of the iron bar above or below the suspended magnet at the time of the experiment; it is about one tenth greater for adjustments in the former position than for those in the latter. In both positions the value given by deflection is the greatest of the two.

* The following are the particulars of the adjustments in Table XXII.: n = number of times the iron bar was reversed, to obtain the values of S and D.

		n	u	S	D	
			° ′	° ′	° ′	
X.	June 23 - -	7	15 9·2	16 36·0	1 19	
XI.	June 24 - -	3	17 55·0	16 34·4	1 21	
XII.	June 24 - -	3	18 38	16 47·6	1 56·6	Collet shifted to other end of bar.
XIII.	June 25 - -	5	14 47·2	16 47·6	1 56·6	

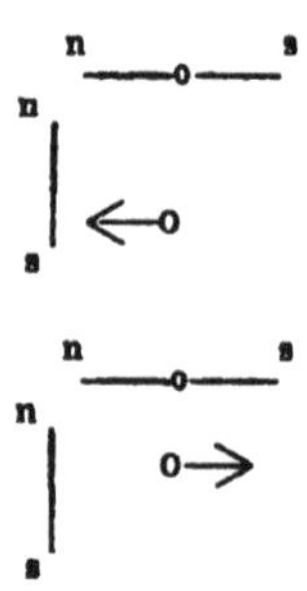

The upper end of the iron bar being always a north pole, the effect of presenting towards it the north and south poles of the deflecting magnet alternately, during its horizontal reversal is always the same as regards its induced magnetism; when, however, the bar is *above*, or the acting end a south pole, this effect concurs with the tendency of the magnet in the same position to deflect the suspended magnet, and the angle of deflection is increased proportionably; when the iron bar is reversed, or the upper end is the acting pole, the contrary is the case, the angle of deflection is diminished; the effect being less, however in the ratio of $\frac{e^3}{e'^3}$ to unity, (e' the value of e when greater than a or the iron bar below.)

Taking the *difference* between observed value of the angle n', in two adjustments, at which the position of the iron bar was different but the distance of deflection the same, to be the sum of the effects produced in each case by the action of the deflector in its *horizontal* positions upon the induced magnetism of the iron bar, it appears, in the case of the experiments numbered VIII. and IX., to have amounted to $(46{\cdot}15-42{\cdot}66)=3{\cdot}49$ scale divisions, when the distance was 33·0 inches; and to $(66{\cdot}26-60{\cdot}84)=5{\cdot}42$ div. when it was only 29·3 inches. Let the effect in the two positions of the iron bar be η and η', where $\eta'=\left(\frac{e}{e'}\right)^3\eta$, then in this case the two values of η are 2·55 and 4·05 div., therefore those of η' are 0·94 and 1·37 div.; that is to say, the double angles of deflection, when the magnet was horizontal and the iron bar *above*, were increased by the two former amounts, at the greater and less distances of deflection respectively, and by the two latter when the iron bar was *below*; in each case in consequence of the effect upon its induced magnetism. The corrected values of n' are therefore 43·6 and 62·2 divisions respectively, which slightly reduces the difference between the resulting co-efficients, but to so small an extent as to prove that it is not caused by the effect in question alone, and that we must look elsewhere for a solution of the difficulty, probably to the introduction of other terms into the expressions involving the distances e and a; but without pursuing this subject any further here, I have concluded that the only way of approximating to the true value of the ratio required is to take the mean between the values found for the same distances under two adjustments, one in which the iron bar is *below*, the other in which it is *above*.

The experiments with the 3·6 deflector supply the following couples:—

TABLE XXIII.

Date.	*a*	*e*	Ratio of Co-efficients.	Mean.
June 6	30·50	25·16	1·290	1·126
9	30·87	36·30	0·962	
12	30·55	25·40	1·411	1·160
13	30·35	36·09	0·910	
13	33·29	28·34	1·311	1·103
13	33·29	38·91	0·896	
16	33·03	27·64	1·359	1·110
16	33·09	38·42	0·862	
16	29·36	23·97	1·312	1·137
16	29·36	34·75	0·963	
Mean -	—	—	—	1·127

next the experiments with the 7·5 inch deflector supply the following couples:—

TABLE XXIV.

Date.	*a*	*e*	Ratio of Co-efficients.	Mean.
June 24	55·09	49·95	1·258	1·171
23	55·09	60·67	1·084	
24	58·83	53·66	1·332	1·230
23	58·83	64·43	1·129	
25	55·09	49·95	1·147	1·105
25	55·09	60·67	1·064	
25	58·83	53·66	1·213	1·159
25	58·83	64·43	1·106	
Mean -	—	—	—	1·166

It would follow, from the whole series, that the scale co-efficient determined for this instrument in the ordinary way will be brought to accordance nearly with the true value, by augmenting it in the ratio 1·146. I have employed, however, the last series alone, the distances of deflection having been more favourable, and the general result less likely to be influenced by any terms involving that quantity beyond what are employed. We have, again, a series of experiments with another instrument, one of those provided with two iron bars. In this case one bar being always above and the other below the suspended magnet, the effect of the deflecting magnet in its horizontal position is neutralized in great measure, being of a contrary sign in the two bars respectively. The following are the particulars:—

Table XXV.

Experiments to determine the Scale co-efficient of a Two-bar Inclinometer, by the method of Deflections, under various adjustments. Length of deflecting magnet 7·5 inches; value of b, or distance of iron bars from suspended magnet, 5·0 inches.

Date 1848-9.		Distances, Inches.		Deflection.		Deduced Scale Value.	Co-efficient by Adjustment.	Ratio.
		a	*b*	*n*	*n'*			
October	3	71·80	4·12	6·48	38·74	0·500	0·357	1·344
"	13	71·80	4·12	5·90	35·15	0·495	—	1·329
"	13	71·80	4·12	5·72	35·59	0·521	—	1·398
"	31	71·80	4·12	6·03	35·48	0·493	—	1·322
"	9	65·74	4·12	7·52	46·75	0·521	—	1·400
"	12	65·74	4·12	8·63	50·08	0·487	—	1·306
"	13	65·74	4·12	8·28	46·31	0·469	—	1·260
November	1	65·74	4·12	8·18	46·44	0·468	—	1·258
April	3	71·91	4·14	6·08	34·21	0·467	0·373	1·251
"	3	65·87	4·14	7·89	44·77	0·499	—	1·338
"	9	71·91	4·14	6·05	33·78	0·494	0·372	1·327
"	9	71·91	4·14	5·75	33·72	0·466	—	1·253
"	9	65·86	4·14	7·34	43·97	0·500	—	1·346
"	9	65·86	4·14	7·49	44·27	0·489	—	1·317

From the deflections at the nearer distance, we find a mean value of 1·332, and from those at the greater distance, a mean value of 1·318, for the ratio in which we must augment the value of the scale co-efficient found in the ordinary way for this instrument, to make it agree with the value deduced from experiment.

Lastly, Dr. Lloyd has found for his instrument a value of about 1·3 for the same ratio.

I conceive that whole evidence warrants the conclusion that the scale co-efficient found by the formula of the magnetical instructions is invariably less than the true value as determined by experiment; that the ratio in which it must be augmented is constant for the same instrument; and that it is nearly the same for all instruments furnished with bars of similar quality. I propose to adopt 1·22 provisionally, for the instrument used at Lake Athabasca, which I consider leaves the changes of inclination under an uncertainty of about one tenth their apparent value; a quantity which, however considerable, does not perhaps greatly exceed the uncertainty of all the observations of the changes of this element thus far; it does not affect their value for many relative purposes, and will not alter the character of any periodical law deducible from the observations.

The great amount of the daily changes of inclination indicated by the scale readings has been shown in Table XIII., in connexion with the corresponding changes of Horizontal Force. Satisfactory proof of the *reality* of these changes, and of the practical value of the instrument, may be given by a comparison of the effect of sudden magnetical shocks, which sometimes occurred, of a very marked character upon the Inclinometer and Bifilar. The following instances have been selected with this view from the observations of Disturbances. The instruments were generally read in succession, with an interval of one minute between them. I have therefore interpolated a value of the Bifilar for the minute of observation of the Inclinometer; the last columns contain the change in scale readings and in terms of the Inclination and Horizontal Force, between each successive observation, usually a space of three minutes. It will be observed, that however great and sudden the changes of Horizontal Force shown by the Bifilar, the Induction Inclinometer never fails to exhibit a corresponding change of scale reading; indeed these changes so much exceed in general what would be inferred from the change of Horizontal Force alone, as to leave an excess sufficiently large to prove that in these cases, making full allowance for probable uncertainty of the scale value of the Inclinometer, the shocks must have been accompanied by large changes of Total Force.

TABLE XXVI.

		Bifilar.			Inclinometer.		
		Readings.	Differences.	$\frac{\Delta X}{X}$	Readings.	Differences.	$\Delta\theta$
	D. H. M.						′
1	Nov. 2 17 10	229·3	—	—	—	—	—
	11	221·8[a]	—	—	113·8	—	—
	15	191·6	—	—	—	—	—
	16	197·7[a]	−24·1	−·0083	126·3	+12·5	+2·1
	20	221·4	+23·7	+·0081	51·7	−74·6	−12·6
	25	175·4[b]	−46·0	−·0157	182·5	+130·8	+22·4
	30	213·4[b]	+38·0	+·0130	132·7	−49·8	−8·5
	35	206·0[b]	− 7·4	−·0025	144·4	+11·7	+2·0
	2 17 40	196·0[b]	−10·0	−·0034	174·8	+30·4	+5·2
2	April 16 18 57	208·9	—	—	—	—	—
	58	239·8[a]	—	—	201·3	—	—
	19 1	271·7	—	—	—	—	—
	2	267·0[a]	+27·2	+·0077	62·5	−138·8	−17·9
	6	248·1	—	—	—	—	—
	7	248·0[a]	−19·0	−·0052	172·6	+110·1	+14·2
	11	247·6	—	—	—	—	—
3	April 16 20 41	203·7	—	—	—	—	—
	42	181·7[a]	—	—	258·8	—	—
	46	93·9	—	—	—	—	—
	47	97·6[a]	−84·1	−·0238	479·3	+220·5	+28·5
	51	112·7	—	—	—	—	—
	52	113·6[a]	+16·0	+·0045	404·5	−74·8	−9·7
	56	117·2	—	—	—	—	—

[a] Readings interpolated.

[b] Readings taken simultaneously with those of the Inclinometer.

TABLE XXVI.—*continued.*

		Bifilar.			Inclinometer.		
		Readings.	Differences.	$\frac{\Delta X}{X}$	Readings.	Differences.	$\Delta\theta$
	D. H. M.						′
4	April 16 23 16	146·8	—	—	—	—	—
	17	101·2	—	—	345·0	—	—
	19	9·9	—	—	—	—	—
	20	21·2	−80·0	−·0226	631·0	+286·0	+37·0
	22	43·8	—	—	—	—	—
	23	56·4	+35·2	+·0099	529·3	−101·7	−13·1
	25	71·5	—	—	—	—	—
	26	80·4	+24·0	+·0068	499·7	−29·6	−3·8
	23 28	108·4	—	—	—	—	—
5	April 17 0 45	176·3	—	—	—	—	—
	46	185·3[a]	—	—	317·9	—	—
	48	203·3	—	—	—	—	—
	49	215·6[a]	+30·3	+·0086	271·5	−46·4	−6·00
	51	240·1	—	—	—	—	—
	52	163·0[a]	−52·6	−·0149	427·6	+156·1	+20·18
	55	−0·8	—	—	—	—	—
	56	0·0[a]	−163·0	−·0461	620·9	−193·3	+24·99
	58	3·2	—	—	—	—	—
	59	−18·1[a]	−18·1	−·0051	807·8	+186·9	+24·17
	17 1 2	−60·7	—	—	—	—	—
	3	−50·0[a]	−31·9	−·0090	837·9	+30·1	+3·89
	5	−30·7	—	—	—	—	—
6	April 17 2 19	179·7	—	—	—	—	—
	20	172·2[a]	—	—	311·6	—	—
	21	157·2	—	—	—	—	—
	22	149·5[a]	−22·7	−·0064	373·6	+62·0	+8·01
	25	131·1	—	—	—	—	—
	26	105·4[a]	−44·1	−·0125	370·3	−3·3	−0·43
	28	54·1	—	—	—	—	—
	29	51·4[a]	−54·0	−·0152	541·3	+170·0	+21·98
	31	46·0	—	—	—	—	—
	32	67·7[a]	+16·3	+·0045	540·9	−0·4	−0·05
	34	111·0	—	—	—	—	—
	35	130·8[a]	+63·1	+·0178	459·3	−81·6	−10·55
	37	170·3	—	—	—	—	—
	38	163·5[a]	+32·7	+·0092	309·6	−149·7	−19·36
	40	150·9	—	—	—	—	—
7	April 25 20 1	239·5	—	—	—	—	—
	2	171·6[a]	—	—	346·3	—	—
	4	65·9	—	—	—	—	—
	5	64·2[a]	−107·4	−·0303	545·6	+199·3	+25·77
	7	60·8	—	—	—	—	—
	8	78·4[a]	−14·2	−·0040	515·3	−30·3	−3·90
	10	113·7	—	—	—	—	—
	11	117·6[a]	+39·2	+·0111	386·2	−129·1	−16·59
	13	125·5	—	—	—	—	—
	14	140·2[a]	+22·6	+·0064	356·7	−29·5	−3·81
	16	169·6	—	—	—	—	—

[a] Readings interpolated.

[b] Readings taken simultaneously with those of the Incl'nometer.

TABLE XXVI.—*continued.*

		Bifilar.			Inclinometer.		
		Readings.	Differences.	$\frac{\Delta X}{X}$	Readings.	Differences.	$\Delta\theta$
	D. H. M.						′
8	April 30 21 22	105·9	—	—	—	—	—
	23	115·9[a]	—	—	475·8	—	—
	25	136·0	—	—	—	—	—
	26	104·0[a]	−14·9	−·0042	346·6	+70·8	+9·15
	28	40·0	—	—	—	—	—
	29	69·9[a]	−34·1	−·0096	525·1	+178·5	+23·08
	31	129·7	—	—	—	—	—
	32	129·4[a]	+60·5	+·0171	455·5	−69·6	−8·98
	34	128·9	—	—	—	—	—
	35	129·9[a]	+0·5	+·0001	343·3	−112·2	−14·47
	37	131·8	—	—	—	—	—
	38	123·7[a]	−6·2	−·0017	382·3	+39·0	+5·04
	40	107·6	—	—	—	—	—
	41	95·0[a]	−28·7	−·0081	457·1	+74·8	+9·65
	30 21 43	69·8	—	—	—	—	—

[a] Readings interpolated.
[b] Readings taken simultaneously with those of the Inclinometer.

I conclude, from the foregoing examples of the action of the instrument in extreme cases, as well as from the close correspondence in the mean diurnal curves of inclination, as derived from the observations, with that of Horizontal Force, which will be pointed out below, that the testimony of the observations themselves is in favour of the opinion that the series, with some uncertainty as to the absolute values assigned, furnishes, as far as it goes, a true representation, and the only one we can at present refer to, of the regular and irregular changes of the Inclination in high magnetic latitudes.

ADJUSTMENTS.

First Adjustment, 14th October 1843.—On conclusion of the experiments of deflection for determination of the absolute Horizontal Force, the instrument was placed on its pedestal, the base levelled, and the telescope adjusted to the meridian. The scale read 420·0, the corresponding reading of the Declinometer being 404·4, Bifilar 282·0, mean reading of Vernier's 199° 56′ 30″. The soft iron bar was now inserted in the socket, the upper or north pole deflecting, and was moved in the collet until the angle of deflection appeared to be a maximum; mean of Vernier's 252° 20′ 10″; when the same division of the scale was on the wire, Declination 406·6. The bar was next reversed, the lower end or south pole deflecting, and the telescope turned in azimuth until the central division was again on the wire; the Vernier's now read 158° 20′ 30″, Declination 406·8. We have then

$$u=41^\circ\ 36' \qquad S=46^\circ\ 59'{\cdot}8 \qquad \theta=81^\circ\ 37'{\cdot}6$$
$$u'=52^\circ\ 23'{\cdot}6 \qquad D=5^\circ\ 23'{\cdot}8 \qquad a=1{\cdot}0007$$

whence the co-efficient for differences of scale reading, when corrected for changes of Declination, Temperature, and Horizontal Force, is

$$aP = a\,\frac{\sin 2\,\theta \cos u}{2 \sin S \cos D} = 0''\cdot 148$$

according to the formula then in use. This value I have augmented in the ratio 1·22, for the reasons already stated in the actual reductions.

Increasing numbers indicated a return of the north end of the magnet towards the north, or a decrease of Inclination; the actual readings have therefore been inverted in the abstracts, by taking the complement of each to 500; and increasing numbers represent increase of Inclination throughout.

It would appear, from the difference between the angles u and u' above, that the iron bar must have possessed a considerable degree of permanent magnetism, or else that the suspended magnet was not on a level with the centre of the collet; it is possible, as there is no record to the contrary, and attention was not directed to this circumstance in the instructions then in use, that there may have been a difference on this account in the position of the acting pole, and, consequently, the amount of its action in the two positions of the iron bar, which would partly account for the difference in question; but the existence of permanent magnetism was afterwards shown by experiments at Fort Simpson. The bar was there employed as a deflector in the horizontal position, its centre at 15·7 inches from that of the suspended magnet, and it was found that there was a regular difference of 29''·8 in the reading, according as one end or the other was presented; this difference gives an angle of deflection of 14''·9, one end acting as a north, the other as a south pole. The angle of deflection produced by a three-inch magnet, the centre at very nearly the same distance, was 612''·0; the relative forces, being expressed nearly by the tangents of these angles, were as 1 to 41·5. I was not aware at that time of the facility with which an iron bar can be deprived of its permanent magnetism, by dipping it, according to Dr. Lamont's suggestion, several times alternately into hot and cold water; but to ascertain whether this circumstance is likely to have had any sensible influence on the results, the experiments of deflection Nos. X. to XIII., at Toronto, above, were purposely made when the bar had contracted a still greater amount of permanent magnetism by being inadvertently placed too near a magnet, but they do not show any difference from those in which the bar was almost entirely free from it, except a slight increase in the value of given changes, which may be due to other causes.

It became necessary to raise the suspended magnet on the 19th October, owing to the difficulty of reading the scale; the effect of thus altering its position with reference to the acting pole of the iron bar was shown by a decrease in the scale readings, indicating an increase in the angle u of about 1° 53′. As the instrument was not otherwise disturbed, and the position of the bar was not altered, this quantity has been added to the angle u in the above formula until the end of the month, making the scale co-efficient

$$a\,\mathrm{P}=0''\cdot1747$$

Second adjustment, 31st October 1843.—The soft iron bar being removed, the telescope was adjusted to the meridian, Vernier's reading 198° 28′ 20″, Declination reading 414·0. The bar was then inserted in its socket, reversed, the telescope being turned in azimuth until the same division of the scale was on the wire, Vernier reading 248° 50′ 30″, Declination 413·0. The bar was lastly inserted, and the telescope again turned in azimuth until the same scale reading was obtained, Vernier reading 150° 0′ 5″, Declination 410·0. We have now

$$u=44^\circ\ 28''\cdot2 \qquad \mathrm{S}=47^\circ\ 25''\cdot2 \qquad \theta=81^\circ\ 37''\cdot6$$
$$u'=50^\circ\ 22''\cdot2 \qquad \mathrm{D}=2^\circ 57''\cdot0 \qquad a=1\cdot0007$$

whence

$$a\mathrm{P}=a\,\frac{\sin 2\theta \cos u}{\sin \mathrm{S} \cos \mathrm{D}}=0''\cdot140$$

And this quantity multiplied by the ratio 1·22 as before, gives for the approximate value of the scale, the co-efficient

$$1''\cdot1705$$

The situation of the instrument made it convenient to have the iron bar in both adjustments in the position in which the permanent magnetism and the induced magnetism were opposed.

Corrections.

Declination Changes.—Each reading has been reduced to the zero of 400 on the Declination scale, by subtracting from it the difference of the corresponding Declination reading from that number. In term days and magnetic disturbances, and whenever observations were made at short intervals, the correction applied was the mean between the Declination reading immediately preceding and following the Inclinometer observation; this was rendered absolutely necessary in many cases, by the rapidity of the changes of the Declination, which not infrequently caused the correction to vary more than a degree from one reading to the next, where the interval between them was only three minutes.

Bifilar Correction.—The correction due to the observed changes in the Inclinometer scale reading, for variations of the horizontal component of the earth force, involves the same quantity (V), of which the determination is the object of the experiments of deflection, and cannot be accurately assigned in the present case; being, however, always very small, compared with the changes of the Inclination itself, the error involved by the application of the original formula can seldom be sensible. That formula is

$$B = \frac{\cos S \sin D}{\cos u \sin 1'} \cdot \frac{k}{a}$$

where k is the co-efficient of the Bifilar, a the value of one division of the scale of the Inclinometer in terms of radius, and the correction to each observed reading is $B.n$, n being the difference in scale divisions between the corresponding Bifilar reading and the standard reading, for the first adjustment $B = 0 \cdot 100$, and for the second $B = 0 \cdot 057$. These corrections have been applied to all the readings; the standard division adopted was the mean of the Bifilar readings for the same day in the month of October, and in the other months the mean of all the observations of each month respectively, to the nearest convenient unit, namely, the division 200 in November, 170 in December, 140 in January, and 160 in February.

The following Table contains the correction in scale divisions of the inclinometer for each value of n from 1 to 100 divisions of the Bifilar, under the adjustment of October 31; also the value of $\frac{\Delta X}{X}$ for the same values of n, for convenience in comparisons of the Bifilar readings.

Table XXVII.

Values of changes of the Bifilar scale reading in parts of the Horizontal Force, also of the Bifilar correction to the Inclinometer. Adjustment of October 31.

n Bif. Div.	$\frac{\Delta X}{X}$	B Incl. Div.	n Bif. Div.	$\frac{\Delta X}{X}$	B Incl. Div.	n Bif. Div.	$\frac{\Delta X}{X}$	B Incl. Div.	n Bif. Div.	$\frac{\Delta X}{X}$	B Incl. Div.
1	·000341	0·06	26	·008871	1·49	51	·017401	2·92	76	·025931	4·35
2	·000682	0·11	27	·009212	1·54	52	·017742	2·97	77	·026272	4·40
3	·001024	0·17	28	·009554	1·60	53	·018084	3·03	78	·026614	4·46
4	·001365	0·23	29	·009895	1·66	54	·018425	3·09	79	·026955	4·52
5	·001706	0·29	30	·010236	1·72	55	·018766	3·15	80	·027296	4·58
6	·002047	0·34	31	·010577	1·77	56	·019107	3·20	81	·027637	4·63
7	·002388	0·40	32	·010918	1·83	57	·019448	3·26	82	·027978	4·69
8	·002730	0·46	33	·011260	1·89	58	·019790	3·32	83	·028320	4·75
9	·003071	0·51	34	·011601	1·94	59	·020131	3·38	84	·028661	4·80
10	·003412	0·57	35	·011942	2·00	60	·020472	3·43	85	·029002	4·86
11	·003753	0·63	36	·012283	2·06	61	·020813	3·49	86	·029343	4·92
12	·004094	0·69	37	·012624	2·12	62	·021154	3·55	87	·029684	4·98
13	·004436	0·74	38	·012966	2·17	63	·021496	3·60	88	·030026	5·03
14	·004777	0·80	39	·013307	2·23	64	·021837	3·66	89	·030367	5·09
15	·005118	0·86	40	·013648	2·29	65	·022178	3·72	90	·030708	5·15
16	·005459	0·92	41	·013989	2·35	66	·022519	3·78	91	·031049	5·20
17	·005800	0·97	42	·014330	2·40	67	·022860	3·83	92	·031390	5·26
18	·006142	1·03	43	·014672	2·46	68	·023202	3·89	93	·031732	5·32
19	·006483	1·09	44	·015013	2·52	69	·023543	3·95	94	·032073	5·38
20	·006824	1·14	45	·015354	2·57	70	·023884	4·00	95	·032414	5·43
21	·007165	1·20	46	·015695	2·63	71	·024225	4·06	96	·032755	5·49
22	·007506	1·26	47	·016036	2·69	72	·024566	4·12	97	·033096	5·55
23	·007848	1·32	48	·016378	2·75	73	·024908	4·18	98	·033438	5·61
24	·008189	1·37	49	·016719	2·80	74	·025249	4·23	99	·033779	5·66
25	·008530	1·43	50	·017060	2·86	75	·025590	4·29			

Temperature Correction.—Experiments were made at Toronto in December 1844, after the return of the instrument from the north-west, to determine the effect of changes of temperature upon the induced magnetism of the soft iron bar. A copper vessel was fixed upon the arm of the instrument itself, surrounding the bar, and provided with a stop-cock for changing the water, a regular adjustment was completed, and the experiments were then made by filling the vessel with water at different temperatures, while the bar was in place. The temperature was carried at once from the lowest to the highest point, the average extremes being 50° and 95° respectively, and the bar allowed 15 minutes to take up the change. Eacn value of the corresponding scale readings employed in the calculation was the mean by three independent observations, with five minutes interval between them. The value of q is found by the formula

$$q = \frac{\Delta u}{\Delta t^{\circ}} \frac{\cos u}{\sin S \cos D}$$

These values and the other particulars of the experiment are given in the following Table.

Table XXVIII.

Experiments to determine the Effect of Changes of Temperature upon the induced Magnetism of the soft iron Bar.

Date.	Adjustment of Inclinometer.			Number of Changes.	Mean $\overset{\circ}{t_{\prime}}-\overset{\circ}{t}$	Mean Δu	q
	$u_{\prime}$	$s_{\prime}$	$D_{\prime}$				
1844:	° ′ ″	° ′ ″	° ′ ″		°	′	
December 23 -	25 43 50	23 59 52	1 43 50	8	46·1	11·22	0·5392
" 24 -	24 22 20	23 35 35	0 45 45	5	43·9	6·12	0·3178
" 24 -	"	"	"	5	46·5	6·44	0·3154
" 26 -	"	"	"	5	41·2	6·61	0·3650
" 26 -	"	"	"	6	41·1	6·45	0·3574
" 27 -	"	"	"	6	47·0	14·55	0·7042
" 28 -	"	"	"	6	46·9	8·63	0·4183
" 28 -	"	"	"	6	46·5	6·87	0·3363
Mean -	—	—	—	—	—	—	0·3737

The observation on the 27th has been rejected.

The correction to the scale readings of the Inclinometer for given changes of temperature is found by the formula—

$$R = \frac{\sin S \cos D}{\cos u} \cdot \frac{q}{a}$$

The values found are R=0·368 for the first adjustment, and R=0·390 for the second. This correction has not been applied to the individual readings. The following table contains the value of R Δt° for each value of $\Delta^{\circ} t$ from 1° to 39°, to be subtracted from the scale readings (as inverted) when the observed temperature is higher than the standard temperature, and the contrary when it is lower.

Table XXIX.

Corrections to reduce the Inclinometer scale readings under the Second Adjustment to a standard temperature.

Δt	RΔt	Δt	RΔt	Δt	RΔt	Δt	RΔt
°	Div.	°	Div.	°	Div.	°	Div.
0	0·00	10	3·90	20	7·80	30	11·71
1	0·39	11	4·29	21	8·19	31	12·09
2	0·78	12	4·68	22	8·58	32	12·48
3	1·17	13	5·07	23	8·97	33	12·87
4	1·56	14	5·46	24	9·36	34	13·26
5	1·95	15	5·85	25	9·75	35	13·65
6	2·34	16	6·24	26	10·14	36	14·04
7	2·73	17	6·63	27	10·50	37	14·43
8	3·12	18	7·02	28	10·92	38	14·82
9	3·51	19	7·41	29	11·31	39	15·21

The adjustment last described remained undisturbed until December 21^{d} 2^{h} Göttingen, when the arm of the instrument was accidentally struck and moved, occasioning a change of 123 scale divisions in the reading; such a movement does not sensibly affect the adjustment, and that quantity has been subtracted from all subsequent readings to correct the series.

Changes of the Inclination.—The approximate amount of the daily range of this element, indicated by the difference between the highest and lowest scale readings, has been alerady given in Table XIII. in connexion with that of the Horizontal Force.

Diurnal variation of the Inclination.—The following Table contains the mean scale readings of the Inclinometer for each month, corrected for changes of the Declination and Horizontal Force, in the manner described above.

TABLE XXX.

Monthly Means of corrected Inclinometer readings, omitting October 20, November 4, January 2–9, which are incomplete.

Civil time -	h. m. 15 55	h. m. 16 55	h. m. 17 55	h. m. 18 55	h. m. 19 55	h. m. 20 55	h. m. 21 55	h. m. 22 55	
Gött. time -	0	1	2	3	4	5	6	7	
October -	54	257·92	234·36	221·11	217·28	218·05	210·97	215·79	
November -	71	130·76	125·94	125·70	126·43	123·46	126·13	125·15	
December -	39	170·59	167·17	168·50	161·39	160·00	163·97	166·39	
January -	264·27	213·92	210·49	200·57	195·92	197·50	203·76	202·25	
February -	[illegible]·58	210·40	200·20	196·86	196·94	202·50	202·67	201·81	
Mean - -	197·34	190·19	182·70	177·80	175·68	176·21	177·92	178·43	
Civil time -	Noon —5m	h. m. 0 55	h. m. 1 55	h. m. 2 55	h. m. 3 55	h. m. 4 55	h. m. 5 55	h. m. 6 55	
Gött. time -	8	9	10	11	12	13	14	15	
October -	·86	218·43	219·20	215·78	215·37	212·78	213·28	213·34	
November -	42	126·84	125·71	123·72	121·88	120·03	119·64	119·22	
December -	36	164·16	162·25	162·25	161·01	160·26	159·89	160·51	
January -	217 57	200·69	201·47	198·82	195·16	195·23	193·68	192·60	
February -	[illegible]·58	196·84	194·31	192·08	191·63	189·16	186·86	185·24	
Mean - -	178·26	177·20	176·22	174·14	173·01	171·25	170·36	169·98	
Civil time -	h. m. 7 55	h. m. 8 55	h. m. 9 55	h. m. 10 55	Midn. —5m	h. m. 12 55	h. m. 13 55	h. m. 14 55	Mean.
Gött. time -	16	17	18	19	20	21	22	23	
October -	212·93	211·89	211·41	222·84	242·82	231·18	248·71	251·07	224·95
November -	120·33	120·60	117·40	120·21	126·39	135·67	137·80	138·99	125·74
December -	158·25	158·65	157·91	158·38	163·03	170·71	182·90	184·46	165·06
January -	191·89	194·48	192·85	193·06	200·16	216·40	209·01	230·05	202·03
February -	187·74	183·50	185·60	184·48	190·32	204·93	205·52	208·73	197·02
Mean - -	169·78	169·53	168·62	170·63	178·88	187·42	191·35	197·80	178·36

The general mean is taken, as in the other Tables, by dividing the sum of all the observations under each hour by the total number, which is 112; and the co-efficient for this curve, which includes observations under the separate adjustment of October, with those of the subsequent months, is

$$aP = 0\cdot1405 \times 1\cdot22 = 0\cdot1714$$

A mean having been taken however for the complete days from November to February inclusive, prior to the application of the Bifilar correction, I subjoin it at the foot of the page*; but as the

---

* TABLE XXXI.

Mean by 98 complete days under the Second Adjustment uncorrected for changes of Horizontal Force.

0	1	2	3	4	5	6	7	8	9	10	11	Gött. time.
186·78	181·75	174·79	172·23	169·94	170·50	172·63	173·41	172·82	171·48	170·14	168·18	
12	13	14	15	16	17	18	19	20	21	22	23	Mean.
166·63	165·15	164·12	163·66	163·40	161·86	162·34	163·28	170·26	181·26	181·02	190·26	171·67

month of October was marked by considerable disturbances, which have probably affected the mean curve of Horizontal Force, it is here included with the other months. In the next Table the small correction necessary to reduce the means to a uniform temperature of 40° has been applied to the general mean; and under each corrected value is given its difference from the mean of the whole in scale divisions of the instrument, together with the approximate value of this difference in terms of the inclination.

Table XXXII.

Corrected Mean Diurnal Curve of the Inclination by 112 *days of observation at Lake Athabasca.*

Civil Time -	h. m. 15 55	h. m. 16 55	h. m. 17 55	h. m. 18 55	h. m. 19 55	h. m. 20 55	h. m. 21 55	h. m. 22 55	
Gött. Time -	Noon.	1	2	3	4	5	6	7	
Scale - -	197·65	190·62	183·09	178·07	176·03	176·76	178·58	$^{+_2}$178·70	
Differences -	+19·22	+12·19	+5·66	−0·36	−2·40	−1·67	+0·15	$^{+_2}$+0·27	
Δ θ	+3′·29	+2′·08	+0′·97	−0′·06	−0′·41	−0′·29	+0′·03	+0′·05	
Civil Time -	Noon −5	h. m. 0 55	h. m. 1 55	h. m. 2 55	h. m. 3 55	h. m. 4 55	h. m. 5 55	h. m. 6 55	
Gött. Time -	8	9	10	11	Mid.	13	14	15	
Scale - -	178·15	176·69	175·75	173·75	172·70	170·98	170·17	169·98	
Differences -	−0·28	1·74	−2·68	−4·68	−5·73	−7·45	−8·26	−8·45	
Δ θ	−0′·05	−0′·30	−0′·46	−0′·80	−0′·96	−1′·27	−1′·41	−1′·44	
Civil Time -	h. m. 7 55	h. m. 8 55	h. m. 9 55	h. m. 10 55	Mid. −5	h. m. 12 55	h. m. 13 55	h. m. 14 55	
Gott. Time -	16	17	18	19	20	21	22	23	Mean.
Scale - -	169·86	169·61	$\overline{168{\cdot}54}$	170·55	178·80	187·77	191·47	$^{+_1}$198·07	178·43
Differences -	−8·57	−8·82	$\overline{-9{\cdot}80}$	−7.88	+0·37	+9·34	+13·04	$^{+_1}$+19·64	
Δ θ	−1′·47	−1′·51	−1′·69	−1′·35	+0′·06	+1′·60	+2′·23	+3′·36	

We see by the foregoing Table that the hour of 3 A.M. is that at which the Inclination deviates most from its mean value, a result precisely similar to what we have found for the other elements; and there is the same reason for attributing the magnitude of the deviation to the effect of disturbance at that hour. If we select the same undisturbed days as before, and take their mean, the result is a signal diminution in the amount of the diurnal change at that period of the 24^h.

Table XXXIII.

*Mean Diurnal Curve of Inclination by 45 Days * selected as free from Disturbance, corrected for Variation of Horizontal Force, and reduced to a uniform Temperature; together with the Difference of each mean from the mean of the whole in Scale Divisions and in terms of the Inclination.*

Civil Time	h. m. 15 55	h. m. 16 55	h. m. 17 55	h. m. 18 55	h. m. 19 55	h. m. 20 55	h. m. 21 55	h. m. 22 55
Gött. Time	Noon	1	2	3	4	5	6	7
Scale	180·52	179·15	177·47	176·29	175·79	175·64	180·02	179·94
Difference	+4·50	+3·13	+1·45	+0·27	−0·23	−0·38	+4·00	+3·92
Δθ	+0'·77	+0'·54	+0'·25	+0'·05	−0'·04	−0'·06	+0'·68	+0'·67

Civil Time	Noon	h. m. 0 55	h. m. 1 55	h. m. 2 55	h. m. 3 55	h. m. 4 55	h. m. 5 55	h. m. 6 55
Gött. Time	8	9	10	11	Midn.	13	14	15
Scale	179·49	177·76	176·61	174·07	174·71	172·36	171·08	172·45
Difference	+3·47	+1·74	+0·59	−1·95	−1·31	−3·66	−4·94	−3·57
Δθ	+0'·59	+0'·30	+0'·10	−0'·33	−0'·22	−0'·63	−0'·84	−0'·61

Civil Time	h. m. 7 55	h. m. 8 55	h. m. 9 55	h. m. 10 55	Midn.	h. m. 12 55	h. m. 13 55	h. m. 14 55	Mean.
Gött. Time	16	17	18	19	20	21	22	23	
Scale	172·19	170·46	170·83	173·33	174·99	178·05	179·38	181·62	176·02
Difference	−3·83	−5·56	−5·19	−2·69	−1·03	+2·03	+3·36	+5·60	
Δθ	−0'·65	−0'·95	−0'·89	−0'·46	−0'·18	+0'·36	+0'·57	+0'·96	

For comparison of the mean diurnal curve of Inclination at Lake Athabasca with that of the same element at Toronto, the only other American station at which it was observed directly, in 1843–4, I subjoin a Table, containing, first, the mean scale readings of a one-bar Inclinometer, observed from the 15th October 1843 to the 10th February 1844; and, secondly, the mean for the same five months, October to February inclusive, of the scale readings of a two-bar Inclinometer for the years 1845, 1846, 1847.

The scale co-efficient given by adjustment for the one-bar instrument was $(a\ P)=0'\cdot723$, which it appears by the experiments contained in Tables XXII.–XXIV., must be augmented in the ratio 1·166† to agree with the value given by experiments of deflection, the instrument there used being the same, giving for the approximate co-efficient $0'\cdot820$. The other mean is related to various adjustments,

* One day less than for the other elements, the 14th February being excluded on account of the omission of an observation.

† I take the experiments with 7·5 inch deflector alone as the best series.

and proportioning the co-efficient according to the number of days under each, the value applicable to it is $(a\ P)=0''\cdot3686$. This must be augmented in the ratio $1\cdot32$, according to the experiments contained in Table XXV., to give the true scale value for this instrument, which is also the one to which those results refer, giving $aP=0''\cdot486$.

TABLE XXXIV.

Mean Diurnal Curve of Inclination at Toronto for 101 *days, October to February* 1843, *by a one-bar Inclinometer; also, Mean Curve for the same period for three years, by a two-bar Inclinometer.*

Civil Time.	Gött. Time.	One-bar Inclinometer, October to February 1843.			Two-bar Inclinometer, October to February, 3 years.		
		Scale.	Daily fluctuation.		Scale.	Daily fluctuation.	
			Scale.	Δ θ		Scale.	Δ θ
h	h			′			′
16	22	157·02	−0·21	−0·17	54·37	−1·25	−0·61
17	23	156·92	−0·31	−0·25	53·81	−1·81	−0·88
18	Noon	156·81	−0·42	−0·34	53·76	−1·86	−0·90
19	1	156·88	−0·35	−0·29	53·99	−1·63	−0·79
20	2	157·23	+0·00	+0·00	54·65	−0·97	−0·47
21	3	157·40	+0·17	+0·14	55·91	+0·29	−0·14
22	4	157·63	+0·40	+0·33	57·43	+1·81	+0·88
23	5	158·12	+0·89	+0·73	58·63	+3·01	+1·46
Noon	6	158·13	+0·90	+0·74	59·04	+3·42	+1·66
1	7	157·61	+0·38	+0·31	58·53	+2·91	+1·42
2	8	157·22	−0·01	−0·01	57·54	+1·92	+0·93
3	9	157·00	−0·23	−0·19	56·58	+0·96	+0·47
4	10	156·86	−0·37	−0·30	55·38	−0·24	−0·12
5	11	156·85	−0·38	−0·31	54·98	−0·64	−0·31
6	Midn.	157·00	−0·23	−0·19	55·07	−0·35	−0·17
7	1	157·12	−0·12	−0·10	55·11	−0·51	−0·25
8	2	157·11	−0·12	−0·10	55·00	−0·62	−0·30
9	3	157·30	+0·07	+0·06	54·93	−0·69	−0·33
10	4	157·23	+0·00	+0·00	55·03	−0·59	−0·29
11	5	157·28	+0·05	+0·04	55·23	−0·39	−0·19
Midn.	6	157·27	+0·04	+0·03	55·19	−0·43	−0·21
1	7	157·21	−0·02	−0·02	55·00	−0·62	−0·30
2	8	157·16	−0·07	−0·06	54·91	−0·71	−0·34
3	9	157·05	−0·18	−0·15	54·73	−0·89	−0·41

It will be remarked, that while the above means give similar diurnal curves, there is a difference of a large proportional amount between the values of the ordinates for the corresponding hours under them; a similar difference being observable between the corresponding mean curves of Horizontal Force, it would appear that the range of both elements was really less for the winter under discussion than its average amount. Taking the difference between the highest and lowest mean scale reading of the Bifilar, and the mean of the whole for each month in the above periods respectively, it appears that the mean of the former, from October 1843 to February 1844, at Toronto, was +5·77 scale divisions, and of the latter −6·12 scale divisions; the corresponding quantities for the period included in the second part of the table are +7·78 and −10·52 scale divisions respectively. The adjustment of this instrument was the same for both periods, but the difference shown is not nearly enough to account for the whole effect. The adjustment of the one-bar Inclinometer was made with every care. I am nevertheless disposed to believe that the co-efficient deduced is too small. I present the result, however, because the instrument was similar to the one used in the north, and the *diurnal law* deduced is independent of the absolute amount of the change.

The mean diurnal curve of Inclination at Lake Athabasca presents two principal maxima and two minima; the first of these occurs at 3 A.M., and corresponds to the minimum of Horizontal Force; the second maximum occurs at 11 A.M., and corresponds to the small relative minimum of the latter element, which has been pointed out as having the effect of creating an undulation in the ascending branch of its daily curve. The principal minimum occurs at 10 P.M., and agrees nearly with the daily maximum of Horizontal Force; the smaller minimum, which is at 8 A.M., agrees in like manner with a subordinate maximum of this element at the same hour. Proceeding in the same way as before, to eliminate, partially, the effect of disturbances, by assembling all those days on which no extra observations were taken, we find that the mid-day maximum becomes more prominent, and the maximum at 3 A.M. considerably less so. The hours of maximum and minimum are but little altered, but a slight increase of inclination is shown at 8 P.M., immediately preceding the lowest value of the day, and answering to a contrary inflexion in the mean curve of Horizontal Force at the same hour. It appears, therefore, that in the minor as well as in the more prominent features of the curves, each maximum of the Horizontal Force corresponds to a minimum of Inclination, and each minimum of the former to a maximum of the latter; and we have, from the independent changes of these two elements, observed, by methods which have nothing in common, a strong mutual support and confirmation.

The mean diurnal curve of Inclination at Toronto, for the period under discussion, consists principally of a single progression, having its maximum at 11 A.M., about one hour after the daily minimum of Horizontal Force. This characteristic is the same, whether we take the mean by 101 days, corresponding to the period of observation at the northern station, or the general mean for the same months. There are indications of a second maximum at 10 P.M. The principal minimum occurs at 5 A.M. There is not the slightest trace of an inflexion corresponding to the daily maximum at 3 A.M., which constitutes the principal feature of the northern curve.

Fort Simpson.—The details of the adjustment of the Inclinometer at Fort Simpson will be given in a future section. The observations at this station are divided into two series; the first of 26 days, five of which are rendered imperfect by the omission of one or more observations; the second of 19 days, three of which are in the same condition. The omissions only affect the means at 0h. and 1h., and the whole 45 days have been combined for a general mean, without excluding the imperfect days, for the reasons assigned in reference to the Declinometer at the same station (p. 11.)

Table XXXV.

Mean scale reading of Inclinometer at Fort Simpson, corrected for changes of Declination and Horizontal Force, for the months of April and May 1844. *To which is added the Mean for the same two months at Toronto, by observations of four years,* 1845—1848.

Scale co-efficients:—

1 April to 1 May	$aP_{,}=0{\cdot}106\times1{\cdot}22=0''{\cdot}1293$
2 May to 24 May	$aP_{,,}=0{\cdot}127\times1{\cdot}22=0''{\cdot}1549$
The general mean	$aP=0{\cdot}115\times1{\cdot}22=0''{\cdot}1402$
Also for Toronto	$aP=0{\cdot}354\times1{\cdot}32=0''{\cdot}4673$

Fort Simpson.										Toronto.				
		Partial Means.				General Means.						Four years.		
		April.		May.		Readings.		Diurnal Variation.					Diurnal Variation.	
Göttingen hour.	Mean time.	Scale.	Temp.	Scale.	Temp.	Scale.	Temp.	Scale.	Δθ	Göttingen time.	Mean time.	Mean.	Scale.	Δθ
noon	15 15	240·61	58°·6	280·41	68′·3	257·11	62°·6	+56·9	+7′·98	21	15	45·49	−1·05	−0′·49
1	16 15	252·18	58·0	294·15	66·8	272·11	62·3	+71·9	+10·07	22	16	45·23	−1·21	−0·56
2	17 15	217·34	57·5	306·84	66·9	255·99	61·6	+55·8	+7·81	23	17	45·01	−1·53	−0·71
3	18 15	199·87	57·1	292·13	66·4	238·92	61·2	+38·7	+5·42	noon	18	44·54	−2·00	−0·93
4	19 15	191·47	56·5	291·47	66·3	233·60	60·8	+33·4	+4·68	1	19	45·03	−1·31	−0·61
5	20 15	175·69	57·0	259·23	65·9	210·96	60·9	+10·8	+1·51	2	20	46·25	−0·29	−0·14
6	21 15	163·74	55·3	250·63	63·5	200·43	58·9	+0·2	+0·03	3	21	48·41	+1·87	+0·88
7	22 15	153·05	56·9	248·23	64·5	193·24	60·2	− 7·0	−0·98	4	22	50·73	+4·29	+2·00
8	23 15	142·27	57·1	249·76	65·4	187·65	60·7	−12·6	−1·76	5	23	51·19	+4·65	+2·17
9	0 15	138·13	58·0	253·21	66·6	186·72	61·3	−13·5	−1·89	6	noon	50·46	+3·92	+1·83
10	1 15	142·45	58·3	251·92	66·1	188·67	61·7	−11·5	−1·55	7	1	49·06	+2·52	+1·18
11	2 15	133·25	58·7	241·54	67·1	178·97	62·4	−21·2	−2·97	8	2	47·20	+0·66	+0·31
midnt	3 15	128·68	58·7	236·41	66·6	174·13	62·0	−26·1	−3·68	9	3	45·77	−0·77	−0·36
13	4 15	123·53	59·0	230·48	67·0	169·84	62·4	−30·4	−3·89	10	4	44·92	−1·62	−0·75
14	5 15	121·51	59·2	225·52	67·3	165·43	62·7	−34·8	−4·87	11	5	44·65	−1·89	−0·88
15	6 15	121·97	59·2	226·15	67·4	165·95	62·8	−34·3	−4·80	midnt	6	44·85	−1·69	−0·79
16	7 15	116·37	59·4	229·58	67·7	164·17	63·0	−36·0	−5·00	13	7	45·41	−1·13	−0·53
17	8 15	120·60	58·8	227·46	68·1	165·73	62·8	−34·5	−4·83	14	8	46·13	−0·41	−0·20
18	9 15	132·04	59·2	230·43	68·7	173·58	63·3	−26·6	−3·72	15	9	46·21	−0·33	−0·16
19	10 15	128·64	59·5	231·25	68·4	171·96	63·4	−28·2	−3·95	16	10	46·06	−0·48	−0·22
20	11 15	151·43	59·7	245·11	68·6	190·98	63·5	− 9·2	−1·29	17	11	46·11	−0·43	−0·20
21	12 15	181·84	59·3	255·47	68·1	212·98	63·1	+12·7	+1·78	18	midnt	43·90	−0·64	−0·30
22	13 15	183·80	59·3	272·91	67·6	221·42	62·9	+21·2	+2·97	19	13	45·91	−0·63	−0·30
23	14 15	190·72	59·1	270·06	67·0	224·22	62·5	+24·0	+3·36	20	14	45·88	−0·66	−0·31
		160·55		254·18		200·20								

The mean diurnal curve of inclination at Fort Simpson deducible from the observations of April and May presents only one well-marked maximum and one minimum, the former at 4 A.M., the latter at 7 P.M.; there is an indication of a very slight maximum at 1 P.M., which coincides with a contrary inflexion in the mean curve of Horizontal Force, but both curves approach more nearly to a single progression than in the winter months. Viewed generally, the mean diurnal curve of inclination at this station, as at Lake Athabasca, is the exact converse of that of the Horizontal Force; the morning maximum of the latter element shown in the winter months is here reduced in amount, proportionably, as much as the mid-day maximum of the former, both features have nearly disappeared. There is no corresponding difference shown at Toronto in the mean curves of the same elements for the same periods respectively; and without grounding too much on observations embracing so short a period, I regard the corroboration afforded by the two elements to one another as giving good grounds for the conclusion that in high latitudes the advance of the season, and the rapid increase in the length of the day, produces changes in the character of the daily course of the magnetic elements which is not experienced in lower ones.

As the Inclinometer was not observed at Toronto in April and May 1844, I have taken the mean for these two months for four years, as a normai curve for that station. It does not differ in any characteristic from that of the five months previously described. It consists of a double progression, having two minima of nearly equal amount, at 6 A.M. and 5 P.M. respectively, and two maxima, of which the first, at 11 A.M., is so strongly marked as to be the great feature of the whole curve, and the second, at 9 P.M., is so small in amount, as to fall on the negative side of the mean line. The curves of this element, therefore, at the two stations have scarcely anything in common.

Total Force.—Owing to the uncertainty in the precise values of the scale co-efficient of the Inclinometer, the inferences we can derive from the observations as to the changes of the Total Force, which depend very much upon those of the inclination, are necessarily somewhat vague; having found, however, by trial, that the *law* of the diurnal changes deducible is not altered by any moderate change in the scale value of the Inclinometer or of the Bifilar readings, but only the *amount*, I think it worth while to present the result.

The changes of Total Force are found by the formula

$$\frac{\Delta R}{R}=\frac{\Delta X}{X}+\tan\theta\Delta\theta$$

in which the quantities $\frac{\Delta X}{X}$ and $\Delta\theta$, are taken from Tables XX., XXXIII., and XXXIV.

Table XXXVI.

Approximate Mean Diurnal Curve of total Magnetic Force at Lake Athabasca from 110 *days of observation, to which are added the corresponding Values from the* 46 *days selected as free from disturbance.*

Mean time -	16	17	18	19	20	21	22	23
Gött. time -	0	1	2	3	4	5	6	7
Whole period	+·00128	—·00013	+·00009	—·00096	—·00091	—·00076	—·00036	—·00069
Selected days	+·00035	—·00005	—·00050	—·00061	—·00074	—·00094	+·00008	+·00008
Mean time -	Noon	1	2	3	4	5	6	7
Gött. time -	8	9	10	11	Midn.	13	14	15
Whole period	+·00029	—·00010	+·00074	+·00044	+·00025	+·00003	—·00026	—·00042
Selected days	+·00021	—·00001	+·00054	+·00002	+·00079	—·00006	—·00043	—·00029
Mean time -	8	9	10	11	Midn.	13	14	15
Gött. time -	16	17	18	19	20	21	22	23
Whole period	—·00046	—·00079	—·00058	—·00041	—·00010	+·00110	+·00090	+·00151
Selected days	—·00026	—·00082	+·00000	+·00106	+·00086	+·00033	+·00004	+·00053

The prevalence of positive and negative signs alternately, in both the mean curves contained in this Table, indicates clearly that the total Magnetic Force at Lake Athabasca in the winter months has two maxima and two minima daily; and on laying down the above quantities upon a sufficiently large scale, the hours of the latter are seen to be 8 A.M. and 8 P.M. nearly, while of the former, one maximum falls by both curves at or near 2 P.M., and the other at or near 3 A.M. if we include the disturbed days, and at or near midnight if we exclude them. There is very little difference in character or amount between the curves given by the whole period and by the selected days respectively in any other than the particular just alluded to. By both curves the Total Force is greater at the maximum in the night than at the one in the day, and less at the minimum, which occurs four hours before noon, than at the one eight or nine hours after noon. The difference between the highest and lowest mean value, or the mean diurnal range of total force, appears to be about ·002. R by both curves. To examine the influence of an error in the co-efficients upon this curve it was assumed that the value of the scale divisions of either instrument might be *one tenth greater* or the same quantity *less* than the value actually employed; this supposition allows of eight combinations, and, having computed and laid down the values upon every one of them, it appears that in each case we have the two daily maxima and minima at nearly the same hours; the differences are chiefly in the amount of the changes and the relative prominence of the two maxima.

Table XXXVII.

Approximate Mean Diurnal Curve of total Magnetic Force at Fort Simpson from 46 *days of observation in April and May* 1844.

Mean time -	15 15	16 15	17 15	18 15	19 15	20 15	21 15	22 15
Gött. time -	0	1	2	3	4	5	6	
	+·0065	+·0097	+·0077	+·0045	+·0050	+·0001	−·0009	−·0009
Mean time -	23 15	0 15	1 15	2 15	3 15	4 15	5 15	6 15
Gött. time -	8	9	10	11	12	13	14	15
	−·0014	−·0024	−·0017	−·0029	−·0027	−·0029	−·0041	−·0038
Mean time -	7 15	8 15	9 15	10 15	11 15	12 15	13 15	14 15
Gött. time -	16	17	18	19	20	21	22	23
	−·0046	−·0043	−·0051	−·0040	−·0009	+·0017	+·0027	+·0037

The above curve presents but one maximum and one minimum, the former at 4^h A.M., the latter at 7^h or 8^h P.M.; the mid-day maximum has disappeared, but the amount of the daily fluctuation is increased fourfold. This change of character in the mean diurnal fluctuation of the total force does not appear to attend the progress of the seasons at Toronto, but would necessarily be inferred from the altered character of the mean curves of all the elements at Fort Simpson; it is evident that the causes preceding the minor fluctuations at Lake Athabasca are here overruled by the more powerful influences to which the principal fluctuation is due, and we see in the case of each of the three elements observed an approach to one great diurnal movement, owing its character almost entirely to the extraordinary constancy and regularity which appears, in these regions, to belong to a class of influences elsewhere denominated *irregular*.

TABLE XXXVIII.

Approximate Mean Diurnal Curve of Total Force at Toronto, from combination of the changes of Horizontal and Inclination Force given by Tables XXXIV. and XXXV.

Mean time - -	16	17	18	19	20	21	22	23
Göttingen time -	22	23	Noon	1	2	3	4	5
October—February -	−·00028	−·00025	−·00042	−·00076	−·00102	−·00109	−·00093	−·00073
April—May - -	−·00063	−·00073	−·00126	−·00087	−·00048	·00025	·00127	·00159
Mean time - -	Noon	1	2	3	4	5	6	7
Göttingen time -	6	7	8	9	10	11	Midn.	13
October—February -	·00043	·00127	·00184	·00181	·00143	·00093	·00017	−·00014
April—May - -	·00153	·00137	·00090	·00051	·00008	·00011	−·00023	−·00026
Mean time - -	8	9	10	11	Midn.	13	14	15
Göttingen time -	14	15	16	17	18	19	20	21
October—February -	−·00001	·00020	−·00038	−·00046	−·00042	−·00039	−·00030	−·00035
April—May - -	−·00013	−·00020	−·00016	−·00031	−·00053	−·00055	−·00055	−·00081

In this Table the observed mean diurnal values of the Horizontal Force for the seasons compared have been combined with the normal curves of inclination given by the observations of three years. They agree in showing that the mean diurnal curve of Total Force at Toronto has one principal maximum, which in the winter seems to occur about 2 P.M. and in the spring about mid-day; the lowest value occurs in the forenoon in the winter at 9 A.M., and in the spring two or three hours earlier. The indications of a secondary maximum and minimum are undecided; in neither case is there any appearance of a maximum of Total Force answering to the nocturnal one in the north. The curve computed from the changes of inclination shown by the one-bar instrument gives also a single maximum at noon, but of a small amount.

ABSOLUTE DETERMINATIONS AND ADJUSTMENTS OF THE MAGNETICAL INSTRUMENTS AT FORT SIMPSON.

Declination.—The Declinometer was adjusted on the 30th March 1844. The Magnet was suspended by a single fibre of silk, of which the force of torsion was quite insignificant; the mean change of scale reading produced by turning the torsion circle 90° was 4·1 div., whence $\frac{H}{F} = \frac{1}{1769}$. The base was levelled, and the fixed wire of the telescope made to cut the central division of the scale, when the instrument was ready for observation. It became evident, however, after a time, that the length of the scale, which was between 12° and 13° of a circle, was insufficient to allow the full range required in some of the great disturbances at this station, if it was to be bisected in the mean position of the Magnet; advantage was therefore taken of a state of entire quiescence of the Magnet, on the 21st April, to move the arm and telescope 1° 40′ to the westward, adding 100 divisions to the range of the scale on the east side of Zero, and diminishing the readings to the same amount; 100 has been subtracted from all readings prior to that date, in order to connect them with the subsequent series.

Absolute Declination.—An observation was made on the 30th March with the azimuth compass. Two sets of azimuths of the sun were observed A.M. and two sets P.M., giving the following values, which are reduced to the mean for 24ʰ by a correction from Table VIII.

At 7ʰ 58ᵐ 1 A.M.	var. east	37° 35′·7	38° 4′·2	38° 0′·0
at 9ʰ 12ᵐ 7 A.M.	„	38° 32′·8		
at 3ʰ 46ᵐ 2 P.M.	„	37° 39′·0	37° 55′·8	
at 4ʰ 55ᵐ 8 P.M.	„	38° 12′·7		

The following observation with a Collimator Magnet was made on the 8th May 1844.

The Theodolite was levelled and directed to the Collimator, mean scale reading 77·84, Declination 416·0, mean of verniers 359° 59′ 30″, deviation from the magnetic axis 3·43 div. = 8′ 30″ to the west. It was then directed to the sun, and the transit of both limbs observed, mean reading of Vernier's 86° 17′ 30″ for the sun's centre, at 9ʰ 0ᵐ 17ˢ apparent time. We have, then, the sun's apparent

Magnetic azimuth - - - - -	86° 18′ 0″
Deviation of telescope - -	0° 8′ 30″
Sun's magnetic azimuth - - -	86° 9′ 30″
Sun's true azimuth at 9ʰ 0ᵐ 17ˢ -	124° 6′ 28″
Absolute Declination - -	37° 56′ 58″ east,

corresponding to 416·0 on the Declination scale. The mean scale reading on the 8th was 399·9, corresponding, therefore, to 37° 40′·9 east Declination. There is a regular and progressive increase of Declination shown by the scale readings in April and May, as already remarked *ante* p. 17, Table IX. The following means were there given:—

			Differences.
April, from 1st to 14th, Mean Declination		358·14	
			27·78
,, ,, 15th to 27th,	..	385·92	
			14·41
,, ,, 28th to 11th May	,,	400·33	
			17·88
May, ,, 12th to 24th	,,	418·21	

The mean for April will be 374·59 = 37° 16′·6 east, and the mean for May 410·01 = 37° 53′·0 east. The mean of the whole, which corresponds to April 27ᵈ·5, will be 390·76 = 37° 31′·7.

The results of the Declination observations have been discussed in connexion with the corresponding series at Lake Athabasca, *ante* p. 15, et seq.

Bifilar.

First Adjustment, 30th March 1844.

1. The telescope was placed in the meridian, by suspending the magnet with unifilar suspension; reading of the azimuth circle 118° 12′.

2. The telescope was next suspended by the same double suspension as was used at Athabasca, and the interval of the threads adjusted by trial to give an angle of about 60°. Scale reading 210.

3. To bring the plane of detorsion of the double suspension into the meridian, the magnet was suspended, and the torsion circle turned until a position was found by trials, in which the scale readings were nearly the same, whether the magnet hung with its marked end to the north, or, by turning the arm 180°, it was reversed and hung with the marked end to the south. The final readings were

Telescope :	Torsion Circle :	Scale :	Bar :
118° 12′	265° 30′	278·0	N. end to N.
298° 12′	265° 30′	300·0	N. end to S.

265° 30′ was therefore taken as the position of the index of the torsion circle which made the plane of detorsion in the meridian.

4. The arm was then turned 90° to 208° 12′, and the torsion circle turned until the scale read 277·0, torsion circle 326° 50′.

We have then $v=(326° 50' - 265° 30' =)$ 61° 20′ when $K = a$ cotan $v = -·000318$. Increasing numbers denote increase of Horizontal Force.*

The Bifilar received an accidental shock after the observation of April 10ᵈ 3ʰ, which produced a change of reading of + 80 div.; this quantity has been added to the readings from that hour down to the re-adjustment of the instrument on the 13th.

Second Adjustment, 13th April.

1. The telescope was placed in the meridian by suspending the magnet with unifilar suspension, reading on the torsion circle 119° 45′, scale 210, Declination 264·0.

2. The double suspension being applied as before, a number of trials were made to bring the plane of detorsion into the meridian. The final readings were as follows:

Telescope :—	Torsion Circle :—	Scale :—
119° 50′	349° 50′	207·5
299° 50′	349° 50′	202·0

Bar :—	Declination :—
N. end to N.	465·0
N. end to S.	460·0

349° 50′ was therefore taken as the position of the index, which made the plane of detorsion in the meridian. The arm was turned 90° to 209° 50′, and the torsion circle turned until the scale read 204·0, when the reading of the torsion circle was 53° 50′, Declination 464·0, whence a cotan $v = -·000283$. Increasing numbers denote increase of force, as before.

The length of the scale of the Bifilar was found insufficient at this station, and a continuation on card was attached to it, on the side of decreasing force; but even this was insufficient on some occasions, when the observed range exceeded both the natural scale and its continuation.

HORIZONTAL FORCE.

In absolute measure.—The determination at Fort Simpson was similar in all respects to that at Lake Athabasca, except that three distances of deflection were employed instead of two. The details were as follows:—

* Decrease of Horizontal Force as actually observed, but the numbers having been inverted, Increase of Force in the register.

TABLE XXXIX.

Determination of the absolute Horizontal Intensity with Bars 30 and 31, at Fort Simpson, Mackenzie's River. Length of suspended Magnet, 2·45 inches.

DEFLECTION.						VIBRATION.								
Date.	Bar.	Dist.	Observed Angle.	Temp.	Corrected Angle.	Date.	Observed Time.	Temp.	Corrected Time.	m	X		Mean.	General Mean.
1844:			° ′ ″	°	° ′ ″		s	°	s					
May 2 -	30	1·0257	23 9 46	37·5	23 6 5	May 3 -	5·1307	39·0	5·1393	0·417	1·960			
„ 2 -	30	1·1757	15 7 46	38·0	15 6 7	„ -	- -	- -	- -	0·417	1·960	1·961		
„ 2 -	30	1·3257	10 28 48	37·5	10 27 0	„ -	- -	- -	- -	0·416	1·964		1·963	
May 2 -	31	1·0257	20 14 12	48·5	20 27 6	May 3 -	5·4616	42·0	5·4691	0·372	1·961			
„ 2 -	31	1·1757	13 14 24	44·0	13 22 9	„ -	- -	- -	- -	0·372	1·970	1·966		
„ 2 -	31	1·3257	9 12 27	48·5	9 19 1	„ -	- -	- -	- -	0·372	1·968			
May 2 -	30	1·0257	23 9 46	37·5	23 6 5	May 3 -	6·3020	47·0	6·3133	0·412	1·935			
„ 2 -	30	1·1757	15 7 46	38·0	15 6 7	„ -	- -	- -	- -	0·412	1·936	1·936		
„ 2 -	30	1·3257	10 28 48	37·5	10 27 0	„ -	- -	- -	- -	0·411	1·939		1·941	1·951
May 2 -	31	1·0257	20 14 12	48·5	20 27 6	May 3 -	6·6707	46·0	6·6867	0·368	1·944			
„ 2 -	31	1·1757	13 14 24	44·0	13 22 9	„ -	- -	- -	- -	0·367	1·948	1·946		
„ 2 -	31	1·3257	9 12 27	48·5	9 19 1	„ -	- -	- -	- -	0·368	1·945			
June 12 -	30	1·0257	22 30 38	57·5	22 26 54	June 12 -	5·2357	57·8	5·2323	0·401	1·952	1·954		
„ 12 -	30	1·3257	10 12 2	57·5	10 10 20	„ -	- -	- -	- -	0·403	1·957		1·949	
June 12 -	31	1·0257	20 5 6	56·0	20 3 18	June 12 -	5·5845	58·8	5·5993	0·360	1·936			
„ 12 -	31	1·0257	9 1 48	57·0	9 2 12	„ -	- -	- -	- -	0·358	1·949	1·943		

In addition to the foregoing values by the standard bars, we have those given by the magnets employed for verification; namely,

By bar 17, May 2, 1844, 1·959	Bar 20, June 2d, 1844, 1·972·
bar 17, June 12 „ 1·956·	bar 23, June 2 „ 1·947·

Including in the present instance the results by bar 17, the general mean is

$$X=1\cdot 952$$

which is the value employed.

Variations of the Horizontal Force.—See Tables XXIII, &c.

INDUCTION INCLINOMETER.

First Adjustment, March 30, 1844.

1. The base was levelled, and the meridian reading of the verniers found to be 297° 30′ 0″.

2. The iron bar was inserted in its collar reversed, the lower or north end deflecting, and the telescope was turned in azimuth until the central division of the scale appeared on the wire, verniers 249° 34′ 10″, hence $u=47° 55′ 50″$.

3. The bar was inverted, the upper or south end deflecting, and the telescope turned as before.

Verniers, 351° 29′ 0″, whence $u=53° 59′ 0″$, also $S=50° 57′ 25″$, $D=3° 1′ 35″$, and $\theta=81° 52′$.

$$\text{Then } aP_{\prime}=a\frac{\sin 2\theta \cos u}{2 \sin S \cos D}=0'\cdot 106$$

which value being multiplied as before by the co-efficient 1·22, gives for the approximate scale value under the first adjustment 0′·1293.

$$B=\frac{\sin D \cos S}{\sin 1' \cos u}\cdot\frac{K}{a}=0'\cdot 0618 \text{ when } K=\cdot 000318.$$

$$=0'\cdot 0551 \text{ when } K=\cdot 000285\cdot$$

$$R=\frac{\sin S \cos D}{\cos u}\cdot\frac{q}{a}=0'\cdot 488$$

The series was broken by an accidental shock to the instrument 2d May, by which the arm carrying the telescope was moved about 5°; advantage was taken of this opportunity to make the series of observations of the Absolute Horizontal Force given at page 72, after which the instrument was re-adjusted.

Second Adjustment, May 2, 1844.

1. The meridian reading of the verniers was 181° 57′ 10″, Declination 400.

2. The iron bar was inserted in its collar reversed, the lower

end deflecting, verniers 135° 59′ 45″, Declination 387·0, whence $u_{,}$=45° 57′′25—13′′0 - 45° 44′ 25″.

$$\text{Then } aP_{\prime\prime}=0''127 \times 1'22=0''1549$$
$$B=0'023$$
$$R=0'388$$

which continue applicable to the end of the series. There are 26 days in the first adjustment and 19 days in the second, giving

$$aP=0''115 \times 1'22=0''1402$$

for the value applicable to the diurnal curve obtained by uniting the whole.

Increasing numbers in every case denote increase of Vertical Force or Dip.

The means of the Inclinometer scale readings at this station will be found in Table XXXV.

IRREGULAR FLUCTUATIONS OF THE MAGNETIC ELEMENTS.

It has been shown in the preceding sections, by an arbitrary selection of days regarded as free from magnetic disturbance, that the portion of the mean diurnal curve of each of the elements, which owes its peculiar character most to the influence of what are usually called the *irregular* movements, is comprised between midnight and 7^h or 8^h A.M, and their effect, as far as can be inferred from the proportionate reduction in the deviation of the magnet from its mean position, made by rejecting them, is much the greatest during that part of the night. As regards the Declination, at least, this result is unexpected, being different from the conclusion to which Colonel Sabine and Dr. Lloyd have been led by their examination of the observations of Toronto and Dublin respectively, and is of so much importance that it will be necessary to investigate it more fully.

If we take the difference between the scale reading (ψ_h) at each hour of observation, and the monthly mean at the same hour $\overline{(\psi_h)}$, the square root of the mean of the squares of these differences $\sqrt{\frac{\Sigma}{n}(\psi_h - \overline{\psi_h})^2}$ is a quantity regarded by Colonel Sabine* as the mean effect of the *irregular disturbing force* at the hour, and is called by Dr. Lloyd the *mean disturbance*†, being analogous to the *mean error* of an observation at that hour; similarly $\sqrt{\frac{\Sigma'}{N}(\psi_h - \overline{\psi_h})^2}$ gives the value for a whole month, or longer period, N being the total number of observations, and Σ′ the sum of all the squares. These quantities have been calculated for each of the elements at

* Preface to Observations during Magnetic Disturbances. Part 1. p. ix. 1842.

† Transactions of Royal Irish Academy. Vol. XXII. part 1.

both the northern stations, and for some of them at Toronto, Philadelphia, and Sitka, with the view of giving stronger prominence to the peculiarity of the others. The work was first done, employing the observations as they stand in the abstracts, but it was evident, while the law is thus deducible, and the quantities for the several hours are relatively correct, their absolute values, when the scale readings of the instrument have undergone any considerable regular change, whether of a periodic or instrumental character, are very much exaggerated, and altogether deceptive. In these cases the scale readings at the beginning and end of the month differ from its mean ($\overline{\psi}$) by a quantity which includes, with the irregular fluctuation, the amount of the regular change in half the month. The differences between the successive fortnightly mean values in Table XI., show how considerable this change was in the Horizontal Force and Inclination at Lake Athabasca; it was equally large in the Declination at Fort Simpson; all these differences ($\psi_h - \overline{\psi_h}$) therefore were exaggerated to an amount proportioned to their interval in time from the middle period. It was thought worth while, in the case of the elements just named, to endeavour to eliminate this change, which was done by assuming that all the daily means of each month would, but for this cause, have been equal, and that the difference of each from the mean of the whole was a measure of the amount of the progressive change to be eliminated in the interval elapsed between that day and the middle day. Each daily mean therefore furnishes the equation $n\,e - (\psi_n - \overline{\psi}) = 0$, where e the value of the daily change required, and n the interval in days; by summing these equations in the usual manner, a mean value of the quantity e was obtained for each month, and this again, being multiplied by n, was added to, or substracted, as the case might be, from all the scale readings of each day, and the differences taken anew from the readings thus corrected. The values of e actually employed, or the approximate change of *mean* scale reading from day to day, were as follows:—Bifilar at Lake Athabasca, for the month of November 2·47 divisions, for the month of February 2·16 divisions; Inclinometer, November 1·29 divisions, December 0·91 divisions, January 0·75; for the other months the change did not appear large enough to call for the correction. Again, for the Declination at Fort Simpson, from the 1st to the 27th April 1·537 divisions, from the 28th April to the 24th May 1·268 divisions. In the following tables the quantities given are those derived from the readings thus corrected.

TABLE XL.

Value of the Mean Disturbance of the Declination, taken without regard to sign, at the several observation hours at Lake Athabasca, to which are added the corresponding quantities for the same period at Toronto and Sitka, the whole expressed in Arc.

Local Mean Time.	Lake Athabasca 1843–4.						Toronto.	Sitka.
	Oct.	Nov.	Dec.	Jan.	Feb.	$\frac{\Sigma}{N}$	$\frac{\Sigma}{N}$	$\frac{\Sigma}{N}$
Midnight	7·84	4·51	8·69	10·82	9·06	8·50	1·72	4·36
1 A.M.	9·55	8·61	7·05	13·54	9·06	9·87	1·33	4·18
2 „	7·80	2·93	7·69	13·05	8·09	8·65	1·00	3·62
3 „	6·28	4·54	6·59	9·75	8·01	7·37	1·59	2·97
4 „	18·90	9·14	5·35	12·76	8·53	10·96	1·52	2·92
5 „	14·12	6·48	7·06	27·04 *a*	10·16	15·25 *a*	1·67	2·84
6 „	18·67	7·21	8·87	11·96	8·36	10·60	1·75	2·75
7 „	7·16	5·96	12·42	8·08	6·83	8·48	1·29	2·82
8 „	5·99	5·10	5·41	5·71	5·72	5·55	1·82	2·88
9 „	4·20	2·41	4·82	4·39	6·16	4·58	1·89	2·80
10 „	4·97	3·65	6·80	5·20	4·56	5·14	1·52	3·10
11 „	3·71	4·51	6·52	4·30	11·62 *a*	6·96 *a*	1·68	2·79
Noon	4·24	3·96	5·21	4·49	4·71	4·56	1·61	2·67
1 P.M.	2·55	3·65	4·67	4·69	4·78	4·27	1·54	2·55
2 „	3·32	2·52	6·40	5·16	4·51	4·68	1·34	2·20
3 „	4·30	2·83	5·17	5·84	4·40	4·66	1·31	2·14
4 „	4·10	2·74	5·47	4·95	4·26	4·45	1·35	1·95
5 „	3·93	3·14	3·67	3·84	3·96	3·67	1·65	1·81
6 „	5·38	4·27	3·47	4·37	4·34	4·28	1·61	1·90
7 „	4·20	3·03	3·36	4·42	3·84	3·76	1·05	3·09
8 „	4·76	3·71	4·34	5·28	4·92	4·62	1·53	2·47
9 „	7·75	2·73	4·69	6·52	8·62	6·26	2·16	2·51
10 „	13·20	15·36	9·97	4·10	4·49	10·18	2·72	3·43
11 „	19·80	8·59	4·37	9·33	5·63	9·71	2·35	3·32

a Irregularity produced by a single unusual observation. See remark below.

It appears that the *mean disturbance* of the Declination at Lake Athabasca, or the *mean disturbing force*, if that term is preferred, is nearly constant, and at its lowest amount from 9 A.M. to 7 P.M.; it is greatest at 5 A.M., but an inferior maximum is presented at 10 P.M., which is the hour at which the greatest value prevails at Toronto and at Dublin. We have consequently an indication of the existence of at least two classes of irregular movements, the one due to causes which apparently act universally, the other due to some cause which only comes into operation in high magnetic latitudes. The characteristics of the curve for the five months in question at Toronto, are a nearly uniform value from 8 A.M. to P.M., when it begins to diminish; it is least at 7 P.M., and then increases regularly until 10 P.M., when it is greatest; it diminishes again until midnight, but appears to vary little during the night. There are indications in each of the three curves of a small increase in the *mean disturbance* about noon. Referring next to the mean for the Russian station at Sitka, the nearest geographically to Lake Athabasca, we find a manifest approximation to the law which prevails at the latter station; the greatest *mean disturbance* is shown at midnight, but high values prevail from 10 P.M. to 2 A.M. The lowest value is here also at 5 P.M.

There are two observations included in the series at Lake Athabasca taken during disturbance, which differ so much from the means for the same hours, as to produce a considerable effect upon the final result. These occurred on January 25^{d} 1^{h}, and February 2^{d} 7^{h}, Göttingen; by omitting them and taking fresh means for those hours, the mean disturbance for 5 A.M. becomes 10′·63, and that for 11 A.M. becomes 5′·28, which are in accordance with the values before and after them. The quantities for the individual months in these cases are then 14′·56 and 6′·37 respectively.

The preceding results are independent of the *direction* of the disturbance. The next Tables are formed by taking the sum of the squares of the easterly and westerly deviations separately, and dividing them by their proper number. It sometimes happens that the scale reading at an individual observation is exactly equal to the mean; in this case the difference being ± 0, the observation is not included in the number with either divisor, but being included for the disturbance without regard to sign, Table XL., occasions that value to be apparently less than, from the values for the same hours under the special signs, it should be; but such is not really the case.

TABLE XLI.

Value of the Mean Easterly Disturbance of Declination at the same stations and for the same period.

Local Mean Time.	Lake Athabasca 1843–4.							Toronto.		Sitka.	
	Oct.	Nov.	Dec.	Jan.	Feb.	Σ(E)/N	Excess.	Σ(E)/N	Excess.	Σ(E)/N	Excess.
Midn. -	4′·0	4′·3	12′·0	7′·8	10′·5	8′·17	- ′ -	2′·16	0′·85	5′·07	1′·37
1 A.M. -	6·2	11·3	4·1	14·2	8·9	9·63	- -	1·66	0·59	4·62	0·63
2 „ -	10·5	2·8	11·6	20·1	6·3	11·16	4·15	1·75	0·27	4·07	1·30
3 „ -	5·6	6·0	8·0	13·2	7·0	8·27	1·71	1·98	0·56	3·18	0·41
4 „ -	32·2	17·7	5·4	6·4	7·2	13·31	3·85	1·59	0·12	3·18	0·51
5 „ -	16·1	8·7	8·3	23·6 *a*	13·8	14·46 *a*	7·58	1·68	- -	2·94	0·20
6 „ -	37·3	10·7	9·2	16·9	11·3	18·06	11·55	1·37	- -	2·60	—
7 „ -	8·7	9·4	18·0	10·3	9·0	12·02	6·62	1·33	0·01	2·47	—
8 „ -	3·3	7·2	6·3	6·5	6·8	6·22	1·33	1·19	- -	2·56	—
9 „ -	4·1	2·1	4·7	4·7	5·9	4·96	0·37	1·35	- -	2·65	—
10 „ -	4·6	4·8	6·7	5·1	4·7	5·34	0·34	1·34	- -	2·45	—
11 „ -	2·5	4·2	7·2	4·7	6·5 *a*	5·28 *a*	- -	1·43	- -	2·31	—
Noon -	2·9	3·4	5·1	4·9	4·1	4·29	- -	1·29	- -	2·35	—
1 P.M. -	2·2	3·4	3·9	5·1	3·7	3·89	- -	1·46	- -	2·15	—
2 „ -	2·2	2·5	5·4	4·3	5·1	4·40	- -	1·18	- -	2·05	0·20
3 „ -	3·8	2·6	4·7	4·2	4·6	4·04	- -	1·24	- -	2·00	—
4 „ -	3·4	2·9	5·8	4·7	4·1	4·33	- -	1·03	- -	1·92	—
5 „ -	2·8	2·5	3·5	3·8	3·6	3·28	- -	1·51	- -	1·96	0·27
6 „ -	5·9	3·2	3·6	5·4	5·0	4·47	0·12	2·34	1·32	1·01	—
7 „ -	3·3	3·6	3·0	5·4	4·6	4·05	0·58	1·33	0·40	4·08	1·91
8 „ -	3·6	4·3	4·6	6·2	5·3	4·93	0·55	2·35	0·47	2·88	0·82
9 „ -	6·1	2·8	5·2	8·2	7·9	6·35	0·47	3·51	2·38	2·85	0·70
10 „ -	16·5	21·0	15·1	5·0	4·0	12·80	4·98	4·93	3·73	4·25	1·68
11 „ -	32·3	5·7	5·1	13·3	6·9	11·77	3·97	1·92	- -	4·17	1·74

a The extreme readings of Jan. 25 and Feb. 2 are here omitted; the mean disturbance is 23′·21 at 1 A.M., and 5′·52 at 11 A.M. if they are retained, the monthly values being then 75′·1 and 7′·1 respectively.

Table XLII.

Value of the mean westerly Disturbance of Declination at the same stations, and for the same period.

Local Mean Time.	Lake Athabasca 1843–4.							Toronto.		Sitka.	
	Oct.	Nov.	Dec.	Jan.	Feb.	$\frac{\Sigma(W)}{N}$	Excess.	$\frac{\Sigma(W)}{N}$	Excess.	$\frac{\Sigma(W)}{N}$	Excess.
Midn. -	18′·2	4′·9	5′·5	11′·9	7′·5	8′·95	0′·78	1′·31	-′ -	3′·70	′—
1 A.M. -	14·9	5·6	10·0	12·[illegible]	9·1	10·14	0·51	1·07	- -	3·97	—
2 „ -	6·5	3·2	4·8	8·0	11·2	7·01	- -	1·48	- -	2·77	—
3 „ -	6·9	2·5	5·2	7·5	8·7	6·56	- -	1·37	- -	2·77	—
4 „ -	9·3	3·3	5·7	14·5	9·6	9·46	- -	1·46	- -	2·67	—
5 „ -	12·2	4·4	5·9	7·1 *b*	5·9	6·88 *b*	- -	1·70	0·02	2·74	—
6 „ -	2·5	4·7	8·5	6·1	5·6	6·51	- -	2·18	0·81	2·81	0·12
7 „ -	5·6	4·0	6·3	6·2	4·8	5·40	- -	1·32	- -	3·19	0·72
8 „ -	9·0	3·8	4·0	4·9	4·6	4·89	- -	2·34	1·05	3·20	0·64
9 „ -	2·2	2·8	5·0	4·2	6·5	4·59	- -	2·08	0·73	2·96	0·31
10 „ -	5·7	2·8	6·9	5·3	4·4	5·00	- -	1·65	0·31	3·73	1·28
11 „ -	7·7	4·8	5·9	3·5	6·2 *b*	5·28 *b*	- -	1·95	0·52	3·07	0·56
Noon -	4·8	4·6	5·4	4·2	5·4	4·84	0·55	2·01	0·72	3·01	0·66
1 P.M. -	2·9	3·9	6·5	4·2	5·8	4·69	0·81	1·65	0·19	2·94	0·79
2 „ -	5·1	2·8	7·5	6·0	4·0	5·34	0·94	1·57	0·39	2·36	—
3 „ -	5·3	3·1	5·7	7·8	4·2	5·37	1·33	1·71	0·47	2·23	0·17
4 „ -	4·8	2·6	5·1	5·1	4·4	4·46	0·13	1·81	0·78	1·98	0·06
5 „ -	5·4	3·8	3·8	3·6	4·4	4·09	0·81	1·86	0·35	1·89	—
6 „ -	4·5	5·8	4·3	3·7	3·9	4·35	- -	1·02	- -	1·95	0·04
7 „ -	5·5	2·5	3·8	3·5	3·1	3·47	- -	0·84	- -	2·12	—
8 „ -	6·0	3·1	4·3	4·5	4·7	4·38	- -	0·88	- -	2·06	—
9 „ -	9·5	2·8	4·2	4·9	9·7	6·06	- -	1·13	- -	2·15	—
10 „ -	10·9	11·4	5·2	3·3	5·1	7·82	- -	1·18	- -	2·57	—
11 „ -	11·4	11·4	3·6	6·8	4·4	7·80	- -	2·60	0·68	2·43	—

b The extreme readings of Jan. 25 and Feb. 2 have been omitted here. The mean disturbance W. is 7′·77 at 5 A.M., and 8′·40 at 11 A.M., if they are retained, the monthly values being 9′·6 and 10′·2 respectively.

In the column headed "Excess" the difference between the mean values E. and W. is shown in connexion with the one which is greatest. Thus, at midnight, Table XLII., the mean westerly disturbance of Declination is 0′·78 greater than the mean easterly.

It appears by the foregoing analysis, that the westerly deviations at Lake Athabasca are, to a small extent, greater than the easterly, from 11 A.M. to 5 P.M., and also at midnight and 1 A.M. During the remainder of the night, and from 6 to 11 P.M., the easterly deviations are the greatest. Individual months present irregularities, as must be expected, but the general law is apparent in each of them; namely, a slight excess of westerly deviation during the day, and a much greater excess of easterly deviation during the night, but with a temporary preponderance of westerly tendencies at midnight; the latter particular agrees with Dr. Lloyd's deductions from the observations at Dublin. There is a considerable difference in the value of easterly and westerly tendencies at 10 P.M., a feature which these observations have in common with those at Dublin and Toronto; but the greatest difference is not as at the last-named stations at that hour, but much later in the night; it is shown from 4 to 7 A.M., and thus proves that the maximum value of the *mean disturbance,* included at those hours, is chiefly occasioned by easterly movements. There is no evidence in the means for five months at Sitka of that westerly tendency about midnight which is, with reference to the physical cause of the phenomena, perhaps, an important feature in the curves at the other stations; in other respects the same general law prevails; and it is interesting to observe, that although this station is in a high latitude, and nearer in point of distance by more than one half to Lake Athabasca than it is to Toronto, yet, being on nearly the same lines of equal magnetic inclination and intensity as the latter station, does not partake to a very much greater extent than Toronto in the great magnetic disturbances shown to prevail so commonly at Lake Athabasca.

LE XLIII.—*Mean Disturbance of the Declination at Fort Simpson in April and May* 1844, *after correcting the scale readings for the progressive change alluded to above, together with the values of the same quantity for Toronto and Sitka. A Mean of the* corrected *readings at Fort Simpson,* excluding incomplete days, *is added; those days having been included in the Mean given Table.*

Local Mean Time.	Scale.	Diurnal Variation.	Mean Disturbance of Declination.					Toronto.					Sitka.				
			Total.	E.	W.	Excess.		Total.	E.	W.	Excess.		Total.	E.	W.	Excess.	
						E.	W.				E.	W.				E.	W.
Midn.	387·7	′ −4·1	′ 14·5	′ 13·9	′ 15·1		′ 1·2	′ 2·89	′ 4·47	′ 2·04	′ 2·43		′ 5·18	′ 7·35	′ 3·32	′ 4·03	—
1 A.M.	389·8	−2·0	11·7	9·6	13·1		3·5	3·16	4·66	2·19	2·47		6·57	8·80	4·14	4·66	—
2	391·6	−0·2	15·8	26·8	8·9	17·9		3·83	6·59	1·83	4·76		5·42	5·56	5·28	0·28	—
3	400·7	+8·8	22·2	33·6	12·1	21·5		4·62	7·97	2·01	5·91		8·01	6·54	9·47		2·83
4	399·0	+7·2	17·3	21·9	11·6	10·3		3·77	5·24	2·18	3·11		8·92	4·84	12·67		7·83
5	410·9	+19·0	25·6	38·1	14·2	23·9		2·42	1·80	2·84		1·04	5·08	5·54	4·62	0·92	—
6	413·1	+21·3	22·9	33·1	14·3	18·8		2·43	2·03	2·86		0·83	3·60	3·50	3·64		0·08
7	414·7	+22·8	22·3	35·8	13·8	22·0		3·97	2·08	5·89		3·87	2·69	2·91	2·41	0·50	—
8	406·5	+16·7	19·7	28·5	9·9	18·6		2·91	1·89	4·06		2·17	2·49	2·16	2·82		0·66
9	402·5	+10·7	12·0	16·4	7·9	8·5		2·16	1·94	2·37		0·43	2·10	1·86	2·38		0·52
10	398·1	+6·2	12·2	16·1	11·4	4·7		2·08	2·17	2·04	0·13		2·46	2·09	2·81		0·72
11	389·5	−2·3	6·2	5·2	7·5		2·3	2·24	2·38	2·10	0·28		2·57	1·86	3·33		1·47
Noon	386·1	−5·8	8·0	8·2	7·8	0·4		2·05	2·06	2·04	0·02		2·58	1·80	3·30		1·50
1	383·3	−8·5	4·6	4·5	4·8		0·3	1·84	1·82	1·86		0·04	2·22	1·59	2·89		1·30
2	380·6	−11·3	7·0	5·2	8·9		3·7	1·71	1·48	2·03		0·55	2·31	1·88	2·84		0·96
3	381·2	−10·6	6·3	5·4	6·9		1·5	1·56	1·48	1·65		0·17	2·81	2·21	3·41		1·20
4	381·0	−10·9	6·2	5·9	6·6		0·7	1·68	1·42	1·98		0·56	2·36	1·78	3·03		1·24
5	381·4	−10·4	7·4	5·9	8·4		2·5	1·88	1·69	2·02		0·33	2·96	1·84	4·31		2·57
6	380·8	−11·1	8·1	6·4	9·6		3·2	2·04	1·90	2·19		0·29	4·08	4·44	3·63	0·81	—
7	383·5	−8·4	8·8	9·8	7·8	2·0		2·78	3·65	2·09	1·56		4·16	5·18	3·06	2·12	—
8	385·9	−6·0	12·3	12·2	12·3		0·1	2·29	2·48	2·15	0·33		2·91	2·65	3·18		0·53
9	385·1	−6·7	15·4	7·8	20·4		12·6	2·94	4·53	1·70	1·83		3·51	3·77	3·25	0·72	—
10	384·7	−7·2	13·1	8·7	16·0		7·3	4·58	7·23	2·69	4·54		3·90	4·55	3·33	1·22	—
11 P.M.	384·9	−7·0	9·2	8·1	10·1		2·0	2·78	3·20	2·35	0·85		3·58	4·81	2·78	1·53	—

The observations at Fort Simpson were taken 15 minutes, and those at Sitka 28 minutes after the hour named.

The diurnal law of mean disturbance of Declination indicated by the observations of April and May at Fort Simpson, appears to differ but little from that of the winter months at Lake Athabasca, as long as we disregard the sign of the movements. There is a maximum of total disturbance at 9 P.M., and another at 5 A.M., or about that hour, the latter being very far the most considerable; on referring, however, to the movements under the respective signs, it appears that the earlier of these two maxima is here caused by westerly and not by easterly movements, the great length of the day at this season having apparently the effect of protracting until a late hour of the night, and of increasing in relative importance the westerly tendency, which was also shown to be characteristic of the hours of the afternoon in the winter months. The very great increase in the amount of the mean disturbance is also apparently a result of the advance of the season, being participated in to a certain degree by all the stations. The apparent anomaly of an excess of westerly movements at Sitka, distant less than 500 miles from Fort Simpson, at two hours (3 and 4 A.M.) when the contrary tendency prevails at the latter station, is most remarkable; but this is not the place to follow out the inquiry it demands.

TABLE XLIV.

Mean Disturbance of the Horizontal Force and Inclination at Lake Athabasca, after applying the corrections specified above (*page* 75), *for the changes of* mean *scale reading; final means for the whole period expressed in scale divisions.*

Hour of Local M. T.	Bifilar k=·000341 X.			Excess.		Inclinometer a P=0'·170.			Excess.	
	–	+	Total.	–	+	+	–	Total.	+	–
16	49·2	22·1	32·9	27·1	—	75·6	21·0	42·9	54·6	—
17	44·6	20·4	30·8	24·2	—	52·2	15·0	31·1	37·2	—
18	25·9	16·1	21·2	9·8	—	29·8	10·8	20·0	19·0	—
19	22·9	14·3	18·5	8·6	—	24·4	9·8	16·7	14·6	—
20	21·6	13·6	17·5	8·0	—	12·0	10·3	11·1	1·7	—
21	18·6	13·1	15·7	5·5	—	14·9	9·4	12·2	5·5	—
22	15·5	12·7	14·1	2·8	—	17·3*	8·2	12·8	9·1	—
23	12·3	12·7	12·5	—	0·4	9·1	8·1	8·6	1·0	—
Noon	12·5	11·7	12·1	0·8	—	8·6	7·9	8·3	0·7	—
1	13·5	13·1	13·3	0·4	—	8·2	6·7	7·4	1·5	—
2	12·5	12·6	12·6	—	0·1	7·9	8·7	8·3	—	0·8
3	13·0	12·9	12·9	0·1	—	6·6	8·0	7·3	—	1·4
4	13·9	12·1	13·0	1·8	—	7·2	8·7	8·0	—	1·5
5	13·1	12·6	12·8	0·5	—	5·8	9·1	7·4	—	3·3
6	14·2	13·0	13·6	1·2	—	7·2	9·3	8·3	—	2·1
7	13·3	12·7	13·0	0·6	—	7·7	10·5	9·2	—	2·8
8	13·8	15·4	14·7	—	1·6	7·9	12·3	9·8	—	4·4
9	16·5	18·4	17·4	—	1·9	12·0	18·6	15·1	—	6·6
10	15·2	16·4	15·8	—	1·2	8·1	17·2	12·8	—	9·1
11	17·3	20·8	19·0	—	3·5	13·6	22·3	17·4	—	8·7
Midn.	29·1	18·6	23·4	10·5	—	37·7	14·2	25·7	23·5	—
13	33·4	17·5	25·1	15·9	—	63·6	17·8	37·2	45·8	—
14	38·9	17·9	26·7	21·0	—	51·5	17·0	30·6	33·6	—
15	55·9	19·7	35·2	36·2	—	94·0	21·7	48·5	72·3	—

* This high value appears to be chiefly produced by the observation of February 2d 6h. Gött.

It appears by the abo·e table that the *mean disturbance* of both the Horizontal Force and Inclination, taken without regard to sign, is greatest at 3 A.M., being two hours earlier than the epoch of greatest disturbance of the Declination, and the subordinate maximum, which is shown by the latter element at 9 or 10 P.M., is wanting in these. The tendency to disturbance is very nearly constant, and at its lowest value, by both elements from 11 A.M. to 8 P.M.; it then begins to increase, but not very rapidly, until 11 P.M.; between this hour and midnight there is a large increase, after which very high values are maintained until 5 A.M. Referring again to the relative values, under the positive and negative signs, we find that there is a very great preponderance of negative movements of force and increasing inclination from midnight to 5 A.M.; but for some hours before midnight the contrary tendency prevails, namely, to increase of Horizontal Force and decrease of Inclination; and this latter appears to be more or less the case throughout the day, but the mean disturbance being comparatively small and the opposite tendencies nearly balanced, the latter conclusion is less certain.

All three elements then agree in supporting the conclusion drawn from the daily mean curves in the preceding part of the volume, that at Lake Athabasca a different periodical law governs the irregular fluctuations from the one established for stations in lower magnetic latitudes, or that the *reaction* succeeding the direct influences preponderating during the day, has its maximum influence at a much later hour. It also appears from the observations, that so regular in their operations in these regions are the so called *irregular* influences, that the denomination might with propriety be reversed, the observations of four or five months being sufficient to show that the mean diurnal curves of all the elements derive their chief characteristics from them. We have also seen that a similar result as far as regards the Declination is deducible from the observations of only 46 days at Fort Simpson. As the other instruments were twice adjusted in this short period, it would be scarcely worth while under ordinary circumstances to refer to them. For the purpose, however, of adding all the confirmation possible to the periodical law in question, I have calculated the *mean disturbance* of the Horizontal Force and Inclination at this station also; the results are contained in the next table, and are in complete accordance with the deductions from the longer period of observation.

Table XLV.

Mean Disturbance of the Horizontal Force and Inclination, as shown by 46 days of observation at Fort Simpson, in April and May 1844, expressed in scale divisions of the instruments.

Local Mean Time.	Bifilar. $k=\cdot000291$ X.			Excess.		Inclinometer a P=0'·140.			Excess.	
	−	+	Total.	−	+	+	−	Total.	+	−
15 15	83·0	35·8	54·1	47·2	—	137·0	70·0	101·6	67·0	—
16 15	82·9	39·3	57·2	43·6	—	192·5	73·4	128·4	119·1	—
17 15	72·8	32·6	47·8	40·2	—	150·7	61·1	98·9	89·7	—
18 15	57·6	28·2	40·5	29·4	—	120·1	47·8	78·4	72·3	—
19 15	67·2	27·1	45·3	40·1	—	131·2	48·8	83·7	82·4	—
20 15	61·6	20·9	38·3	40·7	—	108·6	33·7	64·9	74·9	—
21 15	38·5	13·4	25·1	25·1	—	57·8	23·2	39·2	34·6	—
22 15	15·6	9·5	12·5	6·1	—	35·5	21·4	27·9	14·1	—
23 15	10·9	15·6	13·4	—	5·7	17·3	13·5	15·2	3·8	—
Noon 15	13·7	9·4	11·3	4·3	—	21·7	19·2	20·4	2·5	—
1 15	12·3	13·3	12·8	—	1·0	20·3	26·8	23·3	—	6·5
2 15	12·3	18·6	14·8	—	6·3	18·4	27·6	23·2	—	9·2
3 15	14·0	24·8	19·2	—	10·8	24·6	27·9	26·3	—	3·3
4 15	16·8	23·8	19·5	—	7·0	17·2	38·9	27·5	—	21·7
5 15	17·2	29·4	22·5	—	12·2	20·6	46·5	32·3	—	26·1
6 15	16·6	33·5	23·9	—	16·9	21·1	42·9	32·3	—	21·8
7 15	15·2	23·6	19·2	—	8·4	18·5	13·4	25·6	5·1	—
8 15	17·3	21·0	19·2	—	3·7	17·3	27·1	21·8	—	9·8
9 15	25·4	21·8	23·4	3·6	—	46·8	38·5	43·0	8·3	—
10 15	25·6	16·9	20·6	8·7	—	41·9	31·6	37·0	10·3	—
11 15	31·1	18·1	24·1	13·0	—	71·5	31·6	40·6	39·9	—
Midn. 15	52·3	20·9	37·0	31·4	—	96·0	39·5	68·0	56·5	—
13 15	50·2	20·4	35·0	29·8	—	92·9	35·2	59·5	57·7	—
14 15	48·8	20·1	35·3	28·7	—	112·1	39·0	75·5	73·1	—

The co-efficients are added above for convenience, but the values given are probably somewhat exaggerated by the causes before mentioned. They represent the *relative* amount of the disturbing force affecting the Horizontal Force and Inclination, at the different hours of the day and night; and it must be remarked, that the values themselves are so great, that a reduction, of even a fourth part, would still leave them considerable enough to prove, that the force and activity of the causes producing irregular magnetic fluctuations, in the region to which they belong, must be a matter of not less interesting inquiry than the peculiarity of their epoch. For example, the means of total disturbance of the three first hours of the last table, when reduced by one fourth, represent ·011 X and 12'·4 of Inclination respectively.

The next Table is added for the purpose of completing the comparison attempted in this Report, by bringing into one view the results at all the chain of American stations; and it is of particular interest, as showing a manifest progression, inclining towards the characteristics presented at the northern stations. This is not a proper place to pursue at length the details of such a comparison, or to enter upon the very interesting subject of the annual variations of the mean curves which represent the tendency to disturbance at each hour; but it is desirable to give all the confirmation which can be derived from observations in lower magnetic latitudes, of the special form in which these phenomena present themselves in higher ones, the calculation has been extended to include a full year of observations at the three permanent observatories.

Table XLVI.

Mean Disturbance of the Declination for one year, October 1843 *to September* 1844, *at Philadelphia, Toronto, and Sitka, expressed in arc.*

Local Mean Time.	Philadelphia.			Toronto.			Sitka.		
	E.	W.	Total.	E.	W.	Total.	E.	W.	Total.
Midnight	2′·36	2′·20	2′·21	3′·23	2′·52	2′·80	5′·55	3′·60	4′·57
1 A.M.	2·40	1·83	2·10	3·69	1·82	2·66	5·95	4·00	4·95
2	2·41	2·17	2·29	3·43	1·83	2·58	4·53	3·63	4·07
3	2·37	2·23	2·29	3·38	2·37	2·86	4·16	4·49	4·35
4	2·12	1·70	1·91	2·96	2·53	2·73	4·06	5·30	4·93
5	1·93	2·34	2·10	2·18	3·03	2·63	3·90	3·11	3·51
6	1·77	2·45	2·06	1·88	2·76	2·31	3·26	2·77	3·03
7	1·76	2·65	2·20	1·78	4·18	3·02	2·75	2·96	2·86
8	1·86	2·16	2·00	1·63	3·06	2·21	2·92	2·86	2·91
9	1·90	2·55	2·19	1·65	2·29	1·95	2·41	2·91	2·65
10	1·72	2·14	1·91	1·93	2·17	2·04	2·46	3·79	3·14
11	1·87	2·02	1·93	2·05	2·09	2·05	2·34	3·32	2·83
Noon	1·72	1·89	1·81	1·71	1·95	1·81	2·15	3·06	2·60
1 P.M.	1·60	1·72	1·65	1·67	1·93	1·79	2·09	2·71	2·40
2	1·55	2·00	1·76	1·56	2·08	1·80	2·08	2·36	2·22
3	1·51	2·06	1·77	1·62	1·88	1·75	2·17	2·59	2·38
4	1·61	1·92	1·74	1·60	1·86	1·66	2·21	2·69	2·45
5	1·66	1·70	1·67	2·02	1·80	1·90	2·20	2·83	2·52
6	1·55	1·72	1·62	2·11	1·57	1·80	2·90	2·52	2·71
7	2·48	1·32	1·86	2·58	1·33	1·89	4·92	2·74	3·82
8	3·78	1·49	2·52	5·13	1·75	3·28	5·92	3·03	4·50
9	3·86	1·47	2·61	5·04	1·77	3·17	5·05	3·94	4·06
10	3·81	1·56	2·59	4·90	1·97	3·32	5·89	3·06	4·48
11	2·36	1·56	1·88	3·60	2·10	2·76	4·35	3·04	3·67

The observations at Philadelphia were made 19^m, and those at Sitka 28^m after the hour named.

It appears by this Table that the characteristic of a maximum value of the *mean disturbance* of Declination, extending from 8 to 10 P.M., belongs alike to all the stations, and does not materially differ in relative prominence at any of them; when, however, we compare the course of the mean values during the later hours of the night, a well marked difference presents itself. At each station the maximum just referred to is due entirely to easterly movements, and at each station it is succeeded, in the course of the night, by two other maxima, the first of easterly and the second of westerly movements; it is in the relative prominence of these maxima that the results at the three stations exhibit the difference referred to. At Philadelphia they are too inconsiderable to produce any marked result in the mean taken irrespective of direction; at Toronto they are much more decided, and a comparatively high value of the latter prevails in consequence from 1 to 7 A.M.; lastly, at Sitka they are sufficiently prominent to bear comparison with the maximum at 9 P.M., and so to prepare us for the result of observations in yet higher latitudes, by which, as we have seen above, they are shown to outweigh the latter maximum so much, as to reduce it to comparative insignificance, and to make those hours of the night in which the forces producing these movements predominate, the most important of the twenty-four, as regards their influence on the character of the mean diurnal curves of each of the magnetic elements.

There is a test of a very simple nature which the numerous observations of disturbances enable us to apply, for the purpose of ascertaining whether there is a determinate direction in the movements of the Declination magnet upon these occasions, which is different at different hours of the day. This consists in reckoning the numbers of individual readings taken during disturbances, at which the magnet was east and west of its normal mean position for that day. The latter value is found by making each day of disturbance the centre of a group of five or seven complete days, and finding the mean scale reading of the whole.

Table XLVII.

Showing the total number of Readings during Disturbances at which the Declination magnet was east and west of its mean position.

Mean Time.	Athabasca.		Fort Simpson.		Mean Time.	Athabasca.		Fort Simpson.	
	East.	West.	East.	West.		East.	West.	East.	West.
h 16	272	24	159	15	h 4	—	—	3	2
17	247	11	120	7	5	—	—	—	49
18	174	27	66	1	6	1	3	—	48
19	86	3	55	—	7	27	6	—	27
20	48	14	15	—	8	35	25	4	18
21	27	21	12	—	9	40	50	5	70
22	3	24	2	—	10	72	80	18	52
23	2	13	—	1	11	61	111	16	65
Noon	4	16	—	1	12	57	114	101	38
1	—	2	—	1	13	120	149	119	6
2	—	1	—	2	14	154	103	105	32
3	—	1	—	17	15	237	44	168	27

The observations entered at 16^h mean time at Fort Simpson were taken at 16^h 15^m and so on. It appears from the above Table that from 3 to 7 A.M. the magnet at Lake Athabasca comparatively rarely passes to the westward of its mean position, and from 10 A.M. to noon (and doubtless to 5 or 6 P.M.) comparatively rarely to the eastward of it. Westerly excursions have a preponderance which is not shown by the curve of mean disturbance from 9 P.M. to 1 A.M.

during the remaining hours (8 and 9 A.M. and 7 and 8 P.M.) easterly excursions preponderate. The tendency to westerly movements appears by this test also to be rather greater at Fort Simpson in April and May than at Lake Athabasca in the winter, as has been remarked in connection with Table XLIII. With regard to the apparent difference between the tendencies at 9 P.M. and the succeeding hours to 1 A.M. inferred from this Table and from the curve of mean disturbance (Table X.), it must be remarked, first, that no reference is here made to the relative magnitude of the movements to east and west, which is the subject of the other Table; secondly, that the present Table is derived from disturbances connected, it is probable, in almost every instance with the development of Aurora Borealis, that phenomenon having been visible during some part of every disturbance, on which the state of the sky permitted it to be seen with one exception. And it does appear from a careful comparison of the scale readings during disturbances visibly attended by Aurora, that the preliminary tendency in such instances is to a westerly range, the subsequent tendency to an easterly one. As the Aurora was most frequent at midnight and 1 A.M., the consequence of this distinction, if true, will be a comparative preponderance of westerly movements in the early hours of the night, but unless we admit that the same cause which produces the Aurora Borealis produces the ordinary reactionary or irregular movements, it does not appear to follow that the same law should be manifested by the entire body of observations embracing a great majority of hours on which no Aurora was present.

Shocks.—The following Table contains the total number of readings of the Declinometer which may be denominated shocks, according to the usual definition; that is to say, which differ from the mean scale reading for the same hour by a quantity equal to, or exceeding, twice the amount of the monthly *mean irregular fluctuation* of the element as defined by Colonel Sabine. (Introduction to Toronto Observations, vol. 1. p. xv.) The dates and particulars of these readings will be given in a future section, when we consider the degree of correspondence between the movements at Toronto and the northern stations.

Table XLVIII.

Showing the total number of Shocks or Disturbances of Declination, according to the definition of Colonel Sabine just referred to.

Mean Time.	Athabasca.			Fort Simpson.			Mean Time.	Athabasca.			Fort Simpson.		
	East.	West.	Total.	East.	West.	Total.		East.	West.	Total.	East.	West.	Total.
16[a]	6	1	7	2	1	3	4	—	—	—	—	—	—
17	7	—	7	5	—	5	5	—	—	—	—	—	—
18	8	1	9	1	—	1	6	—	2	2	—	—	—
19	6	1	7	4	—	4	7	—	—	—	—	—	—
20	3	—	3	2	—	2	8	—	—	—	1	1	2
21	1	1	2	1	—	1	9	2	1	3	—	2	2
22	—	—	—	1	1	2	10	5	2	7	—	1	1
23	2	2	4	—	—	—	11	5	3	8	—	—	—
0	—	—	—	—	—	—	12	4	2	6	1	1	2
1	—	1	1	—	—	—	13	4	4	9	1	1	2
2	—	—	—	—	—	—	14	4	2	6	2	—	2
3	—	—	—	—	—	—	15	4	2	6	3	—	3
								61	25	86	24	8	32

a For 0h Gött. Athabasca, and 1h Gött. Fort Simpson.

The total numbers are in the proportion of 1 to every 32 observations at Athabasca, and 1 to every 34 at Fort Simpson. In the five months at Toronto, October to February, we find 116 shocks, being in the proportion of 1 to every 27 observations, and in the two months April and May, we find 50 shocks, being in the proportion of 1 to every 25 observations; thus it appears that the high value of the monthly *mean irregular fluctuation*, caused by the prevalence of a state of disturbance at the northern stations, occasions a smaller proportion of readings to come under the definition of a shock as here applied, than the low value deduced from a comparatively undisturbed series.

It is to be observed that the proportion of easterly shocks is much greater at the northern stations than at Toronto. We have at the latter station in

1841, 70 easterly to 60 westerly.
1842, 77 easterly to 63 westerly.
In eight months, 1843–1844, 99 easterly to 89 westerly.

In all 246 easterly to 212 westerly; whereas we have seen that at the northern stations the easterly deflections are more than double the westerly in number, showing that whatever may be the cause which determines the north end of the magnet to the west, it decreases in activity in the winter season as we proceed to the northward in the American continent.

On the Connexion between the Changes of the Magnetical Elements observed at the Northern Stations and those observed at Toronto and elsewhere.

The circumstance that frequent and very considerable magnetical disturbances were observed at Lake Athabasca and Fort Simpson, during the winter of 1843–4 and the following spring, although the same seasons were remarkable for absence of disturbance at most of the other stations of observation which have been examined, would seem to afford a presumption that some of these disturbances were of a local character, or that their influence, when it extended to Toronto or the European stations, was too slight to attract attention. To this we may add the fact, that the hours of the day most affected by them are apparently not the same in high and in medium latitudes, as has been shown in the previous discussion of the irregular changes; lastly, it has been shown by Colonel Sabine from the Toronto observations of 1841 and 1842 (p. xx.) that the tendency of disturbances at that station is to produce westerly deviations of the declination magnet in the morning hours; and the eight months observations of 1843–4 which are here discussed, (Table XLIX.) lead to the same conclusion; indeed an inspection of any of the more important disturbances will show that the greatest movements of the declination magnet at Toronto are to the westward, whereas at Lake Athabasca and Fort Simpson they are almost invariably to the eastward, particularly in the same morning hours. In the case of this element, therefore, we have an opposite tendency in respect to direction, in addition to the difference of epoch in the two localities. Notwithstanding all these circumstances, however, a careful comparison with other published observations has led to the conclusion, that a state of magnetical disturbance prevailed at one or more other stations upon so many of the occasions upon which it was observed at Lake Athabasca and Fort Simpson, as to leave it doubtful whether, without more positive evidence, any of the disturbances, considered generally, can be considered to have been merely local. It is to be hoped, that the extension of automatic registration by means of photography will soon throw more light upon the question, whether any and what magnetic changes may be regarded as local, and how far we may consider movements which have few or no features of resemblance, to be due to a common cause, because they occur simultaneously in distant localities; meanwhile it would be to neglect a principal purpose of the observations under discussion, to omit to pursue the inquiry as far as they permit.

Extra observations at short intervals were taken at the two northern stations, on sixty-six occasions, exclusive of term days, in a period of about one hundred and sixty days of observation. Upon examination of all the records of magnetical observations for the years 1843–4 which have been published hitherto, namely, those of the four British colonial observations, those of the Russian stations, of Greenwich, Makerstoun, and Philadelphia, there are to be found only twenty-nine instances of corresponding observations elsewhere, and in this number are included five on which the correspondence consists only in one or two extra readings interpolated in the usual series at Greenwich or Makerstoun. This, however, is sufficient to show that in the view of the observers some disturbance existed. The dates to which this remark applies are October 26^d, December 8^d and 26^d 1843, April 30^d, May 3^d 1844. Of the remaining thirty-seven northern disturbances, about ten were of the first order as regards magnitude and duration, namely, those of October 15^d, 16^d, October 17^d commencing 17^h, October 25^d, 26^d, October 30^d, 31^d, December 5^d, 6^d, December 19^d, 20^d, December 29^d 1843, April 9^d, 10^d, April 14^d, 15^d, April 22^d 1844; but on the whole, the coincidences of observation occur generally at the more considerable of the northern disturbances, December 28^d, February 8^d, and February 29^d, being the principal exceptions.

Proceeding next to examine the simultaneous changes of the several elements in detail, we find considerable diversity. Sometimes the movements of one or more elements correspond in epoch and in direction; sometimes in epoch when they are reversed in direction; sometimes the principal movement at one station is represented by the principal movement at the others, and as often by one of secondary prominence; sometimes there is marked agreement during part of a disturbance, no agreement during the remainder; lastly, there are several instances when movements remarkably similar in character occur at remote stations, but separated by a considerable interval of time.

All this may be shown without the aid of diagrams, by selecting the principal features of each disturbance alone for comparison, for which purpose the following brief notes are subjoined. The references are all to Göttingen time.

1843, October 17^d, 1^h to 3^h.—A great reduction of the Horizontal Force at Lake Athabasca between 0^h and 3^h, lowest value — ·064 X *

* The values here given are the differences of the actual readings from the mean for the same hour and month; + or — signs before the change of declination indicate increase or decrease of the absolute value of that element, not simply easterly or westerly movements, as the absolute declination is east at Lake Athabasca, Fort Simpson, and Sitka, but west at Toronto, Philadelphia, Greenwich, Makerstoun, and St. Petersburg; the + sign indicates an easterly movement at the former and a westerly movement at the latter stations. It will be seen that the general tendency in disturbances is to an increase of absolute declination at all these stations, which reduces the difference of direction remarked above to a common principle. In selecting as the characteristic of each principal movement referred to, that one reading which happens to differ most from the mean, it is

at 1^h 46^m, but $-$ ·048 X at 2^h 1^m. Easterly extreme of Declination, $+52''7$ at 1^h 35^m. At Makerstoun observations commenced at 2^h, apparently on account of a low range of Horizontal Force, $-$ ·0019 X, and the same at St. Helena, where they were also commenced at 2^h with a value of $-$ ·0012 X; there is no other obvious correspondence.

October 18–19, Term-day.—The principal movement of the Declination, which occurred between 15^h and 17^h, corresponds in general character at Lake Athabasca, Sitka, Toronto, and Philadelphia, a movement in the contrary direction being presented at the same hour at all the European stations. In the first part of the movement in question the direction is the same at all the American stations; during the latter part it is reversed, and we then have a westerly extreme $-44''5$ at Lake Athabasca corresponding to an easterly one $+10''8$ at Sitka, with which the other stations agree. The changes of Horizontal Force between 15^h and 17^h have a general resemblance at Toronto and Lake Athabasca, save that where the element rises to a high value + ·0172 X at the former station, at 17^h 6^m, it hardly passes its mean value at the latter, and they are very precisely reversed at the European stations at a later period 29^d 4^h to 6^h. The close resemblance of the changes at all the stations in America and Europe is very striking, and they here agree in direction.

October 26^d 20^h to 27^d 2^h.—The principal movement at Lake Athabasca was between 0^h and 2^h, the extreme of Horizontal Force $-$ ·0624 X at 0^h 56^m, and of Declination (easterly) $+41''7$ at 1^h 5^m. a single extra reading at 1^h 10^m at Makerstoun is the only proof that this disturbance, as such, was observed elsewhere; it shows a low value of the Horizontal Force $-$ ·0017 at 1^h 12^m, and an easterly range of Declination $-4''4$ at 1^h 10^m.

November 13^d 4^h to 7^h, and 20^h to 23^h.—A low value of the horizontal force prevails at Lake Athabasca from 4^h to 6^h, lowest $-$ ·0347 X at 4^h 6^m. Declination not particularly affected, but an easterly extreme $+17''6$ at 4^h 0^m. At Philadelphia the minimum of horizontal force, $-$ ·0018 X, occurs at 4^h 11^m, and a low value is also maintained until 6^h. Declination, an easterly extreme $-6''2$ at 4^h 0^m. This was a considerable disturbance at the last-named station, and observations were continued without intermission until 14^d 2^h. They were discontinued from 7^h to 20^h at Lake Athabasca, and then resumed on account of an unusual westerly range of Declination, giving a minimum $-1°$ $6'$ at 22^h 16^m. At Philadelphia we

necessary to remember that from various causes, some of them perhaps instrumental, there may be an apparent difference of many minutes between the epoch of movements at two stations, which, nevertheless, viewed generally are coincident, and this is considered to be the case in all the cases cited unless otherwise stated.

have a small but well marked easterly movement at the same hour, the extreme being only $-1'\cdot 9$ at $22^h\ 22^m$.

November $24^d\ 1^h$ to 2^h.—A minimum of the Horizontal Force $-\cdot 0240$ at Lake Athabasca at $1^h\ 1^m$; no great change of Declination. At Hobarton extra observations also from 1^h to 2^h, the Declination chiefly affected; extreme $-8'\cdot 3$ at $1^h\ 2^m$; the Horizontal Force above the mean $+\cdot 0008$ X at $1^h\ 7^m$.

December $1^d\ 21^h$ to $2^d\ 5^h$.—Unusually large westerly movement of magnet at $22^h\ 15^m$, $-1°\ 37'$, and westerly movements prevailing until $23^h\ 30^m$; afterwards easterly extreme $+59'\cdot 2$ at $3^h\ 27^m$. Lowest value of Horizontal Force $-\cdot 0443$ X at $2^h\ 34^m$, but a very low value about $22^h\ 45^m$, and again from $2^h\ 0^m$ to $3^h\ 30^m$. At Hobarton a low value of the Horizontal Force prevailed from $2^h\ 52^m$, when extra observations commence, to 5^h, the lowest being $-\cdot 0015$ X at $3^h\ 47^m$; Declination most affected about 3^h, extreme $-7'\cdot 0$ at $3^h\ 5^m$.

December $8^d\ 18^h$ to 20^h.—Observations commenced at Lake Athabasca for an easterly range of Declination, maximum $+44'\cdot 5$ at $18^h\ 0^m$; disturbance of Horizontal Force not particularly marked, a minimum $-\cdot 0087$ X at $18^h\ 1^m$, a maximum $+\cdot 0085$ at $19^h\ 10^m$. A few readings were taken at Makerstoun, commencing at 18^h, apparently in consequence of a westerly range of the Declination of no great extent, extreme $+3'\cdot 5$ at $18^h\ 0^m$; the Horizontal Force a maximum $+\cdot 0016$ X at $18^h\ 2^m$.

December $26^d\ 21^h$ to $27^d\ 2^h$.—A low value of the Horizontal Force from 22^h to nearly 0^h, minimum $-\cdot 0354$ X at $23^h\ 28^m$. Declination not particularly affected, maximum $+30'\cdot 4$ at $23^d\ 30^h$. At Makerstoun there is one extra observation only, at $23^h\ 35^m$. Declination $+2'\cdot 8$. Horizontal Force $+\cdot 0001$ X.

December $28^d\ 3^h$ to 4^h.—A great easterly movement of Declination at Lake Athabasca from 2^h to 3^h, extreme $+52'\cdot 3$ at $2^h\ 0^m$; the Horizontal Force not particularly affected, a minimum $-\cdot 0188$ X at $3^h\ 1^m$, a maximum $+\cdot 0105$ X at $5^h\ 1^m$. At Makerstoun this disturbance attracted attention two hours earlier, a westerly extreme of Declination $+9'\cdot 7$ occurs at $1^h\ 52^m$, but the minimum of Horizontal Force occurs at $1^h\ 57^m$, $-\cdot 0018$ X, when this element differed very little from its mean value at Lake Athabasca. At Hobarton observation appears to have been commenced at 2^h, on account of an easterly range of Declination, extreme $+9'\cdot 2$ at $2^h\ 22^m$; the minimum of Horizontal Force coincides with that of Lake Athabasca, being $-\cdot 0013$ X at $3^h\ 2^m$.

1844. January $4^d\ 16^h$ to $5^d\ 9^h$.—Observations commenced at Lake Athabasca on account of the high range of Horizontal Force, which continued from 16^h to 20^h, maximum $+\cdot 0287$ X at $19^h\ 10^m$.

This is succeeded by very low values, giving two marked minima, the first and most important $-\cdot0635$ at $23^h\ 4^m$, the second $-\cdot0342$ at $0^h\ 43^m$; the Declination is also much disturbed, the extremes being $-40'\cdot7$ at $21^h\ 0^m$ and $+1°\ 6'$ at $0^h\ 57^m$, but with many other great inflexions. A state of disturbance was very generally observed on this day. At Makerstoun extra observations were taken with various intermissions through the 4th, 5th, and 6th of January, and we find a maximum of Horizontal Force $+\cdot0012$ X at $19^h\ 12^m$, and the minimum at $23^h\ 17^m$, $-\cdot0021$ X; also the maximum of Declination $+8'\cdot8$ at $5^d\ 0^h\ 50^m$; in the other features no particular coincidence is to be remarked, and the great easterly movement of Declination at Makerstoun between 7^h and 8^h, giving a minimum $-21'\cdot0$ at $7^h\ 35^m$, is entirely wanting at Lake Athabasca, when the observations were resumed at that hour on account of a decrease of Horizontal Force of no great amount, which on the other hand does not appear at Makerstoun. The *mean irregular fluctuation* of the Declination and Horizontal Force at Toronto, on the 4th January, was the highest value of the month; the extra observations embrace but the earlier part of the disturbance, and the minimum of Horizontal Force coincides nearly with the maximum at Lake Athabasca, being $-\cdot0019$ X at $19^h\ 2^m$; the value at the regular observation of $4^d\ 23^h\ 2^m$, which coincides very nearly with the principal minimum at Lake Athabasca, although also a low one $-\cdot0005$ X, is not nearly so low as the reading at the epoch of *greatest* force at the northern station. The Declination appears to have been the most affected at Hobarton, but the minimum of Horizontal Force, $-\cdot0004$ at $23^h\ 12^m$, although very small in amount, coïncides nearly with the principal one at Lake Athabasca.

January $5^d\ 23^h$ to $6^d\ 3^h$.—An extreme depression of the Horizontal Force prevailed at Lake Athabasca from 23^h to almost $1^h\ 30^m$, lowest $-\cdot0627$ X at $23^h\ 1^m$; at the same hour occurs the lowest value of this element at Makerstoun, $-\cdot0021$ X. Again a westerly extreme of Declination, $-19'\cdot4$, occurs at Lake Athabasca at $23^h\ 0^m$; a minor westerly extreme, $+3'\cdot0$, at Makerstoun at the same time, but the principal one, $+4'\cdot5$ at $1^h\ 25^m$, coincides with a considerable easterly inflexion at the former station, $+24'\cdot9$ at $1^h\ 33^m$.

February $1^d\ 0^h$ to 4^h.—Great depression of the Horizontal Force from 0^h to nearly 3^h, lowest $-\cdot0465$ X at $0^h\ 27^m$. At Makerstoun this element was not particularly affected at this time; and extra observations were not commenced until 3^h, when they show a westerly extreme of Declination $+7'\cdot6$, corresponding nearly to an easterly extreme, $+29'\cdot5$, at Lake Athabasca.

Again, from $1^d\ 19^h$ to 21^h an unusual westerly range of Declination prevailed at Lake Athabasca, extreme $-$ 1° 22′ at 19^h 27^m. Extra observations were taken at Makerstoun from 18^h to 19^h, but only one between 19^h and 20^h, which one does not indicate a correspondence of this element, but shows a minimum of Horizontal Force $-$ ·0011 X, coinciding with the one at Lake Athabasca, which is $-$ ·0239 X, also at 19^h 40^m.

February 2^d 6^h to 8^h.—An unusual westerly range of Declination at Lake Athabasca, extreme $-48′·2$ at 7^h 0^m. There is also a minimum of Horizontal Force $-$ ·0258 at 6^h 10^m. Coincident observations were made at Makerstoun and St. Petersburg; at the former also occurs a minimum of Horizontal Force $-$ ·0036 X at 6^h 12^m, and an easterly extreme of Declination $-11′·3$ at 6^h 30^m, being half an hour before the $-$ extreme at Lake Athabasca. At St. Petersburg an easterly extreme of Declination, $+18′·0$, occurs at 6^h 10^m, and a maximum of Horizontal Force at 6^h 30^m, + ·0022 X. Again, 2^d 17^h to 21^h, we have a marked prevalence of westerly Declination, and a high value of the Horizontal Force, at Lake Athabasca; the extremes $-37′·3$ at 17^h 42^m, and + ·0264 X at 18^h 53^m. At Toronto extra observations were commenced at the same time as at the northern station, a westerly extreme of Declination, $+18′·1$, occurs at 17^h 2^m, and another of small extent, $+2′·6$, at 17^h 42^m; a minimum of Horizontal Force accompanies the former, $-$ ·0034 at 17^h 7^m; the lowest of this element value at the northern station is also at 17^h 1^m, but it is still above the mean, + ·0015 X.

February 5^d 0^h to 6^h.—An excessive reduction of the Horizontal Force occurs at Lake Athabasca between 23^h and 0^h, extreme $-$ ·0678 at 0^h 4^m; great changes of Declination accompanied this shock, the extreme values being $+1°$ 4′ at 0^h 21^m, and $-1°$ 23′ at 0^h 30^m. At Makerstoun but a single extra observation is given between 0^h and 1^h, namely, at 0^h 20^m, which shows a considerable reduction of the Horizontal Force also, $-$ ·0029 X, but no particular disturbance of Declination. At Hobarton extra observations were commenced at 1^h, and the lowest value of Horizontal Force is at 1^h 12^m, $-$ ·0017 X, being an hour later than that at Lake Athabasca, a westerly extreme of Declination, $-9′·0$, accompanies it.

Disturbance observations were resumed at 5^d 16^h with a high value of the Horizontal Force, giving a maximum + ·0238 at 16^h 58^m; extra observations were taken at the same time at Makerstoun, showing low values of this element, minimum $-$ ·0034 X at 16^h 32^m. We have for the Declination the easterly extreme $+42′·1$ at 16^h 21^m at Lake Athabasca, and the easterly extreme $-0′·5$ at 16^h 10^m at

Makerstoun; again, the westerly extreme −20′·1 at 17^h 0^m at the former station, and +6′·5 at 17^h 15^m at the latter.

February 8^d 0^h to 1^h and 5^h to 6^h.—Two sudden changes of the Horizontal Force, but of no great extent; the lowest value was that at the regular reading at 5^h 1^m, −·0235 X. Extra observations were made at Makerstoun from 4^h to 6^h, and at Hobarton from 5^h to 6^h; at the former station we find a minimum of Horizontal Force −·0038 X at 4^h 52^m, at the latter a relative maximum of the same element −·0004 at 5^h 2^m. Again we have, of the Declination, an easterly extreme +19′·7 at 5^h 9^m at Lake Athabasca, an easterly extreme −6′·3 at 4^h 55^m at Makerstoun, and a westerly extreme −7′·3 at 5^h 7^m at Hobarton.

February 29^d 0^h to 1^h.—The Bifilar was not in adjustment at Lake Athabasca during this disturbance, which was one of those most generally observed. We find, however, a relative maximum of the Inclination + 3′·1 at 16^h at that station, the value being below the mean at 15^h, 17^h, 18^h, 19^h, and 20^h; the maximum of the same element + 3′·8 being at 16^h 12^m at Toronto, and apparently a little earlier at Philadelphia; on the other hand, the observations at these two stations do not show the minimum which follows at Lake Athabasca at 17^h, −10′·2; they also show a very considerable easterly movement of Declination at 16^h, +25′·1 at Toronto, +22′·2 at Philadelphia, when that element was little disturbed at the other station. Extra observations were discontinued at Toronto at 28^d 19^h; they were continued at Philadelphia until 29^d 3^h, and we find a minimum of the Horizontal Force at 23^h, at which time a maximum of Inclination indicates the same thing at Lake Athabasca. At Makerstoun we find an unusually large easterly movement of Declination between 16^h and 17^h, extreme −21′·3 at 16^h 35^m, half an hour later than at Toronto and Philadelphia, and an equally great reduction of the Horizontal Force, extreme −·0051 X at 16^h 27^m, agreeing, probably, nearly with the maximum of this element at the American stations.

April 2^d 22^h to 0^h.—A very great and sudden shock at Fort Simpson, affecting first the Inclination and Horizontal Force, the former gives + 1° 5′·6 at 22^h 22^m, when the latter was less than the lowest scale reading of the Bifilar (< −0·048 X)* a secondary minimum of the latter element −·0028 X is shown at Makerstoun at 22^h 17^m, but the lowest value −·0042 X is there at 21^h 22^m, when there does not appear to have been any corresponding movement at Fort Simpson. We find the minimum of this element −·0013 X at 22^h 35^m at Hobarton. The regular observations at

* The readings at Fort Simpson are referred to the mean for the 24^h.

Toronto at 21^h and 22^h indicate a great reduction of the same element. The westerly extreme of Declination $+9''2$ occurs at 22^h 25^m at Makerstoun, and at 22^h 27^m at Fort Simpson $-43''0$ at the same time is the westerly extreme $-14''3$ at Hobarton. Of the great easterly range which followed at Fort Simpson, there is no very obvious sign at either of the other stations. Again, the extra observations were resumed at 3^d 4^h, on account of a great easterly range of Declination, with its usual accompaniments, increase of Inclination and Total Force, decrease of Horizontal Force, giving the extremes of each element at or near 4^h 30^m, namely $+ 2° 5'$ of Declination $-\cdot0505$ X $+45''\theta$. Extra observations were resumed at Makerstoun at 5^h, but the extreme movements do not occur until near 6^h, when they are decidedly past at Fort Simpson. We have at the former station the unusual values of $-20''3$ Declination at 5^h 55^m, and $+\cdot0051$ X at 5^h 57^m, a time when no very marked change is shown at Fort Simpson. This second disturbance was not observed at Hobarton, but a few extra observations were made about 6^h at Greenwich, and agree with those at Makerstoun.

April 16–17, 1844.—This great disturbance was recorded at nearly every station of observation, and furnishes at many, probably at all of them, the greatest amount of change of all the elements which had been recorded up to that date. There are, however, several remarkable differences in the character of the changes at different stations: first, as regards the Declination, we have an extraordinary easterly movement at Fort Simpson between 1^h and 2^h on April 17, the extreme range being no less than $+ 6° 40'$ at 1^h 24^m; at this hour no marked feature is presented at Philadelphia or Toronto, whereas at stations in Europe and at Hobarton we find a resemblance. At Makerstoun the westerly extreme observed is $+17''9$ at 1^h 0^m; at Greenwich the westerly extreme is $+16''9$ at 1^h 9^m; at St. Petersburgh the easterly extreme is $-14''6$ at 1^h 25^m, but followed at 1^h 35^m by the westerly extreme $+15''2$; at Hobarton the westerly extreme is $+31''7$ at 0^h 57^m; at St. Helena there is no marked feature at this hour; at the Cape the extra observations were discontinued at 1^h. On the other hand, we have at Toronto and Philadelphia a great easterly movement between 16^d 20^h and 21^h, which answers to a shock of short duration at Fort Simpson; the extreme readings are at Toronto $+39''3$ at 20^h 45^m; at Philadelphia $+26''9$ at 20^h 50^m; and at Fort Simpson $+2° 0''4$ at 20^h 50^m. A similar correspondence appears at 21^h 50^m. Referring next to the Horizontal Force, the greatest reduction of this element at Fort Simpson commenced very suddenly at 17^d 0^h 55^m; the extreme observed was $-0\cdot128$ X at 1^h 28^m, this is, however, but an approximation, the range having passed the limit of the scale;

here are also great negative extremes at or near $17\frac{3}{4}^h$, $19\frac{1}{2}^h$, and $23\frac{1}{2}^h$, but none of them approached the one just cited; this is also the epoch of greatest reduction of this element at Hobarton, the extreme being $-$ ·0069 at 17^h 27^m, and we find it a little later at St. Helena; at Toronto we have at the same hour a considerable negative movement also, giving a minimum $-$ ·0106 X at 0^h 17^m, but at this station and at Philadelphia the lowest value of the day is nearly an hour earlier, and answers to a movement of secondary importance at Fort Simpson; the values are $-$ ·0148 X at 23^h 32^m at Toronto, $-$ ·0080 X at 23^h 27^m at Philadelphia, and $-$ ·0694 at 23^h 19^m at Fort Simpson; about the same time is the lowest value at Sitka and at Greenwich, while at Makerstoun it occurs between 21^h and 22^h. It deserves remark that the Horizontal Force begins to return to its normal value at Toronto and Philadelphia at 17^d 3^h, at Fort Simpson not until 17^d 5^h, in both cases by a regular change which is perfectly similar in other respects.

April 24^d 25^d Term day.—An easterly movement of Declination of very marked character and great extent prevails at Fort Simpson between 25^d 2^h and 4^h, the extreme being $+3°$ $7'$ at 2^h 55^m; we have a corresponding shock at Sitka, but reduced in amount to $+19'$·0, this extreme reading being at 3^h 0^m. A sustained westerly movement of comparatively small extent, but very obviously coinciding in epoch, occurs at Toronto and Philadelphia, giving at the former an extreme westerly reading of $+10'$·9 at 2^h 32^m, at the latter a westerly extreme $+10'$·1 at 2^h 48^m. The European stations do not exhibit any particular movement at this hour.

Referring to the Horizontal Force, we find a very low value of this element at Fort Simpson from 2^h 17^m to 3^h 7^m, the extreme being $-$ ·0585 X at 2^h 57^m; the lowest value at Sitka $-$ ·0101 X occurs at the same reading; the movement generally is well marked at that station. There is a well marked minimum at Toronto about 2^d 17^h being half an hour earlier than the one in question, but the lowest value at this station is $-$ ·0009 X at 7^h 32^m. Viewed generally, there is a marked resemblance in the successive changes of the element at these two stations, but the epochs of maxima and minima do not coincide; they would be made to do so pretty nearly if the whole northern curve were advanced about two hours of time; for example, the first important minimum at Fort Simpson occurs at 0^h, at Toronto at 2^h 10^m, the following maximum, Fort Simpson at 1^h 40^m, Toronto 3^h 10^m. The next minimum Fort Simpson 2^h 57^m, Toronto 5^h 30^m, the next maximum Fort Simpson 4^h 30^m, Toronto 6^h 30^m. An unusually high value of the element prevailed at both stations on the 24th and 25th April.

Extra observations were resumed at Fort Simpson at 20^h on the

25^{h} April, and continued until 26^{d} 2^{h}; they were resumed at 26^{d} 0^{h} simultaneously at Philadelphia, Makerstoun, and Hobarton. Referring them to this part only of the disturbance, we find a very great reduction of Horizontal Force at Fort Simpson between 23^{h} and 0^{h}, the extreme being —·0845 X at 0^{h} 7^{m}; we find a similar movement at Philadelphia, the extreme being —·0016 X at 0^{h} 2^{m}, followed by an immediate return to high values, giving a maximum + ·0010 X at 0^{h} 58^{m}. At Fort Simpson the succeeding maximum is less marked, and not attained until 1^{h} 30^{m}; at Makerstoun we find a minimum at 0^{h} 12^{m} —·0033 X, but the lowest value occurs at 0^{h} 47^{m}, and there is no maximum before 2^{h}. At Hobarton the Declination is the element chiefly affected, but we have a maximum of Horizontal Force at 0^{h} 7^{m} + ·0008, and the minimum —·0019 at 1^{h} 22^{m}.

April 26^{d} 18^{h}.—A great westerly range of Declination prevailed at Fort Simpson from 18^{h} to 19^{h} 20^{m}, extreme —1° 24′·4 at 18^{h} 0^{m}. At Philadelphia extra observations were commenced at 15^{h}; we have an easterly extreme —8′·8 at 17^{h} 54^{m}, but the principal is at 19^{h} 20^{m}, —11′·1, and a westerly movement intervenes, corresponding to an easterly one, or a return towards normal values, at Fort Simpson. Philadelphia at 18^{h} 14^{m}, +1′·9, Fort Simpson at 18^{h} 30^{m}, —23′·8. Again, the Horizontal Force at Philadelphia presents two minima, the greater at 18^{h} —·0013 X, the next from 18^{h} 55^{m} to 19^{h} 20^{m}. At the same periods we have minima at Fort Simpson, but the latter in this case the greater, the extreme being —·0300 X at 19^{h} 31^{m}. At Makerstoun observations occur from 13^{h} to 18^{h}, and indicate, at the latter hour, values of both elements differing very little from their mean.

April 28^{d} 21^{h} to 29^{d} 5^{h}.—The corresponding readings are confined to a couple at Makerstoun, between 2^{h} and 3^{h}. Referring to this period only, we find a great easterly movement of Declination at Fort Simpson, the extreme + 2° 5′ at 2^{h} 3^{m}, accompanied by a very low value of the Horizontal Force, the extreme —·0379 X at 1^{h} 52^{m}, but with no great change for about 20^{m} before and after that hour. The observations at Makerstoun indicate a minimum of this element between 2^{h} 0^{m} and 2^{h} 45^{m}, the lowest value of those recorded being —·0015 X at the 2^{h} 32^{m}, but the Declination differs very little from its mean value.

April 30th.—Extra observations were taken at Fort Simpson every 15^{m} from 14^{h} to 16^{h}, in consequence of an unusually high value of the Horizontal Force, but no marked changes occurred; as usual under similar circumstances, the Declination is westward of its mean. Afterwards, at 21^{h}, a considerable disturbance commences. At Makerstoun a few extra readings were taken from 14^{h} to 18^{h}, showing a small westerly extreme $-7''\cdot 9$ at $14^h\ 5^m$ accompanied by a minimum $-\cdot 0014$ X of Horizontal Force; the subsequent disturbance was not observed.

May $2^d\ 19^h$.—The commencement of extra observations at Fort Simpson coincides with the conclusion of them at Makerstoun.

May 3.—A sudden and great reduction of the Horizontal Force occurs from 3^h to 4^h at Fort Simpson, giving a minimum $-\cdot 0619$ X at $4^h\ 7^m$. At the same time is a considerable easterly movement of Declination, the extreme being $+1^\circ\ 57'$ at $4^h\ 21^m$; at Makerstoun extra observations were also commenced at $4^h\ 35^m$. Apparently in consequence of the easterly movement of the declinometer, since $4^h\ 0^m$ the readings were $+5''\cdot 0$ at $4^h\ 0^m$, and $-4''\cdot 2$ at $4^h\ 35^m$; a maximum of the Horizontal Force $+\cdot 0019$ X appears at $4^h\ 42^m$, but the disturbance was apparently not considered to call for more than occasional readings.

May $8^d\ 13^h$ to 19^h.—An unusual westerly range of Declination prevailed at Fort Simpson, but of no great extent, and without much change. Extra observations were taken at intervals at Makerstoun from 10^h to 20^h on this day, but present no particular features of correspondence or the reverse.

May 13^d, 18^h to 21^h.—Again a westerly range of Declination, attended, as in the last instance, with high value of the Horizontal Force and diminished Inclination. Extra observations were commenced at Makerstoun at 20^h for a small easterly extreme of Declination, but present no marked features.

May 22^d, 1^h to 4^h.—Extra observations were commenced simultaneously at Fort Simpson and at Hobarton. At the former we find an easterly range of Declination prevailing the whole time, extreme $+2^\circ\ 11'$ at $1^h\ 33^m$, accompanied as usual with increased Inclination and a low value of the Horizontal Force, extreme $-\cdot 0442$ X at $1^h\ 43^m$. At Hobarton the disturbances commences with a marked westerly range of Declination, the extreme $-11''\cdot 9$ at $1^h\ 5^m$; the Horizontal Force is also below the mean until the observations ceased at 3^h, extreme $-\cdot 0009$ X at $1^h\ 32^m$. At 5^h extra observations were began at Makerstoun; they were again resumed simultaneously at Fort Simpson and Hobarton at 12^h, and on this occasion, as well as on the 17th April, already described, nearly every stoatin appears to have experienced the disturbance. We find extra observations at Toronto from 13^h to 20^h, Philadelphia 10^h to 20^h, Makerstoun 9^h to 20^h, the Cape of Good Hope 17^h to 22^h, St. Helena 20^h to 0^h, and Hobarton 10^h to 0^h; nevertheless, this disturbance does not appear to have been among the most considerable in point of amount. Referring first to the Declination: the observations commence at Philadelphia at 10^h with an unusual westerly range, extreme $+8''\cdot5$ at 10^h 12^m; the regular observation at 10^h 0^m at Fort Simpson shows no corresponding feature. We have again an easterly inflexion at Fort Simpson at 13^h 6^m, but the general range being westerly, it only reaches $+3''\cdot1$ as referred to the mean, although a change of nearly $56'$ as referred to the preceding westerly extreme at 12^h 24^m; small as it is in amount it answers to an easterly extreme $-9''\cdot9$ at 13^h 7^m at Toronto, or $-3''\cdot9$ at 13^h 0^m at Philadelphia. Again, we have a westerly extreme $-50''\cdot1$ at 13^h 33^m at Fort Simpson, and corresponding to it a westerly extreme $+1''\cdot8$ at 13^h 32^m at Toronto, or of $+0''\cdot8$ at 13^h 24^m at Philadelphia. The easterly extreme of the day, $+12''\cdot4$, occurs at Hobarton at 13^h 22^m. We have then another easterly extreme, $+10''\cdot4$ at 14^h 52^m at Toronto, and $+7''\cdot0$ at 14^h 44^m at Philadelphia, to which there is no corresponding feature at Fort Simpson; then a westerly extreme at both those stations $-0''\cdot5$ at 15^h 52^m at Toronto, to which a return toward mean values at Fort Simpson appears to answer, it gives $-12''\cdot5$ at 16^h 0^m, being still to the westward. So far, therefore, this element has shown several movements apparently common to Fort Simpson and Toronto or Philadelphia, but at the northern station the readings have been chiefly westerly, or of the kind which marks the beginning of a disturbance. At 22^d 23^h we have the more active disturbance, marked by a range of about $2°$ to the eastward, the extreme is $+2°$ $96'$ at 23^h 6^m; disturbance observations had been discontinued at all the stations, except St. Helena, before this hour, but the regular observations at Toronto and Philadelphia give no proof whatever that this shock, or a similar one at 23^d 2^h, extended to them.

Referring next to the Horizontal Force; we find at Fort Simpson a high value prevailing about 12^h, extreme $+\cdot0285$ X at 12^h 31^m; a similar feature is presented at Toronto and at Philadelphia. The regular reading at 12^h 2^m at Toronto gives $+\cdot0010$ X, and at Philadelphia we have $+\cdot0012$ X at 11^h 58^m, which is, however, somewhat less than a value observed an hour earlier. We have next, at the same stations, a minimum between 1^h and 2^h, the whole range being as yet high at Fort Simpson; the minimum in question is only $-\cdot0056$ X at 13^h 7^m at Toronto, the lowest value is $-\cdot0030$ at 13^h 42^m, and at Philadelphia $-\cdot0018$ X at 13^h 18^m; the correspondence ceases with the minimum. At Fort Simpson the element returns very rapidly to its high value, and gives $+\cdot0242$ X at 13^h 46^m. At Toronto and Philadelphia there is no corresponding feature; it returns to its mean value gradually, not attaining a maximum until 15^h. We have next a maximum at Toronto between 16^h and 17^h, and again between 17^h and 18^h, to which there are answering features at Fort Simpson; the lowest value at Toronto is $-\cdot 0040$ at $18^h\ 17^m$; at this hour the values are very slightly below the mean at the northern station, where, on the other hand, we have a very great decrease of this element between 23^h and 0^h, which is indicated by the regular observations at Toronto. The lowest value at Fort Simpson is $-\cdot 0609$ at $23^h\ 25^m$.

The foregoing disturbance completes the list of actual coincidences of observation. In the following cases the observations, although not coincident, correspond so nearly in point of time as to make it probable that the disturbances observed were magnetically the same.

1843, October 25, 26.—Lake Athabasca, $25^d\ 19^h$ to $26^d\ 3^h$; Makerstoun, $26^d\ 4^h$ to 11^h; Greenwich, $26^d\ 7^h$ to 12^h.

October 30, 31.—Lake Athabasca, $30^d\ 21^h$ to $31^d\ 4^h$; Makerstoun, $31^d\ 6^h$ to 10^h.

December 27.—Lake Athabasca, $27^d\ 18^h$ to 20^h; Makerstoun, $27^d\ 22^h$ to 23^h.

1844, January 8.—Lake Athabasca, $8^d\ 15^h$ to 16^h and 20^h to 22^h; Makerstoun, $8^d\ 6^h$ to 13^h and 19^h to 21^h.

February 16.—Lake Athabasca, $16^d\ 21^h$ to 22^h; Makerstoun, $16^d\ 18^h$ to 19^h.

May 2.—Fort Simpson, $2^d\ 19^h$ to 24^h; Makerstoun, $2^d\ 11^h$ to 19^h.

Lastly, we have several instances in which magnetical disturbances were observed at other stations when they were not observed at Lake Athabasca or Fort Simpson, although the attention paid to the instruments was so close as to make it improbable that any considerable disturbance could escape notice. I select those only at which the observations were made at more than one other station, and a few of the more decided disturbances at single stations.

1843, December 8.—Extra observations for disturbance at Makerstoun 6^h to 10^h, at St. Petersburg 7^h to 10^h.

December 10.—Extra observations, Hobarton 5^h to 7^h, St. Petersburg 5^h to 8^h, Makerstoun 13^h to 14^h.

December 11.—Extra observations, Makerstoun 6^h to 12^h, also 18^h to 20^h, Greenwich 8^h to 9^h.

December 12.—Extra observations, Makerstoun 2^h to 11^h, St. Petersburg 3^h to 4^h, and 11^h to 12^h.

1844, January 2.—Extra observations, Makerstoun 7^h to 14^h, St. Petersburg 7^h to 9^h.

January 9.—A state of disturbance appears to have prevailed more or less all day at Makerstoun.

January 10.—The same remark applies, the station the same.

January 23.—Extra observations, Makerstoun 6^h to 8^h, Philadelphia 12^h to 24^h.

February 7.—Extra observations, 6^h to 7^h St. Petersburg, 6^h to 15^h Makerstoun, 9^h to 11^h Greenwich.

February 22.—Extra observations, 7^h to 9^h Makerstoun, faint Aurora being visible.

April 1.—A state of disturbance appears to have prevailed at Makerstoun more or less all day.

April 5.—Extra observations at Makerstoun 12^h to 19^h, at the Cape of Good Hope 13^h to 23^h. Aurora visible at the former station.

May 14.—Extra observations, 14^h to 17^h Makerstoun, 15^h to 17^h Toronto.

The following Table contains the value of the *daily mean irregular* fluctuation of both elements at Toronto and the northern stations, to which is added that of the Declination for the same days at Sitka, the whole calculated according to the method of Colonel Sabine (Observations on days of unusual Magnetical Disturbance, Part I. ix.), which is as follows:—The difference is first taken between the scale reading at each observation (reduced to an invariable temperature in the case of the Bifilar Magnetometer), and the mean of the month for the same hour; these differences are regarded as the effect of the irregular disturbing force at the time of observation, and are represented, by Colonel Sabine, by the symbol $\nabla\psi_n$, n being the number of the Göttingen hour of observation. The *fluctuation* of the element, due to the irregular action between two consecutive hourly observations, is $\nabla\psi_n - \psi_{n-1}$, which is expressed by $F\psi_n$, and the mean irregular fluctuation for a whole day will be $\overline{F\psi} = \sqrt{\frac{1}{24}\Sigma(F\psi_n)^2}$, if the number of observation hours have been 24, as was the case in the present series. Similarly the mean irregular fluctuation for a month or longer period may be found, by dividing the sum of all the squares of $(F\psi)$ by the total number. The scale values employed to convert the mean fluctuation in scale divisions of the Bifilar into parts of the Horizontal Force, were the following: at Toronto $k = \cdot0001056$, at Lake Athabasca $k = \cdot000341$, and at Fort Simpson, from 15th April to 25th May 1844, $k = \cdot000283\cdot$

Table XLIX.

Values of the daily mean Irregular Fluctuation of the Declination and Horizontal Force F (ψ_n), *from October* 1843 *to May* 1844 *inclusive.*

Date.	Declination. Toronto.	Declination. Sitka.	Declination. Athabasca.	Horizontal Force. Toronto.	Horizontal Force. Athabasca.	Date.	Declination. Toronto.	Declination. Sitka.	Declination. Athabasca.	Horizontal Force. Toronto.	Horizontal Force. Athabasca.
1843.						1843.					
Oct. 1	′ —	′ —	′ —	—	—	Nov. 1	1′·0	′ —	3′·1	·0005	·0048
2	2·9	—	—	—	—	2	2·8	—	11·1	·0006	·0049
3	2·0	3·4	—	—	—	3	1·2	—	2·9	·0003	·0070
4	3·2	3·1	—	—	—	4	1·0	0·8	5·2	·0003	—
5	1·5	3·1	—	—	—	5	—	1·3	—	—	—
6	1·3	3·2	—	—	—	6	1·0	1·3	9·1	·0002	·0066
7	1·6	1·3	—	—	—	7	1·1	1·8	4·1	·0003	·0024
8	—	1·8	—	—	—	8	4·5	2·8	16·2	·0004	·0064
9	1·0	1·6	—	—	—	9	1·4	2·3	3·9	·0003	·0047
10	0·9	2·1	—	—	—	10	0·8	1·6	3·4	·0002	·0034
11	0·6	1·3	—	·0003	—	11	1·0	1·1	4·5	·0003	·0024
12	2·2	2·4	—	·0003	—	12	—	1·1	—	—	—
13	1·9	2·2	—	·0005	—	13	2·7	4·1	7·3	·0006	·0082
14	2·0	1·9	—	·0007	—	14	1·7	1·8	5·8	·0007	·0021
15	—	2·7	—	—	—	15	1·5	1·3	6·7	·0003	·0040
16	3·5	6·7	38·0	·0008	·0130	16	1·5	2·1	8·0	·0002	·0026
17	3·6	3·0	47·7	·0012	·0163	17	1·8	2·0	2·6	·0002	·0028
18	2·3	2·3	15·3	·0006	·0052	18	1·2	1·2	2·5	·0003	·0015
19	1·8	2·7	30·4	·0006	·0104	19	—	0·7	—	—	—
20	0·9	1·4	—	·0003	—	20	1·4	1·8	5·4	·0003	·0030
21	0·7	0·8	11·0	·0002	·0037	21	0·7	1·1	2·7	·0003	·0015
22	—	0·9	—	—	—	22	0·9	0·9	4·9	·0002	·0024
23	0·7	1·2	14·2	·0003	·0048	23	0·7	1·1	1·6	·0002	·0016
24	1·0	1·6	17·7	·0004	·0080	24	1·3	1·4	7·0	·0003	·0071
25	1·1	4·8	19·5	·0008	·0066	25	0·6	2·5	3·0	·0001	·0012
26	3·8	4·3	24·1	·0006	·0083	26	—	2·3	—	—	—
27	2·1	2·4	32·5	·0004	·0111	27	0·7	0·8	3·3	·0004	·0016
28	1·6	1·0	13·8	·0003	·0047	28	1·0	1·2	5·9	·0002	·0014
29	—	1·8	—	—	—	29	1·4	1·4	5·1	·0004	·0045
30	1·8	1·8	24·9	·0004	·0085	30	1·2	1·5	5·5	·0003	·0034
31	2·4	2·0	19·9	·0004	·0068						
Mean .	2·07	2·99	11·63	·00054	·00887	Mean .	1·62	1·70	6·28	·00037	·00426

TABLE XLIX.—*continued.*

Date.	Declination.			Horizontal Force.		Date.	Declination.			Horizontal Force.	
	Toronto.	Sitka.	Athabasca.	Toronto.	Athabasca.		Toronto.	Sitka.	Athabasca.	Toronto.	Athabasca.
1843.						1844.					
Dec. 1	1′·0	1′·9	12′·1	·0003	·0071	Jan. 1	0′·8	2′·1	′—	·0003	—
2	2·5	1·0	7·1	·0005	·0068	2	1·3	1·3	7·0	·0004	·0082
3	—	1·5	—	—	—	3	0·6	0·9	4·0	·0002	·0017
4	0·6	0·8	2·8	·0002	·0017	4	3·9	4·8	16·4	·0008	·0189
5	0·8	1·4	4·2	·0001	·0037	5	2·0	2·1	18·0	·0005	·0147
6	0·9	1·1	4·5	·0004	·0039	6	1·4	3·1	10·0	·0004	·0111
7	1·9	0·8	2·0	·0002	·0015	7	—	3·0	—	—	—
8	1·7	2·4	16·2	·0005	·0047	8	2·0	2·5	5·9	·0005	·0034
9	2·1	1·6	5·0	·0006	·0034	9	1·3	1·7	7·2	·0003	·0030
10	—	7·4	—	—	—	10	1·3	2·0	6·2	·0002	·0026
11	2·3	3·2	5·2	·0006	·0032	11	1·4	1·3	5·2	·0004	·0034
12	2·6	1·5	5·6	·0006	·0026	12	1·0	1·3	7·2	·0004	·0017
13	1·8	1·2	3·8	·0004	·0030	13	0·7	1·2	2·7	·0003	·0020
14	0·8	1·5	2·9	·0003	·0014	14	—	0·9	—	—	—
15	0·9	1·0	3·7	·0003	·0015	15	0·8	1·4	4·4	·0002	·0019
16	0·4	0·7	3·3	·0002	·0026	16	0·7	1·0	4·8	·0002	·0030
17	—	0·7	—	—	—	17	0·8	0·9	4·5	·0003	·0027
18	1·1	0·6	3·7	·0002	·0015	18	1·0	1·1	5·8	·0003	·0028
19	1·2	1·8	10·0	·0002	·0051	19	1·0	3·1	5·7	·0002	·0047
20	0·7	1·1	6·3	·0002	·0018	20	0·6	1·0	3·5	·0002	·0027
21	0·4	1·9	4·1	·0002	·0016	21	—	1·4	—	—	—
22	0·5	0·6	4·9	·0002	·0017	22	1·6	1·8	8·9	·0004	·0035
23	0·5	0·7	3·1	·0003	·0013	23	0·9	1·2	6·5	·0003	·0026
24	—	0·9	—	—	—	24	2·0	1·8	15·6	·0006	·0071
25	—	1·0	—	—	—	25	2·3	3·8	23·3	·0005	·0055
26	0·9	1·6	4·9	·0003	·0063	26	0·9	1·2	9·3	·0006	·0089
27	2·1	4·0	6·7	·0004	·0037	27	1·0	1·2	4·0	·0003	·0022
28	1·8	2·5	19·0	·0005	·0068	28	—	1·0	—	—	—
29	0·8	1·2	6·0	·0004	·0071	29	0·8	1·4	5·4	·0002	·0031
30	1·6	1·1	3·7	·0003	·0017	30	1·1	0·9	4·3	·0004	·0025
31	—	1·2	—	—	—	31	1·3	1·9	8·1	·0006	·0071
Mean -	1·42	2·04	7·32	·00042	·00409	Mean -	1·48	2·00	9·29	·00041	·00635

TABLE XLIX.—*continued.*

Date.	Declination.			Horizontal Force.		Date.	Declination.			Horizontal Force.	
	Toronto.	Sitka.	Athabasca.	Toronto.	Athabasca.		Toronto.	Sitka.	Athabasca.	Toronto.	Athabasca. — Fort Simpson.
1844.	′	′	′			1844.	′	′	′		
Feb. 1	2·5	3·5	13·5	·0006	·0078	Mar. 1	2·1	1·8	—	·0003	—
2	7·1	7·1	14·4	·0009	·0083	2	3·1	3·2	—	·0006	—
3	1·6	1·7	4·1	·0005	·0108	3	—	3·2	—	—	—
4	—	1·5	—	—	—	4	3·6	3·0	—	·0007	—
5	2·9	3·6	13·9	·0007	·0148	5	3·0	7·6	—	·0009	—
6	1·8	3·2	7·0	·0004	·0073	6	3·1	7·8	—	·0011	—
7	1·7	5·0	5·3	·0007	·0059	7	4·6	9·2	—	·0011	—
8	2·6	4·5	8·9	·0006	·0057	8	4·2	3·0	—	·0005	—
9	1·2	1·3	4·5	·0003	·0041	9	2·1	7·0	—	·0005	—
10	3·9	1·3	6·2	·0005	·0046	10	—	3·2	—	—	—
11	—	0·8	—	—	—	11	1·3	1·8	—	·0003	—
12	2·0	1·1	3·6	·0002	·0047	12	2·2	2·1	—	·0004	—
13	0·8	1·0	3·5	·0003	·0023	13	1·2	1·4	—	·0003	—
14	0·9	1·9	4·3	·0002	·0022	14	1·6	1·8	—	·0003	—
15	1·1	1·9	4·8	·0003	·0017	15	1·3	1·3	—	·0003	—
16	1·1	1·4	7·3	·0003	·0031	16	2·2	1·5	—	·0003	—
17	1·2	1·0	7·8	·0004	·0017	17	—	1·5	—	—	—
18	—	6·7	—	—	—	18	2·1	3·7	—	·0004	—
19	0·9	1·0	2·7	·0003	·0018	19	2·8	4·0	—	·0005	—
20	1·1	0·9	3·2	·0003	·0016	20	2·1	1·5	—	·0003	—
21	1·0	2·4	5·7	·0003	·0040	21	1·6	2·7	—	·0003	—
22	1·1	2·1	6·1	·0003	·0019	22	1·3	1·4	—	·0003	—
23	0·8	1·2	4·9	·0002	·0017	23	0·9	1·5	—	·0004	—
24	1·3	1·0	7·8	·0003	·0017	24	—	1·4	—	—	—
25	—	1·1	—	—	—	25	1·3	1·7	—	·0003	—
26	1·3	0·9	5·0	·0004	·0073	26	1·3	1·4	—	·0002	—
27	1·1	1·0	4·9	·0002	·0010	27	2·5	3·7	—	·0007	—
28	7·4	7·1	14·3	·0010	—	28	1·4	5·9	—	·0007	—
29	1·6	2·1	—	·0003	—	29	—	15·8	—	·0015	—
						30	6·6	5·4	—	·0016	—
						31	—	5·6	—	—	—
Mean -	2·02	2·80	7·50	·00049	·00570	Mean -	4·05	4·86	—	·00079	—

TABLE XLIX.—*continued.*

Date.	Declination.			Horizontal Force.		Date.	Declination.			Horizontal Force.	
	Toronto.	Sitka.	Fort Simpson.	Toronto.	Fort Simpson.		Toronto.	Sitka.	Fort Simpson.	Toronto.	Fort Simpson.
1844.						1844.					
April 1	4′·9	6′·1	17′·4	·0009	—	May 1	1′·8	3′·4	11′·9	·0007	·0064
2	2·4	3·3	24·5	·0009	—	2	1·7	2·9	10·9	·0006	·0069
3	3·4	4·8	31·0	·0007	—	3	1·2	2·5	12·2	·0005	·0148
4	2·4	1·7	12·1	·0005	—	4	1·3	0·8	5·5	·0002	·0035
5	—	3·3	—	—	—	5	—	2·4	—	—	—
6	1·4	3·3	12·0	·0007	—	6	1·4	1·6	9·5	·0004	·0044
7	—	2·9	—	—	—	7	4·6	1·8	14·2	·0006	·0085
8	2·0	1·1	10·1	·0003	—	8	3·1	1·5	9·4	·0005	·0049
9	1·0	1·3	8·0	·0003	—	9	1·0	1·2	8·2	·0003	·0063
10	1·7	4·7	21·3	·0006	—	10	1·8	1·3	7·2	·0005	·0059
11	1·7	1·6	6·0	·0004	—	11	1·2	1·1	8·1	·0003	·0021
12	1·0	1·0	4·3	·0003	—	12	—	1·4	—	—	—
13	0·9	1·5	9·7	·0003	—	13	2·2	1·8	6·8	·0004	·0045
14	—	2·0	—	—	—	14	5·3	1·8	7·3	·0008	·0048
15	1·1	1·5	27·8	·0004	·0110	15	1·2	1·5	5·2	·0003	·0042
16	8·6	6·1	13·8	·0016	·0112	16	1·0	1·2	7·8	·0003	·0023
17	5·0	6·1	26·6	·0027	·0171	17	0·9	1·2	4·8	·0003	·0025
18	1·3	2·1	4·5	·0006	·0030	18	1·3	1·7	5·8	·0004	·0042
19	1·0	1·5	7·3	·0003	·0064	19	—	1·3	—	—	—
20	1·0	1·1	7·9	·0003	·0069	20	0·7	1·2	5·2	·0003	·0046
21	—	1·7	—	—	—	21	1·7	2·0	6·5	·0009	·0060
22	0·7	1·2	6·6	·0003	·0038	22	5·0	6·6	29·8	·0011	·0116
23	1·7	2·0	12·1	·0004	·0050	23	1·5	11·5	22·5	·0006	·0060
24	2·2	1·9	7·9	·0004	·0033	24	2·9	2·5	9·0	·0004	·0063
25	2·8	6·2	28·3	·0009	·0129	25	1·7	1·4	—	·0006	—
26	4·2	4·4	27·9	·0012	·0193	26	—	1·6	—	—	—
27	3·8	4·8	21·1	·0006	·0104	27	1·8	1·9	—	·0005	—
28	—	1·3	—	—	—	28	1·1	1·2	—	·0003	—
29	1·3	2·2	26·6	·0007	·0072	29	1·4	1·5	—	·0004	—
30	2·6	5·6	24·8	·0007	·0128	30	1·0	0·9	—	·0003	—
						31	1·0	1·1	—	·0006	—
Mean -	2·99	3·45	16·45	·00068	—	Mean -	2·27	2·93	11·48	·00054	·00632

It appears by the foregoing Table that, with a few exceptions, the days of the larger mean irregular fluctuation of the Declination and Horizontal Force in each month coincide at all the stations, and that very low values, indicating a freedom from disturbance, are also generally the same; the exceptions may arise from an actual difference in the relative condition of the elements as regards this characteristic, but may also, in part, be caused by extreme movements concurring with the periods of regular observation at one station and not at another. It is due to this accidental circumstance that the mean fluctuation of both elements has a lower value on the 16th and 17th of April than on the 25th and 26th, although the amount of disturbance was considerably greater in reality on the former than on the latter occasion. In the following Table a few instances of this correspondence are selected, the number against each date being the relative place of that day among the other days of the same month at the same station, as regards the magnitude of the *mean irregular fluctuation* of the elements referred to.

TABLE L.

Extremes of each instrument during Disturbances at Lake Athabasca and Fort Simpson.

Date.	Relative place of each day.					Date.	Relative place of each day.				
	Declination.			Hor. Force.			Declination.			Hor. Force.	
	Toronto.	Sitka.	Athabasca.	Toronto.	Athabasca.		Toronto.	Sitka.	Athabasca. Fort Simp.	Toronto.	Athabasca. Fort Simp.
Oct. 16	3	1	3	2	2	Jan. 8	3	5	11	6	12
17	2	8	4	1	1	24	4	8	4	3	5
25	10	2	2	11	8	25	2	2	1	6	7
26	1	3	6	4	6	Feb. 1	6	6	4	5	3
30	13	18	1	8	5	2	2	1	1	1	6
Nov. 2	2	wants	2	4	6	5	4	5	3	2	1
8	1	2	1	6	5	28	1	1 bis	2	1 bis	wanting.
13	3	1	5	2	1						
14	5	7	9	1	19	Apr. 16	1	2	11	2	7
Dec. 1	12	4	3	13	1	17	2	2 bis	6	1	2
2	2	18	5	6	2	25	7	1	2	5	4
12	1	9	9	4	14	26	4	9	3	3	1
27	4	1	6	10	9	May 1	7	3	5	5	5
28	7	8	1	5	8	14	1	10	13	3	12
Jan. 4	1	1	8	1	1	22	2	2	1	2	3
5	5	6	2	2	2	23	13	1	2	1	8

November 14, December 12, January 8, and May 14, are inserted to show that occasionally a high relative value of the mean irregular fluctuation of one or both elements prevailed at Toronto, with a low one at the northern station; but the general conclusion from this comparison must be, that such a state of things is exceptional, a state of disturbance being more commonly prevalent at the same time over the whole area embraced, which is not an inconsiderable fraction of that of the globe. It has already been shown that there is not in general a correspondence in detail in the movements during disturbances at Lake Athabasca or Fort Simpson and the other stations in America or Europe, although it sometimes exists to a limited extent. The first of these stations is rather less distant, and the latter rather more distant from Toronto, geographically, than Barnaoul in Siberia from St. Petersburg. On referring to the curves of magnetic term days given in the *Annuaire Magnetique et Meteorologique*, &c. for these two stations, it will be seen that distance has apparently less to do with this want of correspondence than difference of magnetical position, for while there are numerous and interesting examples of great movements at St. Petersburg, not shown in the curve for Barnaoul, or only to be identified in some minor inflexion by the aid of the curves given for intermediate stations, yet these cases appear to be rather the exceptions, and in general a correspondence is at once perceptible, notwithstanding the distance, which is about 1,750 geographical miles, and exists both in the character and the precise epoch of the greater movements. For the purpose of showing more fully that this is not the case in the stations we are here comparing, a selection has been made of all the observations at Toronto, Sitka, and the northern stations, which differ from the mean for the month by a quantity exceeding twice the amount of the *mean irregular fluctuation* of the element for the same month. The values of the latter quantity will be found in the preceding Table. The dates of these observations are given in the next Table, together with the amount of the difference in each case from the mean, both at the station at which such difference reaches the given limit or amounts to a shock, and at the remaining stations. To distinguish the latter or corresponding readings from the shocks, they are printed in *Italics*.

TABLE LI.

A List of Shocks of the Declination at Lake Athabasca, or Fort Simpson, Toronto, and Sitka, with the differences of the scale readings at all these Stations from their means respectively at each date. Differences which fall short of $2\sqrt{\frac{\Sigma}{h}F\psi^2}$, *and therefore do not come up to the definition of a Shock, are printed in Italics. A movement of the north end of the magnet to the east is marked with the + sign at all the stations, the contrary movement with the − sign.*

* Observation wanting. S. Sunday, Good Friday, or Christmas Day at the station. d Disturbance observed.

Gött. Date.	Station.		
	Toronto.	Sitka.	Athabasca.
	OCTOBER.		
d. h.	′	′	′
2 19	4·9	*2·5*	—
2 20	9·1	*3·2*	—
2 21	9·2	−7·9	—
2 23	−7·0	−25·6	—
3 0	*−0·8*	−6·9	—
3 16	*−2·5*	6·4	—
3 18	*−1·4*	6·1	—
4 15	8·9	*5·5*	—
5 0	*1·7*	8·1	—
5 1	*2·7*	7·3	—
5 6	−4·8	−10·7	—
5 9	*−1·4*	−6·9	—
5 15	4·6	*−1·9*	—
6 18	*−2·7*	−9·0	—
8 21	−4·1	*−0·6*	—
12 15	5·0	*1·2*	—
12 23	−4·5.	*0·1*	—
13 0	*−1·9*	9·4	—
13 22	−4·3	*−2·4*	—
14 11	−4·2	*−0·4*	—
14 14	*1·4*	−5·9	—
16 0	−11·3	*−5·7*	63·0 d
16 10	−4·5	*−3·3*	*−1·3*
16 20	*+0·1*	26·1	*0·1*
16 21	*−1·5*	26·0	*5·9*
16 22	−6·0	16·3	*3·0*
16 23	*1·5*	10·5	*3·7*
17 2	−11·0	−6·9	*−12·7* d
17 3	*−3·4*	−7·7	*−12·8* d
17 4	−4·1	*−3·9*	*−18·8*
17 18	*−3·2*	5·9	35·6 d
17 20	*−5·5*	*−1·3*	*−17·6* d
18 16	9·3	*+3·0*	*−3·6*
19 16	*−0·9*	−6·0	*4·4*
19 17	−6·8	12·1	*−20·9*
19 18	*−2·1*	7·5	−30·7 d
24 1	*−0·9*	8·4	*5·3* d
25 19	*−0·7*	12·1	62·0 d
26 0	−7·6	*−0·3*	*−8·6* d
26 3	4·2	*−0·8*	*−3·3* d
26 5	*−0·3*	−6·4	*−5·3*
26 10	−4·5	*−3·0*	*1·3*

Gött. Date.	Station.		
	Toronto.	Sitka.	Athabasca.
	OCTOBER.		
d. h.	′	′	′
26 14	*5·2*	*3·4*	*11·7*
26 19	−4·7	6·1	*−29·8*
26 20	−4·1	*−0·5*	*1·3* d
26 22	*−2·8*	−10·2	*1·6* d
27 0	*0·1*	−9·3	*8·1* d
27 1	−6·5	−12·0	*19·8* d
29 22	−4·2	*0·7*	*−1·8*
30 0	*0·7*	*0·5*	28·0
30 17	4·2	*−0·1*	*9·1*
31 14	8·6	*1·2*	*1·7*
	NOVEMBER.		
1 4	3·2	*	*−2·5*
1 5	3·5	*	*0·8*
2 17	6·6	*	*−3·0* d
2 18	*+1·7*	*	−43·8 d
2 19	*1·7*	*	−36·6 d
2 20	5·8	*	*−11·2* d
3 0	−4·0	*	*−2·7* d
5 21	*0·6*	3·7	*−11·5*
5 22	*0·9*	6·	*−0·7* d
5 23	*2·4*	7·	14·3 d
6 0	*1·9*	8·	43·3 d
6 1	*0·4*	6·	*
6 2	*0·4*	4·	22·1
6 9	*0·5*	3·	*−0·6*
7 1	*−0·9*	3·	18·3
7 2	*0·9*	5·	*−10·2*
7 3	*0·2*	4·	*−2·8*
7 4	*1·7*	3·	*−1·1*
7 5	−4·1	*1·*	*2·8*
7 7	−3·6	−3·	−12·5
8 9	*−0·6*	3·	*−2·7*
8 1	*−0·4*	4·	*−5·7*
8 3	*0·2*	4·	*2·3*
8 4	*−0·3*	4·	*1·7*
8 5	*−1·5*	5·	*0·8*
8 15	15·1	3·	*8·9*
8 17	*0·3*	4·	*−1·0*
8 18	*−3·1*	11·	56·6 d
8 20	*−0·7*	8·	*5·6*
8 21	*1·6*	5·	*4·3*
8 22	*1·1*	6·	*11·0*

TABLE LI.—*continued.*

Gött. Date.	Station.			Gött. Date.	Station.		
	Toronto.	Sitka.	Athabasca.		Toronto.	Sitka.	Athabasca.
	NOVEMBER.				NOVEMBER.		
d. h.	′	′	′	d. h.	′	′	′
8 23	0·7	12·5	14·3d	13 4	−4·8	−9·3	17·5d
9 0	2·7	8·1	1·3d	13 5	−4·6	1·9	−5·0d
9 4	1·1	3·5	−0·9	13 6	−4·5	0·6	−8·1d
9 5	0·0	3·8	2·8	13 9	−3·3	−4·7	0·1
9 6	1·7	4·7	−0·1	13 14	7·1	1·2	1·8
9 7	0·1	5·7	4·3	13 15	4·1	1·3	−2·1
9 8	0·0	5·3	4·4	13 21	−0·1	−0·7	−13·1d
9 9	0·5	6·1	2·6	13 22	+0·8	−5·0	−2·2d
9 10	+0·6	5·2	3·0	13 23	−0·8	−8·6	−2·7d
9 11	0·4	5·6	2·5	14 11	−4·2	−5·0	−3·0
9 12	−0·2	5·4	5·5	14 12	−0·8	−6·1	−4·2
9 13	−0·6	5·5	5·1	14 13	−0·4	−3·8	0·4
9 14	+0·6	5·6	1·2	14 15	−0·4	−4·8	−2·1
9 15	−1·1	7·3	3·5	15 12	−0·8	−3·8	−2·1
9 16	+2·6	4·1	1·6	15 14	3·9	−1·7	3·2
9 17	−0·6	3·6	3·0	16 10	−1·1	−3·5	−5·0
9 22	−1·7	−7·3	+5·8d	16 17	3·4	0·6	−1·8
9 23	−0·1	−5·0	1·5d	16 18	2·9	4·4	21·2d
10 0	0·8	−3·9	−0·1d	17 13	0·2	−4·0	1·2
10 6	0·1	7·0	+1·9	17 15	4·4	−3·5	1·9
10 7	−0·2	6·3	−1·3	17 18	−1·0	−4·4	−2·8
10 8	−0·1	5·1	2·4	17 20	0·1	−5·0	1·8
10 9	−0·1	4·3	3·7	17 21	0·0	−3·8	0·3
10 10	−0·2	4·8	1·2	18 6	−1·9	−3·8	−3·1
10 11	+0·1	4·8	0·7	18 8	2·0	−5·2	−6·6
10 12	−0·4	4·0	−1·8	18 9	−0·4	−5·0	−5·8
10 13	−1·1	4·5	−2·4	18 10	−0·2	−4·3	−2·4
10 14	−1·3	3·9	−0·8	19 5	S.	−4·4	S.
10 21	−1·9	7·6	8·1d	19 6	S.	−5·0	S.
10 22	0·1	7·0	2·0d	19 7	S.	−4·4	S.
10 23	1·3	7·2	0·3	20 0	−0·8	−4·1	−2·3
11 0	−0·6	5·3	−3·7	20 1	0·4	−3·8	−0·6
11 1	0·5	4·3	−3·7	20 21	−3·8	+2·9	17·3
11 2	0·6	3·5	−6·2	21 0	0·8	−3·7	−0·7
11 3	0·2	3·7	−3·6	22 2	−2·2	−3·5	14·1
11 4	−0·2	4·1	−1·1	22 3	−1·7	−3·1	14·8
11 5	−0·6	3·9	+1·6	22 18	0·3	−3·9	−4·8
11 6	0·2	4·0	−3·9	23 4	−1·8	−4·4	−3·1
11 12	−1·5	3·8	+1·8	24 1	−3·5	1·5	14·3d
11 13	−0·9	4·5	−0·2	24 2	−1·7	2·5	13·8d
11 14	−0·3	4·8	−6·8	24 4	−2·6	+4·4	9·5
11 15	−1·1	4·4	−2·1	24 19	−0·4	1·7	17·4
11 16	−0·1	3·4	+3·6	26 5	S.	−11·1	S.
11 17	−0·8	4·2	−3·0	26 6	S.	−11·4	S.
11 18	S.	3·5	−2·8	27 4	3·2	−0·8	−0·5
11 19	S.	4·3	−2·6	27 5	3·5	+0·2	2·0
11 23	S.	5·2	S.	28 5	3·0	−3·8	−1·8
12 2	S.	3·6	S.	29 1	0·4	−4·8	−3·7
12 3	S.	4·4	S.	29 2	−1·1	−4·4	−0·4
12 4	S.	4·3	S.	29 4	−0·4	−4·0	−6·5
12 5	S.	4·1	S.	29 14	0·8	−5·0	−13·2
12 6	S.	4·3	S.	30 0	0·1	−4·3	−4·7
12 19	−1·3	−4·8	S.	30 3	−0·8	−4·0	−6·0
12 20	0·0	−4·0	S.	30 4	1·1	−4·0	−8·5
12 21	0·1	−4·2	−5·1	30 5	2·7	−3·5	−5·0
13 2	1·7	−4·9	2·6	30 9	−0·4	−3·4	−3·6
13 3	2·9	−7·5	14·8	30 18	−0·9	1·4	15·2

Table LI.—*continued.*

Gött. Date.	Station.			Gött. Date.	Station.		
	Toronto.	Sitka.	Athabasca.		Toronto.	Sitka.	Athabasca.
		December.				December.	
d. h.	′	′	′	d. h.	′	′	′
1 20	3·8	2·4	-8·9	27 3	2·8	-0·2	-5·6
1 21	2·7	2·8	-28·7 d	27 6	-3·3	-2·8	-3·7
1 22	3·6	-3·8	20·1 d	27 7	-6·3	-2·8	-10·5
1 23	3·0	-0·4	-13·4 d	27 8	-3·4	0·3	3·7
2 0	6·1	2·1	9·0 d	27 22	-2·4	20·0	-2·7
2 1	1·8	1·7	17·3 d	27 23	-4·4	16·4	-0·8
2 2	-2·8	-1·1	-27·3 d	28 0	-0·1	14·8	-3·6
2 3	-8·7	-0·7	20·8 d	28 1	-1·9	12·6	0·3
2 4	-4·5	0·4	20·3 d	28 2	-4 7	11·6	-22·4
3 20	-0·1	-6·5	S.	28 3	-4·2	1·4	52·6 d
5 9	-1·1	-5·5	-7·4	29 8	1·5	4·2	-4·3
5 10	-0·4	-4·9	-10·3	29 20	-1·1	2·2	17·4
7 15	5·5	0·7	2·1	31 9	S.	4·5	S.
8 5	3·1	-5·9	-11·3	31 10	S.	4·2	S.
8 6	-0·2	-4·4	-0·5				
8 7	-1·7	-6·6	+2·7			January.	
8 8	3·5	-13·9	3·9	2 2	-4·5	2·4	12·8
8 9	-1·0	-10·9	4·4	4 12	-3·6	-4·3	-6·2
8 10	1·5	-10·6	-19·3	4 16	14·4	-2·7	3·9 d
8 11	0·3	-4·5	-10·7	4 17	4·8	14·2	18·5 d
8 18	2·4	-1·5	42·0 d	4 19	0·3	6·0	-5·3 d
8 20	4·2	-2·8	-4·5 d	4 20	-0·5	5·6	-8·2 d
9 23	S.	-4·3	S.	4 21	-1·4	3·8	-41·8 d
10 4	S.	-5·6	S.	4 22	1·4	4·4	19·5 d
10 5	S.	-8·4	S.	4 23	-1·4	5·1	32·5 d
10 6	S.	-6·2	S.	5 1	+1·1	1·2	60·0 d
10 7	S.	-7·3	S.	5 10	-5·3	-2·3	-0·4
10 12	S.	-4·2	S.	5 11	-1·0	4·3	-3·7
10 13	S.	-4·8	S.	5 19	1·2	0·3	22·2
10 18	4·2	1·8	S.	5 23	-4·7	0·8	-19·1 d
10 19	0·6	-6·6	S.	6 0	-1·5	-0·7	19·2 d
10 22	-4·2	-22·8	5·9	6 20	S.	9·6	-30·2
11 8	-3·2	-2·9	-3·7	7 0	S.	8·8	S.
11 12	3·5	0·4	-12·6	7 6	S.	4·9	S.
11 14	5·1	-1·7	-7·0	7 19	0·1	4·9	S.
11 18	-4·8	-6·0	-9·8	8 1	0·4	-4·4	-8·9
11 19	-0·8	-5·1	-4·5	8 2	-0·9	-6·8	-3·4
12 3	-2·2	3·9	-14·8	8 3	-0·9	-4·4	-4·3
12 11	10·4	0·2	5·9	8 4	-1·4	-6·5	-4·9
12 12	6·1	2·4	10·8	8 5	0·2	-5·7	-4·9
13 11	3·7	-0·1	2·1	8 6	-0·1	-5·9	1·4
13 20	-3·2	2·9	7·1	8 7	0·4	-6·9	-3·2
14 2	0·3	-5·0	7·6	8 8	1·2	-4·6	-8·3
19 20	-3·0	0·2	30·4 d	8 9	0·4	-4·0	-4·5
19 22	1·9	4·1	24·1 d	8 11	1·9	-7·3	0·3
19 23	1·6	10·3	23·0 d	8 13	4·5	0·7	-2·2
20 0	0·6	7·5	13·2 d	8 16	4·3	2·7	11·3 d
20 1	1·1	4·3	7·3 d	8 22	-3·2	-0·9	-0·1 d
20 2	-0·5	4·9	9·4 d	9 20	-0·1	5·5	3·8
21 5	-0·1	4·9	-11·0	9 21	-2·1	5·2	*
24 9	S.	4·3	S.	10 1	0·6	-4·1	-8·2
25 3	S.	5·7	S.	10 2	0·5	-5·8	-0·2
25 4	S.	4·9	S.	10 3	0·8	-4·4	-1·3
26 22	-2·8	2·1	-5·9 d	10 5	3·2	0·5	-0·1
26 23	-3·5	1·5	7·8 d	10 9	-3·0	-2·2	-4·5
27 0	0·5	3·1	22·0 d	11 3	3·9	+2·2	8·9

TABLE LI.—*continued.*

Gött. Date.	Station. Toronto.	Sitka.	Athabasca.	Gött. Date.	Station. Toronto.	Sitka.	Athabasca.
	JANUARY.				JANUARY.		
d. h.	′	′	′	d. h.	′	′	′
11 4	3·9	+1·0	1·9	27 19	S.	−4·3	−7·8
12 5	0·1	4·4	−5·7	28 3	S.	−[illegible]·3	S.
12 6	−1·4	5·8	4·0	28 4	S.	− ·4	S.
12 7	−2·3	6·5	3·8	28 5	S.	− ·3	S.
12 8	−2·3	6·5	2·3	28 11	S.	− ·8	S.
12 9	−1·5	7·2	−3·0	29 4	0·7	− ·0	4·1
12 14	1·7	4·0	1·9	29 5	0·[illegible]	− ·3	−5·9
12 19	0·5	2·3	23·5	29 8	−0·	− ·3	−6·0
13 5	−0·8	4·4	0·7	29 9	−0·	− ·9	−10·1
13 6	−1·4	4·0	2·6	29 22	−0·	− ·2	−9·3
13 7	−1·4	5 5	2·8	30 17	2·	− ·8	−3·3
13 8	−1·1	4·9	0·9	31 8	0·	− ·6	−0·9
13 9	−0·1	6·8	1·7	31 9	−1·	− ·5	10·7
13 10	0·7	5·0	1·6	31 10	−0	− ·0	−1·0
13 11	0·7	4·1	−0·2	31 17	0	− ·3	−8·7
13 15	0·6	4·2	−0·6	31 18	1·	− ·6	−3·0
13 20	S.	4·0	S.	31 19	1·	− ·3	−18·9
14 5	S.	4·4	S.	31 20	0·	− ·0	−27·8
14 7	S.	4·3	S.	31 21	3·	− ·0	−14·8
14 8	S.	5·7	S.	31 22	3·	− ·3	−11·5
14 9	S.	5·3	S.	31 23	5·	− ·0	−3·5
14 10	S.	4·7	S.		FEBRUARY.		
14 20	−0·4	4·1	S.				
14 22	−1·1	4·1	5·1	1 1	−3·5	2·4	45·9d
14 23	−0·2	5·5	5·5	1 2	−3·2	3·8	35·1d
15 0	0·1	6·4	−1·2	1 3	−0·8	12·4	16·4d
15 1	−0·4	5·3	−3·2	1 4	−0·2	9·9	2·3
15 2	0·4	4·0	−2·2	1 5	−2·2	6·8	6·6
15 8	0·6	4·7	−2·7	1 9	−3·9	−5 8	6·2
15 9	0·1	4·4	11·5	1 10	2·5	−7·3	−5·1
15 10	0·1	5·5	11·6	1 19	0·5	−6·1	+19·4d
18 10	0·7	4·6	3·6	2 6	−5·5	−7·7	−1·1d
19 21	1·2	9·5	22·2d	2 7	−4·8	−9·9	−47·8
19 22	3·3	0·5	7·5d	2 12	6·5	−2·7	5·1
21 9	S.	−4·3	S.	2 13	−0·2	6·7	−3·4
22 3	−6·6	−1·0	13·7	2 17	−18·4	21·0	3·4d
22 4	−4·6	−2·1	13·9	2 18	6·9	3·2	−12·4d
22 5	−3·4	−1·4	5·5	2 22	−0·8	−5·9	−2·6
22 8	−0·8	−5·5	−8·0	3 13	−0·1	−6·3	−7·4
24 16	4·1	−2·5	−7·1	4 21	−2·2	−1·1	−26·4
24 18	7·9	−0·6	−2·8	4 22	0·1	−1·3	−26·2
24 19	6·3	6·2	0·9	4 23	−0·4	−0·4	28·6
24 20	2·2	4·2	15·3	5 0	−0·7	1·0	−28·3d
24 21	5·0	1·5	45·2	5 3	2·2	2·3	18·1
24 22	2·9	11·1	51·9	5 7	−2·3	−5·7	4·6
24 23	2·6	11·8	−11·5	5 8	1·2	−6·7	−9·6
25 0	0·8	4·3	51·4	5 9	−6·1	−2·8	1·0
25 1	−1·9	7·8	113·8	5 14	1·3	−6·4	11·7
25 2	2·5	11·6	55·8	5 17	4·0	6·1	−19·0d
25 3	−5·5	0·2	23·7	6 20	−0·9	10·5	−13·8d
25 4	−5·7	6·3	10·7	7 10	1·0	−8·4	−7·3
25 21	−0·4	−4·3	2·2	7 11	−2·3	−6·9	−4·4
25 22	−0 4	−4·2	−8·7	7 12	−3·5	−6·9	−4·7
26 21	0·1	−4·0	−1·8	7 22	1·3	22·9	−1·2
27 0	0·6	−4·0	*	7 23	3·9	10·3	7·0
27 1	0·6	−4·4	−10·2	8 0	2·0	18·2	22·7d

TABLE LI.—*continued.*

Gött. Date.	Station. Toronto.	Station. Sitka.	Station. Athabasca.
	FEBRUARY.		
d. h.	′	′	′
8 2	5·5	−7·3	16·6
8 3	0·1	−9·5	4·5
8 4	−0·4	−9·5	16·3
8 5	−8·3	−6·2	16·0*d*
9 22	0·1	−6·4	1·2
10 12	11·2	−2·2	−1·1
11 20	0·4	−1·7	28·7
11 21	−2·3	−1·3	17·4
15 5	−1·1	−6·0	−0·6
16 7	1·7	0·0	15·2
18 20	0·1	−1·2	15·4
21 20	1·4	6·9	−12·8
21 21	1·4	6·0	1·2
22 0	1·3	6·5	−4·7
23 5	0·9	1·7	−17·6
26 2	2·4	6·3	1·4
26 3	2·4	5·7	12·5
28 7	2·4	6·1	16·9
28 8	−1·7	−17·1	7·8
28 9	−1·9	1·2	−15·3
28 11	−9·4	−13·1	−10·0
28 16	25·1	−2·2	−4·4
28 17	3·5	−7·8	−20·2
28 22	−1·0	0·1	−21·8
29 17	0·7	4·6	25·6
	MARCH.		
2 19	S.	−12·6	—
2 20	S.	−12·9	—
2 21	S.	−10·9	—
2 22	S.	−13·3	—
2 23	S.	−12·7	—
3 0	S.	−12·5	—
3 1	S.	−10·2	—
4 18	−4·1	9·9	—
5 16	3·3	10·0	—
5 17	5·5	22·7	—
5 23	−8·8	4·7	—
6 0	−10·4	10·0	—
6 2	−5·8	26·2	—
6 3	−7·1	13·3	—
6 4	−8·7	7·6	—
6 17	3·7	17·9	—
6 18	6·4	18·2	—
6 19	9·2	19·3	—
6 20	1·6	18·1	—
6 22	−0·4	9·8	—
6 23	−0·3	10·6	—
7 0	−0·6	9·9	—
7 8	−3·3	11·6	—
7 11	8·1	0·0	—
7 15	13·3	−1·6	—
7 16	13·6	10·7	—
7 18	−4·8	18·4	—
7 21	2·3	10·9	—
8 10	8·1	1·8	—

Gött. Date.	Station. Toronto.	Station. Sitka.	Station. Athabasca / Fort Simps. *
	MARCH.		
d. h.	′	′	′
8 14	10·2	6·1	—
8 19	−1·9	11·5	—
8 20	−1·4	12·6	—
8 21	−0·4	14·1	—
8 22	−3·0	12·8	—
8 23	−0·9	11·9	—
9 0	0·6	11·2	—
9 1	0·1	10·6	—
9 17	3·2	17·5	—
10 22	−1·0	10·0	—
28 23	3·7	24·1	—
29 12	−10·0	−5·1	—
29 14	31·0	−12·2	—*
29 15	8·4	−6·3	—
29 16	11·0	0·6	—
29 17	2·9	−13·7	—
29 18	−25·9	38·6	—
29 19	20·2	−7·9	—
29 20	13·3	−1·9	—
29 21	−10 4	−2·2	—
29 22	9·4	−3·7	—
29 23	13·1	−1·5	—
30 0	−10·0	1·7	—
30 7	3·4	−10·0	—
30 15	8·2	6·3	—
30 18	S.	11·4	—
31 18	5·0	18·4	—
31 23	0·8	−9·7	—
	APRIL.		
1 0	−1·9	−9·3	36·5
1 4	−1·7	−8·9	−11·1
1 6	−1·5	−7·2	10·4
1 15	13·4	1·8	−3·8
1 20	1·0	14·0	−14·7
2 6	0·4	−6·9	8·7
2 16	6·9	−1·7	−5·7
2 20	1·1	7·5	17·5
2 23	−2·2	4·5	90·1*d*
3 16	10·0	8·1	13 8
3 22	−4·7	9·7	4·7
4 0	−2·0	4·4	35·4
4 16	0 1	−7·1	−0·6
5 15	S.	−10·6	S.
5 16	S.	−14·5	S.
5 17	S.	−8·1	S.
5 18	−5·6	−13·2	S.
5 19	1 1	−10·7	S.
5 20	1·7	−8·5	S.
6 15	1·1	−6 9	−15·3
6 23	S.	−7·6	S.
8 7	2·1	1·0	−42·0
9 23	2·7	0·4	42·3*d*
10 1	2·0	−8·3	+23·8*d*
10 3	−0·7	−14·2	+11·7
10 21	1·4	−0·4	45·3*d*

TABLE LI.—*continued.*

Gött. Date.	Station. Toronto.	Sitka.	Fort Simpson.	Gött. Date.	Station. Toronto.	Sitka.	Fort Simpson.
	April.				May.		
d. h.	′	′	′	d. h.	′	′	′
15 1	0·0	1·5	−38·0 d	1 18	−2·3	7·6	17·7
15 2	3·3	0·3	66·1 d	1 23	−3·0	−11·0	−15·0
16 14	2·9	−7·4	−17·0 d	2 0	−2·4	−10·7	*
16 15	2·4	−9·5	−25·3 d	2 1	−0·1	−9·2	−9·6
16 16	−1·0	−7·2	−7·9 d	2 2	−0·9	−8·9	−2·7
16 17	3·0	−8·4	−22·3 d	2 18	5·6	6·9	6·2
16 18	8·4	−2·3	−37·5 d	3 2	2·7	−6·0	−14·6
16 19	10·0	−10·9	−31·4 d	3 3	1·7	−6·6	5·2
16 20	23·3	4·2	−19·1 d	3 4	0·8	−12·6	+35·6 d
16 21	31·0	16·1	24·3 d	7 16	10·8	1·7	−2·0
16 22	23·7	29·1	20·8 d	7 18	5·5	−2·8	−18·0
16 23	−13·5	42·1	25·5 d	7 19	13·6	−1·1	−23·6
17 0	−11·7	25·1	64·2 d	7 22	−4·1	6·2	17·7 d
17 1	−25·2	32·8	40·9 d	7 23	1·3	6·5	−10·5
17 2	−15·4	18·8	65·0 d	8 13	6·7	1·6	−3·0 d
17 3	−1·5	8·4	3·2 d	8 16	4·5	−0·2	−11·7 d
17 4	−5·0	10·3	30·2 d	10 16	5·4	2·6	1·5
17 5	−5·0	2·3	90·9 d	13 18	5·0	−3·6	−17·7 d
17 6	−5·5	−2·8	55·9 d	14 15	5·2	−6·9	−8·5
17 7	−6·3	−6·0	+60·3 d	14 16	21·8	−3·7	−13·4
17 8	−7·4	−9·2	1·4	22 1	3·3	2·8	64·7 d
17 9	−5·5	−13·2	−8·2	22 2	−0·9	7·9	57·6 d
17 10	−2·9	−10·4	0·6	22 3	3·5	7·2	20·6 d
17 11	−7·2	−10·9	−31·7	22 5	6·7	4·6	16·2
17 12	−7·4	−7·3	−17·1	22 10	−6·8	−8·1	−12·8
17 13	−7·9	−11·3	−14·5 d	22 13	8·5	−8·4	−1·5 d
25 2	−4·9	8·7	38·3	22 14	0·9	−6·5	−9·1 d
25 3	−8·5	24·6	116·3	22 15	7·4	−6·1	−13·0 d
25 4	−5·5	7·7	27·5	22 17	8·3	12·8	−9·2 d
26 0	−6·5	−0·2	22·9 d	22 18	8·7	0·2	−23·0 d
26 4	3·5	2·9	49·9	22 19	4·7	0·3	−26·8
26 5	2·7	5·4	63·1	22 20	−2·3	−8·2	−5·2
26 13	11·3	4·5	6·6	22 23	−2·9	−10·6	10·1 d
26 18	−0·5	−0·6	−75·1 d	23 0	−1·3	−22·4	13·6 d
26 19	−7·4	11·7	−47·1	23 1	−1·0	−47·1	7·3 d
26 21	−0·3	5·0	−38·9 d	23 2	0·0	−61·1	−2·8 d
27 11	7·1	3·5	5·4	23 3	0·1	−17·3	−9·0
27 14	7·1	4·9	8·7	23 22	−2·9	−39·1	−39·1
27 16	1·1	18·2	32·9	24 13	7·5	2·8	2·1
27 17	5·2	10·5	33·7	24 14	−5·3	1·7	−12·3
27 20	S.	10·1	4·9 d	24 16	−8·7	3·0	−5·2
29 2	−2·7	6·7	95·5 d	24 17	−6·0	−0·4	−2·8
29 3	−2·0	6·9	26·3 d	24 18	−7·1	2·4	6·0
29 4	−2·4	6·5	68·0 d	24 20	−5·8	4·4	−14·2
29 16	0·9	9·1	−5·5	27 1	0·1	6·4	—
30 17	−1·9	16·7	−42·9	27 9	*	−5·8	—
30 21	−6·3	9·0	16·6 d	27 16	−4·8	−4·4	—
30 22	−5·5	10·3	65·1 d	27 17	−5·0	−0·9	—
				29 21	5·0	0·5	—

It will be seen by the foregoing Table that instances in which the Declination is simultaneously affected, to the extent which is defined as a shock, at Lake Athabasca or Fort Simpson, and Toronto, are comparatively rare; still more so those in which it is so affected at the same observation at Sitka also. Of the latter class we find but eleven instances in a list embracing 623 dates of observation, of which number 473 were common to the three stations. It is to be observed however, as a defect of this mode of comparison, that the number of shocks is generally least where the prevalence of disturbance is greatest, for this circumstance occasions a high value of the mean irregular fluctuation, and the proportion of instances in which the deviation exceeds *double* that amount is not equally augmented; thus the number of so-called shocks is much greater at Sitka than at Lake Athabasca or Fort Simpson, the value of the mean irregular fluctuation being so low at the former station that a temporary prevalence of easterly or westerly ranges without sensible disturbance, and due probably to a different cause than the movements we are particularly investigating, is sufficient to bring nearly all the readings of certain days into the list; of this we have several examples in November and January. The proportion of shocks to the whole number of observations is 1 in 34·4 at the two northern stations taken together, 1 in 26·6 at Toronto, and 1 in 13·3 at Sitka. Referring again to the individual stations, it appears that out of a total number of 83 shocks at Lake Athabasca, 36 coincided with shocks at Sitka; out of 31 shocks at Fort Simpson, 14 coincided with shocks at Sitka; these numbers are in the proportion of 43 and 45 per cent. respectively, whereas the number of dates coincident with shocks at Toronto are only 17 in the former and 7 in the latter number, being in the proportion of 20 and 22 per cent. This difference in the degree of correspondence is further to be seen, on paying regard to the signs of the differences of the coincidences in date at Toronto and the northern stations, 16 have the contrary and 8 the same sign; of those at Sitka and the same stations, 14 have the contrary and 36 the same sign. The evidences of agreement in the movements at Toronto and at Sitka are somewhat greater, but not materially so, than those we have found at the latter station and the temporary ones. In an aggregate of 187 shocks at Toronto, only 61 are coincident with shocks at Sitka, being 32 per cent., and of this number those with like and unlike signs are nearly balanced, the numbers being 29 and 32 respectively. The limit assigned to shocks being arbitrary, we may consider the above list without reference to it, and solely as a selection of simultaneous deviations of the Declination Magnet, from its normal value at the three stations, under circumstances of apparent disturbance at one or more of them. Thus compared, it appears that the magnet was similarly affected, or deviated from its normal position for the same hour, in the same direction at all three stations, in 138 out of 473 instances, or 27 per cent. of the whole; the proportion is the same if we have regard to the sign of the absolute Declination, which is easterly at the three northern stations, westerly at Toronto, and is not greater for movements under one sign than under the other, whether referred to absolute or relative value. The absolute Declination appears to have been increased or decreased simultaneously at all three stations in 140 instances out of the same number. The proportion of instances in which the movements were similar at any two of the stations was as follows:—

Toronto and Sitka, 288 out of 555 selected observations, or 52 per cent.

Toronto and Lake Athabasca, 217 out of 372 selected observations, or 58 per cent.

Toronto and Fort Simpson, 42 out of 109 selected observations, or 38 per cent.

Sitka and Lake Athabasca, 231 out of 370 selected observations, or 62 per cent.

Sitka and Fort Simpson, 75 out of 110 selected observations, or 68 per cent.

As the observations here selected as most favourable for this comparison form somewhat less than one eighth of the whole number at Toronto, and about the same at Athabasca and Fort Simpson, the inference must be that under ordinary circumstances the deviations of the Declination Magnet from its mean position for the hour of observation, at stations so distant as those compared, are not referable to any common cause, and that where apparent resemblance exists, it is entirely casual.

In extending this mode of comparison to the changes of Horizontal Force, it was found that in consequence of the low value of the mean irregular fluctuation of the element at Toronto during the months examined, the proportion of readings which differed from their mean by double that quantity was too large to furnish a real criterion of the state of disturbance prevailing.

The numbers were as follows:—

In October 1843,	number of	observations	432,	shocks	77
In November 1843,	„	„	624,	„	211
In December 1843,	„	„	600,	„	181
In January 1844,	„	„	647,	„	293
In February 1844,	„	„	600,	„	200
In March 1844,	„	„	624,	„	135
In April 1844,	„	„	600,	„	98
In May 1844,	„	„	646,	„	152

To render the view complete, however, as regards the northern stations, a list of the shocks of the Horizontal Force at Lake Athabasca and Fort Simpson is subjoined. The differences are taken in the same manner as those given in disturbances, that is to say, each observation is first referred to the normal mean for that day, and the difference corrected for the mean diurnal change proper to the hour. The latter step has not been taken for the observations of April and May.

Table LII.

A List of Shocks of the Horizontal Force at Lake Athabasca and Fort Simpson, with the difference on each occasion from the mean for the hour, expressed in parts of the Horizontal Force, and the corresponding differences of Inclination; to which are added the differences of the former element from its mean, at Toronto and Sitka, at the same hours. The scale readings at Sitka have been reduced to the temperature of the mean of the month by means of the co-efficient $\frac{q}{k} = 7\cdot0$, *as stated in the "Annuaire Magnetique," &c. 1843, a value which makes* $q = \cdot00038$ *in the English notation.*

* A shock of the Declination at the same hour at one of the stations.

Gött. Date.	Station.				Gött. Date.	Station.			
	Lake Athabasca.		Sitka.	Toronto.		Lake Athabasca.		Sitka.	Toronto.
	Hor. Force.	Inclin.				Hor. Force.	Inclin.		
	October.					October.			
d. h.		′			d. h.		′		
15 21	+·0151	4·4	−·0010	−·0008	27 1*	−·0550	24·9	−·0055	−·0016
15 23	−·0454	+29·8	−·0015	−·0013	30 22	−·0204	+14·7	+·0005	−·0010
16 0*	−·0491	+30·6	−·0012	−·0014	30 23	−·0216	+20·8	+·0008	−·0007
17 1	−·0364	+17·5	−·0020	−·0006		November.			
17 2*	−·0463	+26·2	−·0031	−·0035	1 22	+·0122	+1·5	−·0005	−·0001
17 20*	−·0361	+19·7	+·0006	−·0008	2 18*	−·0170	+1·8	+·0007	−·0015
18 17	−·0206	+20·3	−·0006	−·0009	2 19*	−·0129	+5·9	−·0024	−·0012
19 22	−·0365	+20·3	·0000	−·0002	2 20*	−·0217	+11·4	−·0009	−·0019
24 1*	−·0237	+10·2	+·0008	−·0008	2 21	−·0109	+11·8	−·0008	−·0011
25 19*	−·0231	+9·5	−·0008	−·0006	2 22	−·0139	+6·0	−·0009	−·0010
25 20	−·0242	+12·3	−·0006	−·0004	2 23	−·0392	+22·0	·0009	−·0010
25 21	−·0171	+6·6	−·0008	−·0005	3 0*	−·0345	+16·6	−·0012	−·0008
26 0*	−·0374	+22·5	+·0012	−·0014	5 23*	−·0097	+9·2	−·0001	−·0017
26 20*	−·0286	+6·0	+·0002	−·0002	6 0*	−·0218	+13·4	−·0005	−·0011
26 21	−·0191	6·2	−·0021	·0000	6 4	+·0112	1·2	+·0004	−·0011
26 22*	−·0204	+7·4	−·0010	−·0002	6 5	+·0069	+1·5	+·0005	−·0015
27 0*	−·0265	+13·2	−·0005	−·0004	6 6	−·0083	−0·1	+·0006	−·001

TABLE LII.—*continued.*

Gött. Date.		Station. Lake Athabasca. Hor. Force.	Lake Athabasca. Inclin.	Sitka.	Toronto.
d.	h.				
		NOVEMBER.			
6	9*	+·0062	−0·2	+·0018	−·0008
6	11	+·0083	−0·3	·0000	+·0003
8	22*	−·0148	+10·4	−·0011	−·0003
8	23*	−·0257	+14·2	−·0021	−·0005
9	1	+·0096	+5·8	·0000	−·0001
9	22	−·0176	+10·7	+·0005	−·0002
10	21*	−·0146	+6·9	+·0003	+·0010
10	22*	−·0074	+2·3	−·0016	+·0008
13	4*	−·0333	+21·4	+·0004	−·0033
13	5*	−·0120	+6·2	−·0023	−·0025
13	20	−·0063	+8·7	−·0015	−·0011
15	3	−·0063	+4·8	−·0002	−·0013
24	1*	−·0240	+15·1	+·0005	+·0014
29	1*	−·0068	+4·6	+·0002	·0000
		DECEMBER.			
1	21*	−·0155	+8·2	+·0009	·0000
1	23*	−·0239	+3·4	−·0004	+·0001
2	1*	−·0107	+6·3	−·0013	−·0001
2	2*	−·0279	+16·9	−·0028	−·0011
2	3*	−·0370	+20·5	−·0019	·0000
2	4*	−·0128	+4·8	−·0012	·0010
5	22	−·0239	12·4	·0000	−·0007
5	23	−·0178	+8·9	+·0001	−·0005
6	0	−·0093	+3·6	+·0003	−·0007
6	1	−·0128	+6·9	+·0005	−·0006
6	17	·0107	−3·5	−·0013	−·0011
6	18	+·0120	−4·8	−·0013	+·0003
8	10*	+·0124	−6·8	−·0017	−·0018
8	18*	−·0087	+2·3	−·0002	−·0003
10	21	−·0115	+4·1	·0000	+·0002
19	22*	−·0145	+10·6	+·0007	+·0002
19	23*	−·0181	+11·7	+·0013	+·0004
26	22*	−·0186	+12·1	+·0015	−·0009
26	23*	−·0271	+17·2	+·0005	+·0010
27	0	−·0113	+6·0	+·0001	+·0011
27	19	+·0088	−3·3	−·0007	+·0006
28	3*	−·0188	+9·2	−·0007	+·0100
28	4	·0100	−5·0	−·0007	+·0005
28	5	−·0105	−4·6	−·0008	−·0001
29	22	−·0277	+15·2	−·0008	+·0005
		JANUARY.			
4	16*	+·0192	−10·8	−·0009	−·0002
4	17*	+·0173	−9·0	−·0013	·0000
4	18	+·0141	−6·0	−·0011	−·0004
4	19*	+·0237	−12·7	−·0007	−·0014
4	20*	+·0101	−4·6	−·0005	+·0015
4	21*	−·0300	+26·7	−·0013	+·0013
4	23*	−·0585	+4·0	−·0010	+·0001
5	1*	−·0221	+10·1	−·0016	+·0008
5	23*	−·0626	+29·9	−·0022	−·0004
6	0*	−·0354	+20·3	−·0025	−·0001
6	20*	−·0198	−1·6	−·0003	S.
24	17	−·0196	+18·3	+·0020	−·0025
24	20*	−·0246	+28·8	+·0000	−·0018
24	21*	−·0232	+23·1	+·0003	−·0022
24	22*	−·0191	+17·3	·0010	−·0006
24	23*	−·0258	+24·1	−·0036	−·0005
25	0*	−·0218	+19·5	−·0001	−·0011
25	1*	−·0304	+29·0	−·0024	−·0009
25	2*	−·0183	+15·8	+·0006	−·0006
		FEBRUARY.			
1	0	−·0260	+14·7	−·0021	−·0016
1	1*	−·0450	+24·6	−·0018	−·0015
1	2*	−·0177	+7·6	−·0018	−·0012
1	19*	+·0132	−13·1	−·0024	−·0004
2	6*	−·0255	+13·1	−·0061	−·0028
2	19	+·0189	+8·1	−·0032	+·0019
2	20	+·0202	−5·5	−·0027	−·0004
2	21	+·0125	−2·4	−·0031	−·0011
5	0*	−·0497	+47·9	−·0003	−·0021
5	1	−·0208	+14·5	−·0002	−·0012
5	4	−·0114	−5·2	−·0015	−·0015

TABLE LII.—*continued.*

Gött. Date.		Lake Athabasca. Fort Simpson. Hor. Force.	Lake Athabasca. Fort Simpson. Inclin.	Sitka.	Toronto.
		FEBRUARY.			
d.	h.				
5	16	+·0149	−8′·0	−·0013	−·0004
5	17*	+·0223	−11·3	−·0019	+·0007
5	18	+·0152	−6·1	−·0010	+·0005
5	21	−·0392	+31·2	−·0010	−·0014
5	22	−·0178	+15·6	−·0023	−·0008
8	0	−·0213	+14·2	−·0015	−·0009
8	5*	−·0235	+12·1	−·0017	−·0028
26	23	−·0216	+10·5	−·0003	+·0018
		APRIL. §			
2	23*	−·04	+40·1	−·0067	−·0011
3	4	−·0166	+27·1	−·0029	−·0022
3	5	−·0352	+23·0	−·0031	−·0024
6	17	+·0163	−7·1	−·0013	−·0024
9	23*	−·0166	+15·2	−·0009	−·0005
10	21*	−·0435	+17·1	−·0003	+·0004
15	1*	−·0380	+29·9	−·0006	+·0010
15	2*	−·0361	29·7	−·0005	+·0014
16	14*	+·0280	−20·2	+·0006	−·0009
16	15*	+·0346	−25·5	−·0003	−·0015
16	16*	+·0253	−16·5	−·0009	−·0004
16	18*	−·0173	+9·1	+·0047	−·0019
16	21*	−·0332	+24·2	−·0067	−·0003
16	22*	−·0329	+20·4	+·0021	·0000
16	23*	−·0188	+17·1	−·0038	−·0018
17	0*	−·0272	+21·1	−·0297	−·0119
17	1*	−·0889	+86·6	−·0215	−·0136
17	2*	−·0444	+38·9	−·0145	−·0074
17	3*	−·0367	+33·3	−·0238	−·0068
17	4*	−·0292	+28·6	−·0256	−·0049
17	5*	−·0415	+34·2	−·0085	−·0027
17	6*	−·0358	+23·9	−·0043	−·0008
17	10*	+·0162	+13·0	−·0001	+·0017
17	11*	+·0187	+15·1	+·0004	+·0011
17	12*	+·0222	+15·8	+·0033	·0000
17	13*	+·0252	+19·0	+·0025	+·0007

Gött. Date.		Fort Simpson. Hor. Force.	Fort Simpson. Inclin.	Sitka.	Toronto.
		APRIL.			
d.	h.				
19	23	−·0278	+21′·7	+·0019	−·0066
25	2*	−·0453	+35·3	−·0023	+·0009
25	3*	−·0467	+44·3	+·0068	+·0080
26	0*	−·0638	+59·4	+·0027	−·0022
26	1	−·0449	+35·4	+·0023	−·0023
26	4*	−·0185	+22·1	−·0013	−·0011
26	5	−·0367	+26·4	−·0024	+·0015
26	18*	−·0144	+13·4	+·0021	·0069
26	19*	−·0185	+12·9	+·0019	−·0010
27	0	−·0623	+20·4	+·0002	−·0008
27	20*	−·0229	+15·9	+·0008	S.
29	1	−·0176	+14·3	+·0006	+·0006
29	2*	−·0312	+29·3	−·0001	−·0002
29	3*	−·0209	+18·7	−·0002	+·0006
29	4*	−·0198	+15·8	−·0007	+·0004
30	0	−·0308	+22·7	+·0024	−·0002
30	1	−·0221	+13·7	+·0026	−·0002
30	14	−·0212	+14·2	+·0037	−·0001
30	15	+·0208	+14·1	+·0008	+·0006
30	21*	−·0468	+38·6	−·0019	+·0007
30	22*	−·0377	+29·1	+·0009	+·0007
30	23	−·0175	+13·2	−·0040	+·0003
		MAY.			
3	4*	−·0307	+49·2	+·0008	−·0019
7	22*	−·0257	+20·8	−·0010	+·0081
9	3	−·0245	+26·8	−·0009	−·0006
9	4	−·0274	+24·7	+·0007	−·0001
13	18	+·0188	−16·7	+·0005	−·0017
22	1*	−·0319	+31·0	·0000	−·0004
22	2*	−·0369	+33·7	+·0002	+·0008
22	14*	+·0231	−18·9	+·0002	−·0032
22	23*	−·0290	+36·3	+·0013	−·0022
23	0*	−·0363	+33·2	+·0020	−·0014
23	1*	−·0230	+22·5	+·0032	−·0015
24	2*	−·0235	+24·1	−·0008	+·0001

§ The values at Fort Simpson in April and May are referred to the daily and not the hourly mean.

It does not appear by the foregoing Table that the great and sudden changes to which the Inclination at Lake Athabasca and Fort Simpson was liable, extended in general even to Sitka, the nearest station. These changes were always accompanied by a proportionate change of the horizontal component, and, being generally positive, occasioned those great reductions of the last-named element which are contained in the list of shocks, and form the most conspicuous feature of almost every disturbance. We find that in 105 out of the 170 observations here given, the value of the element had opposite signs at one or more of the stations at the same hour; of these, 74 are observations at Sitka. Of the instances in which the element appears to have been similarly affected, April 16th and 17th are the most decided, and the great reduction appears to have occurred earlier at Sitka on this occasion than at Fort Simpson. There are 106 of these dates at which the deviation of the Declination from its mean, at one or more of the stations, amounted to a shock, and 64 at which the Horizontal Force alone exhibits the effect.

TERM DAYS AND MAGNETIC DISTURBANCES.

Introduction.—The regular magnetical Term Days were observed at Lake Athabasca and Fort Simpson, and, in addition, extra observations were taken whenever the irregularity of the scale readings seemed to call for it, as well as upon some occasions when, from the presence of aurora or from some other cause, it was deemed advisable to commence them in anticipation of magnetic disturbance. During October, and for the first half of November 1843, the customary intervals of five minutes were adhered to; the Declinometer was read at 0^m, 5^m, 10^m, &c. Gött. time, the Bifilar at 2^m, 7^m, 12^m, &c., and the Inclinometer at 3^m, 8^m, 13^m, &c., after the hour; the great rapidity of the changes led afterwards to a general practice of reading the instruments one after the other in recurring succession, with an interval of only one minute between them; thus each element was observed every third minute. Upon Term Days, and a few other occasions, the five minute intervals were retained. The minute entered is, in every case, that of the Declinometer reading.

In computing the simultaneous variations of the three elements of Declination, Inclination, and Total Force, which are given for each observation, as well as those of the horizontal component, the mean diurnal curves given by the whole period of observation were employed, in preference to those of the individual months, for eliminating the regular changes, the latter curves being more or less irregular from the effects of disturbances. These curves correspond in epoch to the middle of the period, or about the 23d December; but since the scale readings of each of the instruments show some progressive change, and those of the Bifilar and Inclinometer a considerable one; the numerical values of the hourly means, by these curves, are not correct for the observations at the beginning or end of the series, and are only correct approximately near the middle of it. In order, then, to deduce true differences, each day of extra observation has been taken as the centre of a group of five, six, or seven days, according to circumstances, and the means of all the readings of that group taken as the simultaneous or co-ordinate means of the three elements for the disturbance observations on the day in question. The difference between each scale reading and the mean thus found for the same element is the whole deviation of the element from its normal value; to this being applied, with the contrary sign, the value of the ordinate of the mean curve for the same hour and minute, the sum is the measure of the irregular fluctuation shown by the observation, which, in the case of the Bifilar and Inclinometer, is expressed in terms of the Horizontal Force and Inclination, by means of the co-efficients given in preceding sections, and stated at the foot of each page. The scale divisions of the Declinometer, being minutes of arc, need no conversion.

The actual mode of proceeding was this; a table was formed for each element and for each month, containing twelve columns and twenty-four lines; in the first column were written the hourly means by the whole number of observations ± the difference between their mean and the mean scale reading for the month; in the succeeding columns the hourly means were repeated + a proportional part of the change of the element from one hour to the next. By taking the difference between the actual readings and the appropriate mean thus found, we have the variation of the element from its normal value, affected by whatever difference there may be between the mean for the month and the true mean for the day of observation, which was sometimes considerable, owing to the progressive changes of scale reading before alluded to. To eliminate this effect a constant correction was next applied to all the readings at each disturbance, being the difference between the mean of the month and the normal mean found as above related.

It must be remarked that the curves, which we have taken as types from 110 days of observation, are still sensibly affected by the irregular observations contained in that period, and consequently can be regarded only as approximations to that character; the degree to which they are affected may perhaps be judged of in some measure by comparing them with the curves given by 46 days selected as free from sensible disturbance, and is shown by the following Table:

Table LIII.

[Difference ($x_{,}-x$) between the hourly ordinates (x) given by the whole period of observation at Lake Athabasca, and the ordinates ($x_{,}$) given by the curves of 46 days selected as free from disturbance.]

Gött. time	0	1	2	3	4	5	6	7
Mean time	16	17	18	19	20	21	22	23
Declination. Whole period	5·83	8·23	5·84	5·56	3·90	3·76	1·33	−2·98
Declination. Selected days	3·15	3·81	3·55	3·81	3·60	3·64	2·35	−1·11
Declination. $x_{,}-x$	−2·68	−4·42	−2·29	−1·75	−0·30	−0·12	+1·02	+1·87
Bifilar. Whole period	−·00522	−·00424	−·00183	−·00083	−·00010	−·00019	−·00042	−·00019
Bifilar. Selected days	−·00117	−·00112	−·00099	−·00091	−·00066	−·00082	−·00131	−·00124
Bifilar. $x'_{,}-x'$	+·00405	+·00312	+·00084	+·00008	−·00056	−·00063	−·00089	−·00105
Inclinometer. Whole period	+3·29	+2·08	+0·97	−0·06	−0·41	−0·29	+0·63	+0·05
Inclinometer. Selected days	0·77	0·54	0·25	0·05	−0·04	−0·06	0·68	0·67
Inclinometer. $x''_{,}-x''$	−2·52	−1·54	−0·72	+0·11	+0·37	+0·23	+0·65	+0·62

Gött. time	8	9	10	11	12	13	14	15
Mean time	Noon	1	2	3	4	5	6	7
Declination. Whole period	−3·93	−4·89	−5·21	−4·19	−3·23	−2·88	−1·80	−1·32
Declination. Selected days	−3·08	−4·55	−3·95	−3·32	−2·37	−1·86	−1·46	−0·88
Declination. $x_{,}-x$	−0·85	+0·34	+1·26	+0·87	+0·86	+1·02	+0·34	+0·44
Bifilar. Whole period	+·00039	+·00049	+·00163	+·00202	+·00218	+·00254	+·00252	+·00242
Bifilar. Selected days	−·00095	−·00060	+·00033	+·00067	+·00122	+·00116	+·00123	+·00091
Bifilar. $x'_{,}-x'$	+·00056	+·00109	−·00130	−·00135	−·00096	−·00138	−·00129	−·00151
Inclinometer. Whole period	−0·05	−0·30	−0·46	−0·80	−0·98	−1·27	−1·41	−1·44
Inclinometer. Selected days	0·59	0·30	0·10	−0·33	−0·22	−0·63	−0·84	−0·61
Inclinometer. $x''_{,}-x''$	+0·64	+0·60	+0·56	+0·47	+0·76	+0·64	+0·57	+0·83

TABLE LIII.—*continued.*

Gött. time -	16	17	18	19	20	21	22	23
Mean time -	8	9	10	11	12	13	14	15
Declination. Whole period -	′ −1·18	′ −0·92	′ −0·23	′ −0·03	′ +1·86	′ +1·54	′ −0·90	′ +1·61
Declination. Selected days -	−0·83	−0·78	+0·49	+0·15	−0·91	−0·08	−0·30	+1·16
Declination. $x_{,}-x$	′ +0·35	′ +0·14	′ +0·72	′ +0·18	′ −2·77	′ −1·62	′ −1·20	′ −0·45
Bifilar. Whole period -	+·00243	+·00219	+·00276	+·00226	+·00022	−·00206	−·00350	−·00513
Bifilar. Selected days -	+·00102	+·00106	+·00176	+·00197	+·00121	−·00038	−·00116	−·00137
Bifilar. $x'_{,}-x'$	−·00141	−·00113	−·00100	−·00029	+·00143	+·00168	+·00234	+·00376
Inclinometer. Whole period -	′ −1·47	′ −1·51	′ −1·69	′ −1·35	′ +0·06	′ +1·60	′ +2·23	′ +3·36
Inclinometer. Selected days -	−0·65	−0·95	−0·89	−0·46	−0·18	+0·36	+0·57	−0·96
Inclinometer. $x''_{,}-x''$	′ +0·82	′ +0·56	′ +0·80	′ +0·89	′ −0·24	′ −1·24	′ −1·65	′ −2·40

The differences of the Declination are in some instances referable to the mean for the same month, and not the mean for the same day; the correction applicable on this account is stated where such is the case, but rarely amounts to 1′.

It appears that at the hours most influenced by disturbances, namely, from about 9ʰ in the evening to 5ʰ in the morning, the simultaneous fluctuations of the three elements are generally less than they would be if referred to diurnal means uninfluenced by disturbances; but since the means actually employed are derived from corresponding observations, the error thus introduced enters proportionably into all of them, and leaves the values actually presented true relatively to one another.

The simultaneous fluctuations at Fort Simpson are measured from the true mean of each element for the day of observation, and include the regular diurnal changes, the period of observation having been so short, and attended by such constant magnetic disturbances that the mean curves deduced can scarcely be regarded as a true representation of the action of those forces which it is the object of the other mode of treatment to eliminate. It was found convenient also to refer the observations in October to the mean for the day, in consequence of the disconnexion of the Inclinometer readings from those of succeeding months.

The approximate changes of Total Force $\frac{\Delta \phi}{\phi}$, have been computed by the formula $\frac{\Delta \phi}{\phi} = \frac{\Delta X}{X} + \tan \theta \Delta \theta$: their accuracy necessarily depends upon that of the scale coefficients of the Bifilar and Induction Inclinometer; the former of these may be considered as sufficiently well determined at Fort Simpson to entitle the values assigned to the changes of the horizontal component to considerable confidence; at Lake Athabasca, where a less delicate method of adjustment was employed, the scale coefficient of the Bifilar may possibly be itself in error in the fifth decimal. The scale coefficient of the Induction Inclinometer at both stations is liable to whatever uncertainty attaches to the correction we have applied in a previous section for the defect of the method formerly employed of determining the coefficient p, in the formula

$$\frac{\Delta Y}{Y} = p \left(\cos u \Delta u + \sin u \frac{\Delta X}{X} \right)$$

by simply inverting the iron bar, and observing the angles of deflection in the direct and inverted position, when it was assumed that $p = \frac{2}{\sin u + \sin u'}$. This method, as was there stated, appears to give a value invariably too low, to the deduced scale coefficient, and from a number of experiments the ratio 1·22 was adopted in which to augment it under all adjustments, and has been applied. It is evident, therefore, that in the face of these admissions, great precision cannot be claimed for the values assigned for the changes of Total Force; but, having carefully considered the subject and examined the results themselves, it appeared to me that, with due explanation, they were entitled to be included in this report. Very little appears to have been determined hitherto with respect to the irregular variations of the Total Force in any quarter; the present observations show at the least that this element does nevertheless undergo in high latitudes very considerable changes; and believing their value also to be here assigned with sufficient accuracy for many purposes, I offer them, as approximations only, but approximations deriving peculiar interest from the remote locality and high magnetic latitude in which the observations were made.

The following Table contains the values of $\tan \theta \Delta \theta$ for each 1′·0 of $\Delta \theta$, at Lake Athabasca, and Fort Simpson.

Table LIV.

Values of tan θ Δ θ for calculating changes of Total Force.

Δθ	Lake Athabas. θ=81° 37′·6	Fort Simpson. θ=81° 52′·3	Δθ	Lake Athabas. θ=81° 37′·6	Fort Simpson. θ=81° 52′·3	Δθ	Lake Athabas. θ=81° 37′·6	Fort Simpson. θ=81° 52′·3	Δθ	Lake Athabas. θ=81° 37′·6	Fort Simpson. θ=81° 52′·3
′			′			′			′		
1	·00197	·00204	26	·05135	·05296	51	·10072	·10389	76	·15010	·15481
2	·00395	·00407	27	·05332	·05500	52	·10270	·10592	77	·15207	·15685
3	·00592	·00611	28	·05530	·05704	53	·10467	·10796	78	·15405	·15889
4	·00790	·00815	29	·05727	·05907	54	·10665	·11000	79	·15602	·16092
5	·00987	·01018	30	·05925	·06111	55	·10863	·11203	80	·15800	·16295
6	·01185	·01222	31	·06122	·06315	56	·11060	·11407	81	·15997	·16500
7	·01382	·01426	32	·06320	·06518	57	·11257	·11611	82	·16195	·16703
8	·01580	·01630	33	·06517	·06722	58	·11455	·11815	83	·16392	·16907
9	·01777	·01833	34	·06715	·06926	59	·11653	·12018	84	·16590	·17111
10	·01975	·02037	35	·06912	·07129	60	·11850	·12222	85	·16787	·17314
11	·02172	·02241	36	·07110	·07333	61	·12047	·12426	86	·16985	·17518
12	·02370	·02444	37	·07307	·07537	62	·12245	·12629	87	·17182	·17722
13	·02567	·02648	38	·07505	·07741	63	·12442	·12833	88	·17380	·17926
14	·02765	·02852	39	·07702	·07944	64	·12640	·13037	89	·17577	·18129
15	·02962	·03055	40	·07900	·08148	65	·12837	·13240	90	·17775	·18333
16	·03160	·03259	41	·08097	·08352	66	·13035	·13444	91	·17972	·18537
17	·03357	·03463	42	·08295	·08555	67	·13232	·13648	92	·18170	·18740
18	·03555	·03667	43	·08492	·08759	68	·13430	·13852	93	·18367	·18944
19	·03752	·03870	44	·08690	·08963	69	·13627	·14055	94	·18565	·19148
20	·03950	·04074	45	·08887	·09166	70	·13825	·14259	95	·18762	·19351
21	·04147	·04278	46	·09085	·09370	71	·14022	·14463	96	·18960	·19555
22	·04345	·04481	47	·09282	·09574	72	·14222	·14666	97	·19157	·19759
23	·04542	·04685	48	·09480	·09778	73	·14417	·14870	98	·19355	·19963
24	·04740	·04889	49	·09677	·09981	74	·14615	·15074	99	·19552	·20166
25	·04937	·05093	50	·09875	·10185	75	·14812	·15278	100	·19750	·20370

METEOROLOGICAL OBSERVATIONS.

METEOROLOGICAL OBSERVATIONS.

THE meteorological observations at Lake Athabasca and Fort Simpson are confined to a register of the temperature of the air, and of the wind and weather; particular attention being paid to the frequent displays of the aurora borealis. Two portable barometers and two thermometers for hygrometric purposes, had formed part of the equipment of the expedition, but were unfortunately rendered unserviceable in the course of the previous journey.

The instrument referred to from the outset as the standard thermometer, was a spirit thermometer by Newman, one of several which had been sent into the Hudson's Bay territory sometime previously by the Royal Geographical Society, and was recommended by that circumstance, as well as by the character of its maker. The tube, however, was so far from being of uniform capacity, that its graduation proved to be many degrees in error at very low temperatures; fortunately, there was on the spot, in the possession of Mr. Colin Campbell, another spirit thermometer, by Dolland, an old instrument, and supposed to be the same which was registered by Mr. Keith in 1825-6*; this proved much the more accurate of the two, and was registered in addition to the other, from the 7th January to the 29th February 1844. These instruments stood as follows in melting snow:—

N (Newman) 33°·5 correction −1°·5
D (Dollond) 31°·0 correction + 1°·0

It was soon observed that the difference here shown between them increased regularly in descending the scale, and much uncertainty was felt as to which was the preferable authority, until an opportunity occurred of testing the thermometer Newman in freezing mercury, which was done as follows:—On the 23d January a portion of mercury was exposed in the open air until it became solid, at the same time another portion was allowed to acquire a temperature very little removed from the freezing point, the solid mass being then added to the fluid, the bulb and about two inches of the stem of the instrument to be tested were immersed in the mixture. The experiment was made in a room having a temperature of 35° Fahrenheit, and the spirit of the thermometer remained steadily at −31°·0 on the scale as long as any of the frozen mercury remained solid. Assuming, then, the solidifying point of mercury to be −40°·9 † Fahrenheit, we have the very large correction of + 9°·9, applicable

* Franklin's second Journey to the Shores of the Polar Sea, Appendix II.

† In adopting this datum instead of the more usual value −39°·5, I am guided by verbal information from M. Regneault, that he has found by accurate experiments with the air thermometer, that the true freezing point of pure mercury is between −40° and −41° centigrade; the mean of these two values is equivalent to −40°·9 Fahrenheit. The mercury employed in the experiments at Lake Athabasca had been purified with nitric acid some time previously, but was dull from exposure and shaking in the constant use it had been put to in the artificial horizon.

to the scale reading, $-31°·0$ on the thermometer N. It appears by the following comparisons that the thermometer D. read $-37°·8$ when N. read $-31°·0$, consequently the error of this instrument at the freezing point of mercury was $+3°·1$, or the correction at the scale reading $-37°·8$ was $-3°·1$.

Comparisons of Dollond's and Newman's thermometers at low temperatures:

Newman Reading.	Number of Observations.	Mean.		Difference.	
		Newman.	Dollond.		
° °		°	°	°	
−30 to −31	18	−30·40	−37·05	−6·65	−6·81
−31 to −32	12	−31·13	−37·89	−6·67	
−32 to −33	7	−32·11	−39·14	−7·03	
−33 to −34	11	−33·34	−40·53	−7·21	
−34 to −35	4	−34·12	−41·77	−7·65	
−35 to −36	3	−35·33	−43·10	−7·77	
−36 to −37	6	−36·20	−44·22	−8·02	
−37 to −38	2	−37·20	−44·70	−7·50	
−38 to −39	1	−38·60	−46·10	−7·50	

These results are confirmed by a number of entries in the meteorological register on the 22d, 23d, and 24th January 1844. A small quantity of mercury being at this time exposed beside the thermometers, it was found frozen at all readings below $-32°·9$ of Newman and $-39°·5$ of Dollond, with one exception as regards the latter instrument, namely, on January 23^d 6^h Göttingen, when it is noted that the mercury was thawing, N. reading $-32°·4$, D. $-39°·6$. There are nineteen hours of observation on the above days at which the fact of the mercury being solid was noted, but it appears from the register that the reading was below $-31°$ by Newman, and therefore the temperature below the freezing point of mercury at forty-six hours of observation in that month; it did not reach it in December or February. Newman's thermometer is entered in the abstracts corrected by subtracting $1°·5$ from each scale reading above $33°·5$, and $1°·5+0·130\,\Delta\,t°$ for each reading below $33°·5$, where $\Delta\,t°=(33°·5-$observed temperature). The coefficient 0·130 is the increase of the correction, which appeared to be uniform for each degree in descending the scale, as shown by the foregoing data. The mean temperature for each day subsequent to January 7th by hourly observations of Dollond has also been corrected independently by subtracting $1°·0$ at each reading above $33°·0$, and $1°·0-0·0596\,\Delta t°$ at each reading below $33°$. The non-agreement of the corrected means of the two thermometers, excepting near the two fixed points on the scale, probably shows that the supposition of a uniform rate of increase of the correction is not in accordance with the fact, for there was nothing in the position of the two instruments to account for one standing permanently about one degree higher than the other, but as neither thermometer was compared with any absolute standard, we appear to have no better resource than to take the mean where both were observed, which has accordingly been done in Table VI.

The position of the thermometer was on the north side of an external porch made to contain the transit instrument; they were attached to a bracket projecting a few inches from the wall, their bulbs about four feet above the soil; the readings were taken through the transit openings.

Table LV.

Mean Temperature of the Air at Lake Athabasca by Newman's thermometer corrected, also mean temperature for the winter quarter, comprising the months of December 1843, *January and February* 1844.

1843–44.	Midn.	1 A.M.	2.	3.	4.	5.	6.	7.
October 16th to 31st	20·08°	18·74°	18·35°	18·10°	18·78°	17·79°	18·37°	19·16°
November - -	9·02	7·89	7·57	7·15	7·58	7·61	7·69	8·29
December - -	−1·16	−0·06	−0·19	−0·14	0·68	0·57	0·58	0·42
January - -	−22·82	−23·02	−23·56	−24·23	−24·45	−25·10	−25·10	−25·55
February - -	3·00	1·44	1·67	1·45	−0·08	−0·03	−0·48	−0·46
Winter Quarter -	−7·00	−7·21	−7·36	−7·64	−7·95	−8·19	−8·33	−8·53

1843–44.	8.	9.	10.	11 A.M.	Noon.	1 P.M.	2.	3.
October 16th to 31st	19·74°	21·18°	22·39°	23·26°	24·69°	25·35°	25·75°	25·52°
November - -	8·48	9·29	10·26	11·31	12·39	12·64	12·82	12·44
December - -	0·46	0·10	1·04	1·77	2·76	2·73	2·19	1·44
January - -	−25·81	−25·30	−23·77	−22·25	−21·52	−20·81	−20·93	−21·11
February - -	0·09	1·54	3·99	6·11	8·65	10·26	10·33	10·76
Winter Quarter -	−8·42	−7·80	−6·58	−4·79	−3·37	−2·61	−2·80	−2·97

1843–44.	4.	5.	6.	7.	8.	9.	10.	11. P.M.	Mean.
October 16th to 31st	24·55°	23·36°	22·37°	22·21°	21·76°	20·96°	21·13°	20·83°	21·44
November - -	11·60	11·04	10·65	10·15	9·93	9·60	9·39	8·95	9·76
December - -	1·02	0·30	−0·32	−0·63	−0·90	−1·09	−0·91	−1·16	0·40
January - -	−21·46	−21·69	−22·53	−21·61	−23·09	−22·30	−22·47	−22·51	−23·00*
February - -	9·41	8·74	8·33	7·42	7·14	6·48	4·98	3·94	4·79*
Winter Quarter -	−3·68	−4·22	−4·86	−4·94	−5·28	−5·64	−6·13	−6·58	−5·94

* See Table LX for the mean by both thermometers.

Mr. C. Campbell, the resident officer of the Hudson's Bay Company, having kindly continued to record the temperature by Dollond's thermometer four times a day after my departure, until he left the station himself, we are enabled to add four months to the foregoing table:

TABLE LVI.

Mean Temperature at Lake Athabasca—continued.

Month.	Sunrise.	9 A.M.	3 P.M.	9 P.M.	Approx. Mean.	Corrected Mean.
1844:	°	°	°	°	°	°
March * -	−2·42	1·13 †	9·13	2·16	1·64	2·4
April - -	27·82	35·48	42·50	33·96	34·72	35·1
May - -	39·14	46·74	50·31	42·07	44·40	44·8
June - -	47 20	57·17	58·83	50·86	53·52	53·9
Spring Quarter -	- -	28·91	35·03	27·06	26·90	27·4

* The first four days of March are wanting.
† 1°·02 by the 31 days, which are complete for this hour only.

The thermometer recorded by Mr. Campbell has been corrected for temperatures above 32°, by the uniform addition of 1°, and for lower temperatures by the scale already given. The approximate mean is that of the observations at 9 A.M. and 9 P.M., which in each of the other months at this station and at Fort Simpson is below the true mean; probably the correction +0°·8, derived from December, January, and February, will be nearly correct for March, and +0°·4, derived from April and May at Fort Simpson, nearly correct for April, May, and June at Lake Athabasca, giving the quantities in the last column.

TABLE LVII.

Mean Temperature of the Air at Fort Simpson, M'Kenzie's River, for April and May 1844.

—	Midn.	1 A.M.	2.	3.	4.	5.	6.	7.	
April . .	° 27·24	° 26·15	° 24·87	° 20·63	° 21·37	° 22·37	° 23·92	° 25·57	
May * .	36·37	36·96	35·36	34·54	35·18	37·91	37·68	39·65	
Mean of 46 days	32·32	31·14	29·88	27·86	28·45	29·24	30·20	32·00	
	8.	9.	10.	11.	Noon.	1 P.M.	2.	3.	
April . .	° 28·22	° 32·41	° 35·85	° 39·30	° 40·69	° 41·94	° 42·14	° 41·46	
May * .	42·07	44·91	48·07	51·44	51·49	53·55	52·92	52·78	
Mean of 46 days	34·54	38·12	41·43	44·84	45·62	47·57	47·06	46·63	
—	4.	5.	6.	7.	8.	9.	10.	11.	Mean.
April . .	° 40·83	° 40·00	° 38·72	° 36·40	° 33·31	° 31·46	° 29·96	° 29·02	° 32·46
May * .	52·61	51·67	50·43	49·16	46·70	43·62	41·21	39·50	44·56 †
Mean of 46 days	46·23	45·33	44·07	42·23	39·48	37·02	35·11	33·81	37·92

* From 1st to 25th May.

† Corrected to the mean of the complete month, at the same average daily increment of mean temperature (0°·30) the mean for May will be 45°·5.

The observations at Fort Simpson were taken 20^{m} after the hours named.

In the observations of Mr. Keith at Lake Athabasca in 1825–6, referred to above, the mean temperature is derived from the mean of the daily extremes; for the sake of comparison, a similar value has been formed for each month in the foregoing tables, and is subjoined, together with other approximations to mean values.

Table LVIII.

Various approximations to the Mean Temperature.

Month.	True Mean by Hourly Observations.	Mean by Daily Extremes.	6 A.M. 6 P.M.	7 A.M. 7 P.M.	8 A.M. 8 P.M.	9 A.M. 9 P.M.	10 A.M. 10 P.M.	11 A.M. 11 P.M.	6 A.M. 2 P.M. 10 P.M.	7 A.M. 3 P.M. 11 P.M.
October -	21·44	21·11	20·37	20·68	20·75	21·07	21·76	22·04	21·75	21·84
November -	9·76	9·51	9·17	9·22	9·21	9·44	9·82	10·13	9·97	9·89
December -	0·40	0·57	0·13	−0·10	−0·22	−0·44	0·06	0·30	0·62	0·23
January -	−23·00	−23·64	−23·84	−23·58	−23·95	−23·80	−23·12	−22·38	−22·83	−23·06
February -	4·79	3·60	3·92	3·48	3·52	4·01	4·48	5·02	4·84	5·05
Winter Quarter - -	−5·94	−6·53	−6·59	−6·73	−6·85	−6·76	−6·35	−5·68	−5·75	−6·08
April - -	32·48	30·90	31·32	30·98	30·76	31·94	32·91	34·16	32·01	32·02
May - -	44·56	42·85	44·05	44·40	44·38	44·26	44·64	45·47	43·94	43·97

It appears, by the foregoing Table, that the best approximation to a true mean, from October to February, is obtained by three equidistant observations, beginning with 6^h or 7^h A.M.; the mean by the daily extremes, that is to say, the highest and lowest hourly observations, is also a good approximation in October, November, and December, but considerably *too low* in the subsequent months; the mean by the homonymous hours from 6^h to 9^h inclusive, is decidedly too low, that of the succeeding homonymous hours 10^h and 11^h is, however, somewhat better. The same remark applies to the months of April and May; the differences from the true mean apparent in the latter are, however, considerably longer than those shown in the previous months, the mean diurnal curves of temperature having themselves a marked difference arising from the change of season.

The mean diurnal curve of temperature for the winter quarter at Lake Athabasca differs but little from that of the corresponding season at Toronto. The mean range is 5°·92 at the former, and 5°·95 at the latter station; the coldest hour is the same, 7^h A.M., and the curve cuts the line of mean temperature in its morning ascent at pretty nearly the same time. By simple interpolation this epoch will be 9^h 53^m at Toronto, and 10^h 21^m at Lake Athabasca. There is a slight difference in the descending branch, which is prolonged above the mean to the latest hour at the more northern station, giving a temperature above the mean at eleven observation hours at Lake Athabasca, and at ten only at Toronto. The more rapid relative increase in the power of the sun with the advance of spring at the more northern station, is evinced by the large amount of the mean daily range at Fort Simpson in April and May, namely,

19°·71, while for the corresponding months at Toronto it is but 15°·76.

Table LIX.

Mean Temperature for December 1843, *January and February* 1844, *at Toronto, also for April and May* 1844, *for comparison of diurnal curves with those given.*

Mean Time -	Midn.	1 A.M.	2.	3.	4.	5.	6.	7.	
Winter Quarter -	25°·2	24°·6	24°·5	24°·3	24°·2	23°·8	23°·5	23°·3	
April—May -	45·9	45·2	44·6	43·8	43·4	43·3	44·0	46·5	
Mean Time -	8.	9.	10.	11.	Noon.	1 P.M.	2.	3.	
Winter Quarter -	23°·8	24°·9	26°·1	27°·5	28°·5	29°·1	29°·5	29°·5	
April—May -	48·9	51·3	53·2	54·9	56·5	58·4	58·8	59·0	
Mean Time	4.	5.	6.	7.	8.	9.	10.	11.	Mean.
Winter Quarter -	28°·9	28°·0	26°·9	26°·3	25°·7	25°·3	24°·9	24°·5	25°·96
April—May -	58·5	58·7	56·4	53·0	50·0	48·5	47·4	46·8	50·73

TABLE LX.—*Highest, Lowest, and Mean Temperature observed on each day at Lake Athabasca and Fort Simpson, during the series of hourly observations. See the columns headed sunrise and 3 P.M. in the Abstracts following, for the Highest and Lowest Temperature from March to June, at the former station, by Mr. Campbell's observations.*

	LAKE ATHABASCA.																FORT SIMPSON.					
	October 1843.			November 1843.			December 1843.			January 1844.			February 1844.				April 1844.			May 1844.		
Day.	High.	Low.	Mean.	High.	Low.	Mean.	High.	Low.	Mean.	High.	Low.	Mean.	High.	Low.	Mean.		High.	Low.	Mean.	High.	Low.	Mean.
1	° —	° —	° —	29°·4	14°·6	21°·86	13°·4	−1°·2	7°·41	° —	° —	° —	−5°·8	−18°·3	−12°·10		25°·0	−2°·8	12°·02	27°·0	12°·6	20°·48
2	—	—	—	32·7	20·5	25·13	23·6	−8·1 }	4·10	−0·6	−17·1	−6·65	−0·1	−9·3	−3·05		32·5	−3·3	14·78	38·0	14·0	38·87
3	—	—	—	30·6	23·8	26·95	—	— }		−7·3	−25·0	−14·65	—	−13·7 }	−5·31		40·3	0·4	24·89	43·3	21·5	34·51
4	—	—	—	29·3	— }	25·25	12·2	2·3	7·52	−5·3	−7·8	−6·45	7·8	— }			46·5	13·1 }	32·07	51·8	29·5 }	40·60
5	—	—	—	—	16·6 }		15·3	0·6	4·71	−6·5	−38·7	−19·56	7·4	−21·7	−7·27		—	— }		—	— }	
6	—	—	—	24·7	13·4	17·83	35·3	6·3	21·34	—	−39·0 }	−36·34	7·6	−23·4	−7·83		17·1	3·6 }	11·87	43·5	28·4	35·46
7	—	—	—	19·1	14·6	16·47	16·7	2·0	7·67	−32·0	— }		9·7	3·1	6·47		—	— }		53·3	28·1	41·51
8	—	—	—	20·2	7·2	14·31	31·4	14·7	21·88	−26·2	−40·1	−30·46*	21·1	6·2	13·46		45·0	9·1	31·58	57·5	34·3	45·38
9	—	—	—	10·0	−1·2	5·67	22·6	— }	10·77	−32·8	−39·8	−36·01	14·8	−2·4	6·80		45·1	14·9	36·60	56·5	35·3	46·32
10	—	—	—	9·6	−1·1	3·99	—	−3·5 }		−18·4	−32·0	−23·80	10·9	— }	−0·25		22·8	6·0	16·74	56·5	36·7	49·00
11	—	—	—	7·3	— }	4·31	7·8	−3·2	0·01	−0·6	−26·8	−8·09	—	−32·1 }			29·4	0·5	18·53	58·4	28·8 }	46·56
12	—	—	—	—	6·8 }		6·4	−12·8	−1·06	−1·4	−17·6	−5·50	−10·8	−30·8	−17·46		38·2	14·0	29·16	—	— }	
13	—	—	—	8·0	−6·8	1·64	−12·6	−24·9	−19·51	−17·1	−30·9 }	−22·35	−2·2	−25·9	−13·10		38·6	20·4 }	30·94	45·5	27·2	37·75
14	—	—	—	19·0	9·1	15·08	0·0	−23·9	8·94	—	— }		7·7	−15·0	−6·31		—	— }		55·3	31·0	45·01
15	—	—	—	13·6	3·4	8·67	0·7	−11·0	−4·22	−11·0	−20·9	−3·41	32·5	4·9	15·97		43·5	19·3	32·74	63·7	34·6	52·12

Table LX.—*continued.*

	LAKE ATHABASCA.															FORT SIMPSON.					
	October 1843.			November 1843.			December 1843.			January 1844.			February 1844.			April 1844.			May 1844.		
Day.	High.	Low.	Mean.	High.	Low.	Mean.	High.	Low.	Mean.	High.	Low.	Mean.	High.	Low.	Mean.	High.	Low.	Mean.	High.	Low.	Mean.
16	44°·5	29°·6	36°·30	17°·8	8°·0	12°·07	° —	−9°·5	° −6·44	−5°·8	−18°·3	−10°·71	34°·4	4°·9	22°·56	51°·5	27°·1	38°·28	72°·5	42°·0	59°·12
17	33·5	27·0	31·87	10·4	4·8	8·33	−3·2	—		−18·4	−30·6	−23·37	37·5	7·5	26·46	50·0	27·5	39·78	71·0	46·4	58·57
18	38·0	30·0	33·08	10·9	3·1	7·25	7·8	−6·5	0·71	−30·5	−37·6	−33·86	—	—		53·1	32·4	42·37	51·0	32·5	44·21
19	34·0	29·0	30·74	—	—		0·3	−11·0	−7·64	−34·0	−39·7	−37·00	21·3	3·0	11·80	56·1	31·3	43·54	—	—	
20	33·3	28·1	30·81	8·4	4·4	6·86	6·8	−11·9	−4·87	−30·4	—	−35·33	32·7	−5·6	15·21	45·7	31·4	38·46	66·1	37·3	55·17
21	49·9	—	36·86	9·3	5·5	6·83	12·4	−11·6	0·31	—	−42·0		22·2	−9·7	11·61	—	—		52·5	28·2	44·65
22	—	14·8		10·0	−3·6	5·68	−2·2	−13·4	−9·25	−35·3	−49·5	−41·42	31·3	11·6	23·94	49·8	28·2	40·16	50·5	29·8	40·36
23	18·0	14·2	15·88	9·0	−5·7	4·40	5·4	−4·9	0·97	−34·5	−44·7	−39·86	32·5	0·4	21·17	37·4	27·1	32·82	62·5	36·5	51·61
24	20·2	12·5	17·27	5·4	2·3	3·69	—	—		−37·5	−46·6	−41·70	3·3	—	−2·07	38·3	25·9	32·19	64·4	44·2	55·66
25	23·2	0·1	15·00	4·1	−0·6	1·92	—	—		−26·6	−47·7	−37·37	—	−9·2		68·0	33·5	53·41	—	—	—
26	6·6	−7·6	−0·65	—	—		−0·2	−6·7	−3·19	−23·9	−38·7	−28·04	34·5	−10·3	17·30	50·0	36·2	42·07	—	—	—
27	9·2	−1·1	4·17	9·9	2·0	5·91	16·8	−5·6	7·24	−10·0	−41·3	−25·40	3·3	−8·0	−0·68	50·1	28·4	40·26	—	—	—
28	—	2·4	6·50	12·3	−2·7	2·38	17·7	−5·9	8·32	—	—		15·7	−2·5	7·43	—	—		—	—	—
29	22·5	—		20·3	−9·4	7·18	−1·4	−11·9	−6·20	14·6	−20·3	−4·10	14·3	−6·9	5·78	56·5	29·3	44·51	—	—	—
30	24·3	19·4	23·45	0·3	−5·8	−2·27	—	−35·3	−21·35	14·6	−19·5	−0·28	—	—	—	51·3	12·9	35·34	—	—	—
31	25·7	14·6	20·26	—	—	—	−6·9	—		−12·8	−22·9	−18·37	—	—	—	—	—	—	—	—	—
	27·35	14·87	21·47	14·68	4·22	9·76	9·02	−7·88	0·40	−15·16	−32·12	−22·74	15·34	−8·13	5·22	—	—	32·46	—	—	44·56

* Mean of both thermometers, corrected, from 7th January to 29th February.

The foregoing Tables show a remarkable prevalence of cold in January 1844 at Lake Athabasca; the mean temperature for that month differs by no less than 23° from that of December, the one being probably above, the other below, the normal mean temperature, and the mean for the hours of 1^h, 2^h, and 3^h P.M. is 31° lower than that for the corresponding hours of February. A similar state of things prevailed at Toronto, where we have the mean temperature for December 1843, 30°·8, and for January 1844, 20°·7, difference 10°·1, the means for those months by the observations of twelve years being respectively 27°·0 and 24°·5, difference only 2°·5. We have also in the same month the extraordinary difference of 62°·3 of temperature in the course of four days, the thermometer having indicated —47°·7 Fahrenheit at 7 A.M. on the 25th, and +14°·6 at 2 A.M. on the 30th January; but this is not the whole range in the month, and is indeed exceeded by a change of 64°·9 between 3 P.M. on 22d March and sunrise on the 25th, when the temperatures observed were 42°·0 and —22°·9 respectively. For the purpose of comparing exactly the fluctuations of temperature at these northern stations with those of Toronto, the situation of which, on a peninsula formed by three of the great lakes, gives its climate somewhat of insular characteristics, I have taken the differences between the highest and lowest observation of each day, and found the mean value of the daily range thus shown, precisely as was done for the corresponding ranges of the magnetical elements, Tables I., IV., and XV. These values, and some other particulars in aid of this comparison, are contained in the next Table.

Table LXI.

Comparison of Range of Temperature.

1843–4.	Mean daily Range.		Extremes in each Month.						—
	Toronto.	Athabasca. / Fort Simp.	Toronto. Max.	Toronto. Min.	Toronto. Diff	Athabasca, &c. Max.	Athabasca, &c. Min.	Athabasca, &c. Diff.	
	°	°	°	°	°	°	°	°	
October	15·0	—	63·4	25·0	38·4	—	—	—	The whole month.
	15·2	15·4	57·8	26·0	31·8	49·9	−7·6	57·5	The 16th to the 31st
November	12·0	11·9	51·6	15·4	36·2	32·7	−9·4	42·1	
December	10·2	18·4	41·4	4·2	37·2	35·3	−35·3	70·6	
January	14·7	10·4	45·0	−6·0	51·0	14·6	−47·7	62·3	
February	15·5	25·4	47·6	1·2	46·4	37·5	−32·1	69·6	
March	25·1	——	50·7	10·9	39·8	——	——	——	
April	23·4	27·0	75·0	21·1	53·9	68·0	−3·3	71·3	
May	22·9	23·3	78·0	29·3	48·7	72·5	12·6	59·9	

WINDS.

THE Direction of the Wind was entered by estimation at each hourly observation, and its force expressed in words. The number of winds from each half quadrant are given in the next Table, where the column North includes N. by W., N.N. by E., and N.N.E., and so on round the circle. The azimuths actually entered were magnetic. This arrangement was adopted to take advantage of the convenient guide furnished by the arrangement of the buildings in Fort Chipewyan, which all ran within half a point of north and south or east and west by compass. They have been converted into true directions, by subtracting two points from each, being the value of the magnetic declination less 5°, the amount of the deviation of the lines to the west of magnetic north. In similar circumstances it will be found preferable to establish a permanent guide of some simple nature to the true directions.

TABLE LXII.

Number of Observations of the Wind from each Direction.

Direction of the Wind.	At Lake Athabasca.												At Fort Simpson.		
	By Hourly Observation.					By Four Observations daily.				Winter Quarter.	Spring Quarter.	The whole Period.	By Hourly Observation.		
	Oct.	Nov.	Dec.	Jan.	Feb.	Mar	Apr.	May	June				Apr.	May	Total.
N. -	0	8	25	32	9	21	34	9	19	66	63	157	73	71	144
N.E. -	68	92	83	90	126	26	17	28	29	299	71	559	5	6	11
E. -	15	22	78	34	14	1	1	3	18	126	5	186	53	56	109
S.E. -	9	54	33	10	8	2	7	1	0	51	10	124	150	154	304
S. - -	1	36	39	7	2	0	8	4	9	48	12	106	42	41	83
S.W. -	26	14	11	23	17	13	6	8	15	51	27	134	24	13	37
W. -	34	8	33	106	7	6	9	6	5	146	21	214	20	6	26
N.W. -	3	33	92	63	55	14	9	26	14	210	23	309	61	32	93
Winds -	156	267	394	365	238	83	91	85	109	997	238	1,789	428	379	807
Calms -	180	357	208	262	338	26	29	30	7	808	94	1,441	172	125	297

Allowing the usual value to each descriptive term (Toronto, vol. 1. xcii.) we have the total pressure from the several quarters, as follows:

Table LXIII.

Sums of the Pressures from each Quarter by estimation, in Pounds upon the Square Foot.

Direction of Wind.	At Lake Athabasca 1843-4.												Toronto, Oct. to June.	
	By Hourly Observation.					By Four Observations daily.				Winter Quarter.	Spring Quarter.	The whole Period, Oct. to June.	No. of Winds.	Total Pressure.
	Oct.	Nov.	Dec.	Jan.	Feb.	Mar.	Apr.	May.	June.					
	lbs.	*lbs.*	*lbs.*	*lbs.*	*lbs.*	*lbs.*	*lbs.*	*lbs.*	*lbs.*	*lbs.*	*lbs.*	*lbs.*		*lbs.*
N.	0·0	25·3	65·4	16·7	8·0	14·7	46·3	2·5	58·1	90·1	63·5	237·0	357	274·5
N.E.	122·6	205·4	293·7	155·4	267·6	50·3	4·9	7·8	84·1	716·7	63·0	1199·8	220	154·8
E.	51·0	31·5	293·2	80·0	31·9	3·0	0·5	0·6	51·6	405·1	4·1	533·3	520	435·4
S.E.	36·0	181·5	71·8	5·3	7·2	0·4	2·0	0·2	0·0	84·0	2·6	304·4	239	96·2
S.	1·5	181·0	171·5	21·0	7·0	0·0	5·9	1·1	18·9	199·5	7·0	407·9	456	247·4
S.W.	92·0	35·5	29·0	58·6	27·8	28·3	1·2	1·9	51·5	115·4	31·4	325·8	407	369·2
W.	126·9	36·5	30·4	64·9	38·0	9·6	2·1	1·3	18·2	133·3	13·0	327·9	426	440·4
N.W.	7·5	18·1	205·4	115·5	81·0	21·5	3·0	7·6	143·0	401·9	32·1	602·6	549	529·7
Sums	437·5	713·8	1160·4	517·4	468·5	127·8	65·9	23·0	425·4	2146·0	216·7	3938·7	—	—

On comparing the results contained in the foregoing Tables with those of the corresponding period at Toronto, it is observable that the proportion of winds from the north-east is much greater at Lake Athabasca than at the more southern stations; at the former the great preponderance, whether we regard number or total pressure, is from that quarter; at Toronto, on the contrary, it is from the north-west. Resolving the total pressures in the four cardinal directions, and obtaining a general resultant, it appears that the following are the equivalents of all the winds at the two stations for the period under comparison:

Table LXIV.

The Wind.	Lake Athabasca.			Toronto.		
	Winter Quarter, 1843–4.	Spring Quarter 1844.	The whole Period.	Winter Quarter, 1843–4.	Spring Quarter 1844.	The whole period.
	lbs.	*lbs.*	*lbs.*	*lbs.*	*lbs.*	*lbs.*
Mean Pressure -	1·19	0·61	1·22	0·56	0·37	0·46
Resultant Direction -	N. 41° E.	N. 4° W.	N. 43° E.	N. 51° W.	N. 10° E.	N 74° W
	lbs.	*lbs.*	*lbs.*	*lbs.*	*lbs.*	*lbs.*
Corresponding or Resultant Pressure -	0·39	0·28	0·28	0·16	0·06	0·08

Thus it appears that the prevalent winds at Toronto and Lake Athabasca belong to different and nearly opposite systems; a north-westerly current preponderates in the lower latitudes (43° 49′), a north-easterly current, inclined to the former at an angle of about 117°, prevails in the higher one (58° 43′.) The general fact of a prevalence of N.E. winds is stated in the Meterological Register by Mr. Keith for 1825–1826, which has been before referred to.

The Mean Force appears to be considerably higher at Lake Athabasca than at Toronto.

There is but little correspondence between the winds observed at Fort Simpson in April and May and those recorded at Fort Chipewyan at the same time; at the former station half of the total number of winds was from the east and south-east, at the latter the same proportion was from the north and north-east, but the period of comparison is too short for any conclusion to be drawn from the observations. A local phenomenon of interest was observed several times at Fort Simpson, in the rapid rise of the temperature of the air when the wind changed to the south-west from an easterly direction. It appeared as if the warmer air of the Pacific Ocean were transferred across the neighbouring ridges of the Rocky Mountains, with little loss of its temperature. Thus we have April 3^d 10^h, wind S.E., temperature 34°·5; 3^d 11^h, wind S.W., temperature 39°·5. Again, April 25^d 5^h, wind S.E., temperature 43°·5; 25^d 6^h, wind S.W., temperature 58°·0. Again, April 29^d 5^h, calm, temperature 39°·5; 29^d 6^h, wind S.W., temperature 48°·5. Lastly, May 8^d 8^h, wind S.E., temperature 53°·7; 8^d 9^h, wind S.W., temperature, 56°·7.

Aurora Borealis, Weather, &c.

Owing to the unavoidable circumstance that only one observer could be on duty at a time during the night, whose attention was required by the magnetic disturbances, which accompanied most of the more active displays of the aurora borealis, the notices of that phenomenon are less full than could be wished, notwithstanding the great desire that was felt to do justice to so favourable an opportunity of studying it. The three magnetometers were commonly observed on these occasions in succession, with an interval of one minute between them; it was considered an object in general to miss a reading as seldom as possible; consequently, although the observation was usually an instantaneous act, as the magnets being suspended in heavy copper boxes were seldom in vibration, there was, notwithstanding, barely time for the observer to step out of doors after one of them, take a survey of the sky, and return to his place before the next, repeating the process if necessary, until the particulars required were collected. The same circumstance led to an abbreviated mode of description, which is to be regretted, but it should be stated that the actual notes taken at the time have been carefully adhered to in the descriptions at the end of this chapter, with no other alteration or expansion than appeared absolutely necessary to make them intelligible. It must be added, that less stress was laid upon particular features and changes of aspect, or position, than would have been suggested by a better acquaintance, on my part, with what had been recommended in observations of the kind. Having been previously employed in the tropics, and passed but one winter in Canada, my attention had not been much given to the subject, nor was I provided with any such invaluable text-book for this class of observations as the report of M.M. Lottin, Bravais, and Martins, of the *Commission du Nord,* 1839, has since supplied to Arctic observers. A definite sense was attached from the first to each descriptive term employed, and I took pains that they should be used as much as possible alike by myself and assistant (Serjeant Henry.) Certain convenient signs or symbols were also fixed on to denote, without verbal description, particular forms of aurora, very much as those of Mr. Howard are used to denote forms of cloud. These were always used in the column of the register appropriated to remarks on the weather, but a place was also provided for more detailed descriptions. Some of the terms and distinctions, such as *striated, serpentine,* although suggested quite independently by the forms of aurora which presented themselves, I have since found more or less used by others, a satisfactory confirmation of their applicability.

In the descriptions which follow, then, the term *streamer* is confined to lines of light in rapid motion, appearing and reappearing suddenly and in any part of the heavens, but directed towards the zenith; they were frequently unconnected with any large body of stationary light. The term *striæ, striated,* refers to a peculiar appearance, sometimes seen in arches and even in detached masses and the larger streamers, an arrangement of the whole body of light in fine parallel lines, directed towards the zenith, and presenting the appearance of wool or cotton combed out; this was considered at the time to indicate a more active state of the electric forces than existed when it was wanting,—to be, in other words, a higher form of development. The terms *beams, streaks, bands,* were used to indicate long narrow portions of light, not sensibly in motion, the last being confined to those which crossed the magnetic meridian nearly at right angles. The term *serpentine* motion was employed to denote changes of outline different in character from the direct rectilineal motion of streamers, but resembling the changes of the folds of a

Plates A. and B. of the admirably faithful Atlas of Auroras, ed in connexion with the report above referred to, exhibit y the features described by the terms striated and serpentine;

save that the vertical divisions are considerably more strongly marked than the writer is conscious of having seen them. The term *cirrous aurora* indicates light detached patches, scarcely distinguishable in form from cirrous clouds. Lastly, *auroral* haze denotes a luminous appearance, usually in the northern quarter, without definite form or boundary, and sometimes also the vanishing light of other descriptions of aurora. The reference (*a*) is given to this appearance in the abstracts as well as to the more definite forms.

The relative brilliancy is generally indicated by figures. Thus, 0·5 represents the faintest description of aurora; 1, faint aurora; 2, moderately bright aurora, and what in low latitudes would be considered bright; 3, decidedly bright; 4, the brightest and most perfect displays. Very few exhibitions were considered to come up to the last class; of these, the principal one was not observed at Lake Athabasca, but at the Painted-stone Portage near Lake Winnipeg, on the 7th August 1843. Much of the comparative brilliancy of the displays however depends on the absence of moonlight, a circumstance which was not sufficiently taken into account at the time. The dates of the changes of the moon are given in the register for reference. The elevation of arches was observed with a wooden quadrant and plummet. The directions given are magnetical, the declination being 28° E. More or less aurora was seen on 49 out of 116 nights of observation at Lake Athabasca. There was no night, properly speaking, at Fort Simpson after the 16th April, the latest date at which the sun sinks in that latitude as far as 18° below the horizon; nevertheless, the aurora was seen on twenty-four out of thirty nights of observation between the 1st April and the 6th May. After the last-named date, as the brightness of the twilight made it very difficult to distinguish between light cirrous clouds and patches of aurora, it did not receive much attention, and was not again seen up to the close of the observations on the 25th May; it is, however, quite possible that close attention would distinguish aurora occasionally, by its motion, to a much later date, as it has been seen in early evening and morning twilight in lower latitudes, and was undoubtedly seen at Toronto *in full daylight*, on the 29th September and 2d October 1851. It appears to have passed the zenith, when last described, on the 6th May, at about 13^h 15^m of local mean time, when the sun's centre was 9° 47′ below the horizon, consequently, if its distance from the earth upon that occasion exceeded fifty geographical miles, it must have been within the sphere of the sun's rays. For want of corresponding observations elsewhere, there are no data for computing the height of any of the displays, but I avail myself of this opportunity of stating, that the impression conveyed to the senses upon many occasions was altogether opposed to the idea of the seat of the display being so distant as it seems to be in lower latitudes. Those who have travelled in mountainous countries must have frequently observed the passage of clouds at a short distance above their heads, and remarked, that without the aid furnished by neighbouring peaks and rocks, there is the most convincing proof of the nearness of the cloud afforded by the manner of its motion, the sensible unfolding of masses of vapour, by the distinctness with which every detail of its form is seen. Precisely of the same nature is the evidence frequently given to the senses in high northern latitudes of the nearness of an aurora, and the universal belief of the fact among those who witness it perpetually, must be allowed to weigh somewhat in the same scale. This inference also is not irreconcileable with the results of actual measurement in Europe and the United States, if we suppose that the circumstances which favour this phenomenon occur nearer the earth in high than in moderate latitudes.

It is important to observe that every night on which aurora was recorded at Christiana in Norway, during the periods of observation under discussion, by M. Hansteen (*Mem. de l'Acad. Royale de Bruxelles, tom.* xx.) coincides with one of the dates of observation at Lake Athabasca or Fort Simpson; there are, however, but eight of the former and one of the latter, namely, October 24th, October 26th, December 8th, 1843, January 8th, January 16th, January 22d, February 17th, and April 17th, 1844. It would appear by the descriptions of M. Hansteen, that the phenomenon was generally more perfectly developed at the American than at the European station, although the latter is about 1° more to the north. There are no observations of aurora recorded at any of the Russian stations or Siberian stations, and but five at Sitka, all, except the last, noted as *faible;* the dates of these are October 15th, October 19th, 1843, and April 16th, 1844, all coinciding with northern observations, and March 7th and 28th, when no observations were made. Of the observations at Makerstoun in the same periods, seven coincide with observations at our two stations, namely, October 16th, October 26th, November 2d, November 13th, April 5th, April 10th, April 17th. Of the remaining Makerstoun dates, twelve in number, observation was impossible from clouds on nine; at two the brightness of the twilight equally prevented it; there is but one (when *traces of aurora* alone are recorded at Makerstoun,) upon which observation seems to have been possible, and no aurora occurred at Lake Athabasca; this date is January 5th. Such a result is the more remarkable, as the displays of the aurora at Lake Athabasca were probably of considerably less than average frequency and brilliancy. The winter of 1843-4, in opinion of residents in that part of America, was remarkable for absence of this phenomenon, as it was in lower latitudes.

The following list of Observations, made in 1850-1 by Mr. J. Anderson, C.F., at the same station, supports such an opinion.

—	Number of Nights observed.	Aurora seen.	Observations impossible.
1850, November - - -	24	11	11
„ December - - -	31	19	9
„ January - - -	31	20	7
„ February - - -	28	21	7
„ March - - -	31	19	11
„ April - - -	30	19	10
	175	109	55

Comparing similar periods, November to February; it appears in this case to have been seen on 89 per cent. of the nights when observation was possible. In the season of 1843-4, the proportion was 71 per cent.

The number of observations at Toronto from the 16th October 1843 to 28th February 1844 was also unusually small. It is only recorded at four hours of observation in that interval, namely, October 19^d 17^h, December 11^d 14^h, and January 24^d 17^h, and 18^h Göttingen. To these may be added October 20th, October 21st, and February 4th, from the reports of the Regents of the University of New York, and December 12th from the record of Mr. Herrick of Newhaven. The latter indefatigable observer, however, recorded suspicions of aurora on November 25th, 26th, 27th, December 10th, January 8th, 13th, and April 24th. The whole number of positive observations, however, in this period amounts to seven only, a number considerably below the average, which for Toronto alone, from 1840 to 1850 inclusive, is between eleven and twelve.

There are only four instances of aurora recorded at Lake Athabasca so soon after sunset or so shortly before sunrise as to come within the period of evening or morning twilight. On the 16th October it was visible at 6 A.M. in the east, very faint, and a portion almost imperceptible in the zenith; at this hour the sun was only 6° 44′ below the horizon, and if the elevation of the aurora in the zenith exceeded twenty-four geographical miles it must have been within the sphere of the sun's rays. It was seen at the same hour on the 23d November as a faint light in the north-west, but at this time the sun was 16° 9′ below the horizon. It was again observed at 6 A.M. on the 29th, the sun being at the time 17° 20′ below the horizon. Lastly, it was seen at 6 P.M. on February 6th as a faint arch, the elevation of which is not recorded, and probably did not exceed 4° or 5°; the sun was at this time 11° 25′ below the horizon.

There are also a few instances of its being seen very soon after twilight ended, or immediately before its commencement. The relative frequency of the phenomenon at different hours of the night is shown by the next Table.

TABLE LXV.—*Showing the Total Number of Observations of Aurora in each month, and the number of entries of clouded, partially clouded, and unclouded sky. An observation between two hours is included under the first of them; thus, an observation at* 7 h. 30 m. *is classed at* 7 h.

Hours of Mean Time.	October.				November.				December.			
	No. of A.	Sky.			No. of A.	Sky.			No. of A.	Sky.		
		Clouded.	Partly Clouded.	Clear.		Clouded.	Partly Clouded.	Clear.		Clouded.	Partly Clouded.	Clear.
6	0	9	1	4	0	15	1	10	0	15	4	6
7	2	7	1	6	1	15	3	8	0	15	3	7
8	2	5	5	4	2	15	2	9	0	14	3	8
9	2	6	2	6	1	16	3	7	0	13	7	5
10	5	6	3	5	1	17	3	6	0	14	6	5
11	4	6	2	6	5	16	3	7	0	15	6	4
Midn.	4	6	2	6	7	14	4	8	1	11	10	4
13	8	8	1	5	5	15	2	9	2	14	5	6
14	5	5	3	6	1	15	2	9	1	14	5	6
15	4	5	3	6	3	15	0	11	2	14	7	4
16	4	7	0	7	0	17	1	8	2	14	6	5
17	4	7	3	4	2	15	2	9	1	13	5	7
18	1	8	2	4	2	16	2	8	0	15	4	6
Total	43	85	28	69	30	201	28	109	9	181	71	73

Hours of Mean Time.	January.				February.				Total.				Sum of the esti-mated Bril-liancy.	April 1 to May 6.			
	No. of A.	Sky.			No. of A.	Sky.			No. of A.	Sky.				No. of A.	Sky.		
		Clouded.	Partly Clouded.	Clear.		Clouded.	Partly Clouded.	Clear.		Clouded.	Partly Clouded.	Clear.			Clouded.	Partly Clouded.	Clear.
6	0	13	3	10	1	18	4	8	1	70	13	33	1·0	—	—	—	—
7	3	10	2	14	2	14	2	9	8	61	11	44	10·5	—	—	—	—
8	3	11	3	12	3	11	5	9	10	56	18	42	18·0	4	30	9	7
9	4	13	2	11	3	11	4	10	10	59	18	39	16·0	7	23	4	14
10	4	10	2	14	2	11	4	10	12	53	18	40	18·0	7	24	4	18
11	2	12	3	11	2	12	3	10	13	61	17	38	24·0	10	26	2	18
Midn.	7	11	3	12	4	10	4	11	23	52	23	41	39·0	9	24	6	16
13	7	12	4	10	6	12	1	12	26	61	13	42	40·0	10	24	5	17
14	6	10	5	11	4	14	0	11	17	58	15	43	18·0	4	24	2	20
15	6	12	3	11	3	14	1	10	18	60	14	42	26·0	0	15	11	20
16	6	11	2	13	1	16	1	8	13	65	10	41	22·0	0	13	14	19
17	3	11	4	11	2	11	4	10	12	57	18	41	15·0	—	—	—	—
18	0	12	1	13	0	13	4	8	3	64	13	39	2·5	—	—	—	—
Total	51	148	37	153	33	167	37	121	166	782	201	525	—	51	208	57	149

All the hours contained in the foregoing Table may be considered as hours of darkness, except 6 A.M. and 6 P.M. for a short period at the beginning and end of the series, and this will be compensated by the longer absence of daylight in mid-winter. The sun rose at Lake Athabasca on the 16th October at 6^h 43^m A.M., and set at 4^h 49^m; morning twilight began at 4^h 6^m, and evening twilight ended at 7^h 22^m; all of mean time. On the 28th February the sun rose at 7^h 6^m, and set at 5^h 19^m; morning twilight began at 4^h 31^m, and evening twilight ended at 7^h 54^m. Hence, if the development of auroral light have no relation to the hour of the night, in other words, no diurnal law, we should expect to find the observations nearly equally distributed throughout that period. It is at once apparent that such is not the case; the number under the several hours increases from 6 P.M. to midnight; there is a great excess at midnight and 1 A.M., after which the numbers diminish down to 6 A.M. The result is the same, if instead of counting the number, we allow weight according to the relative brilliancy of the displays, using the scale already explained; but as the necessary observations were not made on the spot to determine, with any precision, the integral value of the auroral light developed from hour to hour, any estimate founded upon that scale is necessarily vague. However, taking the numbers as they stand, and supplying them by estimation from the descriptions, where they are wanting, they sufficiently confirm the present conclusion. The totals for each hour are added in the last column of the Table for Lake Athabasca.

There is a great difference in the total number of observations recorded in the different months, indicating a less average duration, as well as a minimum of frequency, at the winter solstice. Allowing that the aurora might have been seen upon half the occasions when it is entered as partly clouded, the proportion of observation hours in which it was actually seen, bears the per-centage shown in the following Table, to the number of favourable hours:

—	Total Number of Nights.	Aurora seen.	Observations impossible.	Per-centage of Hours of Observations to favourable Hours.	
1843–4.					
In October -	16	9 Nights.	7 Nights.	0·52	0·266
November -	26	10 ,,	13 ,,	0·24	
December -	25	5 ,,	12 ,,	0·08	
January -	26	15 ,,	10 ,,	0·30	
February -	25	10 ,,	11 ,,	0·24	
April - -	25	21 ,,	7 ,,	0·29	
May 1 to 6 -	6	3 ,,	1 ,,	—	

Thus it appears that aurora was visible, in the winter of 1843–4, at Lake Athabasca, at about one fourth of all the hours of observation when it was possible to see it.

On April 18, at Fort Simpson, which is the mean date of auroral observations, the sun rises at $4^h\ 34^m$ A.M., and sets at $7^h\ 26^m$ P.M., App. T.; and from 8 P.M. to 4 A.M. inclusive, have been taken as the limits of darkness for the whole period.

It may deserve remark that the month of December, when it was least frequent, was a remarkably mild one, and January an unusually cold one.

If the region in which the auroral development takes place be entirely beyond the limits of the atmosphere, as is commonly supposed, it is difficult to conceive any direct connexion between the aurora and the state of that medium, but this question may perhaps be regarded as not finally settled, and it may be worth while to examine the accompanying meteorological features. The first which will be noticed on referring to the meteorological register, is the apparent connexion between the occurrences of aurora and a state of calm. It appears by Table LXII., that the proportion of hours entered as calms, to those of sensible winds, is 1,340 to 1,420 or 94 per cent., whereas the entries accompanying aurora are as follows:

Hourly Observations of Aurora.	Lake Athabasca.	Fort Simpson.	Total.
With High Winds - - - -	21	5	26
With Light Winds - - -	38	10	48
With Calm - - - -	97	36	133

Showing a great preponderance under calm. In order to ascertain whether this could be due to a greater average freedom from cloud under such circumstances, separate abstracts have been formed of the proportion of clear sky accompanying entries of calm and wind. The result is, that the sky was on the average clearer in calm weather than during winds, but in a materially less proportion than is required to account fully for the excess of aurora under calms. The average of clear sky for the thirteen hours included in Table LXV. is—

Under winds, 0·354
Under calms, 0·459

With respect to prevailing winds, it will be noticed, in the same way, that a much larger proportion of the accompanying winds contain easting than westing, and if we admit that the state of the atmosphere may have something to do with the phenomenon, it would follow that the conditions favourable to it, at Lake Athabasca, are derived rather from the side of the Atlantic, or from Hudson's Bay, than from the warmer side of the Pacific.

Table LXVI.

General Statement of the prevailing Winds accompanying Aurora at Lake Athabasca.

Date.	The Six preceding Hours.		Accompanying Aurora.		The Six succeeding Hours.	
Mean Time.	Prevailing Wind.	Description.	Prevailing Wind.	Description.	Prevailing Wind.	Description.
1843:						
Oct. 15	—	Calm.	—	Calm.	—	Calm.
„ 16	—	Calm.	—	Calm.	E.N.E.	Light. Fresh.
„ 20	—	Calm.	—	Calm.	S. by E. E.N.E.	Light. Fresh.
„ 25	W.N.W.	High.	—	Calm.	—	Calm.
„ 26	—	Calm.	— S.S.E.	Calm. Fresh.	— E.S.E.	Calm. Light.
„ 28	E.N.E.	Light.	—	Calm.	No observation.	
„ 31	N.E. S.E.	High. V. High.	S.E.	V. High.	E.S.E. W.N.W.	High.
Nov. 1	S. by W.	Fresh.	E.N.E.	Mod.	E.N.E. S.S.E.	High.
„ 2	S.S.E.	Light.	— S.S.E.	Calm. High.	S.S.E. S.S.W.	High.
„ 9	—	Calm.	—	Calm.	—	Calm.
„ 13	—	Calm.	—	Calm.	E.N.E. S.S.E.	Light.
„ 17	—	Calm.	—	Calm.	—	Calm.
„ 22	— S.S.W.	Calm. Light.	— N.E.	Calm. Fresh.	E.N.E.	Fresh. Light.
„ 27	—	Calm.	—	Calm.	—	Calm.
„ 28	—	Calm.	W.S.W. S.S.W.	Light. High.	—	Calm.
„ 29	—	Calm.	—	Calm.	—	Fresh.
Dec. 9	—	Calm.	—	Calm.	No observation.	
„ 13	E.S.E.	Light.	E.S.E.	Light.	E.S.E.	Light.
„ 15	S.	Light.	—	Calm.	S. by E. N.N.W.	Fresh. Light.
„ 22	S.S.E. E.N.E.	High. Fresh.	E.	High.	E. E.N.E.	High.
„ 26	— S.S.E.	Calm. V. Light.	—	Calm.	S.E.	V. Light.
1844:						
Jan. 8	N.E.	Light.	N.E. E.N.E.	Light.	E.N.E.	Light.
„ 9	W.S.W.	V. Light. Calm.	—	Calm.	—	Calm.
„ 15	S.S.W. W.N.W.	V. Light. Fresh.	W.	V. Light. Fresh.	W.N.W. N.W. by W.	Fresh.
„ 16	N.W.	Mod. High.	N.W.	Mod.	N.N.W.	Fresh. High.
„ 17	W.N.W.	Light. Fresh.	N.N.W.	Mod.	N.N.E. N.	High. Light.
„ 18	N.	Light.	N. W.N.W.	Light.	W.N.W.	Light.
„ 19	—	Calm.	—	Calm.	—	Calm.
„ 21	No observation.		—	Calm.	—	Calm.
„ 22	N.N.E.	Light.	—	Calm.	—	Calm.
„ 24	—	Calm.	—	Calm.	—	Calm.
„ 26	W.N.W. —	High. Calm.	W.N.W. —	Light. Calm.	—	Calm.

TABLE LXVII.—*continued.*

Date.	The Six preceding Hours.		Accompanying Aurora.		The Six succeeding Hours.	
Mean Time.	Prevailing Wind.	Description.	Prevailing Wind.	Description.	Prevailing Wind.	Description.
1844:						
Jan. 28	—	Calm.	—	Calm.	—	Calm.
„ 30	W.N.W. —	Light. Calm.	—	Calm.	—	Calm.
„ 31	—	Calm.	E.S.E.	Light.	E.N.E.	Light. Fresh.
Feb. 5	N.N.W.	Light.	N.N.W.	Light.	—	Calm.
„ 7	—	Calm.	—	Calm.	—	Calm.
„ 11	No observation.		—	Calm.	N.E. —	V. Light. Calm.
„ 12	E.N.E.	Light.	E.N.E.	Light.	—	Calm.
„ 13	—	Calm.	—	Calm.	N.N.W.	V. Light.
„ 15	—	Calm.	—	Calm.	—	Calm.
„ 16	—	Calm.	W.S.W.	Mod. High.	W.S.W.	Light.
„ 17	W.S.W.	Fresh. Mod.	—	Calm.	N.N.W.	High. Mod.
„ 20	N.N.W.	Mod.	—	Calm.	—	Calm.
„ 21	N.E.	High.	E.N.E.	Mod.	—	Calm.
„ 26	W.N.W.	High. Mod.	—	Calm.	—	Calm.
„ 28	—	Calm.	—	Calm.	—	Calm.

It is to be regretted that the observations are deficient as regards the azimuth of arches, and other displays, or their relation to the magnetic meridian; this particular is always too vaguely expressed, and frequently not noted at all. Careful observations of the point of convergence of streamers in the few instances in which they formed a corona, was also overlooked, which however arose chiefly from an exaggerated expectation of something better defined and more regular than ever presented itself.

The most frequent form of the auroral development was the simple arch;* these arches in many instances underwent changes and assumed other forms, but probably in almost every instance the first definite form assumed was of this class. If we classify the entries at

* Mr. Roderick Campbell, an officer in the service of the Hudson's Bay Company, kept, at my request, a meteorological register at Frances Lake, on the west side of the Rocky Mountains, situated about latitude 61° 30′, longitude 129° W., from November 1844 to April 1846. This register comprises 13 months of observations of Aurora Borealis, exclusive of the Midsummer half year, when it could not be distinguished. It was seen on 66 evenings of that period, and is described as an arch in 41 of the entries; it possibly, also, had the same form on some of the nine occasions on which it is not described. There are only five dates on this list coinciding with observations at Toronto, and six more on which coincident observations are found in the Regent's Reports.

the several hours without regard to the subsequent changes, it appears that the numbers are as follow:

Table LXVIII.

Nature of Display.	Before Midnight.	At Midnight.	After Midnight.	Total.
Undefined light, usually in the north	3	3	19	25
A simple arch	34	7	26	67
Arch striated	2	—	2	4
Arch combined with streamers	4	2	3	9
Arch co-existing with transverse bands, which in most cases are probably the remains of earlier arches, advanced to near the zenith	2	—	3	5
Streamers alone, or principally	3	4	8	15
Detached patches alone, or principally	1	6	12	19
Transverse bands alone, or principally	2	1	6	9
	51	23	79	153

This classification rests on rather an arbitrary division, the descriptions not being sufficiently full to enable it to be made satisfactorily, but may serve as an approximation. On comparing it with the register at Toronto, it appears that the more definite forms of aurora occur in much the greater proportional number at the Northern station; a proof, if the more northern region is the nearer to the seat of the display, that the same object cannot be seen at both stations. Thus, we have at Toronto in 8½ years of two-hourly and one hourly observations (January 1840 to June 1848 inclusive), the following number of entries:

Table LXIX.

Nature of Display.	Before Midnight.	At Midnight.	After Midnight.	Total.
Undefined auroral light	147	61	116	324
An arch	35	11	27	73
Arch combined with streamers	64	11	22	97
Other combinations, forming the finer displays	37	13	32	82
	283	96	197	576

The arches at Lake Athabasca form rather the largest proportion at the early hours of the night; the less definable forms, on the contrary, and those which the phenomenon assumes when the display approaches its conclusion, are more numerous in the latter part o.

the night, all tending to show, as already inferred from the numbers in Table LXV., that the luminous display essentially belongs to the night, and that the presence of daylight is not the only reason why it is so very rarely seen when the sun is above the horizon.

A peculiarity may be noticed in the references to the state of the sky accompanying Table LXV., that there are comparatively a small number of entries under the head of "Partially clouded." It may be added, that there are comparatively few observations in the notices which follow, of the definite forms of clouds, the most usual state was a light uniform cloud or haze, covering the entire sky; this prevailed particularly for two or three hours about sunrise and sunset. The sum total of clear sky is considerably less from 6 to 9 A.M., and again from 2 to 5 P.M., than at any other hours. It does not appear, however, that this was the case to a greater extent on the mornings following or the afternoons preceding aurora than on other days, but the reverse. Thus, we have the mean proportion of clouded sky for four hours (6 to 9 A.M.) on mornings following aurora, 0·52, and on the remaining mornings, 0·78. Again, for four hours (2 to 5 P.M.) on afternoons preceding aurora, it is 0·61, and on the remaining afternoons 0·76. The aurora, therefore, would not appear from these observations, either to result from or to tend to produce, circumstances akin to those which produce common cloud, a view which has been sometimes taken. The sums total of clear sky at the different hours are as follows:—

Midnight	51·2	Noon	41·1
1 A.M.	48·1	1 P.M.	41·2
2 ,,	48·1	2 ,,	34·5
3 ,,	47·5	3 ,,	33·7
4 ,,	44·8	4 ,,	29·4
5 ,,	50·4	5 ,,	27·6
6 ,,	45·0	6 ,,	37·9
7 ,,	33·7	7 ,,	48·7
8 ,,	27·2	8 ,,	49·7
9 ,,	24·7	9 ,,	47·4
10 ,,	40·5	10 ,,	47·4
11 ,,	42·9	11 ,,	46·9

With regard to the much disputed question of sound, neither the writer nor his assistant Serjeant Henry, were ever positive of hearing any, but the latter thought he did so upon one or two occasions. The result of inquiries upon the subject was, that opinions were nearly equally divided among the educated residents in the country; a small majority of those the writer consulted, agreed that a sound sometimes accompanied the phenomenon, but among the uneducated and native inhabitants, whose acuteness of sense is probably much superior to that of the other class, a belief in the sound is almost universal, and many individuals assured the writer they had heard it. Similar testimony has been borne very positively by the assistants at the observatory at Toronto, upon one or two occasions of great display.

Connexion of Aurora with Magnetic Disturbances.

A little experience in North America, whether in Canada or in the more northern regions, suffices to correct the impression that every display of aurora, however inconsiderable or distant, is attended by sensible magnetic disturbance. So far as the magnetometers, observed at short intervals, can be taken as a criterion, that is far from being the case, nor does it appear to be so by the more perfect test of photographic registration, as far as it has been applied at Toronto. To this it may be added that the hours at which aurora is most prevalent are midnight and 1 A.M. at Lake Athabasca and Fort Simpson (Table LXV.), whereas the period of greatest mean disturbance at both stations is 3 to 5 A.M.; it is also midnight at Toronto, where the period of greatest mean disturbance is 9 or 10 P.M.; if, therefore, the development of aurora has any immediate relation to the disturbance of the magnetic elements, the latter must precede the former in one region, and follow it in the other, a law which does not appear probable. On the other hand it is unquestionable that the more brilliant displays are almost always attended by magnetic disturbances, as are many of the more moderate ones; exceptions in the first class are very rare, but the writer believes that some can be established; the general conclusion must, however, be that an intimate relation exists between these distinct phenomena, although not that of cause and effect. The general practice of the observers was to read the instruments at intervals of a few minutes, during every aurora; if either of the magnets differed decidedly from its usual position, or was observed to be in vibration, readings were taken as on term days, or more usually they were read in succession, with an interval of one minute only between the observations, each being read, therefore, every third minute. If no sign of disturbance was observed, the remark "no disturbance" was made in the register, but the actual positions at the moments of observation were not thought important, and were not recorded; this omission is to be regretted, since it reduces the amount of proof of the absence of disturbance, which has to be established.

The following are the dates of these entries of "no disturbance;" the character of the aurora on each occasion will be seen by consulting the descriptions appended:

	D.	H.	H.	
October	27	21	22	Gött.
"	28	18	19	
"	31	18		
November	17	19		
"	22	20		
"	27	20		
"	29	19	21	
January	9	16		
"	17	20		

	D.	H.		D.	H.
January	19	16	to	20	1 Gött.
"	20	19	"	20	brilliant.
"	21	21	"	24	
"	22	21	"	22	
"	26	17			
February	12	18	"	20	
"	15	20	"	21	
"	16	17			brilliant.
"	27	1			

The following list contains the dates of the more brilliant and the longest displays of aurora, the number of hours at which they were recorded, and the order or relative place of each day among the other days of the same month, in respect to its "mean irregular fluctuation" of two elements (p. 74), together with the values of those quantities:

TABLE LXX.

Date.	No. of Hours.	Order.		Days in Month.	Date.	No. of Hours.	Order.		Days in Month.
		Dec. F ψ	H. F. F ψ				Dec. F ψ	H. F. F ψ	
1843–4:					1844:				
October 16	9	3	2	14	January 20	7	25	18	26
" 17	9	4	1	"	" 22	4	7	10	"
" 26	10	6	6	"	" 24	6	4	5	"
November 2	5	2	6	26	" 26	4	6	9	"
" 13	4	5	1	"	February 5	4	3	1	24
" 29	6	13	9	"	" 12	4	21	9	"
December 26	3	12	5	25	" 16	5	7	13	"
January 16	4	19	16	26	" 21	4	11	12	"
" 19	9	16	8	"	" 28	5.	2	—	"

By conforming to Göttingen time, the night is divided at 4 A.M. at Lake Athabasca, but as four-fifths of the observations of aurora fall before that hour, its influence on the daily mean irregular fluctuation should be strongly marked in the dates given. It appears that among these days there are several which take a low place in the order of relative disturbance, so far as the quantity referred to is a criterion. Upon the whole, the mean irregular fluctuation of Declination for fifty Göttingen days on which aurora is recorded, is 6′·65, and for the remaining sixty-five days is 7′·10; it is 13·28 scale divisions of the Bifilar, upon forty-nine days of observation with that instrument, when aurora was seen, and 13·38 div. on the remaining sixty-five days, thus being actually less with both instruments on the first than on the second class. The means for the month of January, on which the proportion of auroras to hours of observation (p. 145) was 0·30, are Declination 3′·90, Bifilar 18·66 div.; and in December, when the former quantity is only 0·08, the latter are 7′·32 of Declination and 11·99 div. of the Bifilar, the Bifilar here exhibiting a diminished disturbance, but on the other hand the Declination, as in the other comparison, a greater degree of it.

Although, however, it can be shown that there are instances of aurora, to all appearance unattended by magnetic disturbance, it is remarkable that magnetic disturbances unattended by aurora are very rare; there is but one decided example of it under circumstances of the sky which would have allowed the latter phenomenon to be observed, if it existed; there are also one or two instances in which the disturbance was not observed to commence until some time later than the appearance of aurora, but in every other instance either the sky was clouded or aurora was seen. There appear to be but five instances in which an entire cloudless night passed, without aurora being seen at any time, namely, November 27th, December 20th, January 2d, January 5th, and February 19th. There are also seven half-nights terminating at or commencing from midnight, and some shorter periods, to which the same remark applies, but on only one complete instance of this nature was there any magnetic disturbance observed, namely, January 5th. So far, therefore, as a conclusion can be drawn from such limited data, it would appear that these phenomena are so related, that while the amount of electrical excitement necessary to produce aurora borealis, does not necessarily produce any sensible disturbance of the magnetic elements, yet the latter is almost necessarily attended by the former.

The extra observations on account of disturbance, taken up to the period at which the twilight prevented aurora from being distinguished, are classified with reference to this circumstance in the following list.

[The range of Declination and Horizontal Force is added on each occasion, taking one division of the Bifilar scale = ·0003412 X at Lake Athabasca.]

Table LXXI.

I.—Magnetic Disturbances during which Aurora was visible.

Date.	Mean Time.	Gött. Time.	Range.		$\frac{\Delta X}{X}$	—
			Declination.	Bifilar.		
1843.	D. H. H.	D. H. *	° ′			
October -	15 13 to 21	15 21 to 3+	2 35·5	214·3	·0731	Mem. The differences of Horizontal Force are from the readings, uncorrected for temperature changes.
	16 17 to 19	17 1 to 3	1 3·2	203·0	·0692	
	17 9 to 13	17 17 to 21	1 36·4	135·4	·0461	
	25 11 to 19	25 19 to 3+	2 7·4	233·8	·0798	
	26 12 to 18	26 20 to 2+	0 54·0	160·4	·0546	

* + Indicates the following Göttingen day.

I.—Magnetic Disturbances—continued.

Date.	Mean Time.	Gött. Time.	Range.		$\frac{\Delta X}{X}$	—
			Declination.	Bifilar.		
1843.	D. H. H.	D. H.	° ′			
November	2 08 to 17	2 16 to 1+	2 17·2	197·0	·0672	
	5 15 to 17	5 23 to 1+	0 41·2	85·2	·0290	
	9 14 to 16	9 22 to 0	0 15·2	91·7	·0610	Aurora was visible at midnight, but none during the observations here referred to.
	13 12 to 15	13 20 to 23	1 11·6	91·7	·0310	
	29 13 to 14	29 21 to 22	0 22·0	20·0	·0068	
December -	26 13 to 18	26 21 to 2+	0 42·0	134·0	·0457	
1844. January -	8 12 to 14	8 20 to 22	0 20·2	73·4	·0240	This occasion, although introduced here, extra observations having been taken from 1 to 2 A.M., is perhaps an instance of aurora unattended by disturbance of the magnetic elements.
	19 13 to 14	19 21 to 22	0 22 0	55·2	·0188	
	24 and 25	—	2 15·8	190·5	·0648	Term day.
	26 14 to 15	26 22 to 23	0 31·2	26·2	·0089	
February -	5 8 to 15	5 16 to 23	1 05·0	195·8	·0669	The principal disturbance occurred between midnight and 15h., but no aurora appeared after 9h. although it remained cloudless.
	16 13 to 14	16 21 to 22	0 19·0	33·0	·0113	See remark to January 8th.
	26 15 to 16	26 23 to 0	0 11·2	68·0	·0232	Aurora became visible at 17h; it was clouded during the observations.
	28 16 to 17	29 01 to 1	0 20·8	—	—	Clouded during the observations. See remark to January 8th.
		At Mackenzie's River.				
	2 13 to 22	2 22 to 6+	3 18·4	140·0	·0845	Aurora was seen from 8 P.M. to midnight. This disturbance began after it was quite over.
April -	6 9 to 11	6 18 to 20	0 44·0	61·5	·0195	
	9 12 to 16	9 21 to 1+	1 17·6	75·4	·0240	
	10 9 to 14	10 18 to 23	1 26·2	162·0	>·0515*	
	14 12 to 18	14 21 to 3+	2 16·2	199·0	·0564	
	15 10 to 12	15 19 to 21	0 40·6	78·2	·0221	
	16 5 to 23	16 14 to 7+	8 10·0	570·0	>·1613*	
	19 13 to 14	19 22 to 23	0 45·4	76·1	·0215	
	24 —	—	3 46·0	270·0	·0767	Term day, April 24 and 25.
	25 11 to 18	25 20 to 2+	2 20·4	289·0	·0818	Generally clouded, but auroral light visible.
	28 12 to 20	28 21 to 5+	1 58·5	139·0	·0473	
May -	2 10 to 14	2 19 to 23	0 59·8	66·1	·0187	
	5 22 to 24	5 12 to 14	1 16 5	106·1	·0306	Aurora was not visible after this date.

* Beyond the scale on the negative side.

II.—Magnetical Disturbance with a clear sky, but no Aurora visible.

Date.	Mean Time.	Gött. Time.	Range.		ΔX / X	—
			Declination.	Horizontal Force.		
1843.	H. H.	D. H. H.	° ′			
December 8	10 to 12	8 18 to 20	—	49·2	·0168	It was unclouded during only a portion of this disturbance, which amounted but to a magnetic shock.
" 19	12 to 18	19 20 to 2+	0 34·0+	128·8	·0440	It was clouded during the greater part of this disturbance also, and only became clear at 16h.
" 29	13 to 16	29 21 to 0	1 07·3	126·9	·0433	Perfectly unclouded during the last 2 hours only.
1844. January 5	15 to 19	5 23 to 3+	1 04·0	234·0	·0798	The only complete instance of unclouded sky during a disturbance, and no aurora seen.
February 5	12 to 15	5 20 to 23	0 59·8	229·5	·0783	There was aurora visible down to 9h. as noted above, but none was seen later, although the sky remained unclouded.

III.—Magnetic Disturbance when the sky was clouded over.

Date.	Mean Time.	Gött. Time.	Range.		ΔX / X	—
			Declination.	Horizontal Force.		
1843.	D. D.	D. H. H.	° ′			
October -	18 and 19	—	1 24	70·5	·0241	Term day.
	H. H. 19 10 to 16	19 18 to 0+	0 54·0	184·8	·0630	Snow.
	23 17 to 19	24 1 to 3	0 29·2	80·0	·0273	
	29 18 to 20	30 2 to 4	1 25·0	77·3	·0264	
	30 13 to 20	30 21 to 4+	1 24·0	213·0	·0727	
November	8 9 to 17	8 17 to 1+	1 21·0	124·0	·0424	Cessation from 20h. to 23h. Gött. Snow.
	10 12 to 14	10 20 to 22	0 17·8	83·4	·0285	
	12 19 to 22	13 3 to 6	0 30·8	101·1	·0342	Daylight.
	13 12 to 15	13 20 to 23	1 11·6	91·0	·0310	
	14 11 to 12	14 19 to 20	0 50·6	34·3	·0117	
	16 9 to 11	16 17 to 18	0 37·0	12·9	·0044	Snowing.
	23 16 to 18	24 0 to 2	0 19·3	78·6	·0268	
December -	1 13 to 20	1 21 to 4+	2 40·2	136·5	·0466	
	5 14 to 19	5 22 to 3+	0 57·6	113·6	·0388	
	19 —	—	—	—	—	See List II.
	27 10 to 12	27 18 to 20	0 19·4	19·3	·0066	Snowing towards the close.
	27 18 to 20	28 2 to 4	1 17·4	85·8	·0293	Snowing.
	29 —	—	—	—	—	See List II.
1844. January -	4 8 to 19	4 16 to 3+	1 26·0	292·5	·0998	
	5 0 to 1	5 8 to 9	0 15·2	41·2	·0141	Daylight, an unusual hour for disturbance.

III.—Magnetic Disturbance—continued.

Date.	Mean Time.	Gött. Time.	Range.		$\frac{\Delta X}{X}$	—
			Declination.	Horizontal Force.		
1844.	D. H. H.	D. H. H.	° ′			
January -	16 16 to 17	17 0 to 1	0 7·2	37·5	·0128	The Bifilar Magnet unusual vibration.
	31 16 to 20	Feb. 1 0 to 4	1 12·7	128·0	·0437	
February -	1 11 to 13	1 19 to 21	2 07·2	113·5	·0387	
	1 22 to 24	2 6 to 8	0 52·0	63·0	·0215	Daylight. Snow.
	2 9 to 13	1 17 to 21	0 40·6	49·4	·0169	
	4 0 to 18	5 0 to 2	1 58·4	229·5	·0783	
	6 10 to 12	6 18 to 20	0 45·4	39·4	·0134	
	7 20 to 22	8 4 to 6	0 20·0	58·0	·0198	
		At Fort Simpson.				
April -	26 8 to 18	26 17 to 2+	1 58·5	172·3	·0488	
	29 15 to 17	30 0 to 2	1 36·4	209·5	·0593	
	30 5 to 16	30 14 to 1+	2 15·2	288·3	·0806	

Table LXXII.

Abstract from the Meteorological Journal.

[The entries are given in full when Aurora was visible, but otherwise are given for 3 A.M., 9 A.M., 3 P.M., and 9 P.M. alone. The Göttingen time of particular appearances is retained, for convenience of reference to the Magnetical Observations.]

At Lake Athabasca.

Date.		Wind.		Temp.	Weather.
Gött. Time.	Mean Time.	Direction.	Force.	Newman corrected.	
October 1843.					
D. H.	D. H.			°	
15 21	15 13	—	Calm.	—	Unclouded. Faint aurora, in bands from W. to E.
22	14	—	Calm.	—	Unclouded. Faint diffused aurora both N. and S. of the zenith, and in motion. Observations for disturbance began.
23	15	—	Calm.	—	Unclouded. Aurora brighter, and gathered to a corona near the zenith. Most westerly position of the Declinometer at 23h 5m (−1° 2′·6); most easterly position at 23h 50m (+1° 19′·7); range, 2° 35′·5.*
16 0	16	—	Calm.	31·0	Two parallel arches of aurora in the N.; brightest at the extremities, E. and W., and striated. Lowest reading of the Horizontal Force at 0h 5m (−·056 X.)
1	17	—	Calm.	31·1	Unclouded. Brightest portion of aurora to the S. of the zenith; faint auroral bands in the E. Highest reading of the Horizontal Force at 1h 20m (+·015 X.)
2	18	—	Calm.	30·4	Partially clouded with light cirro-cumuli and cirri. Very faint aurora still visible in the E.
5	21	—	Calm.	33·9	Cirrous haze in the atmosphere.

* The actual difference between the highest and lowest readings during a period of disturbance is here called the range of scale. The deviation E. and W., or + and −, are measured from the mean scale reading for the same hour and minute; thus their sum may be greater or less than the difference of scale reading, by the amount of the mean diurnal change in the interval between them.

Abstract from the Meteorological Journal—*continued.*

Date.		Wind.		Temp.	Weather.
Gött. Time.	Mean Time.	Direction.	Force.	Newman corrected.	
October 1843.					
D. H.	D. H.			°	
16 11	16 3	—	Calm.	43·8	Overcast with cirrous haze; a few clear spaces to the S.E.
17	9	—	Calm.	37·5	Unclouded since 6h.
18	10	—	Calm.	35·4	Unclouded. Bright arch of aurora (3) in the N.
19	11	—	Calm.	34·5	Unclouded. A broad arch or band of aurora (4) extending across the zenith; a brighter arch (3) to the N.
20	12	—	Calm.	33·5	A very faint arch of aurora (5) in the N.
21	13	—	Calm.	31·7	Unclouded. A faint auroral arch extending from N.E. to N.W., brightest at the extremities. A few faint detached patches in the N.W.
22	14	—	Calm.	30·2	Arch less distinct. A broad band of aurora (5) to the S. of the zenith.
23	15	—	Calm.	29·6	Appearance of aurora nearly the same as before.
17 0	16	—	Calm.	27·0	Auroral arch much brighter than before, at its N.W. extremity.
1	17	—	Calm.	28·1	Unclouded. Faint arches or bands across the meridian in the zenith. Observations for disturbance began; at 1h 35m most easterly position of the Declinometer (+0° 46′·6); at 1h 45m lowest value of Horizontal Force (−·062 X); at 2h 40m most westerly position (−0° 15′·5), range 63′·2; at 2h 55m highest value of Horizontal Force (+·005 X).
5	21	N.N.E.	Fresh.	32·6	Hazy. Wind rising with gusts.
11	17 3	N.N.E.	V. light.	33·5	Overcast and hazy.
15	7	N.	Light.	32·4	Unclouded since 6h. At 15¼h a faint auroral arch (4), elevation 12°, extending from N.E. to N.W.
16	8	N.N.E.	Mod.	32·1	Arch stationary. Appearance of aurora little changed.
17	9	N.N.E.	Mod.	31·5	The same as before. Observations for disturbance commenced. At 17h 55m most easterly position of the Declinometer (+0° 39′·9).
18	10	N.N.E.	Mod.	31·5	Arch slightly risen, alt. 17°, and broader. Highest reading of Horizontal Force at 18h 55m.
19	11	N.N.E.	Mod.	32·1	Arch rising; at 19h 16m it extended across the zenith from E. to W.; at 19h 23m began to break up into waves in quick motion, but receded from the zenith to the N. Most westerly position of Declinometer at 19h 30m (+0° 57′·7), range 1° 36′·4.
20	12	N.N.E.	Light.	32·0	Detached masses of aurora, resembling cirrous clouds (4) in the zenith, and to the S. and E., which disappeared before 20h 40m. No aurora at the two following hours. Lowest reading of Horizontal Force at 20h 15m.
23	15	—	Calm.	32·0	Faint cirrous aurora visible.
18 5	21	N.E. by E.	High.	33·7	Wind in gusts. Cir-stratus in the S.E.; remainder clear.
11	18 3	N.E. by E.	Fresh.	36·6	Nearly overcast. Cir-cumulus in the S.E. Term day. Most easterly position of the Declinometer at 15h 40m (+19′·5); at 10h wind fallen.
17	9	N.N.E.	Fresh.	31·8	Overcast since 13h; wind high, and in gusts. Most westerly position of the Declinometer at 16h 35m (−4′·34), range, 1° 2′·4.
23	15	—	Calm.	30·0	A few stars visible near the zenith. Remainder overcast.
19 5	21	—	Calm.	31·8	Dull and overcast. A sprinkling of snow at 16h and 17h.
11	19 3	—	Calm.	32·7	Fine snow mixed with rain since 9h; continued with intervals until past 13h. Considerable disturbance throughout the term observations, with manifest correspondence in some of the principal movements at Lake Athabasca, Toronto, and Greenwich.

Abstract from the Meteorological Journal—*continued.*

Date.		Wind.		Temp.	Weather.
Gött. Time.	Mean Time.	Direction.	Force.	Newman corrected.	
October 1848.					
D. H.	D. H.			°	
19 17	19 9	—	Calm.	30·4	Completely overcast. Observations for disturbance 18h to 19h, and again 21h to 0h. Most easterly position of the Declinometer at 19h (+6° 23'·0); most westerly position at 21h 30m (−24'·4); range, 0° 47'·6. The following entry occurs in the Meteorological Register at Toronto, at 17h Gött.:—"Light cirri and haze "generally over the sky; faint auroral light "in the north; faint streamers." Aurora was seen at three stations in the State of New York on the same evening (Regent's Reports.)
23	15	—	Calm.	29·0	Snowing lightly since 21h.
20 5	21	—	Calm.	30·4	Overcast.
11	20 3	N.N.E.	Light.	32·7	Overcast, with cir-cumulus.
17	9	—	—	31·6	Clear and unclouded.
21	13	—	Calm.	29·0	Unclouded. A faint auroral arch (4) at an elevation of 18°. Aurora was also seen at two stations in the State of New York on the same evening (ib.)
23	15	—	Calm.	29·1	Calm and unclouded.
21 5	21	N.N.E.	V. light.	31·7	Partially clouded. Fleecy cir-cumuli, with clear spaces.
11	21 3	S.W.byS.	Fresh.	47·5	Completely clouded. Wind fresh and gusty.
17	9	S.S.W.	High.	41·5	Dense masses of cir-cumuli to east and zenith, and round horizon; remainder clear.
Sunday.					
22 23	22 15	S.W. by S	V. high.	14·8	Blowing a gale. Sky completely overcast.
23 5	21	W.S.W.	V. light.	16·0	Wind fell at 2h. Snowing lightly since 3h.
11	23 3	S.S.W.	Light.	18·0	Fair, with cirrous clouds. Ceased snowing before 7h.
17	9	W.S.W.	Light.	16·1	Thickly overcast, and very dark.
23	15	—	Calm.	14·8	Dull; calm; a few stars visible; mostly clouded.
24 5	21	—	Calm.	17·1	Overcast since 0h. A few particles of snow at 4h. Extra observations for disturbance, 1h to 3h. A minimum of Horizontal Force at 1h 15m (−·027 X). A maximum of Declination at 1h 25m (+0° 35'·1).
11	24 3	—	Calm.	19·5	Uniformly overcast. A few particles of snow at 12h.
17	9	—	Calm.	16·6	A bank of stratus to east. Remainder unclouded.
23	15	W.S.W.	High.	17·5	Overcast. Wind high and squally.
25 5	21	W.S.W.	Fresh.	21·7	Overcast, beginning to snow lightly. Masses of fog on the lake.
11	25 3	W.S.W.	Fresh.	21·5	Clouds very low. Cir-cumuli.
17	9	W.S.W.	Fresh.	8·2	Densely overcast. Observations for disturbance began at 19h. At 19h 5m most easterly position of Declinometer (+1° 47'·6).
21	13	—	Calm.	4·9	Unclouded. Sky cleared since 20h. An auroral arch (3, 4,) at an elevation of 5°, rising rapidly; at 21h 13m elevation 35°. At 21h 28m aurora in the zenith, and portions to the S.; brightness (2, 3), the brightest portion to the W., but dispersed and broken up. At 21h 33m it appeared as a broad faint arch, 45° above the southern horizon, but dispersed and brighter (3, 2,) to the W.; portions to the E. had a striated appearance. The most westerly position of the Declinometer at 21h 35m (−0° 19'·8); range, 2° 7'·4. At 21h 46m it had nearly disappeared, but a fresh display, of an irregular flexuous form, brightness (3), was rising in the N.
22	14	—	Calm.	3·3	Unclouded. Aurora arch (4) to N. and (5) to S. of zenith; elevation not recorded. At 22h 5m the highest value of Horizontal Force (+·015 X). At 22h 30m aurora faint and stationary. Inclinometer thrown into a small arc of vibration, not usual with this magnet.

Abstract from the Meteorological Journal—*continued.*

Date.				Wind.		Temp.	Weather.
Gött. Time.		Mean Time.		Direction.	Force.	Newman corrected.	
October 1843.							
D.	H.	D.	H.			°	
25	23	25	15	—	Calm.	0·1	A faint striated arch of aurora (4); elevation 16° from southern horizon. Another in the N. (5); elevation 20°. At 23h 10m brilliant annular bands (3, 4) with streamers, and striated to S. and W. At 23h 15m aurora in the form of a thin flexuous ring, striated, in moderate motion, passing from S. to N., and E. of the zenith. At 23h 25m aurora much fainter, and more diffused; Declinometer more disturbed. Easterly movements, and lowest value of Horizontal Force (−·075 X).
26	0		16	—	Calm.	−2·2	Unclouded. Aurora very bright (2, 3,) forming an imperfect circle, striated at 0h 25m. Fainter and more diffused, in detached flexuous portions, striated. At 0h 40m aurora fainter, like cirrous clouds, detached round horizon. Magnets returning to their mean positions.
	1		17	—	Calm.	−5·8	Unclouded. Aurora fainter (4); patches resembling cirrous clouds, and stationary; 1h 30m scarcely perceptible; 1h 45m aurora at an end.
	5		21	—	Calm.	−0·1	Lightly overcast, with cirrous clouds.
	11	26	3	—	Calm.	6·2	Lightly overcast, with cirrous clouds and haze.
	15		7	—	Calm.	−4·7	Unclouded. At 15h 30m a faint auroral arch, at elevation 15°, from N.E. to N.W.
	16		8	—	Calm.	−6·0	Unclouded. Arch rising gradually, and becoming brighter; elevation 18°. A second arch, much fainter, at elevation 13°.
	17		9	—	Calm.	−5·1	Unclouded. A faint auroral arch (5), elevation 5°, to E. of N.
	18		10	—	Calm.	2·2	A faint auroral arch (5), at an elevation of 37° Lightly overcast to the S.; remainder of the sky clear.
	19		11	E.S.E.	Fresh.	0·6	Unclouded. A faint auroral arch (5); elevation 5°. Most westerly reading of the Declination (−27′·9), at 19h 0m.
	20		12	E.S.E.	Fresh.	0·2	Unclouded. Aurora in heavy masses of moderate brightness, (3, 4,) in little motion. Long streamers, extending to the zenith. Observations of disturbance commenced. At 20h 40m two long beams, nearly stationary, alt. 54° and 60°; brightest (2, 4,) to westward of the zenith. Flexuous masses and unconnected streamers in N.N.W. and N.E. As yet little disturbance.
	21		13	E.S.E.	Fresh.	−0·7	Unclouded. Aurora faint (4); diffused like thin vapour in the zenith, with moderate rapid motion to the E. Streamers somewhat brighter (3) in the N. Faint streamers and flexuous masses to the E. and N.W.
	22		14	E.S.E.	Fresh.	−0·1	Becoming clouded. Aurora very faint (5); streamers scarcely perceptible; at 22h 30m no aurora visible. The greatest value of the Horizontal Force (−·028 X) occurred at 23h 30m.
27	0		16	—	Calm.	0·0	Unclouded. A fresh display of aurora, rising rapidly from the northern horizon, and extending itself from an alt. of 20° N. to 40° S. At 0h 30m considerable disturbance. Bright aurora (3, 4,) in flexuous masses, to S. and S.E. of the zenith. An arch (3), at elevation 36°, extending from N.E. to N.W.
	1		17	—	Calm.	0·2	Lightly clouded to S. and E. Patches of aurora of various brightness (3, 5). At 1h 5m the easterly extreme of the Declination (+34′·0); the lowest range of Horizontal Force (−·058 X), about 10m earlier. At this time a large dense mass of aurora passing the zenith, where it became faint, in flexuous streaks, without perceptible motion. At 2h no aurora.
	5		21	E.N.E.	Light.	1·2	Clear and unclouded. Light easterly wind.
	11	27	3	N.N.E.	Fresh.	9·2	Lightly overcast with uniform haze.
	17		9	N.N.E.	Fresh.	5·7	Partly overcast. Snowing lightly.

Abstract from the Meteorological Journal—*continued.*

Date.				Wind.		Temp.	Weather.
Gött. Time.		Mean Time.		Direction.	Force.	Newman corrected.	
October 1843.							
D.	H.	D.	H.			°	
27	21	27	13	N.N.E.	Fresh.	0·4	Lightly clouded in the S. A faint aurora in the N. Observations for disturbance since 20^h. Most westerly position of Declinometer at 20^h 45^m (0° 6′·3).
	22		14	N.N.E.	Fresh.	0·1	Unclouded. A faint aurora, diffusing itself from a point at an elevation of 32°.
	23		15	N.N.E.	—	−0·1	Unclouded. No aurora visible. Horizontal Force greatest at 23^h 30^m, least at 0^h 55^m; most easterly position of Declinometer (+0° 38′·2) at 1^h 5^m; range 0° 54′·0.
28	5		21	N.N.E.	Fresh.	−0·1	Unclouded, but has been generally hazy.
	11	28	3	N.N.E.	Light.	8·0	Unclouded, clear since 5^h; but soon after slightly overcast.
	17		9	N.N.E.	Light.	9·1	Unclouded since 14^h, save a few cirrous and stratus to S.W. at 16^h.
	18		10	N.N.E.	Light.	7·8	A considerable quantity of faint cirrous aurora floating about, with an arch of moderate brightness (4, 5,) and striated to the N.W., but no disturbance.
	19		11	—	Not observed.	6·6	Aurora (4) diffusing itself from a point in the W., with a striated appearance, and as before, in detached patches elsewhere.
	20		12	—	„	5·5	Faint aurora (5). Unclouded.
Sunday.							
29	23	29	15	—	Calm.	22·5	Uniformly overcast. Disturbance observations began at 2^h. Easterly extreme of the Declinometer at 2^h 0^m (+1° 9′·7). Horizontal Force lowest at 2^h 5^m (−·006 X); greatest at 3^h 55^m (+·016 X).
30	5		21	—	Calm.	23·8	Uniformly overcast. Extra observations ended at 4^h.
	11	30	3	—	Calm.	23·8	Uniformly overcast. A few particles of snow at 7^h and 9^h.
	17		9	—	—	21·6	Uniformly overcast.
	23		15	—	Calm.	19·6	Thickly overcast. Observations for disturbance began at 21^h; most westerly position of the Declinometer Magnet at 22^h 5^m (−0° 30′·3); most easterly position at 0^h 10^m (+0° 49′·2); range, 1° 24′·0. Lowest range of Horizontal Force at 22^h 20^m (−·047 X); highest at 2^h 20^m (+·022 X).
31	5		21	—	Calm.	22·8	Thickly overcast.
	11	31	3	N.N.E.	Mod.	24·3	Still overcast.
	17		9	E.	Very high.	16·9	Wind changed since 16^h from N.N.E. to E.S.E. Sky clearing since 14^h.
	18		10	E.N.E.	Fresh.	16·0	Unclouded. A brilliant aurora (3), the light diverging from a focus in the E., extending to the zenith; flexuous, with streamers. At 18^h 15^m it had become diffused and faint (4, 5). No disturbance.
	23		15	W.S.W.	High.	14·6	Wind high and squally. Sky partially covered.
November.							
1	5		21	W.S.W.	Mod.	17·9	Clouded. Wind moderated since 3^h. A little snow at 4^h.
	11	1	3	N.E. by E.	Light.	29·0	Overcast; thick. Wind changed to E. at 6^h.
	17		9	S.E. by S.	Fresh.	21·7	Partially clear. Wind in gusts.
	21		13	N.N.E.	Mod.	22·5	Unclouded. A faint arch of aurora.
	23		15	N.N.E.	Mod.	21·5	Clear; no aurora visible; clouded over from the E. soon after.
2	5		21	E.S.E.	High.	22·9	Lightly overcast. Wind in gusts.
	11	2	3	E.S.E.	Light.	30·2	Sky nearly clear since 6^h.
	16		8	—	Calm.	23·8	A faint auroral arch (5) from N.E. to N.N.W.; elevation 28°. Magnets disturbed, but not much change of position. Greatest value of Horizontal Force (+·008 X) at 16^h 15^m.
	17		9	—	Calm.	23·3	A faint but broad band or arch of aurora (4, 5,) covering the heavens, from alt. 46° to the zenith; 17^h 15^m, aurora more brilliant, the arch breaking up, and masses in motion. 17^h 30^m, the preceding display at an end; a fresh arch (4), elevation 10° in the N.E. Most westerly position of the Declinometer (−1° 29′·9) at 17^h 30^m. 17^h 40^m, the arch risen to 30°, and beginning to break up; 18^h, no aurora.

Abstract from the Meteorological Journal—*continued.*

Date.		Wind.		Temp.	Weather.
Gött. Time.	Mean Time.	Direction.	Force.	Newman corrected.	
November 1843.					
D. H.	D. H.			°	
2 21	2 13	E.S.E.	High.	24·7	Overcast, with dense cirro-cumuli; cleared soon after. At 21^{h} 30^{m} a bright mass of aurora in the N. at elevation 15°. At 21^{h} 45^{m} aurora much fainter (4, 5,) and diminishing in extent. At 21^{h} 55^{m} no longer visible.
22	14	E.S.E.	High.	25·7	No aurora. At 22^{h} 40^{m} a faint (5) annular mass in the N.E.
23	15	E.S.E.	High.	25·5	Clear. A faint arch (5) in the N. At 23^{h} 15^{m} aurora appearing as if it issued, in a stream, from a source in the N., expanding as it extended nearly horizontally to the N.E. at alt. 32°, in outline like the tail of a fox. At 23^{h} 30^{m} a dense but faint mass (5) moving from N.E. to N.W.; most easterly position of the Declinometer Magnet (+0° 40′) at 23^{h} 30^{m}; range, 2° 14′·2. Lowest range of Horizontal Force (−·052 X), a little earlier. The disturbance observations were discontinued at 1^{h}.
3 5	21	E.S.E.	Mod.	23·8	Densely overcast since 0^{h}. Light snow at 4^{h} 5^{h}, and 6^{h}.
11	3 3	E.	Fresh.	30·6	Overcast; occasional sprinkling of snow.
17	9	E.S.E.	Fresh.	27·7	Uniformly overcast, with dense cirro-cumuli.
23	15	—	—	—	Observation omitted.
4 5	21	—	Calm.	27·6	Overcast, with light haze or fog.
11	4 3	—	Calm.	27·7	No change.
17	9	—	Calm.	25·3	No change from 3^{h} to the end of the day at 20^{h}.
Sunday.					
5 ·23	5 15	—	Calm.	16·6	Unclouded, but no appearance of aurora; Magnets slightly disturbed at 23^{h} 30^{m}. A sudden change of moderate extent in all the readings; at the same time aurora visible in streamers in the N.E. and N.W., but faint (4). At 24^{h} no aurora, but considerable disturbance. An easterly extreme of the Declinometer Magnet at 24^{h} (+0° 41′·9), and lowest range of Horizontal Force (−·022 X).
6 5	21	—	Calm.	15·9	Calm and unclouded since 23^{h}.
11	6 3	—	Calm.	24·1	Gradually clouding over since 6^{h}; now completely overcast.
17	9	—	Calm.	17·1	Uniformly overcast since 17^{h}.
23	15	—	Calm.	17·9	No change.
7 5	21	—	Calm.	16·1	No change. A thick fog at 1^{h}.
11	7 3	—	Calm.	15·3	No change. A little snow at 9^{h} and 10^{h}.
17	9	—	Calm.	16·4	No change.
23	15	—	Calm.	19·1	No change.
8 5	21	—	Calm.	19·5	No change.
11	8 3	—	Calm.	15·6	No change. Light snow at 6^{h}.
17	9	—	Calm.	10·0	Light snow, with intervals, since 14^{h}; light N.N.W. wind 18^{h} to 20^{h}; considerable magnetic shock between 18^{h} and 19^{h}; most easterly position of Declinometer (+0° 58′·9) at 18^{h}; most westerly (−0° 12′·6) at 18^{h} 40^{m}; range 1° 11′·4. Lowest Horizontal Force at 18^{h} 45^{m} (−·015 X).
23	15	—	Calm.	7·2	Overcast.
9 5	21	—	Calm.	7·6	Overcast. A few particles of snow at 4^{h}, and again at 9^{h}.
11	9 3	—	Calm.	8·9	Overcast.
17	9	—	Calm.	4·7	Partially clear. Cirro-cumuli.
20	12	—	Calm.	−0·1	Unclouded since 18^{h}. At 20^{h} 15^{m} a bright auroral arch (3) extending from N.W. to N.E., elevation 28°, gathering to focus at the W. extremity. It varied but little until 20^{h} 50^{m}, when it rapidly disappeared. Lowest value of Horizontal Force (−·027 X) at 22^{h} 10^{m}; most easterly position of Declinometer (only +72′·2) at 22^{h} 15^{m}.
23	15	—	Calm.	−1·2	Unclouded. No aurora since 20^{h} 50^{m}. Extra observations were commenced at 22^{h}, on account of a somewhat low range of Horizontal Force, as shown above. The range of Declination was very trifling.
10 5	21	—	Calm.	1·6	Unclouded since 20^{h}, with a short interval of 0^{h} of cirro-cumuli.
11	10 3	—	Calm.	9·6	Unclouded. No change.

Abstract from the Meteorological Journal—*continued.*

Date.		Wind.		Temp.	Weather.
Gött. Time.	Mean Time.	Direction.	Force.	Newman corrected.	
November 1843.					
D. H.	D. H.			°	
10 17	10 9	—	Calm.	5·5	Thickly clouded, with cirro-cumuli gathering since 16^h.
23	15	—	Calm.	3·2	Overcast, and mostly so since 17^h. Slight disturbance 21^h to 22^h. Minimum of Horizontal Force (− ·030 X) at 21^h 10^m; range of Declination 17′·8.
11 5	21	W.N.W.	Light.	5·5	Overcast, with dense cirro-cumuli.
11	11 3	W.N.W.	Light.	6·8	Lightly overcast, with cirro-strati.
17	9	W.N.W.	Light.	6·8	Snowing, with intervals, from 14^h to the end of the day at 20^h.
Sunday.					
12 23	12 15	N.N.E.	Mod.	−6·8	Thickly overcast. Gusts of wind.
13 5	21	N.N.E.	Mod.	0·6	Clouded. Beginning to snow, and continued to do so until 10^h. Slight disturbance, 4^h to 6^h. Minimum of Horizontal Force (− ·035 X) at 4^h 8^m; range of Declination 0° 25′·9.
11	13 3	—	Calm.	6·5	A bank of strata in the S., rising from which, as it appeared, in a vertical direction, was a portion of a rainbow; remainder of the heavens clear.
15	7	—	Calm.	1·6	Unclouded. A faint arch of aurora, at an elevation of 13° in the N. At 15^h 30^m the arch brighter (3), and extending from N.E. to N.W., elevation 18°, and gathering to a focus at the W. end.
16	8	—	Calm.	0·3	Unclouded. Bright auroral band (3), extending from N.W. to N.E., of irregular or serpentine form towards the N.W. extremity, enlarging and becoming brighter at the opposite extremity, elevation 16°; at 16^h.15^m, the band contracting in width and sinking towards the horizon, but retaining its brilliancy.
17	9	—	Calm.	−0·5	Unclouded. No aurora visible.
19	11	—	Calm.	−1·0	Unclouded. Re-appearing, an arch (4), elevation 10°; commenced observation for disturbance at 20^h.
20	12	—	Calm.	2·3	Unclouded. Detached vertical patches of aurora, having a striated appearance at different altitudes, and a faint band (4, 5) across the meridian in the zenith. Immediately afterwards a large circular ring (3, 4) in motion; considerable disturbance of an unusual character as regards the Declination, being chiefly to the westward. Most westerly position (−1° 15′ 2), and minimum of Horizontal Force (− ·031 X) at 20^h 16^m. At 20^h 20^m it clouded over. The magnetic changes were trifling after 21^h.
23	15	N.N.E.	Light.	8·0	Thickly overcast, with the same wind, since 20^h. Most easterly position of the Declinometer (+0° 2′·9), at 22^h 58^m, range 1° 11′·6.
14 5	21	E. by N.	Fresh.	11·2	Overcast. Wind began to rise soon after.
11	14 3	S.E. by S.	Fresh.	17·9	Dispersed cirro-cumuli. Wind somewhat abated, but a violent gale and snow-storm from S.S.E. at 7^h, 8^h, and 9^h.
17	9	E.S.E.	V. light.	18·6	Overcast. At present nearly calm, but a furious gale from S.S.E. has prevailed, with occasional intermission, since 8^h Gött. Slight disturbance 19^h to 20^h. A maximum of Horizontal Force (+ ·008 X) at 19^h 34^m. A westerly extreme of Declination (−32′·4) at 19^h 43^m.
23	15	—	Calm.	10·0	Thickly overcast.
15 5	21	—	Calm.	6·0	Unclouded. A few particles of snow falling. Unclouded at 1^h, 2^h, and 3^h.
11	15 3	—	Calm.	8·0	Overcast.
17	9	—	Calm.	10·4	Overcast. Snow at 14^h, with light E. wind.
23	15	E.N.E.	Light.	11·4	Overcast. Light E. wind since 21^h.
16 5	21	—	Calm.	15·1	Overcast. Began to snow soon after.
11	16 3	—	Calm.	12·3	Cirro-cumuli, with a few clear spaces.
17	9	S.S.W.	Light.	9·1	Overcast. Snowing lightly since 13^h, with a clear interval at 16^h. A slight disturbance prevailed from 18^h to 19^h, marked by Horizontal Force above the Mean (greatest value + ·009 at 18^h 4^m); range of Declination 0° 37′.
23	15	S.S.W.	Light.	10·2	Snowing lightly from 22^h to 1^h.
17 5	21	—	Calm.	8·0	Overcast.

Abstract from the Meteorological Journal—*continued.*

Date.		Wind.		Temp.	Weather.
Gött. Time.	Mean Time.	Direction.	Force.	Newman corrected.	
November 1843.					
D. H.	D. H.			°	
17 11	17 3	—	Calm.	10·0	Unclouded, save a few strati, and cirro-strata near horizon.
17	9	—	Calm.	8·2	Unclouded since 14^h.
19	11	—	Calm.	6·0	A faint arch of aurora in the N., at elevation 22°, first observed at 18^h 30^m, and not changed since.
23	15	—	Calm.	5·7	Unclouded. No trace of aurora since 19^h.
18 5	21	—	Calm.	4·8	Unclouded; lightly overcast with cirrous haze.
11	18 3	N.N.E.	Light.	8·7	Again unclouded since 6^h. Light N.E. wind began at 8^h.
17	9	N.N.E.	Mod.	9·9	Clouded at 13^h; since clear. Wind increasing since 14^h.
Sunday.					
19 23	19 15	—	Calm.	7·0	Thickly overcast.
20 5	21	—	Calm.	7·6	Thickly overcast. Began to snow at 4^h.
11	20 3	—	Calm.	8·4	Still snowing.
14	6	N.N.E.	Light.	5·8	Unclouded. *A light snow was falling from a perfectly clear sky, the stars being visible in every part of it, with little haze. The same at 7^h, but the quantity less, and the stars more hazy.*
17	9	N.N.E.	Mod.	4·7	Overcast, and snowing lightly.
23	15	—	Calm.	6·3	Overcast.
21 5	21	—	Calm.	7·3	Overcast. Snowing since 1^h.
11	21 3	N.N.E.	Light.	7·8	Ceased snowing at 8^h, but now resumed.
17	9	—	Calm.	6·0	Thickly overcast.
23	15	—	Calm.	5·5	Thickly overcast. Snowing from 19^h to 22^h.
22 5	21	—	Calm.	6·9	Thickly overcast.
11	22 3	—	Calm.	9·7	Overcast.
17	9	—	Calm.	4·8	A few stars visible to the N.W. near the zenith. Light air from S. at 14^h, 15^h, and 16^h.
20	12	—	Calm.	0·6	Unclouded. A bright arch of aurora (3) extending from N.N.E. to N.W. at elevation 32°, first observed at 19^h 20^m.
21	13	N.E.	Fresh.	−1·2	Unclouded. A very faint auroral light in the N.
23	15	N.E.	Fresh.	−3·6	Unclouded.
23 1	17	N.E.	Fresh.	−5·7	Unclouded. A faint arch of aurora (5) at elevation 15°.
2	18	N.E. by E.	Fresh.	−3·5	Faint auroral light (5) in the N.W. Sky nearly overcast.
5	21	N.N.E.	Light.	0·2	Overcast since 3^h.
11	23 3	N.N.E.	Light.	6·7	The same.
17	9	—	Calm.	8·1	Overcast. Wind freshened at 13^h, but since fallen off to a calm.
23	15	W.N.W.	Fresh.	5·1	Overcast. A few particles of snow at 22^h, with N.E. wind. Slight disturbance 0^h to 2^h. Minimum of Horizontal Force at 1^h (−·024 X). Most easterly position of Declination (+16′·5) at 1^h 7^m.
24 5	21	W.N.W.	V. Light.	3·2	Overcast. Wind high and squally at 0^h, 1^h, 2^h. Since abated.
11	24 3	W.N.W.	Light.	3·8	Still overcast. Magnetic term day, which wa wholly free from disturbance, commenced a 10^h.
17	9	W.N.W.	Light.	3·0	No change.
23	15	W.N.W.	Light.	3·8	No change.
25 5	21	E.S.E.	Light.	0·9	Overcast, with cirro-cumuli.
11	25 3	—	Calm.	1·2	Unclouded since 9^h.
17	9	—	Calm.	1·3	Partially clear. So continued, with wind from N.E. to the end of the day at 20^h.
Sunday.					
26 23	26 15	E.S.E.	V. high.	2·9	Unclouded, with a strong gale from S.E., since commencement of observation at 21^h.
27 5	21	S.S.E.	Light.	4·2	Wind fell at 3^h. Still unclouded.
11	27 3	S.S.E.	Light.	9·7	Overcast, with light haze.
17	9	—	Calm.	8·0	Unclouded since 12^h.
20	12	—	Calm.	5·1	Unclouded. At 20^h 30^m faint masses of aurora of irregular form to the E. of N.
23	15	—	Calm.	2·0	Continues unclouded, and so on to 3^h.
28 5	21	—	Calm.	−2·3	Lightly overcast, with haze.
11	28 3	—	Calm.	3·6	Unclouded since 6^h.
17	9	—	Calm.	0·4	Unclouded. No change.
19	11	—	Calm.	3·5	Unclouded. An irregular striated arch of aurora (4) at an elevation of 30°.
20	12	—	Calm.	3·2	Unclouded. Floating patches of fain aurora (4.)
21	13	S.S.W.	Light.	7·9	Unclouded. A faint auroral light in the N.

Abstract from the Meteorological Journal—*continued.*

Date.				Wind.		Temp.	Weather.
Gött. Time.		Mean Time.		Direction.	Force.	Newman corrected.	
November 1843.							
D.	H.	D.	H.			°	
28	23	28	15	S.S.W.	High.	12·3	Wind in gusts.
29	1		17	S.S.W.	Fresh.	17·9	Unclouded. Faint streaks of aurora (5) near the zenith.
	2		18	S.	High.	20·3	Unclouded. Patches of aurora faint (5), but striated in the N.W.
	5		21	—	Calm.	14·8	Unclouded, as it has been, with little intermission, for three days.
	11	29	3	—	Calm.	11·0	Unclouded. The same.
	17		9	—	Calm.	0·6	Unclouded. The same.
	19		11	—	Calm.	−3·5	Unclouded. A faint auroral arch (5) from N.E. to N.W. at elevation 37°, gathering to a focus at the N.W. extremity.
	20		12	—	Calm.	−4·7	Unclouded. Aurora visible, but very faint.
	21		13	—	Calm.	−6·5	Unclouded. Arch of aurora gathering to a focus at both E. and W. extremity. Extra observations were taken from 21^{h} to 22^{h}, showing the magnetic elements to be very slightly disturbed.
	23		15	—	Calm.	−9·4	Unclouded. A faint arch (5) in the north.
30	5		23	N.	Fresh.	−4·7	Overcast, with cirrus haze.
	11	30	3	N.N.E.	Mod.	0·3	Overcast, with cirro-cumuli and haze. Wind high, 6^{h} to 10^{h}.
	17		9	N.N.E.	High.	−1·5	Generally clear since 15^{h}.
	23		15	N.	High.	−1·5	Beginning to snow, and continued to do so till 3^{h}.
Dec.							
1	5		21	N.E.	High.	6·4	Overcast. Thick fog at 4^{h}.
		Dec.					
	11	1	3	N.E.	Mod.	5·3	Overcast.
	17		9	N.E. by E.	Mod.	8·9	A few stars beginning to appear.
	23		15	—	Calm.	13·4	Lightly overcast. A great disturbance first observed at 21^{h}. Most westerly reading of Declinometer (−1° 34′·5) at 22^{h} 15^{m}. Most easterly reading (+1° 1′·5) at 3^{h} 27^{m}. Range 2° 40′. Lowest value of Horizontal Force (−·046 X) at 2^{h} 21^{m}.
2	5		21	N.W. by W	High.	7·8	A most violent gale from N.W. at 2^{h}; since abated. Disturbance observations discontinued at 5^{h}.
	11	2	3	W.	Light.	0·6	Snowing lightly since 8^{h}.
	17		9	S.	Fresh.	−2·5	An halo of diameter 40° round the moon. The same at 20^{h}, with two indistinct para selenæ. The thermometer reached 21° 7′ this morning.
Sunday.							
3	23	3	15	S.E. by S.	V. high.	2·2	Overcast. A violent gale since commencement of observations at 21^{h}.
4	5		21	S.E. by S.	High.	3·0	Overcast. Wind at intervals still violent, and so down to 7^{h}.
	11	4	3	—	Calm.	8·9	Cirro-strata and strata generally diffused, but clear from 6^{h} to 9^{h}.
	17		9	—	Calm.	7·8	Lightly overcast.
	23		15	N.W.	Mod.	6·2	A few cirro-strati, otherwise clear.
5	5		21	—	Calm.	3·9	A few light cirro-cumuli overhead, otherwise clear.
	11	5	3	—	Calm.	3·2	Overcast. Snowing, which soon ceased.
	17		9	E.S.E.	Mod.	4·1	Wind in gusts.
	23		15	E.S.E.	High.	15·3	Lightly overcast. Considerable disturbance, 22^{h} to 3^{h}. A minimum of Horizontal Force (−·031 X) at 22^{h} 19^{m}, attended by a westerly range of Declination, and followed by the most westerly reading of the Declinometer (−0° 24′·4) at 22^{h} 25^{m}, a circumstance which was unusual. Most easterly position (+0° 26′·4), at 0^{h} 7^{m}, range 0° 55′·6.
6	5		21	S.E.	High.	27·0	*A heavy shower of rain fell at* 6^{h}; it continued to rain until 11^{h}. The temperature at the surface has only twice been as high as 32° since the 23d October; at 9^{h} and 10^{h} Gött. on the 2d November.
	11	6	3	W.	Mod.	33·7	Dense cirro-cumuli moving rapidly along the S. horizon from the S.W.
	17		9	N.W.	High.	14·6	Wind high all day, violent at 22^{h}. Snow at 15^{h}.
	23		15	N.W.	High.	6·3	Light snow at intervals; a few stars visible, but hazy. Sleet falling.
7	5		21	—	Calm.	−2·0	Unclouded since 1^{h}.
	11	7	3	—	Calm.	7·7	Thickly overcast. Mostly cirro-cumuli since 9^{h}.
	17		9	N.E.	Light.	13·2	Beginning to snow, mixed with rain.
	23		15	E.N.E.	V. light.	16·7	Snow coming to an end.

Abstract from the Meteorological Journal—*continued.*

Date.		Wind.		Temp.	Weather.
Gött. Time.	Mean Time.	Direction.	Force.	Newman corrected.	
December 1843.					
D. H.	D. H.			°	
8 5	7 21	N.N.E.	Light.	15·1	Covered with cirrus haze. A bank of clouds round the horizon.
11	8 3	S.S.W.	Light.	23·4	Overcast, but mostly clear since 5h.
17	9	W.S.W.	Light.	20·7	Again overcast, but unclouded from 13h to 16h. A marked magnetic shock at 18h. Most easterly reading of Declinometer at that hour (+0° 47′·4). Most westerly position at 19h 12m (−0° 11′·3) range 0° 58′·7.
23	15	W.S.W.	Light.	19·8	Nearly overcast, with cirro-cumuli and strati.
9 5	21	—	Calm.	19·6	Overcast, with cirro-cumuli and strati.
11	9 3	W.N.W.	Light.	13·0	Hazy cirri and strati round the horizon.
17	9	—	Calm.	3·1	Unclouded since 13h. After the close of the observations at 20h, a brilliant aurora was witnessed and described by Serjeant Henry; at 20h 22m it appeared in a stream, extending across the zenith in a S.W. direction from a point in the opposite quarter, at an elevation of 54°, like steam escaping from a tube, narrow, very brilliant (2), and of a light but decided green tint. Magnets disturbed at 20h 27m. It broke up into bands crossing the meridian, in the zenith, but equally brilliant. At 20h 30m the bands divided into waves, which dispersed with rapid motion, and soon after disappeared.
Sunday.					
10 23	10 15	—	Calm.	−3·3	Overcast. Sleet at 22h.
11 5	21	—	Calm.	−1·5	Dull and overcast, with occasional sleet since 23h.
11	11 3	N.N.E.	Fresh.	−3·3	Began to blow at 6h, since increasing in force. Occasional snow.
17	9	N.E. by E.	V. high	0·3	From 13h to 16h a most violent gale. Blowing furiously at 16h. *A low bank of auroral light was seen at Toronto this evening at 14h Gött.*
23	15	N.E.	High.	6·5	Gale continuing, but less violent.
12 5	21	E.N.E.	V. light.	4·4	Snowing. Wind abated, and it began to snow at 3h.
11	12 3	—	Calm.	1·8	Snowing, without intermission.
17	9	W.	Light.	−5·7	Snowing, but lightly, and so down to 19h.
23	15	W.	Mod.	−12·8	Occasional snow. A few stars visible S. of the zenith.
13 5	21	—	Calm.	−19·8	Unclouded since 2h.
11	13 3	—	Calm.	−18·3	Unclouded, but hazy.
17	9	E.N.E.	Light.	−24·9	Unclouded since 11h.
23	15	E.N.E.	Light.	−22·9	Unclouded. A faint arch of aurora (5) at elevation 30°.
14 0	16	E.N.E.	Light.	−22·3	Unclouded. The same at elevation 35°.
1	17	E.N.E.	Light.	−20·5	Unclouded. The same; no perceptible change.
5	21	E. by S.	Fresh.	−16·5	Beginning to cloud over, having been clear for 24 hours.
11	14 3	E.S.E.	Fresh.	−6·0	Overcast, with close packed cirro-cumuli.
17	9	E.S.E.	Light.	−2·6	Partially clear. Began soon after to snow.
23	15	S.E.	High.	−4·7	Clear overhead, otherwise clouded.
15 5	21	E.S.E.	Light.	−4·7	Nearly covered with cirri and cirro-strati. A few clear spaces.
11	15 3	S.E.	V. light	−1·3	Light cirri and cirro-cumuli.
17	9	S.E.	Light.	−6·3	Unclouded since 13h.
20	12	—	Calm.	−9·2	An arch of aurora (4) at an elevation of 15°.
23	15	W.N.W.	Light.	−6·8	Overcast since 22h. Snowing fast at 0h.
16 5	21	—	Calm.	−9·5	Overcast.
11	16 3	—	Calm.	−6·8	Overcast. Without break since 22h.
17	9	N.E.	Fresh.	−7·8	Cleared up at 15h. Partially clouded again, and so on to the end of the observations at 20h.
Sunday.					
17 23	17 15	—	Calm.	−3·2	Heavy clouds round the horizon, but clear overhead.
18 5	21	N.N.E.	Light.	4·6	Beginning to snow lightly.
11	18 3	W.SW.	Light.	5·5	Ceased snowing since 10h.
17	9	N.W.	Mod.	−3·3	Continually overcast. Sleet at 18h.
23	15	—	Calm.	−6·5	Overcast. Snowing lightly.
19 5	21	W.N.W.	Light.	−9·4	Overcast. Wind since 2h.
11	19 3	—	Calm.	−11·0	Lightly overcast. Clear at 1h.
17	9	N.E.	High.	−6·9	Overcast. High N.E. wind prevailing since 14h. Began to snow at 18h.

Abstract from the Meteorological Journal—*continued.*

Date.		Wind.		Temp.	Weather.
Gött. Time.	Mean Time.	Direction.	Force.	Newman. corrected.	
December 1843.					
D. H.	D. H.			°	
19 23	19 15	N.E. by E.	High.	0·3	Wind abated a little. Ceased snowing at 22^h. Beginning to clear. A considerable disturbance was first observed at 20^h. Most easterly position of the Declinometer (+0° 32′·4) at 20^h 0^m; the most westerly (−0° 4′·4) at 22^h 36^m, range 0° 34′. Minimum of Horizontal Force (−·039 X) at 22^h 30^m.
20 5	21	W. by S.	Light.	1·1	Lightly clouded, but generally unclouded since 0^h.
11	20 3	W.	V. light	−5·9	Unclouded since 6^h. Magnetic term day, which was wholly free from disturbance, commencing at 10^h.
17	9	—	Calm.	−10·0	Partially clouded to S.E.
23	15	N.N.E.	High.	−11·4	Unclouded since 18^h.
21 5	21	N.N.E.	Fresh.	−9·1	Cirro-strata towards the horizon; remainder clear.
11	21 3	N.N.E.	High.	0·2	Overcast.
17	9	N.N.E.	Light.	12·4	Wind abating. Still overcast.
23	15	N.W.	High.	2·1	Partially clear, and so throughout the night.
22 5	21	W.S.W.	Light.	−9·2	Overcast.
11	22 3	—	Calm.	−12·2	Again overcast, but unclouded from 7^h to 10^h.
17	9	E.S.E.	Fresh.	−13·4	Unclouded since 13^h.
21	13	N.E.	High.	−7·1	Nearly clear. An arch of aurora (5) at elevation 35°.
23	15	N.E.	High.	−5·9	Recently clouded over.
23 5	21	N.N.E.	Fresh.	−0·1	Thickly overcast since 23^h.
11	23 3	N.N.E.	Fresh.	1·2	Overcast again, but partially clear from 8^h to 10^h.
17	9	N.N.E.	Fresh.	2·7	Overcast. It fell calm at 19^h.
Sunday. Christmas Day.					
25 23	25 15	W. by N.	Fresh.	−0·4	Snowing lightly since 21^h.
26 5	21	W.S.W.	V. light.	−3·5	Snowing without intermission all the morning.
11	3	—	Calm.	−3·2	Ceased snowing at 8^h, but still overcast.
17	9	—	Calm.	−4·5	Beginning to clear up.
21	13	—	Calm.	−4·7	Unclouded. A faint arch of aurora ( ) visible but a few minutes before. At 20^h 54^m two brilliant curtains of light ( ), of irregular outline, were observed, place not recorded; they disappeared in a few minutes. At 21^h 30^m aurora again became brilliant ( ), in irregular bands or arches, parallel, striated, most brilliant in the N.E., and much diffused in that quarter. At 21^h 45^m the aurora was very brilliant ( ) and dense near the horizon, but faint in the zenith.
22	14	—	Calm.	−6·1	Unclouded. The aurora very faint, and generally diffused. The most westerly position of the Declinometer (−0° 10′) was at 22^h 3^m. At 22^h 30^m it had passed altogether to the S., and was hardly perceptible; the most easterly position of the Declinometer (+0° 28′·9) was at this time.
23	15	—	Calm.	−6·7	Unclouded. Faint aurora generally diffused. Minimum of Horizontal Force, −·035 X, at 23^h 27^m.
27 0	16	—	Calm.	−5·6	Unclouded. Faint aurora generally diffused. At 0^h 48^m faint bands and streamers crossing the meridian from E. to W., and reaching the zenith. Disturbance inconsiderable, but tending to an increase of Horizontal Force, which gave a maximum (+·007 X) at 1^h 24^m.
5	21	E.S.E.	Mod.	2·3	Unclouded.
11	27 3	E.N.E.	V. light.	8·0	Unclouded, and almost calm, since 6^h.
17	9	—	Calm.	13·2	Gradually clouding over since 13^h. Extra observations were taken from 18^h to 20^h for a disturbance, anticipated from the high value of the Horizontal Force. The range of Declination was 0° 19′·4. Maximum of Horizontal Force (+·012 X) at 18^h 33^m.
23	15	—	Calm.	16·8	Snowing since 20^h.
28 5	21	—	Calm.	16·2	Snowing fast. Occasional light air from N.W. A marked magnetic shock at 3^h. Declination reading (+0° 49′·5); most westerly reading at 3^h 15^m (−0° 12′); range, 1° 2′. This shock was also observed at Makerstoun, in Scotland, but there is no evidence of it in the regular observations at Toronto.

Abstract from the Meteorological Journal—*continued.*

Date.		Wind.		Temp.	Weather.
Gött. Time.	Mean Time.	Direction.	Force.	Newman. corrected.	
December 1843.					
D. H.	D. H.			°	
28 11	28 3	—	Calm.	11·7	Ceased snowing.
17	9	—	Calm.	0·3	Began again to snow at 14ʰ.
23	15	—	Calm.	−5·8	Overcast, but ceased snowing at 19ʰ.
29 5	21	W.N.W.	Fresh.	−6·3	Overcast, snowing at intervals.
11	29 3	W.N.W.	Light.	−2·4	Thick, without decided fog.
17	9	—	Calm.	−6·9	Stars occasionally visible; but thick, with occasional snow.
23	15	—	Calm.	−11·9	Unclouded, and so generally to 4ʰ. Considerable disturbance from 21ʰ to 0ʰ. Most westerly reading of Declination (0° 54′·8) at 21ʰ 18ᵐ. A minimum of Horizontal Force (−·040 X) at 21ʰ 48ᵐ. Most easterly (+0° 10′·6) at 22ʰ 18ᵐ. Range of Declination, 1° 7′·3. A maximum of Horizontal Force (+·006 X) occurred at 23ʰ 33ᵐ.
30 5	21	—	Calm.	−17·4	Overcast.
11	30 3	W.N.W.	V. light.	−22·7	Unclouded from 6ʰ to 9ʰ. At present overcast.
17	9	W.	Light.	−33·7	Unclouded to the end of the observations at 20ʰ.
Sunday.					

Date.		Wind.		Temperature.		Weather.
Gött. Time.	Mean Time.	Direction.	Force.	Newman corrected.	Dollond* as observed.	
January 1844.						
D. H.	D. H.			°	°	
New Year's Day.		—	—	—	—	No observations.
1 23	1 15	E.N.E.	Fresh.	−6·2	—	Overcast.
2 5	21	E.S.E.	Light.	−0·8	—	Overcast, with cirro-strati.
11	2 3	E.S.E.	Light.	−3·3	—	Unclouded, but hazy.
17	9	—	Calm.	−11·0	—	Unclouded since 11ʰ.
23	15	—	Calm.	−17·1	—	Still calm and unclouded.
3 5	21	—	Calm.	−23·4	—	Overcast with a light haze.
11	3 3	—	Calm.	−13·0	—	Overcast.
17	9	—	Calm.	−9·3	—	Overcast.
23	15	—	Calm.	−7·3	—	Overcast.
4 5	21	—	Calm.	−6·2	—	Overcast.
11	4 3	—	Calm.	−5·8	—	Still overcast. A fall of snow from 8ʰ to 10ʰ.
17	9	—	Calm.	−6·9	—	Overcast. A great magnetic disturbance began to be observed at 16ʰ, and continued, with some intermission, to 5ᵈ 3ʰ. It commenced with a range of Horizontal Force above the mean; highest value (+·021 X) at 16ʰ 30ᵐ. Most westerly reading of the Declinometer (0° 40′) at 21ʰ. Lowest value of Horizontal Force (−·063 X) at 23ʰ 3ᵐ. Most easterly reading (+1° 6′·4) at 0ʰ 57ᵐ. Range, 1° 56′.
23	15	—	Calm.	−5·5	—	Snowing slightly from 21ʰ to 23ʰ.
5 5	21	W.N.W.	V. light.	−10·3	—	Snowing slightly since 3ʰ.
11	5 3	W. by S.	Fresh.	−18·2	—	Snowing, but a streak of clear sky in the E.S.E. A slight magnetic shock at 8ʰ Gött.
17	9	W.S.W.	V. light.	−25·2	—	Hazy.

* The *Scale Readings* of Dollond's thermometer are added, to the end of the abstract, as a check upon the true temperature deduced from the Scale Readings of Newman's. Dollond stood at 2−37·8 in freezing mercury. (p. 126.)

Abstract from the Meteorological Journal—*continued.*

Date.		Wind.		Temperature.		Weather.
Gött. Time.	Mean Time.	Direction.	Force.	Newman corrected.	Dollond as observed.	
January 1844.						
D. H.	D. H.			°	°	
5 23	5 15	—	Calm.	−36·7	—	Unclouded, but hazy. Great magnetic disturbance, especially of the Horizontal Force, from 5^{d} 23^{h} to 6^{d} 3^{h}. Most westerly reading of the Declinometer (−0° 18′·7) at 2^{h}, attended by the lowest value of the Horizontal Force (−·063 X). The most easterly (+0° 42′·7) at 23^{h} 48^{m}; range, 1° 4′.
6 5	21	S.S.W.	Light.	−37·5	—	Unclouded.
11	6 3	S.S.W.	V. light.	−34·6	—	Unclouded.
17	9	N.N.E.	V. light.	−37·6	−34·3	Unclouded to the end of the observations at 20^{h}.
Sunday.						
7 23	15	N.E.	Fresh.	−32·0	−28·2	Hazy. Halo round the moon at 22^{h}.
8 5	21	N.N.E.	Fresh.	−32·0	−28·1	Overcast. Partially so since 1^{h}.
11	8 3	N. by W.	Fresh.	−29·3	−25·3	Unclouded again since 7^{h}.
15	9	N.	Light.	−28·6	−26·0	Unclouded. A faint auroral arch ( ), at an elevation of 10°, extending from N.E. to N.W. At 15^{h} 20^{m} the arch was much brighter, gathering to a focus at the N.W. end, and rising gradually. At 15^{h} 28^{m} it broke up into dense masses, which gradually disappeared. At 16^{h} no aurora; a slight degree of disturbance was observed. Most easterly reading of the Declination (+0° 22′·9) at 15^{h} 42^{m}, accompanied by the highest value of the Horizontal Force (+·007 X).
17	9	N.N.E.	Light.	−29·5	−26·0	Unclouded.
20	12	N.N.E.	Light.	−30·4	−27·3	Unclouded. A faint arch of aurora (5), at an elevation of 40°. A slight renewal of disturbance observed. Most westerly reading of the Declination (−0° 14′·6) at 20^{h} 33^{m}; total range, 0° 38′. Lowest value of Horizontal Force (−·018 X) at 20^{h} 30^{m}.
23	15	N.N.E.	Light.	−40·1	−36·6	Unclouded.
9 5	21	—	Calm.	−39·8	−38·0	Unclouded.
11	9 3	—	Calm.	−35·3	−32·1	Unclouded.
16	8	—	Calm.	−34·4	−31·1	A faint arch of aurora, at an elevation of 13°, extending from N.E. to N.W.
17	9	—	Calm.	−34·7	−31·3	A faint arch ( ), at elevation 10°, generally covered with light cirro-strati.
23	15	—	Calm.	−32·8	−29·9	Generally covered with light cirro-strati.
10 5	21	E.S.E.	V. light.	−26·5	−23·6	Lightly overcast.
11	10 3	N.E.	V. light.	−24·1	−21·0	Almost unclouded.
17	9	—	Calm.	−22·6	−18·5	Completely overcast.
23	15	—	Calm.	−24·3	−21·2	Lightly overcast.
11 5	21	W.S.W.	V. light.	−11·4	−9·5	Lightly overcast.
11	11 3	W.S.W.	Light.	−6·9	−4·2	Overcast. Snowing since 9^{h}.
17	9	W.S.W.	Light.	−1·3	−0·5	Overcast.
23	15	W.S.W.	V. light.	−2·6	−1·1	Partially clear at intervals.
12 5	21	W.S.W.	V. light.	−6·0	−3·8	Snowing since 3^{h}.
11	12 3	W.S.W.	V. light.	−2·8	−0·6	Snowing again slightly.
17	9	—	Calm.	−4·7	−2·9	Overcast. Snow 13^{h} to 15^{h}.
23	15	W.S.W.	Light.	−17·6	−14·4	Overcast.
13 5	21	W.S.W.	Light.	−21·8	−19·8	Overcast. Light cirrus haze.
11	13 3	—	Calm.	−20·2	−17·0	Unclouded since 8^{h}.
17	9	—	Calm.	−30·9	−28·3	Unclouded to the close of the observations at 20^{h}.
Sunday.						
14 23	15	—	Calm.	−20·7	−17·3	Unclouded since 21^{h}.
15 5	21	—	Calm.	−16·9	−14·9	Beginning to be lightly overcast.
11	15 3	S.S.W.	Fresh.	−0·7	−1·0	Overcast.
17	9	S.S.E.	Fresh.	−5·8	−7·7	Overcast. Strong gale since 11^{h}; highest at 12^{h} and 13^{h}.
23	15	W.S.W.	Mod.	−0·1	−1·9	Nearly unclouded. Snowing at 19^{h} and 20^{h}, then cleared up.

Abstract from the Meteorological Journal—*continued.*

Date.				Wind.		Temperature.		Weather.
Gött. Time.		Mean Time.		Direction.	Force.	Newman corrected.	Dollond as observed.	
January 1844.						°	°	
D.	H.	D.	H.					
16	0	15	16	W.S.W.	V. light.	−5·8	−3·8	A faint arch of aurora ( ), at elevation of 30°.
	1		17	W.S.W.	Fresh.	−8·0	−6·0	Unclouded. A faint auroral light n the N.
	5		21	W. by S.	Light.	−13·3	−10·7	Overcast, with dense cirro-cumuli.
	11	16	3	W. by S.	Fresh.	−10·6	−8·1	Overcast.
	17		9	W.	High.	−10·3	−8·0	Overcast, began to clear soon after.
	20		12	W.N.W.	Light.	−13·8	−11·2	Faint detached patches of auroral light.
	21		13	W.N.W.	Fresh.	−17·1	−13·6	Partially clear. A faint auroral light in the N.
	23		15	W.N.W.	Fresh.	−18·3	−14·8	Overcast. A slight degree of disturbance at 0h; range of Declination, 19′·8.
17	5		21	W.N.W.	Light.	−23·9	−20·9	Overcast.
	11	17	3	W.N.W.	Light.	−22·4	−19·1	Overcast. Snow at 7h.
	17		9	W.S.W.	Light.	−25·2	−21·6	Unclouded since 15h.
	21		13	W.N.W.	Mod.	−27·8	−25·0	Unclouded. An arch of aurora ( ), crossing the zenith since 20h 45m.
	22		14	W.N.W.	Mod.	−29·7	−27·0	Unclouded. Aurora as before, but fainter ( ); no perceptible motion.
	23		15	W.N.W.	High.	−30·6	−28·0	Unclouded. No aurora.
18	0		16	W.N.W.	High.	−32·5	−29·0	Unclouded. A faint arch of aurora ( ), at elevation 30°.
	5		21	N.W.	Mod.	−37·2	−33·0	Overcast since 3h.
	11	18	3	N.W.	Light.	−33·6	−30·3	Again overcast, but unclouded from 6h to 9h.
	17		9	N.W.	Light.	−32·9	−30·0	Unclouded, mostly so since 13h.
	18		10	N.W.	Light.	−34·0	−31·2	Unclouded. A faint arch of aurora.
	20		12	W.S.W.	Light.	−34·1	−31·2	Unclouded; hazy. At 20h 45m an arch or band across the zenith from W.N.W. to E.S.E., not in motion. No disturbance.
	23		15	W.S.W.	Light.	−36·3	−33·2	Unclouded.
19	5		21	W.S.W.	V. light.	−38·4	−36·2	Still unclouded, but hazy.
	11	19	3	—	Calm.	−34·4	−31·3	Unclouded.
	15		7	—	Calm.	−34·6	−31·5	Unclouded. Faint arch of aurora, at elevation of 10°.
	16		8	—	Calm.	−35·9	−32·5	Unclouded. Aurora, without change.
	17		9	—	Calm.	−37·4	−34·1	Unclouded. Aurora, without change.
	20		12	—	Calm.	−38·6	−36·0	Unclouded. Arch of aurora, at elevation 20°. There have been occasional faint arches and floating patches of aurora from time to time since 16h, but no disturbance of the magnets.
	21		13	—	Calm.	−38·6	−35·5	Unclouded. Faint aurora, diffused generally; brighter portions, of irregular form ( ), in the N. A minimum of Horizontal Force (−·010 X) at 21h; range of Declination from 21h to 22h, 0° 22′·0; most easterly position (+28′·4) at 21h 12m.
	23		15	—	Calm.	−38·3	−35·3	Unclouded. No aurora at this time, but occasional faint arches and patches of light from 21h to 20d 1h.
20	5		21	—	Calm.	−38·3	−35·1	Lightly overcast since 4h; unclouded down to that hour.
	11	20	3	N.N.E.	Light.	−32·2	−28·4	Unclouded again since 6h.
	17		9	N.N.E.	Light.	−34·0	−30·9	Still unclouded.
	19		11	N.N.E.	Light.	−34·0	−30·9	An arch of aurora, elevation 25°, brightest to the E. ( ); detached bands or arches. No disturbance.
	20		12	N.N.E.	Light.	−34·7	−31·5	Arch a little fainter ( ), elevation 26°; striated brilliant detached streamers and patches of light ( ), chiefly in the N.E.
	Sunday.							
21	21	21	13	—	Calm.	−38·4	−35·8	Detached beams or streamers of aurora.
	22		14	—	Calm.	−42·0	−39·0	Unclouded. Aurora as before.
	23		15	—	Calm.	−40·9	−39·1	Unclouded.

Abstract from the Meteorological Journal—*continued.*

Date.		Wind.		Temperature.		Weather.
Gött. Time.	Mean Time.	Direction.	Force.	Newman corrected.	Dollond as observed.	
January 1844.						
D. H.	D. H.			°	°	
22 0	21 16	—	Calm.	−39·6	−37·2	Unclouded. From 21ʰ to 0ʰ a succession of faint arches and detached patches of light with bright streamers ( ), sometimes near the zenith, but no disturbance observed. Exposed some mercury, found it partly frozen at 0ʰ 30ᵐ.
5	21	—	Calm.	−46·5	−44·0	Hazy. The mercury had been solid since 1ʰ.
11	22 3	N.N.W.	Light.	−37·9	−33·9	Still unclouded, but soon after clouded lightly over.
17	9	N.N.W.	Light.	−39·7	−36·9	Unclouded again since 13ʰ.
20	12	—	Calm.	−42·2	−39·0	Unclouded. Faint aurora in patches.
21	13	—	Calm.	−43·6	−40·2	Faint aurora in the N. at 21ʰ 30ᵐ. A great quantity of detached cirrous aurora in different parts of the sky, moderately bright, and not in motion. No disturbance.
22	14	—	Calm.	−43·2	−40·0	Unclouded. Two dense but faint masses of aurora in the N.E. and N.W. Patches of aurora, striated, diffused at various altitudes to the N. round the zenith. No disturbance. At 22ʰ 15ᵐ no aurora in sight.
23	15	—	Calm.	−43·9	−41·1	Unclouded. No aurora visible. At 23ʰ 30ᵐ an arch of aurora extending from N.W. by W. to N. At the western extremity, apparently turned upon itself so as to form a hook, very brilliant; the elevation of the hook was 27°, of the centre 30°, of the N. end 26°. At 23ʰ 40ᵐ the arch broke up into striated masses, of moderate brightness, generally diffused from N.W. to N.E., and disappeared gradually. The Declinometer and Bifilar showed a slight degree of disturbance by the vibration of their magnets, but without change of mean position.
23 5	21	—	Calm.	−43·3	−40·0	Unclouded. Hazy since 0ʰ, and mercury frozen. At 6ʰ the mercury was observed to be partly melted.
11	23 3	—	Calm.	−34·5	−31·1	Lightly overcast, with occasional sprinkling of snow.
17	9	—	Calm.	−39·9	−36·4	Hazy, but unclouded since 14ʰ.
23	15	—	Calm.	−39·9	−36·9	Unclouded. Hazy.
24 5	21	W.N.W.	V. light.	−40·6	−37·6	Overcast, with uniform dense haze.
11	3	—	Calm.	−39·4	−36·0	Hazy. Magnetic term day began at 10ʰ.
15	7	—	Calm.	−43·4	−40·0	Unclouded. No aurora visible. At 15ʰ 30ᵐ a faint auroral haze in the N. near the horizon. Greatest value of the Horizontal Force about this time. At 15ʰ 45ᵐ a mass of aurora in the N.N.W., another in the N.E., both rising vertically to an elevation of 11°, then extending irregularly to an elevation of about 32°, and thence uniting in an arch at an elevation of 60°, from which arch four conspicuous streamers, in violent motion, rose towards the zenith. Brilliancy (2) to (4.)

Abstract from the Meteorological Journal—*continued.*

Date.		Wind.		Temperature.		Weather.
Gött. Time.	Mean Time.	Direction.	Force.	Newman corrected.	Dollond as observed.	
January 1844.				°	°	
D. H. 24 16	D. H. 24 8	—	Calm.	−44·0	−40·8	Unclouded. A heavy and brilliant arch near the zenith ( ), form rather irregular. Width about 5°, and nearly motionless. At 16^h 15^m aurora in the zenith, diverging from points on the E. and W. sides, not in straight beams, but with a wavy or serpentine outline, and breaking up into narrow streaks. At 16^h 30^m a brilliant narrow band, very irregular in outline at about 15° beyond the zenith to the S. Faint detached masses both to the N. and S. of it, at 16^h 45^m. The arch was 55° beyond the zenith, or 35° above the southern horizon, and fainter. There was a little very faint aurora in the N.
17	9	—	Calm.	−44·5	−41·9	Unclouded. Another arch ( ) approaching the zenith, at present 30° from it; broad and diffused. At 17^h 15^m faint detached patches of great extent in the N.E. and N.W. near the horizon, also some detached patches of no regular shape, and streamers. At 17^h 30^m no aurora was visible, save a faint luminous haze near the horizon. *Auroral light was observed at Toronto at* 17^h, but the sky was covered with clouds all night. Aurora was also seen at North Salem, N.Y. (Regent's Reports), but the hour is not named.
21	13	—	Calm.	−44·3	−41·7	Unclouded. Hazy. The most westerly position of the Declinometer (−0° 44′·5) was at 20^h 25^m. At 21^h 30^m vast quantities of faint cirrus aurora ( ) in various parts of the sky, without motion. At 21^h 45^m no aurora was visible.
22	14	—	Calm.	−46·6	−44·0	Partially clouded. No aurora visible. At 22^h 40^m a faint mass of aurora ( ) moving slowly from the N.E. along the eastern horizon (elevation not recorded) gathering to a focus at the N.E. extremity. The lowest value of the Horizontal Force at this time.
23	15	—	Calm.	−46·5	−44·1	Unclouded. A large dense mass of aurora (2) S.E. by S. at elevation 10°. It suddenly broke up into a number of patches, which were scattered at various elevations, and all disappeared before 23^h 15^m.
25 0	16	—	Calm.	−47·0	−44·4	Unclouded. Hazy. No aurora visible. The most easterly position of the Declinometer (+2° 10′·1), at 0^h 15^m, range 3° 4′·8. At 0^h 45^m a faint luminous haze was distinguishable S. of the zenith. The mercury was observed to be solid from 15^h to this hour. No more aurora was seen, but the magnetic disturbance lasted until about 5^h. In that portion which occurred from 15^h to 19^h a tolerably decided correspondence may be distinguished in the movements at Lake Athabasca, and those at Toronto and Greenwich; at the latter stations the greatest degree of disturbance occurred at this time.

Abstract from the Meteorological Journal—*continued.*

Date.				Wind.		Temperature.		Weather.
Gött. Time.		Mean Time.		Direction.	Force.	Newman corrected.	Dollond as observed.	
January 1844.						°	°	
D.	H.	D.	H.					The movements from 21^h to 2^h, which are the most considerable at Lake Athabasca, have no decided counterpart at Toronto or Greenwich, but the disturbance lasted at Toronto and at Greenwich (more decidedly at the former than at the latter station), down to about the same time, viz. 25^d 5^h.
	5	24	21	—	Calm.	−43·7	−41·2	Hazy, but unclouded.
	11	25	3	N.N.E.	Mod.	−40·2	−36·4	Lightly overcast.
	17		9	E.N.E.	High.	−31·8	−28·4	Overcast. Wind high and in gusts since 12^h.
	23		15	E.N.E.	Fresh.	−26·6	−23·3	Overcast. Wind still high and squally, but abated a little.
26	5		21	—	Calm.	−26·0	−22·7	Still overcast. Wind fallen since 2^h.
	11		3	W.S.W.	V. light.	−24·0	−20·9	Overcast.
	17		9	W.S.W.	Light.	−32·9	−29·3	Unclouded since 7^h. A faint arch of aurora extending from N.W. to N.E. at an elevation of 11°.
	20		12	W.S.W.	Light.	−35·7	−32·6	Unclouded. Detached streamers, and part of an arch of moderate brightness from N. to N.E.
	22	26	14	—	Calm.	−37·8	−35·2	Unclouded. Faint and scarcely distinguishable streamers and arches. Horizontal Force slightly disturbed. Range of Declination between 22^h and 23^h 0° 31′. Most westerly (−14′·4) at 22^h 9^m; most easterly (+16′·1) at 22^h 51^m. Lowest value of Horizontal Force (−·013 X) at 22^h 45^m.
	23		15	—	Calm.	−38·7	−36·2	Unclouded. Faint luminous or auroral haze, which continued visible down to 1^h, without assuming any definite form.
27	5		23	—	Calm.	−38·6	−35·4	Lightly overcast since 4^h.
	11	27	3	N.N.E.	Fresh.	−28·0	−24·8	Wind newly risen. Still overcast.
	17		9	N.N.E.	V. light.	−13·8	−10·9	High wind generally since 11^h. At 18^h a little snow. Overcast to the end of the observations.
Sunday.								
28	23	28	15	—	Calm.	−14·8	−12·0	Lightly overcast. Faint auroral light in the N.
29	5		21	N.N.E.	Light.	−17·3	−14·6	Sky recently cleared.
	11	29	13	N.E.	Light.	−5·4	−3·1	Overcast again since 9^h.
	17		9	N.N.E.	Fresh.	1·3	3·0	Snowing thickly since 15^h, and so on to 20^h.
	23		15	E.N.E.	Light.	14·6	15·0	Lightly overcast, with cirro-cumuli. Snow again at 0^h.
30	5		21	W.S.W.	Light.	5·1	6·3	Snowing thickly. Snowing since 2^h.
	11	30	3	W.S.W.	V. light.	2·7	4·7	Overcast, but ceased snowing soon after 5^h.
	17		9	—	Calm.	−8·0	−5·9	Overcast.
	18		10	—	Calm.	−6·0	−8·6	Sky cleared, and slight aurora (not described.)
	21		13	—	Calm.	−13·3	−10·4	Generally clouded. A faint mass of aurora ( ) near the horizon in the N., and streamers in the N.W.
	23		15	—	Calm.	−19·5	−16·2	Partially clear.
31	00		16	—	Calm.	−21·8	−18·2	Partially clear. A faint auroral haze in the N. ( ).
	5		21	—	Calm.	−20·4	−17·9	Unclouded since 2^h.
	11	31	3	—	Calm.	−14·8	−11·9	Light cirri and strati in various parts of the sky.
	17		9	E.N.E.	Light.	−22·8*	−20·2*	Light cirro-strati.
	18		10	E.N.E.	Light.	−22·0	−19·2	Detached streamers in the N.E. ( ), and a faint stationary strip of cirrus aurora at elevation 36° in the N.W.

Abstract from the Meteorological Journal—*continued.*

Date.				Wind.		Temperature.		Weather.
Gött. Time.		Mean Time.		Direction.	Force.	Newman corrected.	Dollond as observed.	
D.	H.	D.	H.			°	°	
January 1844.								
31	23	31	15	N.N.E.	Fresh.	−18·4	−15·9	Uniformly overcast since 19^{h}. A great disturbance, especially of the Horizontal Force, from 0^{h} to 4^{h}. Most westerly reading of the Declination (−0° 14′) at 0^{h} 9^{m}. Lowest value of Horizontal Force (−·046 X) at 0^{h} 27^{m}. Most easterly reading (+1° 1′·7) at 1^{h} 30^{m}. Range 1° 12′·7.
Feb. 1	5		21	N.N.E.	Fresh.	−14·5	−11·8	Overcast, with dense and closely-packed cirro-cumuli.
	11	Feb. 1	3	N.N.E.	Light.	−16·2	−12·7	Unclouded since 8^{h}.
	17		9	N.N.E.	Mod.	−10·5	−8·2	Lightly overcast.
	23		15	N.N.E.	High.	−5·8	−3·4	Overcast. Considerable magnetic disturbance of an unusual character from 19^{h} to 21^{h}, the range of Declination being mostly to the westward. Most easterly reading of Declination (+0° 46′·9) at 19^{h} 12^{m}, followed by the most westerly reading (−1° 19′·8) at 19^{h} 27^{m}. Range 2° 7′·2. Lowest value of Horizontal Force (−·024 X) at 19^{h} 39^{m}.
2	5		21	N. by W.	Light.	−1·1	0·9	Overcast. Slight snow at 7^{h}. Considerable disturbance of the same character as before again observed from 5^{h} to 8^{h}. Range of Declination 0° 51′.
	11	2	3	N.	Light.	−1·5	0·8	Lightly overcast.
	17		9	N.	Light.	−5·1	−3·1	Lightly overcast. Moderate disturbance from 17^{h} to 21^{h}. Most westerly reading of Declination (−0° 35′) at 17^{h} 42^{m}; most easterly (+0° 14′·2) at 18^{h} 48^{m}. Range 0° 49′·6. Horizontal Force above the mean during its continuance. Highest value (+·026 X) at 18^{h} 54^{m}.
	23		15	—	Calm.	−9·3	−6·9	Snowing lightly.
3	5		21	—	Calm.	−10·3	−7·1	Overcast. A few flakes of snow falling occasionally.
	11	3	3	—	Calm.	−1·6	1·0	Very light and fleecy cirro-cumuli, with clear space.
	17		9	—	Calm.	−5·9	−4·0	Overcast since 12^{h}, and so to the end of the observations at 20^{h}.
Sunday. 4	23	4	15	W.N.W.	High.	7·3	8·3	Overcast since resuming observations at 21^{h}. Very great disturbance between 23^{h} and 2^{h}. Most easterly reading of the Declination (+1° 14′·8) at 0^{h} 21^{m}, followed by the most westerly (−1° 22′) at 0^{h} 30^{m}. Range 2° 36′·4. Horizontal Force at 0^{h} 3^{m}, at its lowest value, (−·068 X.)
5	5		21	W.N.W.	Light.	−7·3	−5·0	Overcast.
	11	5	3	W.N.W.	Light.	−1·5	0·2	Unclouded since 6^{h}.
	14		6	W.N.W.	Light.	−8·0	−6·2	Unclouded. A faint arch of aurora.
	15		7	W.N.W.	Light.	−9·6	−7·4	Unclouded. A moderately bright arch of aurora ( ) at elevation 13°.
	16		8	W.N.W.	Light.	−10·3	−8·2	Unclouded; arch as at the last observation. Considerable disturbance, principally of the Declination, from 16^{h} to 19^{h}. Most easterly reading (+0° 48′·3) at 16^{h} 21^{m}, followed by the highest value of the Horizontal Force (+·022 X) at 16^{h} 30^{m}.
	17		9	W.N.W.	Light.	−13·4	−11·0	Unclouded. A double arch of aurora ( ), lower circle at elevation 8°, the upper at 12°, and two detached masses in the N.E., rising vertically to the elevation of 12° and 14°.

Abstract from the Meteorological Journal—*continued.*

Date.		Wind.		Temperature.		Weather.
Gött. Time.	Mean. Time.	Direction.	Force.	Newman corrected.	Dollond as observed.	
February 1844. Sunday.				°	°	
D. H. 5 23	D. H. 5 15	—	Calm.	−21·7	−19·3	Continues unclouded. Disturbance renewed from 20ʰ to 23ʰ. Lowest value of Horizontal Force (−·039 X) at 21ʰ 9ᵐ. Most westerly reading of Declination (−0° 22′·7) at 21ʰ 36ᵐ. Range 1° 3′.
6 5	21	W.N.W.	Light.	−22·0	−19·1	Lightly overcast since 3ʰ.
11	6 3	N.N.E.	Light.	−8·0	−5·8	Generally unclouded since 6ʰ. At present overcast.
17	9	E.N.E.	Light.	4·3	5·6	Overcast. Magnetic shock from 18ʰ to 20ʰ. Most westerly reading of Declination (−0° 23′·6) at 19ʰ 10ᵐ. Most easterly (+0° 21′·2) at 19ʰ 30ᵐ. Range 0° 45′·4.
23	15	—	Calm.	6·6	7·5	Clear at midnight, otherwise overcast since 17ʰ.
7 5	21	—	Calm.	3·8	5·4	A few flakes of snow falling.
11	7 3	—	Calm.	9·7	11·0	Snowing lightly from 10ʰ to 13ʰ.
17	9	—	Calm.	5·9	7·2	Overcast.
23	15	—	Calm.	8·9	9·9	Overcast.
8 1	17	—	Calm.	6·6	7·9	Cirro-strati to the S.; remainder clear. A very faint arch of aurora gathering to a focus in the E. A slight disturbance just over. Range of Declination (0° 37′·4.) Lowest value of Horizontal Force (−·021 X) at 0ʰ 0ᵐ.
5	21	—	Calm.	6·6	7·7	Lightly overcast. A minimum of Horizontal Force (−·023 X) at 5ʰ 0ᵐ.
11	8 3	—	Calm.	19·3	20·0	Light cirrus haze, but otherwise unclouded since 6ʰ.
17	9	—	Calm.	14·3	14·4	Unclouded.
23	15	—	Calm.	12·3	12·9	Clouded over since 22ʰ.
9 5	21	—	Calm.	12·0	12·2	Continues overcast.
11	9 3	—	Calm.	6·8	7·0	Overcast.
17	9	N. by W.	Light.	−0·8	1·2	Wind sprung up at 13ʰ.
23	15	—	Calm.	−2·4	0·0	Overcast.
10 5	21	N.N.E.	V. light.	0·8	2·9	A heavy hoar-frost depositing since 0ʰ.
11	10 3	—	Calm.	7·9	9·7	Snowing lightly. A glimpse of blue sky in the S.
17	9	—	Calm.	8·7	10·0	Overcast, and calm to the end of the observations at 20ʰ.
11 21	11 13	—	Calm.	−30·8	−27·4	Unclouded. An arch of aurora, moderately bright ( ), extending from N.E. to N.W., and rising to an altitude of 54°. A faint patch of aurora ( ) of great extent, near the horizon in the N.E. A slight change in the Horizontal Force and Inclination, but not sufficient to lead to observations for disturbance. At 21ʰ 30ᵐ no traces of aurora.
23	15	N.N.E.	V. light.	−31·8	−27·9	Unclouded.
12 5	21	—	Calm.	−23·1	−20·1	Overcast; cirro-cumuli and strati.

Abstract from the Meteorological Journal—*continued.*

Date. Gött. Time.		Date. Mean Time.		Wind. Direction.	Wind. Force.	Temperature. Newman observed.	Temperature. Dollond as observed.	Weather.
February 1844.								
D.	H.	D.	H.			°	°	
12	11	12	3	E.N.E.	V. light.	−11·9	−8·8	Almost clear; light cirro-cumuli.
	17		9	N.N.E.	Light.	−14·9	−12·6	Unclouded since 12ʰ, with the exception of a few strati in the S.
	18		10	N.N.E.	Light.	−15·3	−12·7	Unclouded. A faint arch of aurora ( ), at elevation 22° from N.E. to N.W.
	21		13	—	Calm.	−14·9	−12·9	A moderately bright arch of aurora ( ).
	22		14	—	Calm.	−18·2	−15·2	Unclouded. No aurora at this hour, but a succession of faint arches and streamers prevailed from 21ʰ to 23ʰ, unaccompanied by any disturbance.
	23		15	—	Calm.	−19·2	−16·9	Unclouded. An arch of aurora ( ), at elevation 20°.
13	5		21	—	Calm.	−22·0	−18·7	Light cirri and strati covering the sky.
	9	13	3	N.N.E.	Fresh.	−15·0	−13·0	Unclouded 6ʰ to 8ʰ; at present lightly overcast.
	17		9	—	Calm.	−7·0	−5·0	Clearing since 15ʰ; at present unclouded.
	19		11	—	Calm.	−10·3	−8·5	Unclouded. A faint aurora visible (undescribed).
	20		12	—	Calm.	−11·2	−8·9	Unclouded. Aurora scarcely distinguishable.
	23		15	—	Calm.	−10·0	−7·3	Unclouded since 17ʰ; continued so to 3ʰ.
14	5		21	—	Calm.	−13·3	−11·0	Overcast at 4ʰ; again unclouded.
	11	14	3	—	Calm.	−4·8	−2·0	Clouded. Light cirro-strati.
	17		9	N.N.E.	Fresh.	−5·7	−2·7	Unclouded, but hazy.
	23		15	—	Light.	7·7	8·3	Overcast since 18ʰ.
15	5		21	—	Calm.	7·9	9·0	Lightly overcast.
	11	15	3	—	Calm.	28·1	29·0	Hazy.
	17		9	—	Calm.	14·9	15·8	Hazy; a few stars visible.
	20		12	—	Calm.	7·7	8·9	Unclouded. No aurora visible. At 20ʰ 15ᵐ a brilliant arch, elevation 25°, extending from N.E. to N.W., and gathering to a focus at the N.E. end No disturbance. At 20ʰ 30ᵐ the arch separated into short narrow portions, which appeared to be suspended vertically, and were dancing up and down, and changing their position, with violent motion. The Bifilar and Inclinometer Magnets in slight agitation, but no change of reading to call for disturbance observations. At 20ʰ 40ᵐ the aurora appeared as two arches, the upper one faint ( ), at elevation 32°, the inner one scarcely perceptible, and at elevation 26°. No disturbance. At 21ʰ no aurora visible.
	23		15	N.N.W.	Light.	5·9	7·2	Unclouded, but hazy.
16	5		21	—	Calm.	12·2	12·9	Unclouded.
	11	16	3	—	Calm.	34·0	32·0	Unclouded, but hazy.
	17		9	—	Calm.	29·2	28·0	Unclouded. A moderately bright arch of aurora ( ), extending from N.E. to N.W. At 17ʰ 30ᵐ the same, at elevation 15°. At 17ʰ 45ᵐ the arch broke up into narrow vertical portions, which were dancing up and down in moderate motion. At the same time two brilliant ( ) masses or foci of aurora to the eastward of N.E.; a slight change of reading in all the instruments.
	18		10	S.S.W.	Light.	29·4	28·4	Unclouded. An arch of aurora of moderate brightness ( ), extending from N.E. to N.W., altitude 35°; readings of the instruments as at 17ʰ 45ᵐ.

Abstract from the Meteorological Journal—*continued.*

Date.		Wind.		Temperature.		Weather.
Gött. Time.	Mean Time.	Direction.	Force.	Newman corrected.	Dollond as observed.	
February 1844.						
D. H.	D. H.					
16 21	16 13	S.S.W.	High.	27°·0	28°·2	Unclouded. A brilliant burst of aurora, of a pale pink colour. At 21h 24m the sky to the S. nearly covered with faint masses and bands of aurora, in slight motion ( ). At 21h 36m a brilliant circle or corona ( ), 20° from the zenith to the N. (This appears to have been caused, not by convergence of streamers, but by a convolution of what may be termed auroral cloud.) The circle in vertical and rotatory motion; large bands of diffused aurora surrounding it. Extra observations from 21h to 22h; range of Declination, 0° 12′·8, completely establishing the absence of disturbance during the finest portion of the above display.
22	14	S.S.W.	Light.	28·9	27·0	An irregular arch in the N.W. near the horizon.
23	15	S.S.E.	High.	34 4	31·8	Unclouded. Faint aurora ( ) still visible.
17 5	21	S.S.W.	Mod.	27·9	26·8	Overcast, with haze.
11	17 3	S.S.W.	Fresh.	37·1	36·0	Overcast.
15	7	W.N.W.	Calm.	35·5	34 6	Partially clouded. Patches of faint aurora or auroral haze in various quarters, with part of an arch, of moderate brightness, from N.E. to N., at elevation 17°.
16	8	—	Calm.	35·1	34·1	An imperfect arch, at elevation 28°. Auroral haze in various parts of the sky to the N. No disturbance.
17	9	—	Calm.	33·5	32·1	Hazy. Clouded over after 18h, and so to the end of the observations at 20h.
Sunday.						
18 23	18 15	—	Calm.	8·0	9·4	Clouded. Snowing lightly since 22h.
19 5	23	—	Calm.	13·8	13·9	Overcast from 21h to 6h.
11	19 3	—	Calm.	19·0	20·0	Unclouded since 7h.
17	9	—	Calm.	4·2	6·2	Continues unclouded.
23	15	—	Calm.	5·5	7·0	Overcast, with dense haze, increasing since 22h. Hoar-frost.
20 5	21	—	Calm.	13·4	14·6	Clouded cirro-cumuli and haze.
11	20 3	—	Calm.	32·5	31·0	Prevalence of cirro-strati and cirro-cumuli, at present unclouded, but hazy.
17	9	W.	Fresh.	8·9	10·3	Unclouded since 15h.
20	12	—	Calm.	2·7	4·0	Unclouded. A faint irregular arch of aurora, at elevation 30°.
21	13	—	Calm.	−1·3	0·8	Unclouded. Faint auroral haze. An irregular arch in the N., at elevation 33°.
23	15	—	Calm.	−5·6	−3·1	Unclouded.
21 5	21	E.N.E.	Light.	2·1	4·6	Overcast since 3h.
11	21 3	N.N.E.	High.	15·6	15·1	Overcast.
17	9	N.	High.	21·5	20·0	Continues overcast.
20	12	N.	Mod.	20·4	20·6	Thick and hazy. Three faint patches of aurora ( ) in the N.W., at the elevations of 54°, 60°, and 62°.
21	13	N.N.E.	Mod.	21·2	21·2	Hazy. A faint arch of aurora (at the elevation of 15°.
22	14	N.N.E.	Mod.	20·1	19·8	Faint aurora visible, through cirro-cumulus clouds.
23	15	N.N.E.	Mod.	19·0	18·9	Hazy. Faint aurora visible, as before.
22 5	21	—	Calm.	19·1	19·5	Calm since 1h. Cirro-cumuli with clear spaces. Hazy.
11	22 3	—	Calm.	24·3	24·4	Unclouded since 9h. Snow at 6h and 7h.

Abstract from the Meteorological Journal—*continued.*

Date.		Wind.		Temperature.		Weather.
Gött. Time.	Mean Time.	Direction.	Force.	Newman corrected.	Dollond as observed.	
February 1844. D. H.	D. H.			°	°	
22 17	22 9	—	Calm.	27·2	26·1	Overcast since 12ʰ.
23	15	—	Calm.	30·8	29·3	Continues overcast.
23 5	21	—	Calm.	32·5	30·9	Unclouded. Clearing since 2ʰ.
11	23 3	—	Calm.	29·9	30·0	Light cirri and strati. Term day commenced at 10ʰ. A day of constant but slight disturbance. Most westerly reading of the Declination (−0° 13′·2) at 4ʰ 40ᵐ; the most easterly (+0° 11′·6); range 0° 23′·6.
17	9	—	Calm.	10·9	11·7	Unclouded since 15ʰ.
23	15	W.N.W.	Light.	0·4	2·7	Snowing slightly since 22ʰ.
24 5	21	—	Calm.	−3·3	−1·2	Unclouded.
11	24 3	—	Calm.	1·3	4·8	Light cirro-cumuli and cirrus haze.
17	9	—	Calm.	−4·9	−2·5	Nearly overcast. Dense cirro-cumuli. Soon afterwards light wind from N.E. to the end of the observations at 20ʰ.
Sunday. 25 23	15	W.N.W.	V. Light.	−9·2	−6·8	Unclouded since the commencement of observations at 21ʰ.
26 5	21	—	Calm.	6·6	9·1	Unclouded.
11	3	E.N.E.	Light.	33·5	33·9	Cirro-strati and cirro-cumuli.
17	9	—	Calm.	27·9	25·9	Hazy. Faint halo round the moon, diameter 40°.
23	15	W.S.W.	Mod.	1·9	3·5	Partially clouded. Snow at 22ʰ. A slight disturbance from 22ʰ to 0ʰ, giving a minimum of Horizontal Force (−·025 X), at 23ʰ 9ᵐ, range of Declination, 0° 11′.
27 1	17	—	Calm.	−2·5	−0·3	Unclouded. Clear bright and broad arch of aurora ( ), at elevation 25°; no disturbance.
5	21	—	Calm.	−5·6	−2·8	Unclouded.
11	27 3	—	Calm.	−0·9	1·9	Hazy.
17	9	E.N.E.	High.	−1·3	1·2	Hazy. Wind from N.E. since 12ʰ; high since 14ʰ.
23	15	N.N.E.	High.	3·3	5·2	Overcast.
28 5	21	N.N.E.	Light.	6·8	8·2	Snowing lightly since 10ʰ.
11	28 3	W.	Light.	15·7	16·7	Ceased snowing at 8ʰ.
16	8	—	Calm.	6·4	6·1	Unclouded. A bright but narrow band of aurora ( ) in the S., at elevation 40°, also an irregular serpentine arch in the N.W., both of a yellowish tint, and closely striated, conveying the impression of being near the earth. At 16ʰ 5ᵐ much yellow and purple colour was developed. At 16ʰ 15ᵐ a bright mass of aurora, extending from W.N.W. to S., at elevation 10° to 17°, with short streamers at an elevation of 50° dancing up and down, appearing and disappearing, with moderate motion. A dense and brilliant stationary body of

Abstract from the Meteorological Journal—*continued.*

Date.		Wind.		Temperature.		Weather.
Gött. Time.	Mean Time.	Direction.	Force.	Newman corrected.	Dollond as observed.	
February 1844. D. H.	D. H.			°	°	aurora in the W.N.W.; detached streamers in the E.; no disturbance. At 16h 30m four faint striated bands or arches crossing the meridian near the zenith. A bright portion of an arch (3) from E. by N. at elevation 6°, to N. at elevation 40°, terminating in a curl at the upper end. At 16h 45m a broad diffused arch of irregular form, at elevation 74°. Five faint and imperfect arches in the N. at various altitudes. Faint streamers in the S.W. No disturbance.
28 17	28 9	—	Calm.	6·4	6·1	Unclouded. A dense and brilliant band, rising in a zigzag form, and extending from N.W. to N.E., much diffused at N.E. end, and portions striated. At 17h 15m streamers of moderate brightness in various parts of the sky. No disturbance. 17h 30m no aurora visible.
19	11	—	Calm.	−0·3	1·9	Unclouded. No aurora. At 19h 30m very faint diffused aurora in the N.; portions striated.
21	13	—	Calm.	−1·5	0·6	Unclouded. Bright aurora ( ) not described. 21h 15m aurora in bright serpentine bands ( ) moderately dense, in rapid motion, and faintly coloured, all in the southern section of the sky.
22	14	—	Calm.	−2·5	−0·2	Diffused bands of aurora ( ) spread irregularly near the zenith, and in the N.W. Slight disturbance.
23	15	—	Calm.	−2·3	0·0	Unclouded. Very faint detached masses of cirrus aurora ( ) floating about in the S. Considerable change of Inclinometer scale reading at 23h, for which extra observations were made for half an hour, then discontinued, on the magnet returning with little irregularity to its mean position.
29 0	16	—	Calm.	−3·8	−2·3	Unclouded. Faint cirrus aurora as before, but no disturbance.

END OF THE OBSERVATIONS AT LAKE ATHABASCA.

ABSTRACT FROM METEOROLOGICAL JOURNAL.

AT FORT SIMPSON.

Date.		Wind.		Temp.	Weather.
Gött. Time.	Mean Time.	Direction.	Force.	Newman corrected.	
1844. April. D. H.	March. D. H.				
1 6	31 21	E. by S.	Mod.	°8·7	Hazy. Snowing from 1^h to 5^h Gött.
12	April. 1 3	S.E.	Mod.	25·0	Overcast, with light cirro-cumuli, interspersed with clear spaces.
17	8	S.E.	Mod.	15·1	A faint mass of aurora, of striated appearance, in the N.N.E., at 50° elevation. At 17^h 30^m an arch of moderate brightness ( ), extending from S.E. to N.N.W., and at 70° of elevation.
18	9	S.E.	Mod.	11·3	Unclouded. The arch has recently separated into faint masses ( ) of striated appearance, diffused generally over the sky, and slightly in motion. At 18^h 15^m the aurora considerably brighter ( ) and nearer the zenith, elevation 82°. At 18^h 30^m still bright and about the same elevation, with moderate serpentine motion.
2 0	1 15	S.E.	Light.	1·6	Hazy, but unclouded.
6	21	S. by E.	Light.	10·3	Unclouded since 0^h.
12	2 3	S.E.	V. light.	29·3	Still unclouded.
17	8	—	Calm.	16·9	Unclouded. A faint arch of aurora ( ) from S.E. to N.N.W., at elevation 74°.
18	9	—	Calm.	15·8	Unclouded. Faint aurora ( ) extending from the E. along the northern quarters; cirrus aurora or haze in various parts of the sky.
20	11	—	Calm.	12·7	Unclouded. Cirrus aurora in various parts of the sky.
21	12	—	Calm.	11·2	Unclouded. Faint aurora from E. to N.W., and auroral haze in various parts of the sky.
22	13	—	Calm.	8·6	Unclouded. No aurora. At 22^h 10^m a brilliant burst of aurora, attended by great disturbance of the magnets. It appeared in the W.N.W., rising rapidly in vertical streamers, which were highly coloured, exhibiting tints of pink, green, and yellow, and in violent pulsating or dancing motion; sometimes, also, changing position by serpentine development, and presenting themselves in different parts of the sky, both N. and S. of the zenith. It disappeared at 22^h 30^m. Most easterly position of the Declination (+2° 37′) at 22^h 40^m; most westerly (−0° 50′·3) at 22^h 50^m; range 3° 27′. The movements of the Inclinometer and Bifilar Magnets exceeded the range of their respective scales.
3 0	15	—	Calm.	1·5	Unclouded. No aurora visible; but little disturbance. Observations were discontinued at 23^h 30^m, but resumed at 4^h, giving another maximum to the E. of +1° 39′·8 at 4^h 24^m.
6	21	—	Calm.	12·7	Light cirrus clouds generally.
12	3 3	S.W.	Fresh.	39·8	Cirro-cumulus clouds. Wind since 8^h began S.E.
17	8	S.	V. light.	28·9	Unclouded. Faint arch of aurora from E. to N., at elevation 40°. The magnets not disturbed.
18	9	—	Calm.	30·4	Two faint arches ( ) from S.E. to N.N.W., at 57° and 65° elevation, and a broad diffused band crossing the zenith from S. to N.
4 0	15	—	Calm.	22·6	Overcast.
6	21	S.E.	Light.	29·3	Lightly clouded to the E.; remainder clear.
12	4 3	S.	Fresh.	46·5	Nearly overcast.
18	9	S.	V. light.	33·8	Unclouded, but hazy.
19	10	S.E.	V. light.	31·5	The same. Faint auroral haze in the N. and S.W.
Good Friday. 5 18	5 9	—	—	—	Casual observation. An aurora of moderate brightness ( ), chiefly confined to the N.W., annular and irregular in form and motion.
21	12	N.N.W.	High.	22·0	Unclouded. Faint auroral haze in S. and W. Wind squally, increased to a gale at 22^h.
6 0	15	W.N.W.	High.	9·5	Clouded since 22^h, wind somewhat abated. Snow mixed with rain at 23^h.

Abstract from the Meteorological Journal—*continued.*

Date.		Wind.		Temp.	Weather.
Gött. Time.	Mean Time.	Direction.	Force.	Newman corrected	
April 1844. D. H.					
6 6	5 21	N.W.byN.	Mod.	° 9·3	Unclouded. A sprinkling of snow at 2h, 3h, and 4h.
12	6 3	—	Calm.	16·4	Light cirrous clouds, 0·9 of blue sky.
18	9	—	Calm.	8·9	Nearly unclouded. No aurora. At 18h 15m an arch of aurora ( ) extending from E. to W.N.W., extremities at 20° elevation, but rising to 78° in the centre, rather irregular above the elevation of 45°, width 11°, in motion; also, two faint arches from E. to N., at elevation 25° and 33°. Extra observations were commenced, but the disturbance manifested was slight. At 18h 30m the arch was in the zenith, where it separated into three bands or arches; the general form was regular at the N.W. end, but at the opposite extremity was curved almost to a circle, at the elevation of 18°. Most westerly reading of Declination (−0° 27′·4) at 18h 30m. At 18h 45m the arch appeared in three detached portions, extending from the zenith to the N.W., slightly in motion. There was also a bright patch in the N., at elevation 20°, and faint haze ( ) in various parts of the sky.
19	10	—	Calm.	6·2	Nearly unclouded. A faint arch of aurora (5), at elevation 34°. The most easterly reading of the Declinometer was observed at 19h 45m (+0° 16′·5); range only 0° 43′·8.
20	11	—	Calm.	3·6	Unclouded. A double arch of aurora ( ) of striated appearance, at an elevation of 27°, rather brighter than before.
Sunday.					
7 23	7 14	—	Calm.	15·3	Hazy. A faint arch of aurora extending from E. to N.W., elevation 56°.
8 0	15	—	Calm.	13·2	Unclouded.
6	21	—	Calm.	23·7	Unclouded.
12	8 3	S. by W.	Mod.	42·5	Nearly overcast, with light cirro-cumuli, 0·3 of blue sky.
18	9	S.	Light.	38·5	Overcast since 13h.
9 0	15	S. by W.	Mod.	37·1	Unclouded. Wind gusty.
6	21	W.	High.	42·4	Unclouded.
12	9 3	N.N.W.	Brisk.	36·8	Overcast, and fresh N.N.W. wind since 7h. Heavy snow soon after 12h.
18	9	W. by N.	High.	23·7	Hazy, but only partially clouded. Light snow at 15h and 16h.
20	11	W. by N.	Brisk.	22·8	Unclouded. No aurora. At 20h 45m the aurora appeared as a broad arch, crossing the zenith from E. to W. by N., brightest and most regular in the S.E., up to an elevation of 40°, from thence about 9° wide to a distance of 30° from the zenith towards the W. It was extending itself with moderate motion. No disturbance.
21	12	W. by N.	Mod.	17·1	Aurora was extending in the N.N.W. with violent whirling motion, from an elevation of 25° to 68°, exhibiting faint tints of pink and yellow, with several long streaks rising towards the zenith in the N. and S. The most westerly reading of the Declinometer (−0° 4′·4) was at 21h 6m. At 21h 15m a faint band of aurora extended across the zenith, and a very bright mass of streamers ( ) rose to the zenith from an elevation of 14° E. by S., so closely arranged as to resemble a single striated beam, but in violent motion, and beautifully tinted. At 21h 30m four faint transverse bands ( ) extended across the meridian near the zenith, from an elevation of 30° above the horizon in the S.E. to an elevation of 25° in the N.W., having a serpentine outline and a moderate motion. At 21h 45m vertical beams of aurora and detached haze in various parts of the sky.

Abstract from the Meteorological Journal—*continued.*

Date.		Wind.		Temp.	Weather.
Gött. Time.	Mean Time.	Direction.	Force.	Newman corrected.	
April 1844. D. H.	D. H.			°	
9 22	9 13	—	Calm.	16·4	Unclouded. The aurora as last described. At 22h 15m the same. At 22h 30m it had nearly disappeared, leaving only a few vertical beams, and streaks of light moderately bright near the horizon. The most easterly reading of the Declinometer (+1° 10′) was at 22h 45m, range 1° 10′·6.
10 0	15	W. by N.	Mod.	14·9	Unclouded, but hazy.
6	21	N.W.byN.	Fresh.	20·4	Nearly overcast.
12	10 3	—	Calm.	22·8	Nearly unclouded.
18	9	—	Calm.	12·7	Unclouded. A vertical mass of aurora rising in the S.E., elevation 24°. At 18h 15m faint vertical beams extending from N.N.E. to E. near the horizon. Imperfect arches rather brighter ( ) rising from the S.E. and N.W., in the latter quarter approaching the zenith, in the former rising to 60° elevation. At 18h 30m vertical beams still stationary, and ranging from N.N.E. to E., also an imperfect annular body of aurora in the N.N.E., at an elevation of 40°, and about 23° in diameter. Imperfect arches, and cirrus aurora or haze in various parts of the sky. At 18h 45m the aurora had nearly disappeared. The instruments were watched and scale readings taken every 15m during this display, but no disturbance was manifested.
20	11	—	Calm.	11·9	Unclouded. A faint arch of aurora from E. to N. at elevation 28°.
21	12	—	Calm.	7·1	Unclouded. A heavy band of aurora, moderately bright, crossing the zenith from S.E. to W. by N., of irregular form, with rapid serpentine changes, faintly coloured; it vanished in a few minutes, and at 21h 6m nothing but a few faint streamers and cirrus aurora was visible; this continued with little change to 21h 45m, and at 22h there was no aurora visible. The magnets considerably disturbed. Most easterly reading of Declinometer (+1° 27′·3) at 21h 6m, most westerly reading (−0° 0′·9) at 22h 3m, range 1° 26′·2.
11 0	15	—	Calm.	2·7	Unclouded.
6	21	E.	V. light.	13·8	Unclouded.
12	11 3	E. by S.	V. light.	28·7	Unclouded.
18	9	E.	Brisk.	21·5	Unclouded, but hazy.
19	10	E.	Brisk.	21·5	Still unclouded. A faint double arch from N. to E. at an elevation of 26°, and auroral haze in detached masses.
23	14	E.	Brisk.	17·1	Long and faint vertical beams, and auroral haze extending from N. to E.; also faint bands ( ) across the meridian near the zenith.
12 0	15	E. by S.	Brisk.	17·1	Nearly unclouded.
6	21	E.S.E.	Light.	23·7	Unclouded since 1h.
12	12 3	—	Calm.	37·0	Still unclouded.
18	9	—	Calm.	32·9	Haze gathering since 13h. At present overcast.
13 0	15	—	Calm.	30·4	Thickly overcast, and snowing. Very dark.
6	21	N.W.	Mod.	33·5	Still snowing slightly.
12	13 3	N.	Brisk.	34·5	Partially clouded, but with clear spaces.
18	9	W.N.W.	Brisk.	27·5	Overcast since 13h, but now clearing again. Still hazy.
19	10	N.W.	Light.	26·9	Unclouded. An arch of aurora ( ) at an altitude of 33°, which separated at 19h 40m into faint serpentine bands of little density, in moderate motion, and apparently at a considerable elevation in the atmosphere. No disturbance.
20	11	N.W.	Light.	26·2	Dispersed portions of aurora still visible.

Abstract from the Meteorological Journal—*continued.*

Date.		Wind.		Temp.	Weather.
Gött. Time.	Mean Time.	Direction.	Force.	Newman corrected.	
April 1844. D. H.	D. H.			°	
14 21	Sunday. 14 12	S.E.	Mod.	22·6	Unclouded. A bright arch of aurora ( ) from E. to N. at 35° of elevation, terminating in a curve or hook at the E. end, at 14° elevation. Long vertical streamers in slight motion from N. to N.E. Dense masses of aurora in the zenith, and in various parts of the sky. The magnets slightly disturbed. Most westerly reading of Declination (−0° 19′·1) at 21^h. At 21^h 15^m the features of the aurora were much the same, but considerably fainter. At 21^h 30^m it had nearly disappeared, excepting faint haze, and a faint imperfect arch from E. to N. at elevation 18°. At 21^h 45^m there was a faint arch from N.E. to N. at an elevation 27°. At 22^h no aurora was visible.
15 0	15	—	—	20·6	Unclouded. At 1^h the same. No aurora visible, but a considerable degree of disturbance, for which the extra observations were resumed, and continued to 3^h. Most easterly reading of the Declinometer (+1° 38′·7) at 1^h 42^m, range since 21^h, 2° 16′·0.
6	21	S.E.	V. light.	27·1	Unclouded.
12	15 3	S.E.	V. light.	43·5	Clouded, light cirrus and haze.
18	9	S.E.	V. light.	32·5	Uniformly overcast, with light haze.
19	10	—	Calm.	33·7	Unclouded since 18^h. A faint arch crossing the meridian near the zenith from E. to W.N.W., form irregular in the highest portion, expanding with moderate motion. No disturbance, but readings taken every 15^m. At 19^h 15^m the aurora in bright serpentine bands, faintly coloured, crossing the meridian near the zenith, and extending rapidly, with winding motion. At 19^h 30^m the bands were extending with rapid motion to the S. of the zenith, but much fainter than at the last observation. At 19^h 45^m about half of the sky was covered with aurora in vertical streamers, and cirrus masses of moderate brightness. The magnets were now observed at intervals of one minute, and showed a slight degree of disturbance, the Declination ranging to the W. of its mean position. Most westerly reading (−0° 23′·9) at 19^h 54^m.
20	11	—	Calm.	33·5	Unclouded. A long range of vertical streamers, faintly coloured, extended from E. nearly to the N., at elevation of 24°, pulsating or dancing with rapid motion. Cirrus aurora, or haze, in various parts of the sky. Most easterly reading of the Declination (+0° 16′·7) at 20^h 3^m, range, 0° 40′·6; a remarkably small quantity for so considerable a display of aurora. At 20^h 15^m there were no traces of the aurora visible.
23	14	—	Calm.	26·4	Unclouded since 20^h. No observation from 23^h to 3^h.
16 6	21	N.W.	V. light.	41·4	A few light cirrus, but nearly unclouded since 3^h.
12	16 3	N.W.byN.	Mod.	49·5	Unclouded.
14	5	—	Calm.	46·5	Light cirro-cumulus and haze. The Bifilar and Inclinometer began to give evidence of a state of disturbance; the Declination also ranged considerably to the westward of its mean position at this hour. Extra observations were commenced and continued for sixteen hours.
17		—	Calm.	38·5	Hazy. Disturbance still continuing, the Declination exclusively to the west of its mean position. At 17^h 38^m aurora was first observed rising in the S.E. in two curved streams, moderately bright; elevation above the horizon, 35° and 40°. *At this same hour the following entry occurs in the Meteorological Register at Toronto:—"Clear overhead; "clouded round the horizon. A faint light "apparent behind the clouds in the N.W. "horizon."*

Abstract from the Meteorological Journal—*continued.*

Date.		Wind.		Temp.	Weather.
Gött. Time.	Mean Time.	Direction.	Force.	Newman corrected.	
April 1844.					
D. H.	D. H.			°	
16 18	16 9	—	Calm.	37·7	Unclouded. No aurora described, but at Toronto an entry similar to the last is found: "*A faint auroral light behind the clouds in* " *the northern horizon.*" After which it was clouded at that station, and began to rain at $20^h 40^m$. At $18^h 45^m$ imperfect striated arches of aurora extended from E. to N., and from S. to N.N.E., moderately bright. A very faint double arch or band crossed the zenith from S. to N.W. Bright cirrus aurora, or haze, in various parts of the sky.
19	10	—	Calm.	33·9	Unclouded. The aurora nearly the same as at $18^h 45^m$. At $19^h 15^m$ scarcely any traces of aurora, but considerable disturbance, the Declination ranging to the eastward. At $19^h 45^m$ two narrow serpentine bands of aurora crossed the sky, near the zenith, from N.W. by N. to S., moderately bright, and with slight motion. The most westerly reading of the Declinometer (−1° 17′·2) was at $19^h 50^m$.
20	11	—	Calm.	30·4	Unclouded. A broad band of aurora in violent motion, extending from the S.E. to the zenith, with vertical streamers pulsating or dancing in the N. Extra observations were commenced at Toronto at this hour, and continued for sixteen hours. At $20^h 15^m$ long vertical streamers in the S.E. and N.E., in moderate motion. At $20^h 45^m$ aurora extending from E. to W., and passing near the zenith, in the form of an arch of moderate brightness, faintly coloured at the eastern end. The extreme easterly reading at Toronto (+0° 39′·3) was at this time.
21	12	—	Calm.	28·2	Unclouded. The aurora rising from the E. and W., in fine vertical streamers, in rapid motion. They all united to form a star or corona near the zenith, having a diameter of 7°, its exact position not recorded. At $21^h 15^m$ no traces of the aurora remained, except a few fine streamers in the N. At $21^h 30^m$ the same. At $21^h 45^m$ all trace had disappeared.
22	13	—	Calm.	28·6	Unclouded. Aurora reappeared, of irregular form, moderately bright, and extending from N. to S.W., in slight motion. Extra observations were commenced at Greenwich at this hour, in consequence of a change of Declination of 8′ 45″ between 20^h and 22^h. Soon after the same hour the range of Declination at Toronto passed to the westward of the mean, and did not rise to the mean value during the remainder of the observations. At $22^h 15^m$ no traces of aurora were visible; the same at $22^h 30^m$ and $22^h 45^m$.
23	14	—	Calm.	28·0	Unclouded. Aurora visible, in vertical streamers and dense cirrous patches, both N. and S. At $23^h 45^m$ a slender serpentine band crossing the zenith. The most westerly reading of the Declination at Toronto (−0° 54′·5) was observed at $23^h 25^m$; range, 1° 14′·3.
17 0	15	—	Calm.	27·6	Aurora no longer visible; day dawning. The disturbance still continued, and soon after 1^h exceeded the limits of the scales of all the instruments. Most easterly reading of the Declinometer (+6° 32′·0 nearly) at $1^h 24^m$; range not less than 8° 10′, possibly somewhat greater. The more active part of this disturbance, which was the greatest observed, appears to have terminated about 8^h Gött., or 11 A.M. of mean time at the station. It is remarkable that its relative extent was by no means so great at Toronto, where the range observed has been often exceeded, and was quite inconsiderable at Greenwich. Disregarding minor changes, it may be described

Abstract from the Meteorological Journal—*continued.*

Date.				Wind.		Temp.	Weather.
Gött. Time.		Mean Time.		Direction.	Force.	Newman corrected.	
							to have consisted at Toronto, as regards the Declination, of a great easterly excursion, having a maximum at 21^{h}, and followed by a westerly excursion, giving one minimum at or near $23^{h}\ 30^{m}$, and another at $0^{h}\ 15^{m}$, the two being separated by a marked return to the eastward at 0^{h}. The succeeding maximum is at 3^{h}, after which the changes of this element are unimportant. Referring to the observations at Fort Simpson, there is no general feature corresponding to either of these. Great and rapid changes of Declination prevailed during the whole continuance of the observations, but the most important of these, between 1^{h} and $1^{h}\ 30^{m}$, when the element reached the very large deviation of 6° 32′ from its normal value at the same hour, has no corresponding feature at Toronto, where at that period the changes of Declination were moderate. As regards the Horizontal Force, we have at Toronto a minimum soon after 20^{h}, succeeded by a very decided increase of force, having a maximum two hours later; this is followed by two minima, the most considerable between 23^{h} and 0^{h}, and the other soon after 1^{h}, after which there is a very gradual return towards the mean value of the element for several hours. It is curious to observe that a feature very much resembling the first of these, namely, a maximum between two minima, occurs at Fort Simpson three hours earlier, at 19^{h} instead of 22^{h}, but in relative extent is not so great. To the minimum in question at Toronto, there is no feature corresponding at Fort Simpson; on the other hand, each of the two succeeding minima at Toronto has correspondence which cannot be regarded as accidental with a minimum at Fort Simpson; the important difference being, that the first, which is by far the most considerable at Toronto, is the least at Fort Simpson, and the second, which at the latter station exceeds any other observed, is but moderate at Toronto.
April 1844.							
D.	H.	D.	H.			°	
17	6	16	21	—	C m	40·5	Nearly overcast, with cirro-cumuli and cirro-strati.
	12	17	3	—	C m.	50·0	0·7 of blue sky. A few cirro-cumuli and cirro-strati. A slight degree of disturbance was observed from 13^{h} to 14^{h}, the Declinometer ranging to the westward of its mean position. Most westerly reading (−0° 20′·9) at $13^{h}\ 9^{m}$.
	18		9	—	Calm.	38·3	Unclouded, but hazy.
	22		14	—	Calm.	33·2	Almost unclouded. A few vertical beams of aurora in the N.
18	0		15	—	—	—	No observation. At 23^{h} and at 2^{h} calm and unclouded.
	6		21	—	Calm.	42·2	Nearly unclouded.
	12	18	3	E. by S.	Light.	51·7	Unclouded.
	18		9	—	Calm.	42·1	Nearly unclouded.
	20		11	—	Calm.	38·0	Unclouded. Auroral haze in the S.E.
	21		12	—	Calm.	36·3	Unclouded. A faint arch of aurora from E. to W., at an elevation of 79°. No disturbance.
	22		13	—	Calm.	33·9	Unclouded. A long range of slender vertical beams, ranging from E. to N., of moderate brightness, and showing but little motion.
19	0		15	—	Calm.	33·3	Unclouded.
	6		21	—	Calm.	44·4	Unclouded.
	12	19	3	N.W.	Light.	54·5	Still unclouded, but somewhat hazy.
	18		9	—	Calm.	40·7	Partially clouded.
	20		11	—	Calm.	37·5	Unclouded again. A dense mass of aurora in E. by S., at elevation 22°, moderately bright.

Abstract from the Meteorological Journal—*continued.*

Date.				Wind.		Temp.	Weather.
Gött. Time.		Mean Time.		Direction.	Force.	Newman corrected.	
April 1844.							
D.	H.	D.	H.			°	
19	21	19	12	S. by E.	V. light.	36·5	An arch of aurora ( ) extending from N.N.E. to W., elevation 39°, the end nearest the N. terminating in a curve or hook, at an elevation of 36°; also a faint streak from N. to E. No disturbance. At 21ʰ 15ᵐ no traces of the aurora were visible.
	22		13	S.E.	Light.	35·0	Faint broad bands of aurora crossing the zenith from E. to W. At 22ʰ 45ᵐ faint striated masses of aurora ( ) in the N.W. At 23ʰ no aurora visible. Extra readings were taken from 22ʰ to 0ʰ, showing a slight degree of disturbance. Most westerly reading (+0° 6′·3) at 22ʰ 45ᵐ; most easterly (+0° 50′·5) at 23ʰ 6ᵐ; range, 0° 45′·4.
20	0		15	—	Calm.	32·8	Unclouded.
	6		21	—	Calm.	36·8	Unclouded.
	12	20	3	—	Calm.	45·7	Still unclouded.
	18		9	S.E.	V. light.	37·6	Unclouded, but hazy.
	20		11	—	Calm.	32·6	A faint irregular arch of aurora, elevation 40°, with slender vertical beams or striæ in slight motion. No disturbance.
Sunday.							
21	22	21	13	E.	Light.	38·0	Unclouded. Auroral haze in various parts of the sky.
	23		14	E.	Light.	36·8	A very faint arch of aurora, at elevation 63°, extending from E. by N. to N.N.W.
22	0		15	—	Calm.	35·5	Unclouded.
	6		21	S.E.	V. light.	44·7	Unclouded.
	12	22	3	N. by W.	Light.	46·5	Unclouded.
	18		9	N.	Brisk.	36·7	Unclouded. Wind in gusts.
23	0		15	N.	Mod.	28·1	A few light cirrous clouds.
	6		21	N.	Fresh.	33·4	A few cirri and cirro-cumuli.
	12	23	3	N.	Brisk.	37·0	About 0·2 of well defined cirro-cumuli, ranging from W. to S.W.; remainder of the sky unclouded.
	18		9	N.	Mod.	31·0	Densely overcast, with close packed cirro-cumuli.
24	0		15	E.S.E.	Mod.	25·9	Nearly overcast. Heavy snow at 22ʰ.
	6		21	S.E.	Fresh.	30·4	Overcast.
	12	24	3	S.E.	High.	37·2	Overcast. Magnetic term day began at 10ʰ.
	18		9	S.E. by S.	Fresh.	33·5	Overcast, with little change since the last observation.
25	0		15	—	Calm.	33·0	Overcast. Thick cirro-cumuli. It was partially clear at 21ʰ and 22ʰ. No aurora seen.
	6		21	S.W.	Light.	58·0	Light uniform haze. A considerable degree of disturbance prevailed during this term day, especially from 2ʰ to 4ʰ Gött., at which period both Declinometer and Bifilar exhibited its maximum effect. The movements of the Bifilar have their counterpart very decidedly marked at Toronto. Those of the Declinometer, which were equally great at Fort Simpson, have no corresponding movement whatever at Toronto, but, on the contrary, there is at that time a marked absence of movement there. On the other hand, the Declinometer at Toronto was disturbed from 12ʰ to 20ʰ Gött., during which time there was no disturbance of that element at Fort Simpson. Most easterly reading of the Declinometer (+2° 54′·5) at 2ʰ 55ᵐ Gött.; most westerly (−0° 27′·5) at 8ʰ 15ᵐ; range, 3° 40′·0.
	12	25	3	S.W. by S.	Brisk.	68·0	Overcast, with fleecy cirro-cumuli.
	18		9	W. by S.	High.	54·9	Overcast. Wind increasing, in gusts.
	22		13	N.	V. high.	42·1	Sky clearing; a few stars visible since 21ʰ. Faint auroral haze. Blowing a gale since 20ʰ. A considerable disturbance began at 20ʰ, with a westerly range of the Declinometer, and prevailed down to 26ᵈ 2ʰ. Most westerly reading (−1° 16′·7) was actually at 0ʰ 24ᵐ. The most easterly reading (+1° 2′·1) preceded it at 0ʰ 9ᵐ, but a westerly deviation, amounting to 0° 48′·1, was previously attained at 20ʰ 45ᵐ, and the general character of the disturbance of Declination was westerly; range 2° 20′·4.

Abstract from the Meteorological Journal—*continued.*

Date.				Wind.		Temp.	Weather
Gött. Time.		Mean Time.		Direction.	Force.	Newman corrected.	
April 1844.							
D.	H.	D.	H.				
26	0	25	15	N. by W.	Light.	37°·3	Unclouded. Wind fallen since 23^h.
	6		21	N.W.	V. light.	44·5	Overcast again since 1^h.
	12	26	3	E.	V. light.	46·5	Overcast.
	18		9	E.	V. light.	40·7	Overcast with dense cirro-cumuli; began to rain soon after. A considerable disturbance, which commenced with a westerly range of Declination, prevailed from 18^h to 2^h. Most westerly reading (−1° 15′·1) at 18^h; most easterly (+0° 26′·7) at 27^d 0^h 18^m; range, 1° 59′·5. This disturbance was marked by a constant state of vibration in the magnets, which was not usual. No aurora was visible at 0^h, when it was for a short period unclouded.
27	0		15	—	Calm.	34·5	Continued rain since 18^h, at present mixed with sleet.
	6		21	—	Calm.	41·9	Overcast. Ceased rain and snow after 1^h.
	12	27	3	N. by W.	V. light.	50·1	Overcast, with close cirro-cumuli.
	18		9	—	Calm.	36·8	Dense cirro-cumuli, closely packed.
Sunday.							
28	21	28	12	—	Calm.	29·6	Unclouded. A broad and diffused but faint ( ) arch of aurora, extending through the zenith from S.E. to N.W. At 21^h 15^m faint auroral haze alone. A considerable disturbance prevailed from 21^h to 8^h, commencing with a westerly range of the Declinometer. Most westerly reading (−0° 28′·6) at 21^h 46^m; most easterly reading (+1° 46′·5) at 2^h 3^m; range 2° 36′·8. No aurora was visible at 22^h, when it was unclouded.
29	0		15	E.	V. light.	29·6	A few light cirro-strati.
	6		21	S.W.	High.	48·5	Wind high, with gusts. Sky covered with cirro-cumuli.
	12	29	3	S.S.W.	Light.	53·8	Covered with cumuli and cirro-cumuli.
	18		9	—	Calm.	48·7	Hazy, but unclouded.
30	0		15	—	Calm.	34·6	Overcast with uniform haze. A moderate disturbance was observed from 0^h to 2^h. The range of Declination easterly. Most easterly reading (+1° 27′·9) at 0^h 18^m; most westerly (−0° 7′·3) at 1^h 3^m; range, 1° 36′·4. The sky was overcast during its continuance.
	6		21	S.E.	V. light.	48·6	Lightly overcast.
	12	30	3	N.W.	Mod.	43·3	Fine drizzling rain commencing. Observations at intervals of 15^m were taken from 14^h to 16^h, in consequence of high range of Horizontal Force and prevailing westerly range of Declination, but no disturbance was observed. Most westerly reading (−0 12′·9) at 14^h 30^m. This amount of deviation from the mean is large for that hour, having been exceeded only four times.
	18		9	N.	High.	23·7	Rain from 0^h to 15^h, which changed to snow, and so continues.
May.							
1	0		15	N. by W.	V. high.	14·6	Strong northerly gale prevailing since 19^h, attended by snow down to 20^h; overcast the whole time. At 21^h a considerable disturbance commenced, and was observed down to 1^d 1^h. Most easterly reading (+2° 18′·3) at 21^h 36^m. Most westerly (−0° 26′·7) at 22^h 51^m; range, 2° 44′·8.
	6		21	N. by W.	High.	18·9	Nearly unclouded.
		May.					
	12	1	3	N.	Brisk.	26·8	Covered with close cirro-cumuli.
	18		9	N.	V. light.	22·1	Thickly overcast.
2	0		18	—	—	—	No observation. At 1^h it was calm and unclouded.
	6		21	S.E.	V. light.	29·0	Still unclouded.
	12	2	3	S.E. by E.	V. light.	37·7	Still unclouded.
	18		9	S.E. by E.	Brisk.	32·0	Clouding over since 16^h. At present overcast.

Abstract from the Meteorological Journal—*continued.*

Date.		Wind.		Temp.	Weather.
Gött. Time.	Mean Time.	Direction.	Force.	Newman corrected.	
May 1844.					
D. H.	D. H.			°	
2 19	2 10	S.E.	Brisk.	23·2	Again unclouded. Two arches or bands of aurora extending from E. and S.E. through the zenith to N.W., moderately bright, and in slight motion. At 19ʰ 15ᵐ a bright narrow band extended through the zenith from E. to N.W., in moderate serpentine motion. Considerable disturbance prevailing from 19ʰ to 0ʰ, commencing with a westerly range of Declination. Most westerly reading (−0° 31′·2) at 19ʰ 36ᵐ.
3 0	15	S.E.	Fresh.	21·5	Generally unclouded since 19ʰ.
6	21	S.E.	Light.	31·5	Light cirro-cumuli and cirro-strati. The disturbance observations were resumed at 4ʰ, the range of Declination being now easterly. Most easterly reading (+1° 33′·1) at 4ʰ 21ᵐ; range, 2° 33′·4.
12	3 3	S.E.	Light.	42·5	Light cirri.
18	9	S.E.	Mod.	36·7	Overcast since 15ʰ. Slight rain from 15ʰ 40ᵐ to 17ʰ.
4 0	15	S.E.	Mod.	31·2	Cirri and strati.
6	21	S.E.	Light.	42·5	Unclouded, save light cirrus haze.
12	4 3	N.W.	Light.	40·8	Nearly overcast, with cirro-cumuli. Slight rain at 13ʰ.
18	9	—	Calm.	38·7	Overcast, with dense cirro-cumuli.
19	10	—	Calm.	35·7	Unclouded. Irregular serpentine bands of aurora moderately bright ( ) crossing the zenith, and extending to the S., with rapid motion. A very slight degree of disturbance was shown by the magnets, for which extra readings were not taken.
20	11	—	Calm.	33·4	Unclouded. Aurora still visible, but not described.
Sunday.					
5 22	13	N.N.W.	V. light.	30·4	Unclouded. Aurora crossing the meridian from E.N.E. to N.W. in an irregular narrow band, with slight serpentine movement. Aurora was also observed on the evening of the 5th at two stations in the State of New York (Regent's Reports.) A slight disturbance was observed from 22ʰ to 23ʰ 30ᵐ. Most easterly reading of Declinometer (+1° 4′·4) at 22ʰ 9ᵐ, followed by the most westerly reading (−0° 11′·7) at 22ʰ 21ᵐ; range, 1° 16′·5. This is the last observation of aurora recorded, the nights being already sufficiently light to permit out-door pursuits even at midnight, and to make it nearly impossible to distinguish that appearance from light cirrus clouds.
6 0	5 15	—	Calm.	29·1	Unclouded.
6	21	N.N.W.	V. light.	38·5	Overcast. It began to snow at 6½ʰ, and so continued to 9ʰ.
12	6 3	N.W.	Light.	39·8	Strati and cumuli, with 0·6 of clear sky.
18	9	—	Calm.	32·0	Overcast, with uniform haze.
7 0	15	—	Calm.	28·1	Cirri and cirro-strati scattered over the sky.
6	21	—	Calm.	42·4	Unclouded since 4ʰ.
12	7 3	S.E. by E.	Light.	48·9	Scattered cumuli and cirro-cumuli.
18	9	—	Calm.	40·0	Closely packed cirro-cumuli. Auroral light was observed at Toronto at 16ʰ.
8 0	15	E.	Mod.	34·4	Thickly overcast.
6	21	S.E.	Light.	44·4	Unclouded since 3ʰ.
12	8 3	S.	Light.	55·7	Detached cumuli.
18	9	—	Calm.	43·0	Overcast entirely since 16ʰ. *Auroral light was observed at Toronto* at 15ʰ.
9 0	15	—	Calm.	35·7	Light cirri, with haze.
6	21	S.E.	V. light.	47·1	Nearly unclouded.
12	9 3	S.	V. light.	56·5	0·4 of blue sky, with cumuli over the remainder.
18	9	—	Calm.	44·7	Closely overcast since 15ʰ.

Abstract from the Meteorological Journal—*continued.*

Date.				Wind.		Temp.	Weather.
Gött. Time.		Mean Time.		Direction.	Force.	Newman corrected.	
May 1844.							
D.	H.	D.	H.			°	
10	0	9	15	—	Calm.	36·7	A few stars visible since 21^h, and now nearl unclouded.
	6		21	S.	V. light.	54·0	Unclouded from 1^h to 4^h. At present hazy.
	12	10	3	S. by W.	Mod.	55·5	Overcast, with uniform haze. Wind in gusts.
	18		9	S.	Light.	48·3	Overcast, with uniform haze.
11	0		15	—	—	—	No observation. At 23^h and at 1^h calm and unclouded.
	6		21	W. by S.	Fresh.	54·8	Unclouded.
	12	11	3	S.W.	Brisk.	55·5	Cirri and cirri-cumuli scattered over the sky.
	18		9	N.N.W.	Mod.	37·4	Heavy rain, which began at 15^h 45^m, and continued to 19^h. Heavy gale from N.N.W., with snow, nearly all day.
Sunday.							
13	0	12	15	N.N.W.	Fresh.	27·2	Unclouded.
	6		21	N. by W.	Light.	36·5	Unclouded.
	12	13	3	E.	V. light.	44·5	Cirri-cirro-strata, and haze.
	18		9	E. by S.	V. light.	37·3	Overcast, with thick cirro-cumuli. Extra observations were made from 18^h to 21^h, in consequence of a westerly range of the Declination, but no decided disturbance was observed. Most westerly reading (−0° 29′·7) at 18^h 12^m; most easterly, which was still to westward of the mean (−0° 7′·3) at 19^h 9^m.
14	0		15	S.E.	Mod.	31·0	Unclouded. A slight magnetic shock was observed from 0^h to 1^h, the range of the Declination being easterly. The most easterly reading (+0° 39′·1) at 0^h 15^m; the most westerly (−0° 7′·1) at 0^h 48^m; range, 0° 47′·2.
	6		21	S.E.	Light.	43·4	Still unclouded.
	12	14	3	E.S.E.	V. light.	54·6	Light cirri and cirro-strati since 11^h.
	18		9	E.S.E.	Light.	45·1	Covered with light cirro-cumuli. *A faint auroral light was visible at Toronto* at 16^h and 17^h, connected with streamers at 16^h. Aurora was also observed at four stations in the state of New York (Regent's Reports.)
15	0		15	S.E.	Light.	34·6	Cirri, with general haze.
	6		21	—	Calm.	47·9	General haze.
	12	15	3	N. by W.	Light.	63·0	Scattered cirri.
	18		9	—	Calm.	53·3	Overcast with cirro-cumuli.
16	0		15	S.E.	V. light.	42·0	A few cirri but nearly unclouded.
	6		21	—	Calm.	61·4	Unclouded.
	12	16	3	—	Calm.	69·3	A few cumuli, 0·8 unclouded.
	18		9	—	Calm.	58·7	Covered with cirro-cumuli.
17	0		15	—	Calm.	46·4	Unclouded since 23^h. Rain at 20^h.
	6		21	S.S.E.	V. light.	58·5	A few cirri, but nearly unclouded.
	12	17	3	S.	Light.	67·6	Still unclouded.
	18		9	S.E.	Mod.	58·7	Thick cirro-cumuli, increasing since 14^h. Distant thunder at 17^h 45^m. Loud thunder followed by heavy rain at 19^h.
18	0		15	N.	V. high.	47·5	Wind increasing since 21^h. At present blowing a gale.
	6		21	N.	High.	48·4	Detached cumuli, but sky nearly clear.
	12	18	3	N.	High.	47·0	Overcast.
	18		9	N.N.E.	Mod.	36·7	Still overcast.
Sunday.							
20	0	19	15	S.E. by S.	Light.	37·3	Cirri and haze.
	6		21	—	Calm.	53·5	Unclouded since 1^h.
	12	20	3	N.	V. light.	65·9	Still unclouded.
	18		9	E.	V. light.	56·7	Light uniform haze.
21	0		15	S.E.	V. light.	43·5	Cirro-cumuli and strati.
	6		21	S.E.	V. light.	46·3	Unclouded since 2^h.
	12	21	3	S.E.	Light.	52·4	Unclouded.
	18		9	S.	Brisk.	41·7	Cirro-cumuli and haze.

Abstract from the Meteorological Journal—*continued.*

Date.				Wind.		Temp.	Weather.
Gött. Time.		Mean Time.		Direction.	Force.	Newman corrected.	
May 1844.							
D.	H.	D.	H.			°	
22	0	21	15	—	—	—	No observation. At 23h wind S. and fresh. Overcast, with thick cirro-cumuli
	1		16	S.E.	Fresh.	31·1	Light cirri prevailing. A great disturbance began to be observed at 21d 1h, and prevailed throughout this day and part of the next. The most easterly reading of the Declinometer on the 22d (+1° 56′·8) was at 1h 33m; the most westerly (−0° 43′·4) at 12h 24m; range, 2° 40′.
	6		21	S.E.	Fresh.	37·9	Light cirri prevailing.
	12	22	3	S.E.	Fresh.	50·5	Overcast, with cirro-cumuli.
	18		9	S.E.	Fresh.	42·7	Uniform light haze. *Auroral light was visible in the N. at Toronto* at 17h, 18h, and 19h Gött., accompanied at 17h by an arch of small streamers, extending from N.W. to N.E., about 1½° wide in the centre, with an elevation of 40°. The extra observations, which were discontinued at Fort Simpson at 4h Gött., were resumed at 12h, and continued to 19h. One hour later a disturbance began to be observed at Toronto, the observations being there continued to 20h.
23	0		15	—	Calm.	36·5	Overcast. Light rain at 22h. Extra observations were resumed at 23h, and continued to 3h. The most westerly reading of the Declinometer being (−0° 44′·5) at 0h 51m; the most easterly (+1° 40′·0) at 1h 45; range, 2° 33′·4. Of the separate portions of the disturbance observed at Fort Simpson, the middle one only has an imperfect correspondence with the disturbance at Toronto, principally shown by the changes of Horizontal Force; the first and last portions, which comprise the greatest changes, do not appear to have extended to Toronto.
	6		21	—	Calm.	48·6	Uniform light haze.
	12	23	3	S.E.	V. light.	61·2	Uniform light haze.
	18		9	S.E.	Light	53·8	Cirro-cumuli, with haze.
24	0		15	—	Calm.	44·4	Overcast, with cirri and cumuli. A slight disturbance was observed from 0h to 4h. The most westerly reading (−0° 8′·9) at 0h 57m; the most easterly (+0° 54′·2) at 1h 33m; range, 1° 3′·0.
	6		21	N.W.	V. light.	57·7	Uniform light haze.
	12	24	3	—	Calm.	63·5	Uniform light haze. Magnetic term day began at 10h.
	18		9	—	Calm.	56·4	Unclouded. Occasional cirro and cirro-strati since 14h.
25	0		15	E.	Light.	45·0	Unclouded, but hazy.
	6		21	S.E.	V. light.	58·7	Overcast since 5h.
	12	25	3	—	Calm.	—	Uniform light haze. A constant but moderate degree of disturbance prevailed during the whole of the term day, which terminated at 10h, and also characterizes the term observations at Toronto. There is no decided correspondence in the changes of the Declination at the two stations; but there is some correspondence in the changes of Horizontal Force between 12h and 14h Gött. on the 24th, and again at 9h Gött. on the 25th.

END OF THE OBSERVATIONS AT FORT SIMPSON.

FORT CHIPEWYAN.

Abstract of Hourly Observations made during the month of October 1843.

Spirit Thermometer by Newman, corrected.

Noon.	1.	2.	3.	4.	5.	6.	7.	8.	9.	10.	11.	12.
—	—	—	—	—	—	—	—	—	—	—	—	—
—	—	—	—	—	—	—	—	—	—	—	—	—
—	—	—	—	—	—	—	—	—	—	—	—	—
—	—	—	—	—	—	—	—	—	—	—	—	—
—	—	—	—	—	—	—	—	—	—	—	—	—
—	—	—	—	—	—	—	—	—	—	—	—	—
—	—	—	—	—	—	—	—	—	—	—	—	—
—	—	—	—	—	—	—	—	—	—	—	—	—
—	—	—	—	—	—	—	—	—	—	—	—	—
—	—	—	—	—	—	—	—	—	—	—	—	—
—	—	—	—	—	—	—	—	—	—	—	—	—
—	—	—	—	—	—	—	—	—	—	—	—	—
—	—	—	—	—	—	—	—	—	—	—	—	—
—	—	—	—	—	—	—	—	—	—	—	—	—
—	—	—	—	—	—	—	—	—	—	—	—	—
31·0	31·1	30·4	31·7	31·7	33·9	36·5	39·0	42·1	43·6	44·5	43·8	42·9
27·0	28·1	27·4	33·5	32·8	32·6	33·0	32·5	32·7	32·7	33·5	33·5	32·7
31·9	31·5	30·8	31·1	32·5	33·7	34·4	35·0	37·5	37·8	38·0	36·6	36·0
30·0	29·4	29·2	29·8	30·4	31·8	32·3	32·3	34·0	31·7	31·9	32·7	31·7
28·1	—	29·8	29·2	30·1	30·4	31·5	31·7	32·3	32·3	31·7	32·7	32·5
29·2	29·2	29·6	29·6	30·6	31·7	36·0	37·7	43·5	47·5	48·5	47·5	47·5
—	—	—	—	—	—	—	—	—	—	—	—	—
14·9	14·9	15·1	15·5	15·6	16·0	15·9	16·6	16·6	17·9	17·7	18·0	16·1
14·6	12·7	12·5	13·0	13·8	17·1	17·1	19·8	20·0	20·2	20·2	19·5	19·1
17·7	19·6	20·9	21·3	21·6	21·7	22·0	22·3	22·7	23·2	22·9	21·5	17·0
−2·2	−5·8	−7·6	−6·5	−5·1	−0·1	1·2	1·9	3·3	5·1	6·6	6·2	4·4
0·0	0·2	−0·1	−1·1	−0·5	1·2	3·3	4·2	5·9	8·2	8·5	9·2	8·9
−1·6	−2·0	−2·4	−2·4	−2·2	−0·1	2·5	4·2	5·0	6·5	7·8	8·0	7·0
—	—	—	—	—	—	—	—	—	—	—	—	—
22·5	22·6	22·5	22·9	22·9	23·8	23·8	23·8	24·8	24·0	24·0	23·8	23·9
19·8	19·8	19·1	20·6	22·1	22·8	23·9	24·7	25·7	24·7	24·7	24·3	24·0
262·9	231·3	257·2	268·2	276·3	296·5	313·4	325·7	345·6	354·9	360·5	357·3	343·7
18·78	17·79	18·37	19·16	19·74	21·18	22·39	23·26	24·69	25·35	25·75	25·52	24·33
−2·66	−3·65	−3·07	−2·28	−1·70	−0·26	0·95	1·82	3·25 N	3·91	4·31	4·08	3·11

8^h Göttingen time = noon of local mean time.

FORT CHIPEWYAN.

Abstract of Hourly Observations made during the month of October 1843.

Spirit Thermometer by Newman, corrected.

13.	14.	15.	16.	17.	18.	19.	22.	23.	Sums.	Means.
—	—	—	—	—	—	—	—	—	—	—
—	—	—	—	—	—	—	—	—	—	—
—	—	—	—	—	—	—	—	—	—	—
—	—	—	—	—	—	—	—	—	—	—
—	—	—	—	—	—	—	—	—	—	—
—	—	—	—	—	—	—	—	—	—	—
—	—	—	—	—	—	—	—	—	—	—
—	—	—	—	—	—	—	—	—	—	—
—	—	—	—	—	—	—	—	—	—	—
—	—	—	—	—	—	—	—	—	—	—
—	—	—	—	—	—	—	—	—	—	—
—	—	—	—	—	—	—	—	—	—	—
—	—	—	—	—	—	—	—	—	—	—
—	—	—	—	—	—	—	—	—	—	—
—	—	—	—	—	—	—	−1·5	−1·5	−4·5	−1·5
41·5	37·3	38·5	39·3	37·5	35·4	34·5	30·2	29·6	871·2	36·30
31·9	32·4	32·4	32·1	31·5	31·5	32·1	32·5	32·0	764·9	31·87
34·5	31·7	31·9	31·8	31·8	31·7	31·4	30·4	30·0	793·7	33·08
31·3	30·4	30·4	30·4	30·4	30·1	29·9	29·0	29·0	737·7	30·74
32·2	32·3	33·3	32·2	31·6	31·3	30·4	28·1	29·1	647·1	30·31*
46·0	40·9	47·5	45·0	41·5	41·3	41·4	—	—	884·6	36·86
—	—	—	—	—	—	—	15·0	14·8		
15·7	14·2	14·4	15·9	16·1	16·5	16·4	14·6	14·8	381·1	15·88
19·1	18·3	18·3	18·2	16·6	17·2	18·3	17·2	17·5	414·4	17·27
14·6	12·3	10·7	9·2	8·2	7·8	7·8	3·3	0·1	360·1	15·00
0·0	−3·4	−4·7	−6·0	−5·1	2·2	0·6	−0·1	−0·1	−15·7	−0·65
8·2	8·0	7·4	7·2	5·7	5·5	5·3	0·1	−0·1	100·0	4·17
6·3	6·8	8·0	8·0	9·1	7·8	6·6	—	—	155·9	6·50
—	—	—	—	—	—	—	23·5	22·5		
23·1	22·5	22·5	22·3	21·6	21·5	21·5	19·5	19·6	538·9	23·45
22·7	20·5	20·3	19·1	16·9	16·0	15·5	14·6	14·6	466·2	20·26
327·1	313·2	310·9	304·7	293·4	295·8	291·7	256·9	253·4	7184·0	300·64
23·36	22·37	22·21	21·76	20·96	21·13		18·35	18·10	514·42	21·44
1·92	0·93	0·77	0·32	−0·48	−0·31		−3·09	−3·34	—	—

* Sum and mean of triplets.

FORT CHIPEWYAN—*continued.*

Abstract of Hourly Observations made during the month of November 1843.

Date. Gött. Mean Time.	Noon.	1.	2.		4.	5.	6.	7.	8.	9.	[illegible]
1	14·8	14·6	14·6	14·8	15·7	17·9	18·2	20·6	21·9	26·7	[illegible]·1
2	21·3	20·6	20·5	21·1	21·1	22·9	26·0	28·1	30·4	32·0	32·7
3	24·7	24·9	23·8	23·8	24·1	23·8	25·1	25·1	27·4	27·0	29·4
4	—	—	—	27·9	27·0	27·6	28·6	29·3	29·1	29·3	29·2
5	—	—	—	—	—	—	—	—	—	—	—
6	14·8	14·8	14·6	13·4	13·6	15·9	15·0	16·8	22·3	24·7	24·5
7	16·9	16·8	15·4	15·7	15·7	16·1	16·8	17·0	17·5	16·8	15·7
8	19·1	19·0	18·1	17·6	18·9	19·5	20·1	19·3	20·2	18·9	18·9
9	6·9	6·8	6·6	6·4	6·4	7·6	8·0	8·9	10·0	10·0	9·5
10	−1·1	−1·0	−0·2	−0·9	−0·9	1·0	3·5	8·3	8·5	9·6	9·1
11	2·0	2·1	2·3	4·4	6·4	5·5	7·2	7·3	6·8	5·5	6·3
12	—	—	—	—	—	—	—	—	—	—	—
13	−6·8	−6·0	−5·0	−8·3	−0·1	0·6	2·3	4·4	4·9	5·7	6·0
14	9·1	10·2	13·2	11·5	10·3	11·2	13·3	14·1	15·9	18·4	18·9
15	8·0	6·6	5·1	3·4	4·2	6·0	8·1	8·3	9·8	4·2	7·9
16	11·0	10·1	10·1	11·9	12·7	15·1	15·6	17·8	17·3	15·1	13·2
17	10·4	10·0	10·0	8·8	7·2	8·0	8·3	8·7	9·2	9·1	8·9
18	4·5	3·3	3·3	3·1	3·6	4·8	6·7	8·0	8·9	9·0	9·2
19	—	—	—	—	—	—	—	—	—	—	—
20	7·0	7·7	7·3	7·7	7·7	7·6	7·8	7·8	8·1	8·4	8·4
21	6·7	6·9	6·4	6·2	7·0	7·3	7·2	9·0	9·3	9·3	7·7
22	5·8	5·8	5·6	5·7	5·6	6·9	8·3	9·3	9·8	10·0	10·0
23	−5·5	−5·7	−3·5	−3·1	−1·3	0·2	2·7	4·5	5·5	5·7	6·5
24	5·0	4·3	3·8	3·2	3·0	3·2	3·1	4·7	5·3	5·4	4·7
25	4·1	3·8	4·1	3·3	2·2	0·9	−0·6	−0·6	1·2	1·9	2·1
26	—	—	—	—	—	—	—	—	—	—	—
27	2·7	3·2	3·1	3·3	2·3	4·2	4·6	5·1	7·1	6·9	8·9
28	0·3	−0·7	−1·5	−2·6	−2·7	−2·3	−0·2	0·3	3·0	4·2	4·3
29	13·4	17·9	20·3	17·8	10·4	14·8	14·6	15·3	15·0	14·9	13·0
30	−5·5	−5·8	−5·8	−5·6	−5·5	−4·7	−3·6	−3·4	−2·3	0·0	0·1
31	—	—	—	—	—	—	—	—	—	—	—
Sums	189·6	190·2	192·2	215·5	220·6	241·6	266·7	294·0	322·1	328·7	333·2
Hourly Means	7·58	7·61	7·69	8·29	8·48	9·29	10·26	11·31	12·39	12·64	12·82
Diurnal Variation	−2·18	−2·15	−2·07	−1·47	−1·28	−0·47	0·50		2·63 N	2·88	3·06

8^h Göttingen time = noon of local mean time.

Fort Chipewyan—*continued.*

ber 1843.

Abstract of Hourly Observations made during the month of November 1843.

Spirit Thermometer by Newman, corrected.												
13.	14.	15.	16.	17.	18.	19.	20.	21.	22.	23.	Sums.	Means.
29·2	28·1	26·1	24·0	21·7	20·2	21·1	21·5	22·5	22·5	21·5	524·7	21·86
26·3	25·4	24·8	23·8	23·3	23·2	23·2	22·1	24·7	25·7	25·5	603·0	25·13
29·8	29·4	27·7	27·7	27·7	27·7	27·9	27·2	—	—	—	404·2	26·95*
27·0	26·8	27·0	26·4	25·3	24·1	23·3	23·3	—	—	—	378·7	25·25*
—	—	—	—	—	—	—	—	19·3	19·3	16·6		
17·3	17·3	16·5	16·8	17·1	17·5	17·9	18·1	18·1	18·2	17·9	427·9	17·83
14·6	14·8	15·5	15·7	16·4	16·6	17·3	17·8	18·1	18·6	19·1	395·2	16·47
13·9	12·3	11·3	10·2	10·0	8·9	8·2	8·0	7·7	7·5	7·2	343·4	14·31
7·6	6·6	5·7	5·5	4·7	2·2	1·1	−0·1	−0·2	−0·3	−1·2	136·1	5·67
6·5	5·7	4·3	3·8	5·5	4·6	2·5	3·3	2·0	1·2	3·2	95·7	3·99
6·4	6·7	6·6	7·3	6·8	6·8	6·8	6·8	—	—	—	103·5	4·31
—	—	—	—	—	—	—	—	−6·0	−6·5	−6·8		
2·3	1·4	1·6	0·3	−0·5	0·4	−1·0	2·3	5·9	5·9	8·0	39·3	1·64
16·5	16·4	16·2	17·0	18·6	19·0	17·8	17·8	17·9	13·6	10·0	362·0	15·08
8·1	8·9	9·4	9·9	10·4	11·2	11·9	12·8	13·1	13·6	11·4	208·1	8·67
12·5	12·5	12·4	11·1	9·1	8·7	8·0	9·2	10·7	10·7	10·2	289·6	12·07
9·5	9·2	9·8	7·9	8·2	7·5	6·0	5·9	6·5	4·8	5·7	199·8	8·33
8·9	9·3	9·5	10·9	9·9	8·8	7·2	6·9	—	—	—	174·1	7·25
—	—	—	—	—	—	—	—	7·0	6·8	7·0		
6·6	5·8	4·8	4·4	4·7	5·9	6·3	6·6	6·2	6·4	6·3	164·7	6·86
6·0	5·9	5·7	5·8	6·0	6·3	6·6	6·5	6·0	5·7	5·5	163·8	6·82
9·1	8·8	7·7	7·2	4·8	2·7	0·8	0·6	−1·2	−2·7	−3·6	136·3	5·68
8·0	8·1	8·0	8·1	8·1	8·6	8·8	9·0	8·0	5·7	·1	105·5	4·40
2·5	2·5	2·8	3·3	3·0	3·2	3·6	4·2	4·1	3·8	3·8	88·6	3·69
0·6	0·7	0·7	2·4	1·3	1·7	2·5	2·6	—	—	—	46·1	1·92
—	—	—	—	—	—	—	—	3·1	3·2	2·9		
9·9	9·3	8·8	8·2	8·0	7·1	4·7	5·1	4·3	3·0	2·0	141·9	5·91
3·5	3·5	1·3	1·2	0·4	0·8	3·5	3·2	7·9	10·4	12·3	57·0	2·38
5·3	2·2	0·9	0·8	0·6	0·4	−3·5	−4·7	−6·5	−6·1	−9·4	172·2	7·18
0·1	−1·3	−1·2	−1·6	−1·5	0·0	0·2	−1·4	2·4	−2·3	−1·5	−54·4	−2·27
—	—	—	—	—	—	—	—	—	—	—	—	—
287·0	276·8	263·9	258·1	249·6	244·1	232·7	234·6	197·3	189·2	178·7	6031·3	257·90
11·04	10·65	10·15	9·93	9·60	9·39	8·95	9·02	7·89	7·57	7·15	233·74	9·7
1·28	0·89	0·39	0·17	−0·16	−0·37	−0·81	−0·74	−1·87	−2·19	−2·61	—	—

* Mean by triplets.

FORT CHIPEWYAN—*continued.*

Abstract of Hourly Observations made during the month of December 1843.

Date. Gött. Mean Time.	Spirit Thermometer by Newman, corrected.											
	Noon.	1.	2.	3.	4.	5.	6.	7.	8.	9.	10.	11.
1	−1·2	0·6	2·1	3·4	5·1	6·4	7·3	8·0	8·2	7·2	5·8	5·3
2	23·6	19·1	18·9	17·6	15·2	7·8	5·1	4·7	4·7	3·0	1·9	0·[illegible]
3	—	—	—	—	—	—	—	—	—	—	—	—
	4·0	3·1	3·0	2·3	3·0	3·0	4·6	6·7	10·8	11·5	10·1	8·[illegible]
5	3·7	3·3	4·4	4·4	3·3	3·9	5·3	0·6	1·5	2·1	2·7	3·[illegible]
6	17·9	19·1	21·1	22·9	26·1	27·0	29·9	31·6	34·2	34·2	35·3	33·[illegible]
7	4·6	3·2	1·8	0·9	−1·3	−2·0	−0·4	3·5	4·1	5·2	6·5	7·[illegible]
8	17·0	16·4	15·7	14·7	15·3	15·1	22·5	27·0	30·4	31·4	29·4	28·[illegible]
9	19·4	19·4	21·2	22·0	21·4	19·6	21·5	21·9	21·5	19·3	15·7	13·[illegible]
10	—	—	—	—	—	—	—	—	—	—	—	—
11	−2·9	−1·3	−1·2	−1·6	−1·4	−1·5	−2·0	−2·1	−1·9	−1·9	−1·9	−3·[illegible]
12	6·4	4·9	4·4	4·2	4·4	4·4	3·1	2·3	2·2	2·0	2·4	1·[illegible]
13	−13·7	−12·6	−13·3	−14·7	−16·3	−19·8	−18·5	−16·3	−15·2	−15·9	−17·6	−18[illegible]
14	−22·3	−20·5	−22·4	−23·9	−19·4	−16·5	−12·7	−10·1	−9·3	−7·2	−7·0	−6[illegible]
15	−4·7	−3·3	−4·7	−4·0	−4·6	−4·7	−3·4	−1·1	−0·7	−0·1	−0·6	−1[illegible]
16	−5·8	−6·5	−7·8	−8·7	−8·7	−9·5	−7·1	−6·4	−4·6	−5·8	−5·9	−6[illegible]
17	—	—	—	—	—	—	—	—	—	—	—	—
18	−2·8	−1·4	−0·1	2·0	4·0	4·6	5·5	7·8	6·8	7·2	6·8	[illegible]
19	−7·2	−8·0	−8·2	−8·4	−9·0	−9·4	−9·2	−9·1	−7·8	−9·8	−10·3	−1[illegible]
20	5·3	5·3	6·8	5·5	2·7	1·1	1·1	−2·8	−2·8	−2·7	−4·1	−[illegible]
21	−11·6	−10·0	−10·1	−9·2	−9·0	−9·1	−7·0	−7·3	−4·7	−2·3	−3·2	
22	−2·2	−3·4	−4·9	−7·1	−8·0	−9·2	−9·2	−9·2	−8·4	−9·0	−10·3	−1[illegible]
23	−4·9	−4·7	−3·1	−1·7	−0·1	−0 1	0·3	2·0	3·3	3·9	5·4	
24	—	—	—	—	—	—	—	—	—	—	—	
25	—	—	—	—	—	—	—	—	—	—	—	
26	−0·5	−2·3	−3·4	−2·7	−3·7	−3·5	−3·5	−3·5	−2·9	−3·2	−3·0	
27	−5·6	−4·6	−3·5	−1·1	−0·1	2·3	3·3	4·8	5·9	7·0	7·3	
28	17·7	17·7	17·2	15·8	15·8	16·2	16·8	17·7	17·7	16·1	12·5	
29	−5·9	−7·1	−8·4	−8·4	−8·0	−6·3	−4·6	−3·5	−2·3	−1·4	−1·5	
30	−11·4	−12·1	−10·9	−13·8	−15·3	−17·4	−20·5	−23·0	−23·0	−22·5	−22·0	
31	—	—	—	—	—	—	—	—	—	—	—	
Jan. 1	—	—	—	—	—	—	—	—	—	—	—	
Sums -	16·9	14·3	14·6	10·4	11·4	2·4	26·0	44·2	69·1	68·3	54·4	
Means -	0·68	0·57	0·58	0·42	0·46	0·10	1·04	1·77	2·76	2·73	2·18	
Diurnal Variation	0·23	0·17	0·18	0·02	0·06	−0·30	0·64	1·37	2·36 N	2·33	1·78	

8[h] Göttingen mean time = noon of local mean time.

FORT CHIPEWYAN—*continued.*

Abstract of Hourly Observations made during the month of December 1843.

	16.	17.	18.	20.	21.	22.	23.	Sums.	Means.
7·7	7·8	8·9	9·3	10·0	11·6	12·8	13·4	167·8	7·41
−2·6	−1·8	−2·5	−3·5	−8·1	—	—	—	98·5	4·10
—	—	—	—	—	0·3	1·6	2·2		
9·4	9·4	7·8	12·2	9·9	9·6	6·6	6·2	180·4	7·52
4·3	4·3	4·1	4·3	7·2	7·8	11·2	15·3	113·0	4·71
17·2	16·4	14·6	12·0	9·7	7·7	6·6	6·3	512·2	21·34
10·9	11·7	13·2	13·9	13·5	14·6	15·6	16·7	184·2	7·67
22·1	22·2	20·7	21·4	23·7	21·9	22·0	19·8	525·1	21·88
4·4	3·3	3·1	0·3	−2·4	—	—	—	258·6	10·77
—	—	—	—	—	−3·3	−3·5	−3·3		
−0·2	1·1	0·3	0·3	6·6	7·8	6·1	6·5	0·2	0·01
−2·4	−4·7	−5·7	−6·6	−9·4	−10·0	−11·8	−12·8	−25·4	−1·06
−21·9	−23·7	−24·9	−23·9	−23·9	−23·8	−23·5	−22·9	−468·2	−19·51
−3·5	−3·3	−2·6	0·0	−2·4	−2·6	−2·8	−4·7	−214·6	−8·94
−3·4	−4·6	−6·3	−6·9	−9·2	−11·0	−8·6	−6·8	−101·2	−4·22
−8·0	−8·0	−7·8	−5·9	−5·1	—	—	—	−154·6	−6·44
—	—	—	—	—	−3·5	−3·5	−3·2		
−2·0	−2·9	−3·3	−3·7	−5·0	−5·5	−5·9	−6·5	17·0	0·71
−9·9	−9·9	−6·9	−6·1	−3·5	−2·8	−1·5	0·3	−183·4	−7·64
−11·9	−10·3	−10·0	−10·1	−9·2	−10·3	−11·4	−11·4	−116·8	−4·87
8·2	8·9	12·4	11·1	6·5	5·9	5·9	2·1	7·4	0·31
−11·6	−13·4	−13·4	−12·5	−9·9	−7·1	−7·2	−5·9	−222·0	−9·25
2·6	2·3	2·7	2·1	4·3	—	—	—		
—	—	—	—	—	—	—	—	23·3	0·97
—	—	—	—	—	−0·1	0·1	−0·4		
−0·4	−0·2	−4·5	−4·7	−5·4	−4·7	−6·1	−6·7	−76·5	−3·19
8·1	10·0	13·2	14·6	16·6	16·4	16·1	16·8	173·7	7·24
3·1	1·8	0·3	−1·1	−2·4	−4·7	−5·9	−5·8	199·8	8·32
−8·0	−8·0	−6·9	−5·9	−5·8	−8·6	−10·3	−11·9	−148·9	−6·20
−28·7	−30·9	−33·7	−33·3	−35·3	—	—	—		
—	—	—	—	—	—	—	—	−512·4	−21·35
—	—	—	—	—	−7·1	−7·3	−6·9		
−15·8	−22·5	−27·2	−22·7	−29·0	−1·5	−4·7	−3·6	+237·2	+10·29
−0·63	−0·90	−1·09	−0·91	−1·16	−0·06	−0·19	−0·14	+9·49	+0·49
−1·03	−1·30	−1·49	−1·31		−0·48	−0·59	−0·54	—	—

Fort Chipewyan—*continued.*

Abstract of Hourly Observations made during the month of January 1844.

Date. Gött. Mean Time.	Spirit Thermometer by Newman, corrected.												
	Noon.	1.	2.	3.	4.	5.	6.	7.	8.	9.	10.	11.	12.
1	—	—	—	—	—	—	—	—	—	—	—	—	—
2	−5·4	−3·5	−1·5	−2·5	−1·6	−0·8	−0·6	−0·8	−1·1	−0·8	−0·8	−3·3	−6·5
3	−18·1	−20·5	−23·5	−24·8	−25·0	−23·4	−22·7	−18·4	−17·1	−13·9	−14·8	−13·0	−13·3
4	−7·6	−7·3	−6·0	−6·7	−6·6	−6·2	−6·0	−6·2	−6·9	−5·6	−5·3	−5·8	−5·9
5	−6·5	−6·9	−6·9	−7·8	−9·2	−10·3	−10·3	−11·5	−12·2	−14·4	−15·6	−18·2	−20·[illegible]
6	−36·3	−35·9	−35·7	−38·9	−39·0	−37·5	−38·6	−36·3	−33·8	−34·8	−35·1	−34·6	−36·[illegible]
7	—	—	—	—	—	—	—	—	—	—	—	—	—
8	−31·8	−32·0	−32·9	−32·9	−32·5	−32·0	−29·9	−27·8	−26·3	−26·2	−27·4	−29·3	−29·[illegible]
9	−39·7	−39·3	−38·1	−38·0	−38·8	−39·8	−37·5	−38·8	−38·0	−36·7	−35·2	−35·3	−35[illegible]
10	−32·0	−30·6	−29·7	−29·5	−28·4	−26·5	−22·3	−20·7	−22·1	−21·6	−23·7	−24·1	−25[illegible]
11	−25·0	−26·8	−22·9	−18·3	−17·4	−11·4	−10·1	−10·3	−9·6	−9·2	−9·0	−6·9	−[illegible]
12	−2·3	−2·8	−4·7	−5·8	−5·8	−6·0	−6·1	−1·5	−1·5	−1·4	−1·4	−2·8	−[illegible]
13	−17·6	−17·9	−17·1	−18·9	−20·2	−21·8	−22·7	−21·0	−18·4	−18·1	−18·2	−20·2	−[illegible]
14	—	—	—	—	—	—	—	—	—	—	—	—	[illegible]
15	−20·7	−20·9	−19·2	−19·9	−19·1	−16·9	−14·4	−11·0	−8·8	−6·6	−6·6	−0·7	[illegible]
16	−5·8	−8·0	−10·3	−11·2	−12·8	−13·3	−10·6	−10·7	−10·3	−10·3	−10·1	−10·6	[illegible]
17	−18·4	−20·9	−21·6	−21·8	−22·8	−23·9	−23·0	−24·1	−23·2	−23·2	−23·1	−22·4	[illegible]
18	−32·5	−32·9	−35·6	−36·3	−37·6	−37·2	−33·4	−32·5	−30·5	−30·6	−32·2	−33·6	[illegible]
19	−38·5	−39·0	−38·5	−38·7	−39·7	−38·4	−37·8	−36·5	−35·6	−35·4	−35·3	−34·4	[illegible]
20	−37·6	−37·4	−38·6	−38·8	−38·6	−38·3	−35·6	−34·3	−32·8	−30·6	−30·4	−32·2	[illegible]
21	—	—	—	—	—	—	—	—	—	—	—	—	[illegible]
22	−39·6	−46·7	−47·2	−49·5	−48·2	−46·5	−40·8	−37·4	−36·1	−35·4	−35·3	−37·9	[illegible]
23	−44·1	−44·7	−43·3	−43·2	−43·3	−43·3	−42·5	−39·5	−37·8	−35·6	−35·9	−34·5	[illegible]
24	−40·9	−39·9	−40·9	−40·9	−41·3	−40·6	−38·5	−37·5	−38·8	−38·3	−38·5	−39·4	[illegible]
25	−47·0	−45·6	−46·3	−47·7	−45·4	−43·7	−42·0	−38·8	−39·7	−39·6	−40·6	−40·2	[illegible]
26	−26·1	−25·5	−24·7	−24·2	−25·9	−26·0	−25·0	−23·9	−25·0	−24·1	−24·3	−24·0	[illegible]
27	−38·9	−39·3	−41·3	−41·3	−40·9	−38·6	−34·2	−33·1	−31·8	−30·3	−28·4	−28·[illegible]	[illegible]
28	—	—	—	—	—	—	—	—	—	—	—	—	[illegible]
29	−16·0	−16·0	−16·4	−16·4	−16·1	−17·2	−17·3	−15·4	−13·1	−9·7	−7·2	−5·[illegible]	[illegible]
30	14·6	9·9	9·1	10·1	6·6	5·1	4·5	4·2	4·5	4·2	3·1	2·[illegible]	[illegible]
31	−21·8	−22·3	−18·8	- 20·5	−21·5	−20 4	−17·6	−14·8	−13·5	−12·8	−12·9	−14[illegible]	[illegible]
Sums -	635·6	652·7	652·6	64·4	671·1	654 9	615·0	578·6	559·5	541·0	544·2	54[illegible]	[illegible]
Means -	−24·45	−25·10	−25·10	−25·55	−25·81	−25·18	−23·65	−22·25	−21·52	−20·81	−20·9[illegible]	−21[illegible]	[illegible]
Diurnal Variation -	−1·45	−2·10	−2·10	−2·55	−2·81	−2·18	−0·65	0·75	1·48 N	2·19	2·07	1[illegible]	[illegible]

8h Göttingen mean time = noon of local mean time.

Fort Chipewyan—*continued.*

Abstract of Hourly Observations made during the month of January 1844.

			18.	19.	20.	21.	Mean by Dollond's Thermo.
—	-	—	—	—	—	—	—
−7·0	—	−11·0	−12·8	−14·1	−14·1	−14·8	by trip-lets.
−11·9	−11·6	−9·3	−9·3	−8·3	−8·0	−8·0	—
−6·2	−7·1	−6·9	−7·1	−6·9	−6·4	−6·1	—
−23·1	−24·1	−25·2	−27·5	−28·9	−30·6	−34·4	
−38·5	−36·3	−37·6	−37·9	−38·6	−39·0	—	
—	—	—	—	—	—	−34·3	
−30·3	−29·5	−29·5	−29·8	−29·8	−30·4	−34·4	−29·9
−34·1	−34·4	−34·7	−35·2	−35·2	−34·5	−33·8	−35·9
−24·3	−24·3	−22·6	−19·3	−18·6	−18·4	−22·0	−23·3
−4·5	−3·5	−1·3	−1·2	−1·3	−0·9	−0·6	−7·7
−4·5	−4·9	−4·7	−7·4	−9·7	−11·5	−15·0	−4·8
−24·8	−28·3	−30·9	−30·9	−29·4	−29·6	—	−21·9
—	—	—	—	—	—	−20·5	
4·3	5·5	5·8	5·5	7·8	8·5	11·0	−2·9
−9·9	−9·8	−10·3	−11·8	−12·2	−13·8	−17·1	−10·0
−21·8	−23·0	−25·2	−26·1	−26·3	−26·6	−27·8	−22·8
−34·9	−35·7	−32·9	−34·0	−34·5	−34·1	−35·4	−33·5
−34·0	−34·9	−37·4	−38·4	−38·6	−38·6	−38·6	−36·9
−34·2	−34·0	−34·0	−34·0	−34·0	−34·7	—	−35·2
—	—	—	—	—	—	−38·4	
−39·4	−40·3	−39·7	−40·9	−42·0	−42·2	−43·6	−41·4
−37·4	−37·6	−39·9	−40·6	−41·3	−42·0	−40·9	−39·7
−40·9	−43·2	−44·5	−44·3	−43·6	−42·2	−44·3	−41·8
−37·4	−36·1	−31·8	−30·9	−30·4	−27·8	−27·3	−37·2
−25·5	−26·3	−32·9	−34·2	−35·2	−35·7	−36·7	−27·6
−23·9	−21·6	−13·8	−11·5	−10·2	−10·0	—	−25·0
—	—	—	—	—	—	−14·6	
−1·6	−1·5	1·3	3·3	5·7	5·9	11·2	−3·6
−0·1	−1·3	−8·0	−6·0	−6·9	−13·7	−13·3	+0·04
−18·1	−20·2	−22·8	−22·0	−22·7	−22·9	−18·7	−17·9
		579·8	584·3	585·2	593·3	596·4	—
−21·69	−22·58	−22·30	−22·47	−22·51	−22·82	−23·02	—
1·31	0·48	0·70	0·53	0·49	0·18	−0·02	

Fort Chipewyan—*continued.*

Abstract of Hourly Observations made during the month of February 1844.

Date. Gött. Mean Time.	Spirit Thermometer by Newman, corrected.											
	Noon.	1.	2.	3.	4.	5.	6.	7.	8.	9.	10.	11.
1	−18·3	−16·1	−16·4	−16·0	−14·8	−14·5	−15·5	−16·0	−14·1	−14·0	−15·1	−16·2
2	−2·4	−1·7	−1·3	−0·7	−1·5	−1·1	−0·6	−0·3	−0·1	−0·1	−0·5	−1·5
3	−11·3	−13·7	−13·5	−11·6	−11·2	−10·3	−7·1	−5·9	−4·5	−3·5	−2·0	−1·6
4	—	—	—	—	—	—	—	—	—	—	—	—
5	7·4	3·8	1·0	−2·6	−3·8	−7·3	−6·3	−4·5	−4·2	−2·2	−2·2	−1·5
6	−22·3	−20·7	−23·4	−23·7	−22·5	−22·0	−17·4	−15·2	−3·3	−10·5	−10·3	−8·0
7	4·6	3·3	3·1	3·8	4·3	3·8	4·6	6·3	7·8	8·7	8·9	9·7
8	9·8	6·6	6·4	6·2	6·2	6·6	8·9	12·1	16·9	18·6	21·1	19·3
9	12·1	9·7	10·4	11·0	11·2	12·0	13·5	14·6	14·8	12·3	9·4	6·8
10	−3·1	−3·3	−2·8	−1·3	−1·1	−0·8	2·3	6·4	7·5	10·9	8·3	7·
11	—	—	—	—	—	—	—	—	—	—	—	—
12	−30·8	−27·2	−26·2	−26·0	−24·4	−23·1	−21·0	−16·5	−15·2	−12·9	−10·8	−11·
13	−21·4	−22·6	−25·0	−25·9	−25·9	−22·0	−19·7	−19·5	−17·0	−15·0	−12·5	−8·
14	−10·3	−14·2	−13·8	−15·0	−11·8	−13·3	−12·6	−7·8	−5·3	−5·4	−5·8	−4·
15	8·8	7·0	4·9	5·0	5·3	7·9	13·2	15·6	21·3	25·8	30·1	28·
16	5·1	4·9	9·8	8·7	9·2	12·2	19·9	24·5	27·2	29·3	21·5	34·
17	23·8	27·2	28·5	27·0	24·4	27·9	30·4	32·7	35·3	37·5	37·4	37·
18	—	—	—	—	—	—	—	—	—	—	—	—
19	8·2	8·9	11·2	11·1	11·5	13·3	13·4	15·9	16·6	19·3	21·3	19
20	6·7	7·8	8·1	11·0	11·2	13·4	15·5	16·8	22·4	27·9	30·3	32
21	−5·8	−6·0	−9·7	−5·0	−3·8	2·1	4·7	6·6	7·4	9·9	12·3	15
22	15·9	15·6	12·1	11·6	15·3	19·1	23·6	24·7	27·4	28·0	26·1	24
23	30·6	30·9	28·1	28·4	30·3	32·5	30·4	30·4	30·9	30·4	29·4	29
24	1·0	−1·4	−2·5	−6·8	−5·3	−3·3	−1·4	0·0	3·3	3·3	2·6	1
25	—	—	—	—	—	—	—	—	—	—	—	—
26	−10·3	4·2	5·5	4·5	5·3	6·6	14·6	19·3	28·1	33·0	33·3	3
27	0·6	−2·5	−6·2	−6·7	−8·0	−5·6	−2·7	−3·1	1·9	3·2	2·9	−
28	3·1	3·4	6·6	6·4	6·8	6·4	11·7	13·4	14·6	15·1	15·1	1
29	−3·8	−4·7	−6·9	−5·8	−4·7	−3·5	−2·6	2·3	6·5	6·8	7·5	
Sums	−2·1	−0·8	−12·9	−11·4	+2·2	+38·6	+99·8	+152·8	+216·2	+256·4	+258·3	+2
Means	−0·08	−0·03	−0·48	−0·46	+0·09	+1·54	+3·99	+6·11	+8·65	+10·26	+10·33	+1
Diurnal Variation	−4·87	−4·82	−5·27	−5·25	−4·73	−3·25	−0·80	1·32	3·86 N	5·47	5·54	5

8^h Göttingen mean time = noon of local mean time.

FORT CHIPEWYAN—*continued.*

Abstract of Hourly Observations made during the month of February 1844.

21.	22.
−6·1	−6·1
−8·2	−9·3
—	—
7·8	7·5
−18·3	−17·6
5·7	7·6
6·6	7·7
13·4	12·7
−1·5	−0·7
—	—
−30·8	−32·1
−14·9	−18·2
−11·9	−9·7
−0·2	4·4
6·6	6·0
27·0	28·9
—	—
7·8	8·3
3·0	4·2
−1·3	−2·8
21·2	20·1
30·4	31·1
1·7	1·0
—	—
−2·4	−5·8
1·0	2·1
−0·6	3·3
−1·5	−2·5
—	—

+1·44	+1·67
−3·35	−3·12

LAKE ATHABASCA.

Spirit Thermometer by Dollond, corrected.

Date. Mean Time.	March 1844.					April 1844.			
	Sunrise.	9 a.m.	3 p.m.	9 p.m.	Mean.	Sunrise.	9 a.m.	3 p.m.	9 p.m.
*1	—	0·7	—	—	—	14·4	19·6	22·7	12·3
*2	—	−12·5	—	25·0	—	−2·2	2·0	14·4	10·2
*3	—	27·9	34·0	3·5	—	11·3	17·5	39·0	33·0
*4	—	−14·9	−8·4	−15·6	—	31·0	35·0	46·0	39·0
5	−20·8	−15·6	−7·4	−13·6	−14·35	25·8	37·0	46·0	39·0
6	−10·5	−10·5	−5·8	−12·5	−9·82	33·0	40·0	45·0	35·0
7	−11·5	−15·6	−6·3	−11·5	−11·23	31·0	38·0	38·0	28·9
8	−7·4	−7·4	0·9	−1·1	−3·75	28·9	36·0	41·0	33·0
9	−22·9	−21·9	−7·9	−17·7	−17·60	36·0	46·0	45·0	35·0
10	−17·7	−8·4	−4·3	−12·5	−10·72	28·9	37·0	39·0	28·9
11	−14·6	−10·5	−1·1	−6·3	−8·13	18·5	24·7	29·9	19·6
12	−20·8	−15·6	3·0	4·0	−7·35	17·5	23·7	25·8	22·7
13	12·3	15·4	21·6	11·3	15·15	24·7	28·9	41·0	35·0
14	8·2	10·2	16·5	4·0	9·72	28·9	29·9	31·0	27·9
15	2·0	4·0	12·3	7·1	6·35	18·5	28·9	33·0	29·9
16	4·0	5·1	8·2	2·0	4·82	31·0	41·0	53·0	43·0
17	7·1	12·3	28·9	10·2	14·63	33·0	45·0	53·0	37·0
18	35·0	37·0	31·0	19·6	30·65	28·9	29·9	35·0	28·9
19	17·5	20·6	19·6	18·5	19·05	31·0	48·0	56·0	44·0
20	24·7	35·0	32·0	23·7	28·60	16·5	23·7	33·0	24·7
21	8·2	10·2	32·0	37·0	21·85	22·7	32·0	39·0	39·0
22	31·0	38·0	42·0	27·0	34·50	33·0	39·0	61·0	43·0
23	−2·2	−3·2	4·0	−6·3	−1·92	41·0	53·0	65·0	41·0
24	−12·5	−6·3	−2·2	−12·5	−8·38	22·7	21·6	25·8	22·7
25	−22·9	−15·6	−5·3	−8·4	−13·05	24·7	36·0	53·0	49·
26	−4·3	−2·2	3·0	−6·3	−2·45	45·0	59·0	55·0	41·
27	−16·7	−10·5	2·0	−5·3	−7·62	40·0	51·0	55·0	45·
28	−15·6	−10·5	2·0	−11·5	−8·90	41·0	48·0	59·0	39·
29	−22·9	−16·7	0·0	−2·2	−10·45	37·0	45·0	43·0	43·
30	−4·3	−3·2	8·2	8·2	2·23	41·0	55·0	52·0	49·
31	12·3	16·5	19·6	14·4	15·70	—	—	—	—
Sums -	−65·3	30·6	246·5	58·3	67·53	834·7	1064·4	1274·6	1018·
Means -	−2·42	1·13	9·13	2·16	2·50	27·82	35·48	42·49	33·
Diurnal Variation }	—	—	—	—	—	—	—	—	—

* Not included in Sums and Means.

LAKE ATHABASCA.

Spirit Thermometer by Dollond, corrected.

9 a.m.	3. p.m.
56·0	49·0
28·9	27·9
32·0	36·0
42·0	52·0
53·0	56·0
41·0	39·0
41·0	40·0
45·0	56·0
51·0	55·0
53·0	56·0
56·0	62·0
49·0	55·0
37·0	33·0
31·0	39·0
41·0	47·0
49·0	45·0
53·0	64·0
35·0	45·0
33·0	40·0
51·0	59·0
31·0	35·0
34·0	36·0
46·0	51·0
49·0	44·0
55·0	69·0
73·0	66·0
59·0	64·0
61·0	56·0
57·0	63·0
47·0	59·0
59·0	63·0
1448·9	1561·9
46·74	50·38

9 a.m.	3 p.m.	9 p.m.
—	—	—
35·0	39·0	36·0
33·0	34·0	29·9
34·0	35·0	31·0
41·0	45·0	33·0
37·0	46·0	41·0
45·0	40·0	38·0
51·0	56·0	45·0
63·0	64·0	55·0
68·0	70·0	58·0
71·0	75·0	55·0
46·0	46·0	38·0
45·0	55·0	46·0
59·0	65·0	59·0
61·0	66·0	58·0
55·0	58·0	53·0
74·0	74·0	68·0
55·0	54·0	46·0
65·0	64·0	55·0
84·0	88·0	69·0
88·0	88·0	71·0
76·0	74·0	66·0
66·0	66·0	58·0
58·0	61·0	55·0
65·0	63·0	58·0
65·0	64·0	59·0
65·0	66·0	61·0
56·0	53·0	49·0
48·0	44·0	42·0
49·0	53·0	47·0
—	—	—
1658·0	1706·0	1474·9
57·17	58·83	50·86

FORT SIMPSON.

Abstract of Hourly Observations made during the month of April 1844.

Date. Gött. Mean Time.	Spirit Thermometer by Newman, corrected.											
	Noon.	1.	2.	3.	4.	5.	6.	7.	8.	9.	10.	11.
1	−2·8	−1·0	0·1	0·8	3·1	5·7	8·7	13·7	17·3	20·4	23·7	23·9
2	1·6	−0·7	−2·1	−3·3	−1·3	2·9	10·3	17·2	23·8	24·8	32·5	31·6
3	1·5	0·5	0·4	2·1	7·3	15·1	12·7	27·1	36·0	35·5	34·5	39·5
4	23·6	20·4	19·4	19·4	19·4	23·6	29·3	33·4	41·5	44·5	44·6	46·3
5	—	—	—	—	—	—	—	—	—	—	—	—
6	9·5	7·1	6·0	6·0	7·3	8·2	9·3	11·6	16·2	16·5	16·0	16·2
7	—	—	—	—	—	—	—	—	—	—	—	—
8	13·2	—	9·1	10·4	12·3	13·8	23·7	29·2	37·2	42·5	44·4	45·0
9	37·1	37·5	35·5	37·4	38·3	40·4	42·4	44·3	45·1	41·0	41·5	40·4
10	14·9	14·0	13·8	16·9	16·0	16·3	20·4	21·5	22·4	22·5	22·6	22·6
11	2·7	0·5	0·5	2·8	4·9	6·4	13·8	19·3	24·6	28·1	29·4	29·3
12	17·1	14·0	14·7	15·0	16·0	19·3	23·7	28·2	31·5	34·6	36·4	38·2
13	30·4	31·0	31·5	31·5	31·1	30·5	33·5	34·3	35·7	36·8	38·6	35·9
14	—	—	—	—	—	—	—	—	—	—	—	—
15	20·6	19·3	21·4	22·6	22·9	26·1	27·1	33·5	37·5	41·2	33·5	43·3
16	—	—	—	27·1	29·1	36·7	41·4	45·5	49·5	50·0	51·5	51·3
17	27·6	27·5	27·6	30·4	34·5	36·7	40·5	42·4	45·4	48·5	49·4	50·0
18	—	—	33·4	32·4	34·5	37·4	42·2	45·7	49·9	51·3	51·2	51·2
19	33·3	31·3	31·3	33·6	37·2	39·6	44·4	48·2	52·5	54·5	54·6	55·1
20	32·8	31·4	31·5	32·4	33·3	34·9	36·8	39·4	42·3	44·1	44·5	45·5
21	—	—	—	—	—	—	—	—	—	—	—	—
22	35·5	—	35·3	36·5	39·1	40·6	44·7	47·5	49·6	49·8	48·5	47·4
23	28·1	28·1	28·4	28·5	29·3	32·4	33·4	35·3	37·0	36·7	37·3	37·4
24	25·9	—	26·0	27·2	28·1	29·0	30·4	31·7	32·8	32·8	34·1	35·7
25	33·0	32·5	35·2	37·8	40·6	43·5	58·0	58·3	58·7	61·5	65·5	66·5
26	37·3	36·2	36·5	40·0	40·5	42·4	44·5	45·6	48·1	47·6	50·0	46·5
27	34·5	34·4	34·8	37·9	39·5	40·0	41·9	42·5	43·5	46·5	49·4	50·0
28	—	—	—	—	—	—	—	—	—	—	—	—
29	29·6	29·3	32·5	34·2	35·7	39·5	48·5	53·3	54·6	54·3	56·4	56·5
30	34·6	34·2	35·2	38·5	40·5	45·5	46·6	46·6	49·7	51·3	49·4	48·3
Sums -	520·6	427·5	537·0	598·1	639·2	705·5	810·2	896·3	962·4	1017·3	1048·5	1053·6
Means -	22·63	21·37	22·37	23·92	25·57	28·22	32·41	35·85	39·30	40·69	41·94	42·1
Diurnal Variation -	−9·85	−11·11	−10·11	−8·56	−6·91	−4·26	−0·07	3·37	6·82	8·21 N	9·46	9·6

9h Göttingen mean time = 0h 18m local mean time.

FORT SIMPSON.

Abstract of Hourly Observations made during the month of April 1844.

15.	16.		23.	Sums.	Means.
21·5	19·7	15·1	4·2	288·5	12·02
26·5	19·7	16·9	7·8	354·7	14·78
37·0	33·0	28·9	24·1	597·4	24·89
41·5	38·9	35·7	—	} 769·8	32·07
—	—	—	13·1		
14·9	11·4	8·9	—	} 284·9	11·87
—	—	—	15·3		
41·1	40·6	37·5	37·0	747·0	31·76
30·5	26·2	26·4	14·9	782·3	26·60
20·4	18·2	16·0	6·0	401·8	16·74
20·3	24·2	21·7	17·1	444·7	18·53
34·8	35·9	33·9	31·9	699·8	29·16
33·4	31·1	30·4	—	} 742·6	30·94
—	—	—	29·4		
41·0	41·7	38·7	26·4	785·8	32·74
44·5	42·5	38·5	23·0	833·1	37·89
49·5	47·0	41·9	30·4	954·7	39·78
52·1	50·6	45·4	33·5	954·9	42·09
50·7	50·5	43·8	34·4	1045·0	43·54
41·8	41·0	38·9	—	} 923·0	38·46
—	—	—	36·8		
46·0	41·0	38·5	33·2	938·4	40·00
35·8	34·5	32·3	27·1	787·8	32·82
34·8	33·6	32·5	32·6	745·2	32·38
63·0	66·5	56·1	39·5	1257·9	52·41
48·5	43·1	40·9	36·7	1009·6	42·07
46·3	45·7	39·3	—	} 966·2	40·26
—	—	—	28·5		
53·0	47·5	46·5	34·8	1068·3	44·51
35·0	30·6	28·0	12·9	848·1	35·34
967·9	910·1	832·7	631·6	19226·5	812·74
38·72	36·40	33·31	24·87	775·92	32·46
6·24	3·92	0·83	−7·61	—	—

Fort Simpson—*continued.*

Abstract of Hourly Observations made during the month of May 1844.

Date. Gött. Mean Time.	Spirit Thermometer by Newman, corrected.										
	Noon.	1.	2.	3.	4.	5.	6.	7.	8.	9.	10.
1	14·6	12·6	16·0	14·3	14·7	17·1	18·9	20·1	24·8	24·2	27·0
2	—	14·0	17·8	20·6	23·6	26·1	29·0	31·5	34·6	35·5	36·3
3	21·5	22·0	—	23·7	24·2	28·2	31·5	34·4	37·6	39·5	42·4
4	31·2	32·7	33·6	35·3	37·3	39·8	42·5	45·0	51·8	50·6	48·3
5	—	—	—	—	—	—	—	—	—	—	—
6	29·1	30·6	33·5	33·7	35·1	36·5	38·5	38·6	40·0	43·1	43·5
7	28·1	30·4	31·5	33·4	36·0	39·7	42·4	46·0	51·5	48·5	53·3
8	34·4	34·3	34·5	36·6	39·5	40·8	44·4	46·5	53·7	56·7	57·5
9	35·7	35·3	39·5	40·3	42·0	45·5	47·1	51·2	54·3	55·5	53·3
10	36·7	38·3	43·5	42·1	44·0	46·5	54·0	54·6	56·3	50·8	56·5
11	—	48·5	47·4	47·4	49·5	52·4	54·8	58·3	58·4	56·5	56·9
12	—	—	—	—	—	—	—	—	—	—	—
13	27·2	29·3	31·3	31·0	32·3	33·8	36·5	40·0	44·1	43·3	44·5
14	31·0	33·5	36·0	36·5	37·6	40·0	43·4	49·3	52·5	49·9	51·3
15	34·6	37·6	41·2	40·4	42·8	43·8	47·9	52·5	58·3	60·3	62·4
16	42·0	44·6	49·0	50·1	52·2	55·5	61·4	64·2	66·0	66·5	67·7
17	46·4	50·6	53·5	51·1	54·0	56·9	58·5	61·0	64·5	66·3	67·5
18	47·5	46·5	46·7	46·4	48·5	47·5	48·4	50·0	51·0	49·5	48·6
19	—	—	—	—	—	—	—	—	—	—	—
20	37·3	40·5	44·4	44·7	47·5	50·5	53·5	60·9	63·5	64·4	65·5
21	43·5	44·5	46·3	43·6	43·5	44·5	46·3	49·1	52·4	52·4	52·5
22	—	31·1	29·8	30·7	32·6	35·2	37·9	42·1	44·0	46·4	47·5
23	36·5	37·7	44·8	41·3	45·5	46·6	48·6	52·9	57·0	56·9	57·2
24	44·4	44·2	—	48·0	50·3	54·5	57·7	61·2	63·9	64·4	63·9
25	—	—	—	—	—	—	—	—	—	—	—
26	—	—	—	—	—	—	—	—	—	—	—
27	—	—	—	—	—	—	—	—	—	—	—
28	—	—	—	—	—	—	—	—	—	—	—
29	—	—	—	—	—	—	—	—	—	—	—
30	—	—	—	—	—	—	—	—	—	—	—
31	—	—	—	—	—	—	—	—	—	—	—
Sums	621·7	738·8	720·3	791·2	832·7	883·4	943·2	1009·4	1080·2	1081·2	1103·6
Means	34·54	35·18	37·91	37·68	39·65	42·07	44·91	48·07	51·44	51·49	52·55
Diurnal Variation	−10·02	−9·38	−6·65	−6·88	−5·91	−2·49	0·35	3·51	6·88	6·93	7·99
Two Months Σ	1142·3	1166·3	1257·3	1389·3	1471·9	1588·9	1753·4	1905·7	2062·6	2098·5	2188·1
Two Months $\frac{\Sigma}{N}$	27·86	28·45	29·24	30·20	32·00	34·54	38·12	41·43	44·84	45·62 N	47·57

9h Göttingen mean time = 0h 18m local mean time.

FORT SIMPSON—*continued.*

Abstract of Hourly Observations made during the month of May 1844.

	15.	16.	17.
	25·0	23·7	23·7
	35·1	33·9	32·8
	41·3	42·0	41·4
	46·2	45·5	41·0
	—	—	—
	39·5	37·7	35·9
	46·9	44·2	43·7
	53·2	53·1	46·9
	52·7	49·7	47·8
	53·9	51·9	50·6
	45·7	43·7	41·1
	—	—	—
	45·0	42·5	39·7
	55·3	54·5	52·2
	62·7	61·8	58·2
70·6	67·9	63·6	62·1
	64·1	64·1	57·8
	43·2	41·2	39·5
	—	—	—
	65·1	63·8	61·5
	49·7	48·5	44·9
	44·8	46·5	44·8
	59·1	57·9	56·7
	62·7	62·5	58·4
	—	—	—
	—	—	—
	—	—	—
	—	—	—
	—	—	—
	—	—	—
	—	—	—
1085·1	1059·1	1032·3	980·7
51·67	50·43	49·16	46·70
7·11	5·87	4·60	2·14
	2027·0	1942·4	1813·4
	44·07	42·23	39·42

MAGNETICAL ABSTRACTS.

Note.—Regular Observations which were followed by Extra Readings, in consequence of disturbance, are distinguished by *Italic* figures throughout the following Abstracts. The Daily Means of imperfect days are derived from the 8-hourly series that may be complete. At Lake Athabasca 0^h Gött. $= 15^h\ 55^m$ M.T., or $3^h\ 55^m$ A.M. At Fort Simpson 0^h Gött. $= 15^h\ 14 \cdot 6^m$ M.T., or $3^h\ 14 \cdot 6^m$ A.M.

LAKE ATHABASCA.

Abstract of Hourly Observations made during the month of October 1843.

Day. Gött. Mean Time.	Declination Magnetometer.										
	Noon.	1.	2.	3.	4.	5.	6.	7.	8.	9.	10.
1	—	—	—	—	—	—	—	—	—	—	—
2	—	—	—	—	—	—	—	—	—	—	—
3	—	—	—	—	—	—	—	—	—	—	—
4	—	—	—	—	—	—	—	—	—	—	—
5	—	—	—	—	—	—	—	—	—	—	—
6	—	—	—	—	—	—	—	—	—	—	—
7	—	—	—	—	—	—	—	—	—	—	—
8	—	—	—	—	—	—	—	—	—	—	—
9	—	—	—	—	—	—	—	—	—	—	—
10	—	—	—	—	—	—	—	—	—	—	—
11	—	—	—	—	—	—	—	—	—	—	—
12	—	—	—	—	—	—	—	—	—	—	—
13	—	—	—	—	—	—	—	—	—	—	—
14	—	—	—	—	—	—	—	—	—	—	—
15[b]	—	—	—	—	—	—	—	—	—	—	—
16	401·6[a]	417·0[a]	423·3[a]	422·6	412·1	420·0	416·5	413·6	415·0	406·0	408·4
17	425·8	446·0[a]	413·0	410·0	405·0	414·0	408·0	411·0	408·4	410·0	410·0
18 (Term Day)	414·0	418·2	415·2	418·8	420·0	424·4	412·0	408·0	403·0	406·0	408·0
19 (Term Day)	418·4	416·6	419·6	422·0	424·0	422·2	418·2	411·8	415·0	410·8	411·8
20[b]	430·0	—	418·0	426·0	421·5	418·0	418·3	415·4	406·0	406·6	412·6
21	417·0	418·0	418·0	417·2	420·0	425·4	426·6	414·0	411·0	410·0	412·0
22	—	—	—	At Fort Chipewyan.				—	—	—	—
23	420·0	419·0	419·0	422·0	418·6	418·0	416·0	416·0	419·4	411·0	414·0
24	425·0	436·0	429·6	432·8	419·2	419·2	416·0	415·0	412·0	410·0	410·0
25	419·8	420·0	420·0	424·0	421·0	412·0	406·4	410·2	408·0	406·0	407·2
26	420·0[a]	434·6[a]	431·8	419·5	424·6	414·5	420·5	401·0	410·0	414·0	411·0
27	436·7[a]	450·5[a]	414·0	423·0	419·0	426·0	417·8	411·8	416·4	412·0	410·0
28	425·0	421·4	419·6	423·6	422·0	424·0	420·0	412·5	414·5	407·8	400·0
29	—	—	—	At Fort Chipewyan.				—	—	—	—
30	419·6	443·0	490·0	441·6	424·6	419·8	420·0	414·4	413·0	417·0	412·0
31	439·0	457·8	424·0	417·4	421·0	418·0	416·0	413·0	412·0	412·0	412·0
Sums -	5571·9	5599·8	5533·7	5496·2	5451·3	5457·5	5414·0	5352·3	5357·7	5322·6	5326·4
Means -	428·62	430·75	425·67	422·78	419·33	419·81	416·46	411·72	412·13	409·43	409·72
Diurnal Variation	19′·19	21′·32	16′·24	13′·35	9′·90	10′·38	7′·03	2′·29	2′·30	0′·0	0′·29

[a] Visible aurora.

[b] The 15th and 20th are excluded in forming the means, as imp

LAKE ATHABASCA.

Abstract of Hourly Observations made during the month of October 1843.

Declination Magnetometer.													
13.	14.	15.	16.	17.	18.	19.	20.	21.	22.	23.	Sums.	Means.	Fortnightly Means.
—	—	—	—	—	—	—	—	—	—	—	—	—	
—	—	—	—	—	—	—	—	—	—	—	—	—	
—	—	—	—	—	—	—	—	—	—	—	—	—	
—	—	—	—	—	—	—	—	—	—	—	—	—	
—	—	—	—	—	—	—	—	—	—	—	—	—	
—	—	—	—	—	—	—	—	—	—	—	—	—	
—	—	—	—	—	—	—	—	—	—	—	—	—	
—	—	—	—	—	—	—	—	—	—	—	—	—	
—	—	—	—	—	—	—	—	—	—	—	—	—	
—	—	—	—	—	—	—	—	—	—	—	—	—	
—	—	—	—	—	—	—	—	—	—	—	—	—	
—	—	—	—	—	—	—	—	—	—	—	—	—	
—	—	—	—	—	—	—	—	—	—	—	—	—	
—	—	—	—	—	—	—	—	—	—	—	—	—	
—	—	—	—	—	—	—	—	390·0a	*401·6*a	*386·0*a	—	392·53	
408·0	409·7	407·0	413·8	405·1	409·0a	407·3a	410·0a	416·5a	420·0a	422·0a	9967·2	416·13	414·24
411·6	410·0	409·0a	411·0a	*410·2*a	*416·5*a	*414·6*a	*392·5*a	410·0	415·2	419·5a	9921·2	413·38	
410·0	412·0	407·6	409·6	409·0	411·6	412·5	414·5	414·0	410·5	413·2	9889·9	412·06	
410·0	412·2	412·6	417·6	392·0	*389·0*	436·0	410·0	*398·0*	*408·2*	*420·0*	9901·4	412·56	
413·0	424·0	414·0	414·8	414·6	412·0	414·4	413·4	414·0	414·0a	414·6	9867·2	416·48c	
412·0	412·4	411·4	412·0	416·0	403·0	411·0	412·0		Sunday.		9954·9	414·79	
—	—	—	—	—	—	—	—	418·4	419·4	419·6			
414·0	412·9	418·0	416·6	417·0	416·4	413·0	414·0	414·0	417·0	422·0	10001·9	416·75	
409·8	410·4	414·0	412·0	416·0	418·0	412·0	416·0	426·0	420·0	421·8	10018·2	417·42	
413·0	414·0	414·0	411·5	412·8	411·4	*410·6*	*414·0*	*412·0*a	*417·0*a	*422·5*a	10014·0	417·25	
406·0	424·0	412·0a	416·4a	420·0a	414·0a	388·0a	*411·2*a	*414·0*a	*418·6*a	*416·4*	9972·6	415·52	
413·0	412·4	416·0	415·8	423·6	412·0	414·6	414·0	408·0a	411·5a	415·0	10015·8	417·32	
398·4	402·0	400·8	399·4	408·0	407·8a	412·0a	391·0a		Sunday.		9851·8	410·49	
—	—	—	—	—	—	—	—	412·8	415·2	416·0			415·84
414·0	414·0	414·0	420·0	422·0	420·0	420·2	416·0	*397·5*	*440·0*	*403·6*	10100·1	420·84	
407·2	414·0	410·6	415·6	416·0	412·0a	406·0	414·0	412·4	412·8	414·4	9999·2	416·63	
5329·0	5360·0	5347·0	5371·3	5367·7	5365·5	5432·0	5329·0	5337·6	5420·4	5437·4	129635·4	—	
409·92	412·31	411·31	413·18	412·90	412·73	417·85	409·92	410·58	416·96	418·26	9964·27	415·50	
0′·49	2′·88	1′·88	3′·75	3′·47	3′·30	3′·72	0′·49	1′·15	4′·61	3′·31	—	6′·07	

Increasing numbers denote a movement of the north end of the magnet towards the East.

LAKE ATHABASCA—*continued.*

Abstract of Hourly Observations made during the months of November and December

Date. Gött. Mean Time.	Declination Magnetometer.											
	Noon.	1.	2.	3.	4.	5.	6.	7.	8.	9.	10.	1
1	417·0	423·6	423·0	418·0	418·0	420·0	417·1	412·0	413·1	416·6	413·8	41
2	417·0	414·4	417·2	414·8	419·8	418·0	423·5	410·0	409·0	409·0	411·0	40
3[b]	*418·0*	425·0	420·0	414·0[c]	416·0	418·2	415·8	411·2	408·6	412·0	414·0	41
4[b]	—	—	—	415·2	418·0	418·0	416·0	418·0	414·0	415·2	414·5	41
5	—	—	—	At Fort Chipewyan.				—	—	—	—	
6	*464·0*	*430·0*	444·5	426·0	424·0	420·0	418·0	414·0	410·0	412·0	412·0	41
7	414·2	416·0	412·2	418·4	419·4	422·0	413·0	400·2	401·5	408·0	410·0	41
8	418·0	422·0	418·9	424·0	422·2	420·0	416·0	413·6	407·8	414·0	409·6	41
9	*422·0*	419·8	419·0	420·4	419·6	422·0	418·0	417·0	416·0	415·2	416·0	41
10	420·6	418·0	421·0	418·0	420·0	418·4	420·0	411·4	414·0	416·3	414·2	41
11	417·0	419·4	416·2	417·6	419·4	420·8	414·2	405·0	405·0	404·0	410·0	41
12	—	—	—	At Fort Chipewyan.				—	—	—	—	
13	422·2	414·0	425·0	436·0	*438·0*	*414·8*	*410·0*	408·0	416·0	412·7	413·0	41
14	416·0	420·0	426·0	416·0	422·2	416·0	420·0	412·0	410·5	410·6	410·4	41
15	420·0	428·0	426·0	433·2	411·0	419·6	416·0	416·0	414·4	412·0	414·0	4
16	414·0	419·2	417·4	416·6	419·4	420·4	419·2	414·6	410·0	408·0	408·0	4
17	418·2	418·0	420·0	420·0	422·0	418·2	418·4	412·4	414·8	412·0	414·0	4
18	418·0	421·1	418·0	418·4	420·2	420·0	415·0	410·0	405·0	406·8	410·6	4
19	—	—	—	At Fort Chipewyan.				—	—	—	—	
20	418·4	422·4	427·0	422·0	421·0	420·0	418·0	416·2	412·2	418·0	416·2	4
21	420·0	416·4	418·0	417·6	420·0	420·4	419·0	417·0	412·2	415·2	412·0	4
Term Day. 22	419·8	434·0	436·5	436·0	422·4	425·0	419·6	415·7	412·8	411·8	413·7	4
23	422·0	419·1[a]	420·0[a]	418·9	417·4	416·0	417·0	415·0	412·5	414·0	414·0	4
{24	421·2	*436·0*	436·2	424·0	430·0	420·0	417·8	408·2	412·0	410·4	413·0	4
{25	420·0	419·0	419·0	418·4	419·0	420·0	418·0	418·4	420·0	418·0	415·2	4
26	—	—	—	At Fort Chipewyan.				—	—	—	—	
27	420·0	422·1	422·0	418·0	420·0	421·2	423·0	420·8	414·2	416·0	415·8	4
28	421·0	418·0	418·4	420·0	420·4	417·4	430·0	408·0	408·8	414·0	419·0	4
29	419·6	425·5[a]	422·0[a]	421·0	414·0	416·2	418·0	417·4	416·7	418·4	416·2	4
30	416·0	415·2	414·0	415·2	412·0	414·2	416·5	411·0	410·4	409·0	411·0	
Sums -	10096·2	10121·2	10137·5	10108·5	10031·4	10060·0	10035·3	9903·9	9878·9	9902·0	9912·7	9
Means -	420·67	421·72	422·40	421·19	420·47	419·17	418·14	412·66	411·62	412·58	413·03	4
Diurnal Variation	9′·05	10′·10	10′·78	9′·57	8′·85	7′·55	6′·52	1′·04	0′·0	0′·96	1′·41	
1	422·4	430·8	417·0	428·0	416·0	412·2	406·4	408·2	406·0	406·4	410·0	
2	*434·0*	*442·0*	*430·4*	*446·0*	*448·8*	422·0	416·0	414·2	420·0	417·6	417·0	
3	—	—	—	At Fort Chipewyan.				—	—	—	—	
4	421·2	420·0	419·6	420·2	420·0	420·4	421·8	417·0	413·0	415·0	414·0	
5	416·1	424·0	427·0	424·4	421·0	421·6	417·0[b]	410·0	412·3	408·2	405·0	
6	*429·8*	*436·5*	*443·8*	419·8	420·0	422·8	421·2	418·2	418·0	414·0	413·0	
7	421·4	419·2	415·6	417·0	416·4	417·0	416·0	416·6	414·0	414·4	413·9	
8	419·2	416·0	417·8	419·2	420·0	410·4	420·0	420·0	420·0	420·0	396·0	
9	414·4	417·8	424·0	422·0	418·0	420·0	420·0	410·2	406·4	410·0	408·2	
10	—	—	—	At Fort Chipewyan.				—	—	—	—	
11	422·2	423·2	420·8	416·4	423·8	427·0	408·8	410·0	412·4	406·0	410·0	
12	425·8	414·4	413·6	410·4	416·6	420·0	425·4	426·0	414·0	419·6	415·0	
13	424·2	428·0	427·5	420·8	421·0	420·2	428·0	424·2	420·0	416·2	419·6	
14	429·0[a]	430·0[a]	430·4	426·0	427·8	430·0	434·0	428·8	426·2	424·2	429·0	
15	426·0	422·4	426·0	431·0	422·0	428·4	426·0	427·4	418·0	416·4	420·0	
16	425·0	424·7	425·2	420·6	420·4	423·8	426·4	426·0	420·8	418·0	418·0	
17	—	—	—	At Fort Chipewyan.				—	—	—	—	
18	429·4	422·2	429·5	429·0	426·2	418·0	416·7	416·2	418·0	413·4	420·0	
Term Day. 19	425·0	426·0	426·3	426·0	422·0	420·0	421·2	421·0	422·4	421·0	420·0	
{20	*438·2*	432·0	432·2	427·0[h]	423·8	422·0	408·4	410·0	408·0	418·0	421·0	
{21	426·0	415·8	412·6	414·2	415·0	418·1	417·4	412·2	412·0	411·0	410·6	
22	420·0	419·0	419·2	419·0	420·0	420·0	420·0	412·0	412·6	411·3	416·0	
23	415·5	416·0	417·2	420·0	419·7	422·8	421·0	414·0	416·0	417·0	416·5	
24	—	—	—	At Fort Chipewyan.				—	—	—	—	
25	—	—	—	At Fort Chipewyan.				—	—	—	—	
26	422·0	428·0	429·5	426·0	426·0	426·5	424·6	426·0	422·0	420·2	420·0	
27	*447·0*[a]	*438·0*	421·2	419·6	419·0	428·0	416·8	406·8	415·8	419·0	422·1	
28	421·4	421·4	400·4	*477·8*	424·0	426·0	426·6	415·5	419·8	418·0	418·2	
29	425·5	425·0	419·5	424·0	424·2	423·0	433·4	424·0	422·0	420·0	413·8	
30[k]	426·4	424·0	429·0	425·8	429·0	430·0	426·0	419·7	411·8	412·0	418·8	
Sums -	10200·7	10192·4	10147·0	10204·4	10125·1	10120·2	10093·1	10014·5	9987·7	9974·9	9966·9	
Means -	425·03	424·68	422·79	425·18	421·88	421·67	420·55	417·27	416·15	415·62	415·29	
Diurnal Variation	9′·74	9′·39	7′·50	9′·80	6′·59	6′·38	5′·26	1′·98	0′·86	0′·33	0′·0	

[a] Visible aurora. Hourly readings taken during disturbances are distinguished by a difference o
[b] Omitted from the Means, as not being complete days. [c] Eight minutes late.
[d] By the five complete 8-hourly series. [e] Taken at 23^h 30^m. [f] Twenty minu

LAKE ATHABASCA—*continued.*

Abstract of Hourly Observations made during the months of November and December 1843.

Declination Magnetometer.

13.	14.	15.	16.	17.	18.	19.	20.	21.	22.	23.	Sums.	Means.	Fortnightly Means.
413·2	415·0	414·0	412·4	415·5	413·7	418·4	418·2	416·8[a]	416·0	416·0	9992·9	416·37	
403·4	406·0	410·0	*412·4*[a]	*414·0*[a]	*375·0*	*378·0*	*401·0*	*415·4*[a]	*416·2*[a]	*416·0*[a]	9838·1	409·92	
415·0	416·0	415·0	417·5	416·2	414·0	413·8	414·4	—	—	—	8725·7	415·57[d]	
418·0	416·4	414·4	414·2	417·0	424·0	416·5	412·0		Sunday.		8739·8	416·06[d]	
—	—	—	—	—	—	—	—	404·2	410·0	*433·0*[a]			
416·0	423·9	414·2	414·0	412·0	416·0	416·2	415·2	411·2	414·8	414·0	10078·0	419·92	
415·2	415·8	414·2	412·0	416·0	416·6	417·0	416·0	415·2	414·4	415·0	9932·5	413·85	
411·2	411·0	425·0	426·0	416·0	*475·4*	*481·4*	420·8	420·0	428·0	*432·0*	10091·9	420·50	
419·5	416·0	419·6	418·0	420·0	416·5	416·0	418·0[a]	416·0	*422·8*	*419·2*	10042·1	418·42	
412·0	414·0	414·0	411·5	410·0	408·2	411·4	416·0	*483·8*	419·0	418·0	9976·2	415·67	
414·2	408·0	414·0	420·0	420·0	416·0	412·0	410·8[f]		Sunday.		9928·9	413·62	
—	—	—	—	—	—	—	—	410·8	407·0	419·4			416·61
414·0	416·6	414·0[a]	418·2[a]	419·0	419·0	416·0[a]	*407·0*[a]	*408·6*	*414·8*	420·4	9996·2	416·51	
414·8	410·6	414·0	411·6	416·0	414·0	*422·0*	407·0	408·8	416·6	418·4	9953·8	414·74	
410·2	418·0	416·8	412·8	420·0	420·6	416·2	417·0	418·0	416·0	415·0	10017·8	417·41	
416·0	418·0	415·2	414·0	415·2	*440·0*	408·0	414·6	417·5	418·2	420·0	9993·3	416·39	
415·6	413·0	418·0	422·0	418·0	416·0[a]	414·2[a]	417·0	416·0	415·0	416·0	9999·8	416·66	
416·2	416·0	418·2	414·8	417·0	415·5	415·0	419·6		Sunday.		9976·2	415·67	
—	—	—	—	—	—	—	—	415·0	419·8	419·6			
417·0	415·0	417·0	415·0	414·4	417·6	413·4	420·0	433·0	414·2	416·0	10085·4	418·14	
414·7	417·8	422·0	418·4	419·0	419·1	414·0	412·0	419·2	418·8	419·4	10006·7	416·95	
417·0	417·2	417·0	421·0	419·6	414·0	416·4[a]	414·2[a]	414·8[a]	421·0	418·0	10072·5	419·69	
416·4	416·2	414·5	415·0	416·0	416·0	414·0	414·0	416·2	417·7	417·6	9987·6	416·15	
416·5	414·6	418·0	423·2	415·0	423·0	432·0	415·8	415·6	419·0	415·8	10061·3	419·22	
414·0	414·3	416·4	417·2	420·2	415·0	417·0	414·0		Sunday.		10010·1	417·09	
—	—	—	—	—	—	—	—	414·4	416·0	417·2			417·58
419·2	417·6	418·0	418·0	416·8	414·0	410·0	414·0[a]	418·0	417·0	416·0	10021·7	417·57	
412·6	416·4	414·6	418·0	418·6	418·0	415·5[a]	425·6[a]	416·0[a]	414·7	418·7	10014·7	417·28	
413·0	401·6	412·0	416·2	416·8	417·4	421·6[a]	419·0[a]	*410·0*[a]	417·0	412·0[a]	9985·4	416·06	
414·0	420·2	414·8	412·6	422·0	434·0	414·1	414·0	411·8	414·6	416·4	9959·6	414·96	
9945·9	9955·0	9985·5	9994·3	10007·1	10050·6	9949·8	9963·8	9975·9	10008·6	10026·1	239970·7	—	
414·41	414·79	416·06	416·43	416·96	418·77	414·57	415·16	415·66	417·02	417·75	9998·75	416·61	
2′·79	3′·17	4′·44	4′·81	5′·34	7′·15	2′·95	3′·54	4′·04	6′·40	6′·13	—	4′·99	
417·0	422·0	414·0	416·0	416·5	420·0	416·6	412·2	*390·0*	*448·0*	*407·8*	9959·1	414·96	
416·0	418·4	416·0	416·6	419·0	421·8	416·0	417·0		Sunday.		10146·3	422·85	
—	—	—	—	—	—	—	—	420·0	418·0	416·4			
412·8	414·0	419·6	420·4	419·5	422·0	417·2	418·0	421·8	420·6	418·6	10033·0	418·04	
418·4	419·0	417·8	417·2	415·4	414·4	416·2	416·0	*418·8*	*418·0*	*426·0*	10009·4	417·06	
411·8	414·8	412·6	414·0	416·0	415·0	418·4	415·2	417·8	420·0	419·7	10040·2	418·34	
416·4	419·0	421·5	416·2	416·0	419·2	416·4	417·0	418·0	419·5	418·8	10009·5	417·06	
416·4	418·0	418·5	410·2	410·0	*464·2*	*418·0*	416·6	414·0	417·6	416·4	10017·1	417·38	
415·2	412·4	417·0	416·8	414·6	416·7	416·0	410·4[a]		Sunday.		9969·0	415·37	
—	—	—	—	—	—	—	—	419·6	416·0	419·4			419·28
410·0	412·0	413·0	409·8	412·3	412·4	414·6	412·7	412·0	414·6	413·5	9929·3	413·72	
422·0	420·0	419·8	419·8	424·5	423·0	423·0	421·5	422·4	423·6	423·0	10093·8	420·57	
420·6	421·0	420·8	421·2	419·0	425·8	426·0	428·2	426·0	425·0	427·0[a]	10153·3	423·05	
422·6	426·0	426·2	426·2	426·0	427·0	423·0	428·2	424·0	425·6	424·8	10246·3	426·93	
418·0	420·8	421·8	419·2	418·0	420·0	424·0	423·8[a]	421·8	422·5	423·0	10137·3	422·39	
420·1	421·6	420·0	419·6	421·4	418·0	415·0	417·0		Sunday.		10113·7	421·40	
—	—	—	—	—	—	—	—	418·0	423·4	428·2			
419·5	419·0	421·0[i]	421·0[i]	424·2	423·9	421·0	420·0	418·0[g]	421·1	422·8	10109·9	421·25	
419·6	419·0	420·0	416·8	419·8	417·0	414·0	*451·5*	*483·3*	*446·0*	*444·2*	10183·3	424·30	
417·0	420·0	424·0	418·6	419·0	413·8	416·7	416·2	421·6	417·2	418·2	10077·0	419·87	
413·2	416·0	417·0	417·6	418·6	416·0	418·0	419·0	412·0	413·6	416·4	9964·6	415·19	
421·0	419·5	418·0	416·5	415·2	420·0	421·8	420·0	425·6[a]	418·0	416·0	10035·4	418·14	
419·6	419·0	422·0	420·4	420·0	418·0	420·0	417·5		Sunday.		10034·7	418·11	421·26
—	—	—	—	—	—	—	—		Christmas-day.				
—	—	—	—	—	—	—	—	416·3	416·0	420·4			
425·6	426·0	421·5	426·0	424·4	420·4	423·4	425·2	*483·0*[a]	*416·0*[a]	*489·0*[a]	10179·1	424·13	
421·2	419·8	419·0	420·0	426·4	*430·4*	*419·4*	424·0	424·0	419·2	420·4	10141·1	422·55	
416·4	418·0	422·8	423·0	424·0	423·8	421·4	421·2	420·0	421·5	422·4	10138·0	422·42	
422·4	422·0	421·8	423·8	426·8	429·2	428·2	438·5	*486·0*	*430·2*	*416·4*	10164·7	423·53	
421·2	423·7	424·5	428·0	428·0	433·4	428·6	430·0		Sunday.		8911·6	424·48	
10052·8	10057·3	10065·7	10046·9	10066·6	10132·0	10058·3	10106·9	10050·4	10125·2	10108·7	241887·1	—	
418·03	419·05	419·40	418·62	419·44	422·17	419·10	421·12	418·75	421·88	421·20	10078·61	419·94	
2′·74	3′·76	4′·11	3′·83	4′·15	6′·88	3′·81	5′·83	3′·46	6′·59	5′·91	—	4′·65	

[g] Fifteen minutes late. [h] Five minutes late. [i] Twelve minutes late.

[a] Not included in the mean, the day being incomplete.

Increasing numbers denote a movement of the north end of the magnet towards the East.

Lake Athabasca—*continued.*

Abstract of Hourly Observations made during the months of January and February 18

Date. Gött. Mean Time.	Declination Magnetometer.											
	Noon.	1.	2.	3.	4.	5.	6.	7.	8.	9.	10.	11.
1	—	—	—	—	—	—	—	—	—	—	—	—
2[b]	429·1	430·0	443·0	428·0	438·0	428·4	422·0	416·7	415·8	414·4	416·8	418·
3	424·8	419·0[c]	420·4	421·8	421·0	424·0	426·1	429·5	428·4	426·0	424·2	422·
4	425·2	424·9	426·6[d]	426·4	423·0	431·2	427·5	422·0	424·8	418·2	419·6	407·
5	*449·8*	*497·2*	*406·0*	419·0	424·2	432·0	420·0	420·0	*419·0*	422·6	420·0	419·
6	*448·0*	*448·2*	*434·0*	444·0	434·6	436·0	432·8	430·6	424·8	417·4	422·0	426·
7	—	—	—	At Fort Chipewyan.				—	—	—	—	—
8	432·0	423·3	426·8	426·0	425·4	425·6	429·0	422·0	414·2	416·0	408·0	423·
9[b]	429·2	428·7	433·0	430·0	432·0	424·0	436·5	424·0	420·4	420·0	416·2	428·
10	428·0	429·0	430·0	429·0	430·0	430·4	435·6	425·0	420·2	416·0	412·5	405·
11	425·1	428·0	432·5	434·2	432·2	436·0	436·0	429·8	430·0	422·0	422·0	420·
12	424·0	419·0	417·2	420·0	421·2	424·8	431·6	420·0	424·8	417·5	424·2	419·
13	423·6	428·0	426·0	432·2	433·0	431·2	430·2	428·0	423·4	422·2	422·0	422·
14	—	—	—	At Fort Chipewyan.				—	—	—	—	—
15	431·6	434·6	423·0	429·0	429·0	429·5	429·0	424·0	419·8	432·0	432·0	426
16	421·0[a]	427·5[a]	426·2	426·7	427·0	426·4	424·8	421·8	433·0	424·8	423·0	420
17	*436·0*	430·8	438·0	438·0	435·5	432·0	431·0	430·2	426·0	419·8	419·4	424
18	432·0[a]	436·4	430·6	434·8	436·6	440·0	422·8	426·4	426·0	422·4	427·2	424
19	427·4	423·8	429·0	432·6	430·0	430·0	423·8	426·0	424·0	425·3	424·0	424
20	434·0[a]	430·0[a]	427·0	429·1	426·0	426·4	430·0	425·2	417·5	417·4	422·5	426
21	—	—	—	At Fort Chipewyan.				—	—	—	—	—
22	428·8	423·4	430·8	444·0	444·2	436·0	416·5	412·0	414·5	424·5	423·6	421
23	430·0	434·0	432·0	435·0	435·0	434·8	433·0	427·0	421·0	420·4	421·0	430
24 } Term Day.	434·0	424·0	430·3	422·2	428·0	432·4	432·4	424·4	421·0	423·0	415·8	428
25 } Term Day.	484·2[a]	551·0	494·0	454·0	441·0	427·0	423·0	424·2	422·0	422·0	420·0	430
26	421·4	423·0	426·0	426·8	429·8	436·2	423·5	426·0	423·0	420·0	421·4	430
27[b]	—[k a]	427·0[a]	429·6	428·0	427·2	429·8	433·0	427·0	420·4	421·8	419·8	424
28[b]	—	—	—	At Fort Chipewyan.				—	—	—	—	—
29	432·4	434·0	423·0	424·2	434·4	424·6	420·0	418·0	416·5	410·4	413·0	42
30	426·0	419·0	418·4	417·5	427·5	429·0	426·0	426·0	421·4	421·0	421·8	42
31	432·0[a]	432·2	431·0	430·0	423·0	425·0	424·8	425·8	421·6	409·8	410·0	42
Sums -	9954·3	10056·7	9893·8	9896·5	9896·6	9900·5	9834·4	9778·9	9716·9	9670·7	9669·2	972
Means -	432·80	437·25	430·17	430·27	430·29	430·46	427·58	425·17	422·47	420·47	420·40	422
Diurnal Variation	12′·40	16′·85	9′·77	9′·87	9′·89	10′·06	7′·18	4′·77	2′·07	0′·07	0′·0	2
1	*430·2*	*476·0*	*468·8*	*444·8*	432·2	437·0	424·0	411·4	412·4	425·5	412·4	41
2	424·0	426·0	424·2	434·0	439·0	435·8	*405·0*	*373·0*	411·0	414·0	412·8	41
3	422·8	425·2	426·0	425·7	425·2	428·0	420·0	418·6	424·4	412·4	416·8	4
4	—	—	—	—	—	—	—	—	—	—	—	
5	*399·1*	*436·0*	430·0	449·6	420·0	436·0	420·0	425·4	411·0	420·3	410·8	4
6	434·0	425·8	436·0	428·0	424·8	437·8	428·0	421·0	416·0	414·4	413·6	4
7	*490·2*	432·8	430·0	437·2	432·0	432·0	425·8	418·0	423·7	420·6	410·2	4
8	*450·6*	445·2[e]	449·2	436·0	*416·2*	446·4	414·2	408·6	415·8	418·0	416·4	4
9	424·2	428·6	426·0	422·0	426·0	424·0	425·2	420·6	418·8	419·8	421·2	4
10	424·4	424·0	427·7	424·0	426·4	429·6	424·2	428·0	425·0	421·6	417·0	4
11	—	—	—	—	—	—	—	—	—	—	—	
12	434·0	433·0	430·4	431·0	426·0	430·6	429·0	424·6	420·3	425·0	422·0	4
13	430·2	434·4	430·6	428·2	434·3	432·0	438·0	424·2	422·8	418·8	417·2	4
14	427·9	426·0	428·0	425·8	428·0	431·0	430·3	424·0	424·7	424·8	424·0	4
15	428 8	430·0	419·6	421·6	427·1	429·8	423·6	422·8	424·0	421·6	414·0	4
16	425·0	425·4	429·6	428·0	428·0	429·6	425·3	436·0	418·0	416·1	413·8	4
17	432·0	430·0	430·2	427·1	424·8	423·4	419·2	416·0	419·0	419·0	417·8	4
18	—	—	—	—	—	—	—	—	—	—	—	
19	431·4	436·4	434·0	434·0	432·0	429·2	427·6	422·4	422·4	422·0	424·0	
20	422·4	426·0	426·6	426·2	426·0	429·8	424·0	424·6	420·0	422·0	416·6	
21	427·4	434·0	436·8	438·6	439·6	429·8	424·2	419·6	416·8	415·7	412·4	
22	423·2	426·0	430·4	428·2	431·7	435·2	432·8	430·0	424·0	417·0	415·5	
23 } Term Day.	426·0	427·8	428·8	430·4	424·6	412·8	424·3	422·2	420·6	419·7	422·0	
24 } Term Day.	433·8	425·0	434·2	432·0	434·2	423·0	430·4	422·8	428·2	425·8	424·0	
25	—	—	—	—	—	—	—	—	—	—	—	
26	432·0	430·4	434·0	444·0	434·0	424·2	426·2	425·0	422·4	420·4	419·0	
27	432·8	435·4[a]	432·6	430·0	425·2	431·0	430·8	423·0	424·8	424·0	426·0	
28	432·4	428·3	437·0	429·0	430·0	430·8	434·0	437·7	428·4	404·0	420·0	
29	*437·0*[a]	428·2[a]	430·0[f]	438·0	427·2	432·3	431·0	413·0	420·8	418·2	422·6	
Sums -	10268·8	10383·7	10383·7	10355·4	10317·3	10328·8	10227·1	10099·5	10094·5	10062·5	10019·5	
Means -	427·87	432·65	432·65	431·47	429·80	430·37	426·13	420·81	420·60	419·27	417·48	
Diurnal Variation	10′·39	15′·17	15′·17	13′·99	12′·41	12′·89	8′·65	3′·33	3′·12	1′·79	0′·0	

[a] Visible aurora. [b] Days omitted from the Mean, not being complete. [c] Seven
[d] Fifteen minutes late. [e] At 21h 30m 418·0. [f] Seventeen minutes late. [g] Ten mi
[h] Nine minutes late. [k] At 0h 20m 436·0. [l] Eight minutes la

LAKE ATHABASCA—*continued.*

Abstract of Hourly Observations made during the months of January and February 1844.

Declination Magnetometer.

13.	14.	15.	19.	20.	21.	22.	23.
—	—	—		—	424·8	425·2	425·0
428·0	—	427·0		424·4	421·5	424·2	421·2
422·0	420·2	424·0		425·6	427·0	428·0	426·1
417·0	416·0	426·2		414·0	*384·0*	*448·0*	*460·0*
420·0	423·0	423·0		423·8	427·2	420·0	*408·4*
428·4	425·0	432·5		302·0	Sunday.		
—	—	—		—	428·4	434·0	435·0
430·0	424·3	*437·6*[a]		*420·4*[a]	*424·4*	428·4	427·0
420·8	419·0	428·0		428·0	—[a]	428·0	421·8
416·8	423·8	424·6		426·0	427·0	425·0	426·0
416·0	420·0	421·8		427·8	422·0	424·5	422·2
421·0	426·2	430·0		428·4	427·8	423·4	420·4
420·2	424·0	425·5		426·2	Sunday.		
—	—	—		—	434·0	433·6	433·0
420·0	421·8	424·0		422·0	428·0[e]	428·0	427·0
420·3	421·2	418·9		419·0[a]	424·2[a]	424·8	426·0
422·6	424·0	421·2		412·2	418·2[a]	424·2[a]	424·0
426·4	438·0	437.6		428·0	420·0	426·0	425·2
427·4	423·0	423·0[a]		440·4[a]	*448·0*[a]	436·0[a]	436·2[a]
424·4	421·3	424·0		414·8[a]	Sunday.		
—	—	—		—	426·0[a]	432·2[a]	430·0[a]
422·4	421·8	420·4		434·8[a]	422·0[a]	406·0[a]	412·2[a]
427·0	427·8	432·0		432·0	426·4	430·6	436·5
425·6	430·0	424·1[a]		437·5	471·0[a]	440·4[a]	416·0[a]
426·0	428·0	426·0		425·0	428·0	419·8	419·0
425·0	428·0	430·0		427·0[a]	424·0	*474·0*[a]	415·0[a]
418·2	418·4	421·8		422·4	Sunday.		
—	—	—		—	426·0	426·9	428·0[a]
424·4	426·0	417·0		424·0	420·4	419·3	424·0
420·0	423·2	420·0		413·0	424·0[a]	432·0	431·4
422·2	422·0	424·0		304·4	411·0	417·0	424·0
9728·1	9758·6	9773·4		9711·3	9793·0	9855·1	9832·6
422·96	424·29	424·93		422·23	425·78	428·48	427·50
2′·56	3′·89	4′·53		1′·83	5′·38	8′·08	7′·10

13.	14.	15.	16.	17.	18.	19.	20.	21.	22.	23.	Sums	Means	
415·4	420·8	417·8	425·0	423·8[m]	421·8	*444·0*[a]	*412·0*	417·0	420·2	424·0	10229·5	426·23	
418·0	420·0	424·0	409·6	*435·6*	*410·0*	*420·0*	*412·0*	421·2	420·2	420·0	10003·9	419·33	
414·0	426·0	423·2	421·8	421·8	420·0	424·0	420·4	—	—		10046·8	418·62	423·32
—	—	—	—	—	—	—	—	396·0	396·6	397·6			
414·4	434·0[a]	432·0[a]	*428·2*[a]	*403·2*[a]	*419·6*	424·0	424·8	*435·0*	*475·0*	423·0	10159·9	423·33	
418·2	437·0	423·0	422·6	428·0	422·0	*417·2*	409·0	418·6	421·2	424·8	10154·8	423·12	
426·8	422·0	420·1	429·2	426·0	423·8	419·6	421·3	417·5	421·6	433·2	10183·4	424·31	
419·7	421·0	432·2	426·0	429·0	419·3	422·2	422·0	411·4	420·6	423·4	10225·4	426·06	
421·6	421·6	426·0	426·0	418·4	424·0	423·0	417·9	430·6	424·0	422·0	10158·9	423·20	
422·8	423·2	424·0	425·3	425·3	423·0	422·5	431·5	—	—		10253·2	426·38	
—	—	—	—	—	—	—	—	439·8[a]	426·3	430·6			
426·0	423·5	424·0	422·0	421·8	416·0[a]	419·4	420·6	430·2[a]	450·4[a]	430·0[a]	10220·8	426·24	
424·3	424·0	422·0	424·2	423·4	423·8	427·7[a]	424·0[a]	425·0	423·5	420·7	10233·1	426·38	
426·0	424·0	420·2	422·0	422·0	428·0	431·0	423·8	410·5	418·2	424·2	10187·9	424·50	
419·9	422·0	420·0	420·0	420·0	421·5	421·0	414·2[a]	419·8	423·0	421·6	10121·9	421·73	
419·6	418·8	420·0	416·4	434·8[m]	420·6[a]	420·0	417·2	*400·0*[a]	416·2[a]	420·0[a]	10137·2	422·38	
414·6	415·4	418·0[a]	413·2[a]	417·2	417·6	420·0	438·2	—	—		10130·4	422·10	
—	—	—	—	—	—	—	—	423·3	426·4	429·0			
424·4	423·2	423·8	422·4	426·0	424·0	420·4	419·6	420·0	419·8	417·8	10206·8	425·28	
420·7	420·3	422·0	428·0	426·0[e]	425·2	426·0	428·0[a]	420·6[a]	426·8	429·0	10175·3	423·97	424·63
420·2	418·0	422·0	422·0	422·4	416·0	426·7	410·0[a]	423·6[a]	423·8[a]	426·0[a]	10171·6	423·82	
424·0	421·4	426·0	424·0	424·4	424·4	428·2	428·0	416·0	435·2	423·2	10218·5	425·76	
421·8	420·4	421·0	423·8	423·2	430·4	428·4	427·9	428·0	428·2	427·6	10186·3	424·49	
423·6	424·6	428·0	422·0	423·2	426·0	428·4	425·5	—	—		10269·7	427·90	
—	—	—	—	—	—	—	—	433·0	428·6	430·8			
424·6	426·4	420·0	426·0	427·0	423·2	428·2	422·6	424·2	430·0	*437·0*	10245·4	426·89	
423·5	418·4	420·2	420·0	425·0	426·0	430·4	434·0	420·4	429·8	440·0	10248·1	427·00	
429·0	412·0	416·6	418·2[a]	402·0[a]	425·4	417·1[a]	424·0	434·0[a]	401·0[a]	431·0	10148·7	422·86	
423·0	419·8	420·0	436·2	447·8	423·0	425·2	428·4	—	—	—	—	428·80	
10113·1	10135·2	10146·1	10142·9	10133·7	10138·8	10189·4	10148·5	10137·7	10148·2	10229·5	244361·4	—	
421·38	422·30	422·75	422·62	422·24	422·45	424·56	422·85	422·40	422·84	426·23	10181·70	424·24	
3′·90	4′·82	5′·27	5′·14	4′·76	4′·97	7′·08	5′·37	4′·92	5′·36	8′·75	—	6′·76	

[m] Twelve minutes late. [a] Three minutes late. [e] Twenty minutes late.

Increasing numbers denote a movement of the north end of the magnet towards the East.

LAKE ATHABASCA—*continued.*

Abstract of Hourly Observations made during the months of January and February 1844.

Date. Gött. Mean Time.	Second Declination Magnetometer, 2-inch bar.											
	Noon.	1.	2.	3.	4.	5.	6.	7.	8.	9.	10.	11.
1	—	—	—	—	—	—	—	—	—	—	—	—
2[b]	216·6	217·0	241·8	220·6	230·0	221·0	216·0	212·0	208·2	208·0	209·4	211·0
3	222·0	219·6[c]	220·6	220·0	219·6	220·6	218·0	219·6	219·0	217·2	218·4	216·0
4	220·3	220·0	221·0[d]	222·5	218·6	224·2	220·0	216·0	216·5	213·6	216·5	210·
5	*252·5*	*300·0*	*219·8*	214·6	218·8	223·8	215·8	219·6	*219·0*	216·0	218·0	217·
6	*259·4*	*248·4*	*291·5*	234·5	232·0	226·0	220·4	222·2	219·6	212·0	213·0	222·
7	—	—	—	—	—	—	—	—	—	—	—	—
8	224·0	223·4	222·0	221·8	220·0	219·3	224·0	215·0	213·0	214·0	212·6	216·
9[b]	229·0	228·0	228·0	222·8	226·0	217·0	223·0	213·6	210·0	209·6	207·4	218·
10	227·0	226·0	228·0	226·4	226·0	226·0	229·0	226·5	222·5	218·2	214·5	206·
11	225·0	230·8	232·0	236·0	235·5	238·0	236·0	232·0	229·1	222·8	222·4	222
12	228·4	229·0	232·0	228·4	230·0	236·0	241·0	238·0	229·6	225·4	222·5	224·
13	230·7	234·4	233·8	234·8	238·3	237·0	236·0	232·0	226·4	225·2	225·0	226·
14	—	—	—	—	—	—	—	—	—	—	—	—
15	228·4	232·0	232·5	233·0	234·2	237·5	238·2	229·2	224·8	224·3	225·0	229·
16	236·1[a]	235·0[a]	236·2	235·8	235·7	234·4	233·0	235·2	234·0	232·0	230·0	230·
17	*244·8*	236·8	242·2	238·0	238·2	236·4	238·0	234·2	231·2	226·4	224·2	227·
18	234·2[a]	232·4	234·7	240·0	242·1	250·0	228·0	228·4	230·0	228·0	231·4	233
19	234·2	233·0	233·2	234·0	237·0	234·5	230·8	226·0	228·0	227·4	226·0	232
20	233·0[a]	231·5[a]	232·0	231·0	230·4	230·0	230·0	229·8	226·8	227·0	226·8	229
21	—	—	—	—	—	—	—	—	—	—	—	—
22	230·0	229·6	226·0	241·4	241·4	234·0	211·0	210·0	214·5	227·8	230·0	230
23	230·0	236·0	236·4	235·0	232·2	233·0	234·0	226·0	224·4	222·8	220·0	226
24	232·0	229·0	231·0	230·6	233·0	232·0	228·2	223·8	220·2	220·0	220·4	224
25	310·0[a]	368·0	300·0	254·4	238·0	232·7	230·0	228·2	226·0	222·6	226·4	22
26	225·4	227·0	228·0	229·2	230·0	234·0	231·2	227·7	227·2	227·2	226·0	22
27[b]	— [a h]	233·2[a]	232·0	230·4	230·0	229·4	230·4	231·0	226·0	225·2	227·8	23
28	—	—	—	—	—	—	—	—	—	—	—	
29	242·4	238·2	232·0	234·2	235·0	236·4	232·0	231·8	228·2	222·6	223·2	22
30	233·0	236·0	235·2	234·4	239·3	240·8	234·6	233·0	232·0	232·0	232·0	23
31	241·0[a]	241·4	236·0	236·0	235·8	222·0	230·2	230·0	227·0	222·0	222·4	23
Sums -	5441·8	5532·5	5376·1	5346·0	5341·1	5338·6	5269·4	5214·2	5169·0	5126·5	5126·7	516
Means -	236·60	240·54	233·74	232·43	232·22	232·11	229·10	226·70	224·74	222·89	222·90	22
Diurnal Variation	13′·71	17′·65	10′·85	9′·54	9′·33	9′·22	6′·21	3′·81	1′·85	0·0	0′·01	

Date. Gött. Mean Time.	Noon.	1.	2.	3.	4.	5.	6.	7.	8.	9.	10.	11.
1	*245·0*	*308·1*	*269·8*	*250·4*	*240·6*	250·0	229·6	220·0	222·4	230·0	220·4	2
2	231·6	237·8	232·0	243·5	248·0	*249·6*	*238·2*	179·8	218·4	224·2	226·0	2
3	234·4	240·8	239·0	238·2	236·0	240·4	231·0	226·0	229·0	219·0	222·8	2
4	—	—	—	—	—	—	—	—	—	—	—	
5	*218·0*	*262·0*	252·0	272·0	243·0	258·0	242·0	246·0	230·0	244·0	232·2	2
6	267·4	248·0	264·2	253·2	250·0	258·4	251·0	243·4	242·4	241·8	239·8	
7	*250·8*	256·2	253·2	257·0	260·0	259·4	250·6	242·2	252·8	242·2	230·8	
8	*296·0*	272·2[a]	268·0	263·0	*270·2*	270·0	242·4	236·0	242·0	241·8	210·0	
9	250·0	252·0	250·8	240·4	252·2	250·8	252·2	246·0	244·0	246·0	248·2	
10	253·4	256·0	253·2	230·0	252·2	254·8	254·0	250·4	248·2	246·0	242·2	
11	—	—	—	—	—	—	—	—	—	—	—	
12	244·0	238·8	232·0	244·8	241·8	247·0	246·2	240·6	236·0	241·0	238·8	
13	247·8	252·2	250·2	248·0	249·8	247·4	247·0	242·2	238·8	238·8	236·9	
14	244·4	246·0	243·6	245·6	246·0	246·0	245·4	240·6	242·0	239·6	240·0	
15	256·0	261·5	244·2	246·2	252·8	250·8	242·8	241·8	239·8	240·4	234·6	
16	248·6	248·6	253·8	252·4	230·8	246·2	244·6	254·0	245·4	243·8	240·2	
17	259·0	256·8	256·0	253·2	252·0	243·2	243·8	240·4	243·0	241·8	241·2	
18	—	—	—	—	—	—	—	—	—	—	—	
19	251·2	254·4	252·2	252·4	250·8	250·0	247·2	243·0	242·8	242·0	243·8	
20	244·6	246·0	246·6	247·1	248·0	247·2	242·8	238·4	238·0	239·4	235·8	
21	248·4	256·0	256·8	251·2	250·8	247·6	243·8	238·4	233·8	232·0	229·2	
22	238·0	244·0	248·2	245·8	247·4	250·4	248·2	244·0	237·0	230·0	230·6	
23	245·0	246·0	248·5	251·4	242·2	246·8	244·4	241·8	242·0	240·7	239·6	
24	245·0	245·6	246·2	247·0	247·1	242·4	241·0	234·0	235·5	236·0	237·2	
25	—	—	—	—	—	—	—	—	—	—	—	
26	246·0	240·2	253·8	258·0	250·2	242·0	242·0	241·6	242·0	235·9	235·0	
27	252·6	251·6[g]	249·4	246·3	246·0	248·4	243·8	240·8	240·8	238·8	238·2	
28	240·0	218·8	244·5	223·5	248·0	252·0	250·4	257·7	240·4	217·6	236·0	
Sums -	5947·2	6029·6	6013·3	5980·6	5984·9	5998·8	5864·4	5729·1	5726·5	5692·8	5630·1	
Means -	247·80	251·23	250·55	249·19	249·37	249·95	244·35	238·71	238·60	237·20	234·59	
Diurnal Variation	13′·21	16′·64	15′·96	14′·60	14′·78	15′·36	9′·76	4′·12	4′·01	2′·61	0·0	

[a] Visible aurora.
[b] These days are omitted from the Means, not being complet
[c] Seven minutes late.
[d] Fifteen minutes late.
[e] At 21h 32m
[f] Seventeen minutes late.
[g] Ten minutes late.
[h] Nine minutes

LAKE ATHABASCA—*continued.*

Abstract of Hourly Observations made during the months of January and February 1844.

Second Declination Magnetometer, 2-inch bar.													
13.	14.	15.	16.	17.	18.	19.	20.	21.	22.	23.	Sums.	Means.	Fortnightly Means.
—	—	—	—	—	—	—	—	217·2	214·0	214·4	—	—	
216·3	—	221·6	221·0	217·0	220·8	220·0	223·0	218·0	220·0	219·2	—	218·55	
215·0	214·2	216·8	217·2	219·0	219·0	217·1	217·8	220·0	222·0	221·4	5230·1	218·75	
208·4	212·0	220·0	*217·0*	*226·0*	*214·0*	*216·2*	209·4	*184·0*	*252·0*	*274·0*	5206·2	219·42	
218·0	220·0	220·0	226·0	221·0	216·0	240·0	223·0	223·2	218·0	211·0	5389·7	224·57	
222·2	220·0	230·0	234·5	218·5	219·8	218·0	195·0ᵃ	—	—	—	5380·7	224·20	
—	—	—	—	—	—	—	—	219·5	226·0	228·0			
223·2	221·8	220·0ᵃ	232·4	232·0	218·0	222·4	*216·0*ᵃ	*238·0*	228·4	228·2	5312·3	221·35	224·46
218·0	216·4	225·8	230·4ᵃ	225·6ᵃ	224·4	224·2	224·0	—ᵉ	221·0	222·2	—	221·42	
220·0	225·2	226·4	226·0	227·2	214·2	244·2	230·4	226·2	226·0	226·4	5376·3	224·01	
220·0	214·0	228·8	228·4	226·0	228·0	230·0	234·2	227·0	232·6	226·0	5479·1	228·30	
229·6	232·2	232·0	232·4	232·2	233·0	259·0	232·8	234·2	230·4	228·0	5566·7	231·95	
228·2	230·0	232·0	233·6	232·2	228·4	228·7	230·2	—	—	—	5538·1	230·75	
—	—	—	—	—	—	—	—	230·2	229·6	229·0			
228·5	230·0	229·5	232·0	232·6	230·2	232·0	232·2	222·4ᶠ	224·0	235·2	5526·5	230·27	
231·0	230·0	230·0	234·0	231·0	230·0	230·8	228·2ᵃ	232·0ᵃ	231·8	234·0	5578·2	232·42	
228·4	233·2	233·5	234·0	234·0	234·0	232·2	222·4	223·0ᵃ	230·0ᵃ	230·4	5581·1	232·55	
232·6	236·0	236·0	236·0	235·0	239·0ᵃ	235·0	232·0ᵃ	225·0	231·2	230·0	5612·4	233·85	
226·0	224·1	220·0ᵃ	229·6ᵃ	227·8ᵃ	230·0ᵃ	230·6ᵃ	234·4ᵃ	*246·0*ᵃ	237·0ᵃ	234·2ᵃ	5546·6	331·11	
230·0	229·8	230·0	200·0	228·0	229·0	226·0ᵃ	229·0ᵃ	—	—	—	5500·7	229·20	
—	—	—	—	—	—	—	—	226·0ᵃ	228·4ᵃ	228·0			
229·4	230·0	227·3	226·2	230·0ᵐ	229·0	222·0	228·4ᵃ	216·0ᵃ	206·0ᵃ	212·0ᵃ	5414·0	225·58	
229·0	230·8	230·0	231·0	230·0	232·0	232·0	223·0	230·8	231·2	233·8	5517·8	229·91	231·84
228·4	224·8	225·0	233·0ᵃ	218·0ᵃ	219·4ᵃ	236·4	246·6	282·2ᵃ	300·0ᵃ	233·4ᵃ	5605·2	233·55	
226·4	228·0	227·3	226·4	226·0	234·4	237·0	225·2	225·2	226·0	225·2	5801·4	241·72	
227·0	230·5	235·0	227·4	240·0ᵃ	230·0	232·0	230·0ᵃ	229·0	*219·5*ʰᵃ	242·0	5510·3	229·60	
229·0	228·0	229·2	227·4	230·0	231·0	226·0	231·0	—	—	—	—	230·56	
—	—	—	—	—	—	—	—	234·4	236·0	237·4ᵃ			
229·0	232·4	233·0	233·0ˡ	232·8	230·0	232·4	230·0	234·0	234·8	234·0	5568·9	232·04	
233·2	234·7	232·8	231·2	233·2	230·0ᵃ	229·2	224·6	236·0ᵃ	240·0	240·2	5611·0	233·79	
232·2	232·0	231·5	232·2	228·0	230·0ᵃ	226·0	204·0	217·0	230·0	236·0	5498·7	229·11	
5195·7	5215·7	5246·9	5282·9	5269·5	5217·4	5309·2	5178·8	5236·9	5335·7	5318·4	126432·0	5268·00	
225·90	226·77	228·13	229·69	229·11	226·84	230·83	225·17	227·69	231·99	231·23	5497·08	229·04	
3′·01	3′·88	5′·24	6′·80	6′·22	3′·95	7′·94	2′·28	4′·80	9′·10	8′·34	—	6′·15	
226·4	230·0	228·8	230·0	224·0ᵐ	226·0	*258·0*ᵃ	*228·3*	220·0	226·2	231·4	5663·0	235·96	
219·8	224·0	230·0	215·2	*230·0*	*216·4*	*227·0*	*216·4*	216·6	230·0	232·6	5447·3	226·97	
216·0	229·7	227·4	223·0	230·9	230·0	231·0	231·0	—	—	—	5473·6	228·07	
—	—	—	—	—	—	—	—	212·6	214·0	217·2			
232·8	255·0ᵃ	253·0ᵃ	*240·8*ᵃ	*220·0*ᵃ	*252·0*	243·0	246·0	*262·8*	*259·6*	248·8	5871·0	244·62	
244·0	246·8	243·0	244·0	250·0	245·2	235·0	244·6	242·0	244·0	246·8	5925·8	246·91	240·40
240·0	244·0	241·1	252·8	253·8	243·2	241·6	239·2	233·2	246·4	251·2	5920·3	246·68	
245·3	244·0	256·4	247·2	250·0	248·0	249·8	249·8	236·6	252·0	250·6	6022·6	250·94	
249·0	251·2	249·8	242·2	250·0	250·3	248·2	239·0	253·2	249·0	251·5	5966·6	248·61	
247·9	253·6	251·4	254·0	253·6	252·6	251·0	283·0	—	—	—	6033·1	251·38	
—	—	—	—	—	—	—	—	254·0ᵃ	242·0	243·8			
242·0	243·2	243·0	243·0	242·0	235·6ᵃ	240·4	239·0	243·8ᵃ	249·5ᵃ	248·0ᵃ	5801·7	241·74	
240·2	240·0	240·4	243·2	242·2	242·0	240·6ᵃ	240·3	239·2	242·4	243·8	5836·2	243·17	
242·6	242·0	245·0	243·2	242·0	247·6	254·0	243·8	226·0	242·4	248·2	5841·0	243·37	
242·0	242·0	242·4	242·2	242·6	243·2	245·0	237·4ᵃ	242·2	244·0	245·8	5850·1	243·75	
246·0	246·0	249·2	245·8	262·2ᵃ	259·0ᵃ	244·8	246·0	*247·6*ᵃ	243·0ᵃ	252·4ᵃ	5956·2	248·17	
237·3	238·2	240·0ᵃ	240·0ᵃ	240·4	243·6	243·8	264·0	—	—	—	5900·0	245·83	
—	—	—	—	—	—	—	—	244·8	246·4	249·7			
242·0	242·0	244·2	243·0	246·0	243·0	240·2	236·0	236·6	242·2	240·0	5875·4	244·81	
240·6	240·4	241·6	250·0	242·0ᵒ	241·8	240·6	242·0ᵃ	242·0ᵃ	245·8	248·0	5829·4	242·89	242·77
234·2	235·0	236·4	236·0	238·2	237·2	241·2	222·4ᵃ	241·0ᵃ	244·2ᵃ	243·6ᵃ	5764·4	240·18	
240·0	240·6	241·6	244·0	243·2	244·0	245·0	244·2	233·2	254·0	245·6	5813·5	242·23	
241·0	240·4	240·6	242·6	242·0	242·8	244·0	244·0	246·0	246·0	246·0	5844·9	243·54	
240·0	240·6	240·0	240·0	239·0	241·5	241·8	241·0	—	—	—	5784·9	241·04	
—	—	—	—	—	—	—	—	241·2	240·0	243·4			
241·0	243·0	240·0	242·4	245·2	241·8	246·0	241·8	242·8	247·0	*259·0*	5859·7	244·15	
242·0	238·6	240·0	220·2	217·2	231·4	222·0	230·0	240·5	260·0	260·4	5778·8	240·78	
242·2	222·0	224·5	235·0ᵃ	209·8ᵃ	240·0	228·0ᵃ	239·0ᵃ	ᵈ254·0ᵃ	220·0ᵃ	248·0	5635·4	234·81	
5733·4	5772·3	5789·8	5759·8	5756·3	5788·2	5802·0	5788·2	5752·1	5830·0	5902·8	139694·9	—	
238·89	240·51	241·24	239·99	239·85	241·17	241·75	241·17	239·67	242·92	245·95	5820·59	242·52	
4′·30	5′·92	6′·65	5′·40	5′·26	6′·58	7′·16	6′·58	5′·08	8′·33	11′·36	—	7′·93	

ʰ At 0ʰ 20ᵐ 232·2. ˡ Eight minutes late. ᵐ Twelve minutes late.
ᵃ Three minutes late. ᵉ Twenty minutes late.

Increasing numbers denote a movement of the north end of the magnet towards the East.

FORT SIMPSON.

Abstract of Hourly Observations made during the months of April and May 1844.

Date. Gött. Mean Time.	Declination Magnetometer.										
	Noon.	1.	2.	3.	4.	5.	6.	7.	8.	9.	10.
1	396·0	354·0	382·4	398·0	362·4	376·0	371·7	348·0	352·0	343·5	334·0
2	346·0	353·6	358·0	361·4	377·2	371·0	371·6	364·0	348·4	347·0	336·9
3	354·6	344·0	352·2	360·0	450·0	376·6	382·0	378·0	346·0	345·0	341·6
4	399·6	368·0	368·0	368·4	368·0	351·0	362·0	360·8	361·0	353·0	348·6
5	—	—	—	—	—	—	—	—	—	—	—
6	358·4	358·8	354·0	354·0	361·5	363·0	363·0	352·0	354·0	375·0	352·0
7	—	—	—	—	—	—	—	—	—	—	—
8	358·2	—	358·4	358·4	359·8	362·8	356·0	326·0	346·0	354·0	347·0
9	357·1	357·0	361·5	362·0	362·8	365·0	364·8	365·0	358·0	352·5	354·0
10	374·8	396·8	379·8	397·8	370·8	368·0	366·0	362·0	358·0	351·5	355·5
11	363·1	374·0	380·0	372·0	374·6	374·0	371·0	364·8	365·0	363·0	356·0
12	364·7	372·2	—[b]	372·4	377·0	373·0	364·0	374·0	368·0	362·0	360·5
13	368·4	390·8	376·8	376·0	377·0	378·0	376·8	375·0	371·2	369·0	366·0
14	—	—	—	—	—	—	—	—	—	—	—
15	380·8	348·0	458·0	491·0	388·2	385·0	382·0	380·4	374·0	368·0	364·0
16	—	—	—	388·0	398·0	405·5	389·0	378·4	378·0	373·2	372·6
17	449·0	484·0	460·0	400·4	489·0	483·5	448·5	442·5	375·0	362·0	368·0
18	—	—	388·0	381·0	391·8	393·0	383·5	380·0	378·0	374·0	364·8
19	386·0	396·0	387·6	392·2	386·8	389·2	399·5	392·0	388·4	384·8	382·6
20	394·6	416·4	410·0	396·2	394·2	391·0	388·4	391·0	384·0	381·0	377·5
21	—	—	—	—	—	—	—	—	—	—	—
22	400·8	—	401·8	398·6	398·5	398·6	395·2	386·1	389·8	385·0	382·6
23	422·0	400·6	398·0	416·0	394·4	388·0	382·0	380·0	378·6	374·0	371·9
24 (Term Day)	385·8	—	389·8	392·0	393·2	390·0	395·0	383·0	385·8	380·0	378·0
25 (Term Day)	414·2	403·7	446·0	526·2	439·0	406·3	407·0	409·8	368·2	383·0	384·0
26	488·0	489·0	423·6	—[c]	463·0	470·0	482·0	380·0	378·0	376·0	378·2
27	378·4	396·4	399·0	417·8	402·0	398·0	405·0	408·6	394·0	404·4	386·8
28	—	—	—	—	—	—	—	—	—	—	—
29	398·0	416·0	509·0	448·0	485·3	426·0	404·0	394·0	390·0	385·0	380·0
30	420·4	397·8	402·2	402·0	408·8	402·2	400·0	385·6	385·5	366·7	377·0
Means	386·02	382·72	399·03	397·97	396·22	398·43	385·63	380·75	369·52	367·13	363·13
Diurnal Variation	25·56	22·26	38·57	37·51	35·76	27·97	25·17	20·29	9·06	6·67	2·07

Date.	Noon.	1.	2.	3.	4.	5.	6.	7.	8.	9.	10.
1	406·4	398·8	413·5	396·0	410·0	407·0	415·0	404·0	396·6	394·6	396·0
2	—	395·8	420·0	412·2	403·6	418·0	407·5	392·0	403·6	400·0	390·0
3	397·2	405·2	404·0	426·0	458·0	444·0	408·4	406·0	399·0	386·2	387·5
4	398·0	400·0	404·0	411·0	409·2	411·8	409·0	408·0	401·2	396·5	392·0
5	—	—	—	—	—	—	—	—	—	—	—
6	408·8	419·6	402·8	407·8	425·4	426·0	419·0	402·0	401·6	396·0	393·9
7	415·8	400·0	411·6	418·0	421·6	420·0	412·2	417·0	402·0	398·0	397·0
8	408·4	—[a]	434·0	435·0	420·6	438·0	418·0	412·2	404·0	386·0	391·5
9	404·0	405·0	427·8	431·8	459·0	442·2	413·6	402·0	403·9	404·0	401·0
10	397·8	403·0	407·2	414·0	421·4	434·0	413·0	408·0	400·8	400·0	400·0
11	—	412·0	412·4	435·2	420·0	417·0	416·0	412·2	409·6	402·0	400·8
12	—	—	—	—	—	—	—	—	—	—	—
13	409·0	412·0	411·8	428·0	421·2	418·0	412·0	408·0	407·0	406·5	400·0
14	419·0	421·0	441·0	446·0	428·0	430·0	426·0	414·8	412·2	410·0	410·0
15	419·2	423·0	430·2	430·0	438·0	438·8	425·0	420·0	414·0	417·0	413·6
16	438·4	428·0	436·5	454·4	451·0	433·2	425·0	413·0	412·6	403·4	411·[illegible]
17	416·0	422·0	422·0	430·0	430·8	423·0	431·0	426·0	418·0	416·0	409·[illegible]
18	417·2	422·6	427·0	428·0	425·8	432·0	420·8	422·0	407·5	411·0	412·[illegible]
19	—	—	—	—	—	—	—	—	—	—	—
20	424·8	422·0	424·6	426·0	430·0	429·0	430·0	428·0	424·0	419·0	418·[illegible]
21	419·0	426·0	441·8	441·0	437·0	436·0	434·2	430·0	428·0	404·2	413·[illegible]
22	—	496·0	500·8	465·0	444·6	457·0	436·0	426·0	417·6	401·0	402·[illegible]
23	448·0	440·0	441·8	437·8	405·0	432·8	413·8	436·0	426·2	415·2	414·[illegible]
24 (Term Day)	439·2[d]	434·0	448·0	432·0	432·2	442·4	444·0	427·0	417·0	423·0	421·[illegible]
25 (Term Day)	See term-day observations.				—	—	—	—	—	—	—
26	—	—	—	—	—	—	—	—	—	—	—
27	—	—	—	—	—	—	—	—	—	—	—
28	—	—	—	—	—	—	—	—	—	—	—
29	—	—	—	—	—	—	—	—	—	—	—
30	—	—	—	—	—	—	—	—	—	—	—
31	—	—	—	—	—	—	—	—	—	—	—
Sums	414·91	415·14	421·88	427·86	433·21	428·80	419·25	415·30	409·66	405·08	403·[illegible]
Diurnal Variation	14·49	14·62	21·38	27·34	32·69	28·28	18·73	14·78	9·14	4·56	3·[illegible]

[a] Visible aurora.

[b] At 2h 25m 361·8. Whence at 2h 0m 360·9.

[c] At 1h 30m 438·0. Whence at 1h 0m 452·1.

[d] At 0h 21m 439·[illegible] Whence at 0h [illegible]

FORT SIMPSON.

Abstract of Hourly Observations made during the months of April and May 1844.

Declination Magnetometer.													
13.	14.	15.	16.	17.	18.	19.	20.	21.	22.	23.	Sums.	Means.	Fortnightly Means.
339·8	332·0	335·8	330·2	360·2[a]	326·2[a]	320·2	329·0	324·2	322·0	330·4	8343·6	347·65	
344·2	340·4	334·1	338·2	334·2[a]	361·4[a]	345·6	362·8[a]	354·4[a]	331·6[a]	448·4	8505·3	354·39	353·09
346·2	344·2	343·6	359·2	344·1[a]	352·1[a]	351·4	350·2	349·6	356·4	360·2	8568·4	357·02	
333·8	341·2	344·2	346·4	329·2	341·8	345·2[a]	340·4	—	—	—	} 8467·9	352·83	
—	—	—	—	—	—[a]	—	—	343·1[a]	350·0	350·2			
334·1	330·0	332·2	347·4	331·4	356·0[a]	338·0[a]	342·8[a]	—	—	—	} 8383·0	349·29	
—	—	—	—	—	—	—	—	348·0	349·4	351·2[a]			
350·0	346·2	348·0	348·0	348·4	347·8	354·2	350·0	338·4	352·0	356·1	—	350·38[b]	
354·2	356·0	352·6	352·1	355·4	352·0	356·0	359·4[a]	359·8[a]	370·4[a]	405·4	8641·2	360·05	
354·2	354·0	350·2	356·0	360·2	356·5[a]	384·3	350·0[a]	406·0[a]	363·2[a]	365·1	8787·6	366·15	
359·8	359·2	360·0	360·0	362·4	369·2	361·8[a]	356·0	366·0	361·4	359·2[a]	8752·7	364·70	
363·4	364·0	365·1	366·0	367·8	368·0	367·2	364·0	367·2	366·4	364·7	—	367·09[b]	
366·2	368·2	368·0	359·4	368·0	362·2	368·0[a]	374·0[a]	—	—	—	} 8879·8	369·99	
—	—	—	—	—	—	—	—	348·0[a]	372·0	368·2			
364·0	363·2	368·0	369·0	376·0	369·8	379·6[a]	373·7[a]	375·9	366·0	371·6	9048·6	377·02	
360·1	347·0	338·4	358·2	346·8[a]	330·8[a]	335·9[a]	348·4[a]	394·6[a]	393·2[a]	409·7[a]	—	374·76[b]	377·02
350·6	350·2	369·8	370·4	368·2	374·0	370·4	374·3	372·3	376·2[a]	373·2	9366·7	390·28	
382·2	373·4	378·2	379·1	378·3	378·0	378·2	380·7[a]	378·4[a]	368·0[a]	380·0	—	379·53	
370·7	377·2	373·0	380·7	380·2	380·2	379·9	379·6[a]	393·0[a]	386·6[a]	392·3	9241·0	385·04	
383·8	376·4	380·2	381·8	381·2	380·0	378·0	384·6[a]	—	—	—	} 9278·3	386·60	
—	—	—	—	—	—	—	—	382·4	384·6[a]	388·6[a]			
380·2	378·0	379·2	388·3	382·0	381·8	364·4	379·9	382·0	383·4	382·4	—	386·03[b]	
373·4	374·0	375.2	379·8	376·0	371·9	376·0	398·2	378·4	374·4	376·8	9203·1	383·46	
376·0	376·0	381·2	379·4	381·0	394·8	376·5	386·0	380·0	382·8	402·3	—	385·08	
371·8	376·2	384·0	357·3	385·0	381·4	372·2	374·0	388·0	350·0[a]	386·2	9504·5	396·02[b]	
386·0	360·0	380·4	386·1	390·0	308·4	396·0	369·9	347·2	384·5	400·0	—	392·59[b]	
386·0	390·1	362·4	416·4	419·6	374·3	395·0	380·9	—	—	—	} 9472·4	394·68	
—	—	—	—	—	—	—	—	408·0[g]	400·2	394·0			
386·6	388·0	379·2	380·4	388·4	386·1	388·2	394·0	390·0	390·0	392·0	9663·0	402·62	394·73
381·8	396·0[f]	380·0	376·0	346·9	386·0	386·0	386·2	406·2	458·8	377·0	9395·9	391·50	
											Mean -	374·59	
361·51	362·03	360·69	364·48	364·81	365·61	366·44	367·18	371·96	371·86	376·84	—	373·85	
1·05	1·57	0·23	4·02	4·35	5·15	5·98	6·72	11·50	11·40	16·38	—	13′·30	
390·0	384·4	372·0	386·0	384·2	408·0	404·0	384·0	394·4	396·6	381·8	9475·3	394·80	
390·2	381·3	392·0	384·0	391·0	397·7	397·8[a]	386·2	386·0	385·8	396·7	—	395·61	
386·0	393·7	394·3	398·0	391·0	389·8	397·3	396·2	406·0	396·1	396·0	9641·9	401·75	
394·4	384·0	386·6	390·0	394·4	390·0	387·0[a]	385·8	—	—	—	} 9523·2	396·80	
—	—	—	—	—	—	—	—	372·0	416·8[a]	409·8			
396·4	396·1	396·2	390·6	403·0	398·2	395·2	404·0	396·6	397·8	397·8	9666·0	402·75	
392·0	392·2	392·2	394·4	424·6	380·0	374·0	378·0	390·2	490·4	394·0	9627·4	401·14	
392·2	390·8	384·4	386·0	392·2	373·6	380·3	390·0	374·0	376·4	390·2	—	390·08	
390·2	392·0	392·2	396·1	398·0	404·4	398·2	403·6	411·4	400·3	401·8	9800·5	408·35	
400·0	400·2	396·0	401·8	401·0	401·2[g]	404·0	404·0	402·8	410·2	398·3	9712·5	404·60	
396·0	398·3	390·0	401·7	404·6	401·8	399·5	400·0	—	—	—	} —	404·83	
—	—	—	—	—	—	—	—	402·1	402·0	404·0			
402·2	408·1	408·0	404·2	404·0	388·0	390·4	398·4	398·0	414·0	408·2	9750·4	406·27	410·64
403·2	398·0	394·2	392·0	404·0	408·0	400·2	416·0	416·4	409·2	409·7	9918·0	413·25	
402·0	412·3	412·0[h]	416·2	425·0	414·0	412·0	410·0	410·6	408·0	412·3	10027·2	417·80	
410·2	414·2	414·0	413·6	415·1	416·0	418·4	410·0	409·7	419·6	413·9	10081·1	420·05	
408·0	410·2	414·2	412·6	416·7	410·3	417·0	414·2	418·0	416·2	410·2	10025·4	417·72	
403·2	408·3	414·2	414·0	418·2	414·0	412·4	417·0	—	—	—	} 10018·2	417·42	
—	—	—	—	—	—	—	—	416·0	426·0	426·6			
407·6	409·2	410·0	408·6	410·2	410·6	410·6	410·0	409·9	407·4	420·0	10008·9	417·04	
407·7	403·2	404·0	406·2	410·0	410·0	412·0	410·2	412·6	412·0	410·0	10028·0	417·85	
411·8	404·6	400·1	410·2	409·0	394·4	390·2	412·0	410·0	406·6	384·0	—	428·91	
416·0	414·0	414·3	415·6	416·2	414·0	412·6	404·0	396·0	384·4	406·3	10088·2	420·34	
418.0	404·0	400·4	413·2	418·0	426·0	420·4	405·6	410·0	418·4	434·0	10167·8	423·06	
—	—	—	—	—	—	—	—	—	—	—			
—	—	—	—	—	—	—	—	—	—	—	Mean -	410·01	
—	—	—	—	—	—	—	—	—	—	—			
—	—	—	—	—	—	—	—	—	—	—	Mean of the whole	390·76	422·69
—	—	—	—	—	—	—	—	—	—	—			
—	—	—	—	—	—	—	—	—	—	—			
—	—	—	—	—	—	—	—	—	—	—			
400·57	401·26	400·90	402·49	407·22	404·28	402·83	402·46	403·79	406·37	406·61	—	409·88	
0′·05	0′·74	0′·38	1′·97	6′·70	3′·76	2′·31	1′·94	3′·27	7′·85	6′·09	—	9′·36	

[f] Six minutes late. [g] Five minutes late. [h] Four minutes late.
Increasing numbers denote a movement of the north end of the magnet towards the East.

LAKE ATHABASCA.

Abstract of Hourly Observations made during the month of October 1843.

Date. Gött. Mean Time.	Bifilar Magnetometer.										
	Noon.	1.	2.	3.	4.	5.	6.	7.	8.	9.	10.
1	—	—	—	—	—	—	—	—	—	—	—
2	—	—	—	—	—	—	—	—	—	—	—
3	—	—	—	—	—	—	—	—	—	—	—
4	—	—	—	—	—	—	—	—	—	—	—
5	—	—	—	—	—	—	—	—	—	—	—
6	—	—	—	—	—	—	—	—	—	—	—
7	—	—	—	—	—	—	—	—	—	—	—
8	—	—	—	—	—	—	—	—	—	—	—
9	—	—	—	—	—	—	—	—	—	—	—
10	—	—	—	—	—	—	—	—	—	—	—
11	—	—	—	—	—	—	—	—	—	—	—
12	—	—	—	—	—	—	—	—	—	—	—
13	—	—	—	—	—	—	—	—	—	—	—
14	—	—	—	—	—	—	—	—	—	—	—
15[b]	—	—	—	—	—	—	—	—	—	—	—
Term Day. 16	*77·6*[a]	*244·6*[a]	*277·4*[a]	*234·8*	*278·3*	273·0	287·3	275·6	285·7	283·9	289·3
Term Day. 17	261·3[a]	*147·6*[a]	*118·3*	261·2	253·1	261·8	256·9	284·9	280·1	277·1	290·1
18	231·7	260·2	254·5	259·9	261·9	263·6	262·1	271·6	260·2	278·8	288·1
19	268·7	273·3	270·1	265·1	267·1	257·4	275·3	279·2	250·7	258·4	267·0
20[b]	225·6	—	266·1	260·1	265·9	260·4	258·6	260·9	257·3	262·3	258·6
21	272·3	208·1	270·9	268·5	260·3	264·6	269·8	268·9	263·9	271·9	277·5
22	—	—	—	—	—	—	—	—	—	—	—
23	269·6	265·6	262·7	255·3	261·6	266·7	265·0	253·7	247·3	267·1	276·9
24	214·1	*153·6*	*217·1*	223·3	244·2	227·5	239·0	236·8	233·2	229·7	237·3
25	229·6	237·0	231·3	221·0	220·1	227·9	222·5	227·1	232·5	235·0	240·2
26	*111·8*[a]	*208·7*[a]	217·1	223·7	217·5	207·8	230·3	207·6	230·2	221·7	234·0
27	*145·1*[a]	*63·4*[a]	211·0	206·5	218·9	213·7	216·7	218·7	209·4	210·5	221·4
28	160·2	206·1	212·8	214·1	218·4	212·3	225·4	229·7	231·3	227·2	257·8
29	—	—	—	—	—	—	—	—	—	—	—
30	248·1	195·0	*230·1*	*276·7*	274·3	280·6	279·6	279·9	282·0	268·6	276·[illegible]
31	*183·6*	*178·3*	*242·1*	*279·7*	268·3	270·0	262·3	261·1	261·5	264·9	258·[illegible]
Sums -	2679·1	2701·5	3012·4	3185·8	3253·0	3226·9	3292·2	3294·8	3273·0	3294·8	3413·[illegible]
Means -	206·08	207·81	231·72	245·06	250·23	248·22	253·25	253·45	251·77	253·44	262·5

[a] Aurora visible.

[b] Omitted from the Hourly Means, as being imperfect days.

LAKE ATHABASCA.

Abstract of Hourly Observations made during the month of October 1843.

	14.	15.	16.	17.	18.	19.	20.	21.	22.	23.	Sums.	
	—	—	—	—	—	—	—	—	—	—	—	—
	—	—	—	—	—	—	—	—	—	—	—	—
	—	—	—	—	—	—	—	—	—	—	—	—
	—	—	—	—	—	—	—	—	—	—	—	—
	—	—	—	—	—	—	—	—	—	—	—	—
	—	—	—	—	—	—	—	—	—	—	—	—
	—	—	—	—	—	—	—	—	—	—	—	—
	—	—	—	—	—	—	—	—	—	—	—	—
	—	—	—	—	—	—	—	—	—	—	—	—
	—	—	—	—	—	—	—	—	—	—	—	—
	—	—	—	—	—	—	—	—	—	—	—	—
	—	—	—	—	—	—	—	—	—	—	—	—
	—	—	—	—	—	—	—	—	—	—	—	—
	—	—	—	—	—	—	—	—	—	—	—	—
	—	—	—	—	—	—	—	295·1[a]	*204·8*[a]	*190·0*[a]	608·9	302·97
	300·1	300·6	290·3	275·3	293·5	268·0	210·8	236·5	227·3	234·5[a]	6286·2	261·92
	265·8	267·0[a]	286·6[a]	*264·3*[a]	*264·4*[a]	*273·4*[a]	149·6[a]	254·5	263·7	256·1[a]	6115·3	254·80
	260·6	268·6	265·4	311·0	285·6	249·9	251·0	266·1	260·0	278·7		266·77
276·0	274·3	268·7	240·9	223·7	*277·2*	280·2	264·2	*260·2*	*143·8*	*218·3*		258·32
	273·1	265·3	269·3	269·6	272·0	259·9	271·7	268·3[a]	274·1	271·9		263·27
	272·4	274·1	273·7	263·1	266·2	271·1	272·5	—	—	—		270·45
	—	—	—	—	—	—	—	268·0	273·3	270·7		
	252·4	232·6	236·7	239·6	240·0	230·1	247·3	244·4	240·3	236·1		252·61
	242·3	241·1	234·9	242·7	246·8	234·8	200·4	222·2	222·5	223·1		228·46
	230·8	234·9	234·8	226·6	230·1	*175·3*	*164·4*	*180·4*[a]	*230·7*[a]	*190·4*[a]		222·05
	221·6	236·0[a]	229·4[a]	230·9[a]	228·7[a]	194·9[a]	*152·0*[a]	*175·7*[a]	*167·1*[a]	*197·1*		210·45
216·8	215·1	216·3	216·1	222·1	231·5	216·7	216·1	214·7[a]	187·9[a]	217·7		206·09
	244·0	251·4	244·4	241·8	243·8[a]	244·2[a]	217·1[a]	—	—	—		233·48
	—	—	—	—	—	—	—	270·0	272·2	258·8		
283·7	288·5	283·5	283·6	272·6	280·8	286·6	275·4	*222·1*	*163·8*	*186·4*		260·28
275·3	268·8	266·2	268·0	278·6	251·3[a]	223·6	263·5	257·6	252·3	251·7		254·57
·9	3336·7	3341·0	3294·8	3312·3	3340·3	3148·8	2884·3	3072·4	2694·9	2979·6	—	—
	256·67	257·00	253·45	254·79	256·95	242·22	221·87	236·34	222·68	229·20	—	

[a] Mean of the Hourly curve (from which the 20th is excluded) 244·63. [d] Including October 20th.

Increasing numbers denote increase of Horizontal Force.

Lake Athabasca—*continued.*

Abstract of Hourly Observations made during the months of November and December

Date. Gött. Mean Time.	Horizontal Force Magnetometer.										
	Noon.	1.	2.	3.	4.	5.	6.	7.	8.	9.	10.
1	252·8	239·0	222·5	252·0	254·4	257·0	252·1	247·3	246·6	256·3	253·3
2	247·4	255·1	249·3	246·3	245·7	241·1	234·2	239·6	242·1	237·9	244·6
3	*190·9*	211·4	232·4	232·6[c]	237·3	238·9	233·1	231·5	229·3	230·8	231·7
4[b]	—	—	—	243·0	241·7	244·3	243·2	243·7	241·7	240·7	237·8
5	—	—	—	—	At Fort Chipewyan.			—	—	—	—
6	*156·8*	*242·2*	225·0	256·5	267·1	258·4	260·0	255·4	255·6	260·1	263·3
7	233·8	246·9	231·9	234·6	233·0	234·1	236·9	227·1	237·5	239·2	234·6
8	222·9	217·3	207·4	225·1	218·3	231·6	232·4	228·6	233·0	226·1	221·3
9	*197·5*	238·6	240·9	238·7	234·5	226·6	216·5	217·7	213·5	217·1	221·0
10	219·6	220·6	223·6	209·8	204·3	204·6	209·4	208·7	206·4	205·5	207·6
11	185·5	203·0	207·4	213·6	213·8	201·3	198·4	204·4	204·0	216·5	206·5
12	—	—	—	—	At Fort Chipewyan.			—	—	—	—
13	205·0	209·0	208·5	194·4	*199·6*	*173·0*	*194·4*	206·0	200·4	197·9	206·0
14	202·2	198·7	190·9	207·4	197·2	209·4	201·0	200·4	199·0	196·2	200·3
15	209·0	189·5	200·9	176·6	208·3	211·6	206·7	199·2	204·0	202·9	209·0
16	207·6	201·7	199·9	203·5	206·5	202·4	198·8	195·9	196·2	198·9	202·0
17	201·3	195·5	221·3	197·6	198·9	197·3	197·6	199·9	195·7	197·9	203·2
18	194·4	197·2	199·0	197·0	194·6	196·1	187·2	189·9	188·2	200·7	197·0
19	—	—	—	—	At Fort Chipewyan.			—	—	—	—
20	205·5	200·6	194·3	203·0	216·3	202·4	201·2	197·6	198·4	193·3	197·4
21	202·1	203·0	201·0	201·6	202·5	201·4	190·3	194·1	193·7	194·4	197·8
22	198·2	178·1	202·0	198·0	196·8	195·6	192·7	199·4	198·9	201·3	203·0
23	191·6	193·4[a]	190·6[a]	193·7	195·2	196·2	183·2	183·1	187·2	188·9	192·0
Term Day { 24	183·6	*196·6*	167·4	182·3	192·1	188·0	187·1	188·4	187·3	186·0	186·1
Term Day { 25	185·2	184·8	185·4	187·8	186·0	186·4	184·8	185·4	182·7	184·6	186·9
26	—	—	—	—	At Fort Chipewyan.			—	—	—	—
27	189·8	181·0	187·4	185·7	187·1	184·0	182·9	185·4	184·6	186·6	185·4
28	185·2	186·0	186·0	185·1	182·7	183·9	175·9	184·0	181·5	186·0	182·1
29	177·7	141·6[a]	154·5[a]	165·4	168·0	187·0	186·3	183·3	190·9	191·0	186·6
30	180·8	182·6	180·4	180·3	182·0	182·0	171·2	175·0	171·9	160·3	168·9
Sums -	4965·4	5022·4	5109·9	5166·6	5132·2	5190·3	5114·3	5127·3	5132·6	5165·4	5187·6
Means -	198·62	200·90	204·40	206·66	205·29	207·61	204·57	205·09	205·30	206·62	207·50
Differences	4·59	6·87	10·37	12·63	11·26	13·58	10·54	11·06	11·27	12·59	13·47
$\frac{\Delta X}{X}$ }	·001629	·003439	·003681	·004484	·003997	·004821	·003742	·003928	·004001	·004469	·004782
Nov. 30	—	—	—	—	—	—	—	—	(These three observation		
Dec. 1	168·1	153·4	174·9	168·1	176·8	176·5	170·4	168·8	170·6	172·0	174·7
2	*141·8*	*130·9*	*87·6*	*63·8*	*136·4*	178·9	184·1	181·6	178·4	177·2	176·7
3	—	—	—	—	At Fort Chipewyan.			—	—	—	—
4	167·4	168·4	169·1	170·1	169·5	169·4	169·9	171·1	174·6	175·4	177·9
5	171·1	172·8	169·4	182·7	178·2	175·7	174·2[f]	176·3	174·6	169·0	175·8
6	*127·1*	*126·5*	*171·8*	†185·7	186·3	181·9	179·0	179·2	182·9	186·7	192·2
7	180·6	181·3	186·6	183·0	182·2	180·7	180·3	178·7	180·7	179·3	180·5
8	182·4	186·6	186·3	183·9	178·8	171·7	164·1	168·7	170·6	201·2	220·4
9	187·3	184·0	157·5	151·8	185·7	186·9	185·8	182·9	185·8	187·0	188·3
10	—	—	—	—	At Fort Chipewyan.			—	—	—	—
11	179·6	169·1	179·3	178·6	163·0	164·2	176·4	175·4	169·8	165·2	184·2
12	165·2	179·0	171·4	174·6	170·6	166·4	178·4	187·4	177·9	191·7	189·2
13	173·4	170·4	168·3	167·1	171·4	168·1	165·8	158·6	162·5	155·5	171·4
14	143·0[a]	143·3[a]	148·7	154·1	159·1	158·1	157·8	158·7	160·7	157·9	161·0
15	165·7	162·6	160·6	152·4	165·7	173·0	169·5	168·4	166·7	165·4	167·3
16	167·3	168·5	169·1	171·6	169·8	160·0	163·6	163·8	166·4	167·5	164·7
17	—	—	—	—	At Fort Chipewyan.			—	—	—	—
18	161·2	165·2	162·9	168·0	166·3	172·0	171·0	165·4	169·4	170·3	172·2
19	167·5	162·8	163·4	164·2	168·7	170·2	164·2	162·4	172·3	166·6	165·7
Term Day { 20	*155·6*	*154·8*	169·3	179·3[e]	182·2	180·0	179·2	178·4	180·6	182·0	183·1
Term Day { 21	168·7	167·6	170·7	170·4	167·1	169·8	164·0	163·0	166·3	165·8	168·
22	173·0	177·1	176·8	170·8	169·4	169·0	167·6	173·3	182·3	173·9	175·
23	164·0	164·8	166·1	166·9	164·7	165·9	162·7	162·6	166·9	168·9	172·
24	—	—	—	—	At Fort Chipewyan.			—	—	—	—
25	—	—	—	—	At Fort Chipewyan.			—	—	—	—
26	169·3	173·7	187·3	187·3	181·5	184·3	181·0	178·1	174·7	178·8	180·
27	*123·0*[a]	169·8	187·0	183·0	180·0	170·4	167·0	158·6	183·8	178·5	180·
28	174·7	175·1	164·6	*112·6*	199·1	200·4	179·6	180·6	181·3	180·7	183·
29	170·5	175·6	170·3	183·0	183·8	177·1	172·6	169·4	168·5	166·8	174·
30	170·4	167·4	162·8	167·9	165·4	163·0	163·1	160·0	165·0	166·2	162·
31	—	—	—	—	At Fort Chipewyan.			—	—	—	—
Sums -	4117·9	4150·7	4190·8	4140·9	4321·7	4347·6	4291·3	4271·4	4333·3	4349·5	4441·
Means -	164·72	166·03	167·63	165·64	172·87	173·90	171·65	170·86	173·33	173·98	177·6
Differences	7·84	9·15	10·75	8·76	15·99	17·02	14·77	13·98	16·45	17·10	20·7
$\frac{\Delta X}{X}$ }	·002783	·003248	·003816	·003110	·005676	·006042	·005243	·004963	·005840	·006070	·00737

[a] Visible aurora. [b] Omitted from the Means, as being incomplete. [c] Eight

[d] Twenty minutes late. [e] Fifteen minutes late. [f] Five minutes late. [g] Twelve

Lake Athabasca—*continued.*

Abstract of Hourly Observations made during the months of November and December 1843.

Horizontal Force Magnetometer.

18.	19.	20.	21.	22.	23.	Sums.	Means.
268·0	252·0	246·6	219·4[a]	268·7	249·1	6026·3	251·10
206·8	215·9	182·8	168·0[a]	195·9[a]	117·9[a]	5534·5	230·60
234·6	231·3	234·0	—	—		—	231·11
243·6	242·9	236·8	Sunday.			—[b]	238·40
—	—	—	239·2	230·4	198·0[a]		
255·8	253·5	246·6	246·9	248·9	228·6	5979·1	249·13
231·0	232·0	229·5	236·3	235·8	232·3	5643·3	235·14
248·0	226·9	237·8	203·6	169·6	133·1	5363·8	223·49
221·0	219·0	219·9[a]	211·3	156·7	198·1	5229·7	217·90
213·5	213·9	205·9	169·8	186·7	204·1	5018·2	209·09
233·8	223·1	222·3[d]	Sunday.			5078·9	211·62
—	—	—	222·5	220·1	210·6		
203·6	202·9[a]	182·1[a]	182·2	188·8	207·7	4777·6	199·07
207·8	216·1	215·1	206·5	205·3	203·2	4896·6	204·02
201·9	206·6	205·9	196·9	208·0	193·1	4903·5	204·31
237·5	230·5	210·3	206·2	204·6	203·3	4965·4	206·89
195·9[a]	198·1[a]	200·3	200·1	197·8	197·0	4785·9	199·41
200·0	200·3	198·3	Sunday.			4741·6	197·57
—	—	—	203·6	200·0	203·0		
209·3	205·2	194·9	176·0	203·9	204·1	4846·3	201·93
198·4	209·6	197·8	196·0	196·0	195·5	4777·8	199·07
199·9	202·0[a]	203·8[a]	194·1[a]	189·6	183·3	4743·5	197·65
193·3	194·3	200·2	194·2	192·6	191·4	4620·5	192·52
190·5	187·0	183·3	183·5	182·6	185·0	4398·3	183·26
183·3	182·9	183·5	Sunday.			4463·7	186·82
—	—	—	194·7				
183·5	193·0	191·8[a]	187·5			4479·8	186·66
180·0	187·0[a]	183·1[a]	179·0[a]			4406·2	183·59
188·6	179·9[a]	191·4[a]	154·0[a]			4308·8	179·53
184·0	182·7	180·2	165·8			4226·1	176·09
5269·0	5248·3	5147·4	4937·3			123676·5	—
210·76	209·93	205·90	197·49			4947·07	206·13
16·73	15·90	11·87	3·46			—	—
·005832	·005644	·004214	·001225			—	—

					18.	19.	20.	21.	22.	23.	Sums.
(… on this sheet to complete the days.)					—	—	—	165·8	168·9	171·7	—
46·3	176·1	182·6	181·0	188·0	191·0	190·8	192·6	183·6	141·9	90·4	4073·8
76·0	174·0	180·2	185·0	177·3	183·7	176·6	174·9	Sunday.			3913·4
—	—	—	—	—	—	—	—	165·3	165·3	167·2	
183·1	181·6	175·4	177·2	180·4	179·9	180·5	180·6	179·8	182·3	173·7	4234·5
174·2	178·3	177·7	177·3	176·9	176·3	177·6	176·0	171·3	96·6	110·0	4061·0
199·9	191·7	194·5	197·4	214·6	220·0	191·7	188·9	183·9	180·0	179·6	4424·2
181·6	181·3	175·0	178·3	179·5	183·4	184·4	182·6	174·6	175·9	180·6	4329·3
192·9	191·4	189·9	195·9	190·3	161·5	203·8	187·0	190·3	181·3	183·9	4473·4
195·6	197·1	194·8	193·3	194·0	191·0	192·2	182·0[a]	Sunday.			4456·0
—	—	—	—	—	—	—	—	205·5	168·6	172·8	
191·3	191·5	179·6	180·9	191·7	194·1	195·4	192·6	178·7	176·8	178·6	4328·9
195·4	183·8	180·0	181·7	174·2	177·1	175·7	177·4	167·4	167·5	169·6	4294·4
163·1	165·3	172·8	166·9	165·2	169·2	167·4	151·5	149·2	153·2	153·5[a]	3964·0
166·4	165·5	167·8	170·2	168·0	169·4	171·3	163·6	162·7	165·1	162·5	3863·5
170·0	170·0	167·6	169·1	170·6	171·5	169·5	170·2[a]	167·7	165·6	166·6	4018·8
166·8	166·3	166·5	165·4	161·5	161·7	162·0	161·1	Sunday.			3977·0
—	—	—	—	—	—	—	—	169·4	162·8	162·8	
173·6	175·2	173·5[a]	174·4	179·0	173·4	170·9	169·0	165·9[a]	166·9	163·4	4079·5
174·4	173·8	174·6	170·2	169·1	167·4	169·7	152·3	170·0	116·4	101·2	3918·1
177·4	177·5	178·1	176·4	176·4	175·9	172·0	170·4	169·2	171·2	171·1	4198·1
171·9	168·5	167·1	168·4	173·8	180·6	179·3	179·6	165·3	167·9	170·8	4074·3
170·7	170·2	169·0	167·9	168·6	163·4	163·9	165·9	161·7[a]	164·2	164·4	4062·4
173·9	175·0	169·8	174·8	174·5	174·4	175·6	174·8	Sunday.			—
—	—	—	—	—	—	—	—	Christmas Day.			—
—	—	—	—	—	—	—	—	167·8	168·0	168·0	—
182·6	185·3	182·6	178·4	179·3	187·1	176·4	169·5	150·6[a]	106·0[a]	77·5[a]	4115·7
184·0	183·0	182·0	184·4	186·0	105·0	206·5	195·3	186·6	179·3	174·4	4298·3
192·1	190·4	183·8	184·2	183·9	183·1	179·6	181·1	176·8	156·0	165·2	4281·1
172·4	180·3	173·6	178·6	172·8	168·7	171·5	173·2	140·0	73·8	152·0	4023·5
162·8	162·1	158·6	157·7	160·5	156·1	166·8	154·2	Sunday.			—
											96468·2
—	—	—	—	—	—*	—	—	New Year's Day.			—
478·4	4455·3	4417·1	4435·0	4456·1	4454·0	4468·1	4364·3	4209·1	3922·1	3931·3	103466·8
79·14	178·21	176·69	177·40	178·24	178·16	178·72	174·57	168·36	156·88	157·25	4138·65
22·26	21·33	19·80	20·52	21·36	21·28	21·84	17·69	11·48	0·00	0·37	—
007902	·007572	·007629	·007235	·007563	·007554	·007753	·006290	·004075	·000000	·000131	—

Increasing numbers denote increase of Horizontal Force.

Lake Athabasca—*continued.*

Abstract of Hourly Observations made during the months of January and February 1

Date. Gött. Mean. Time.	Horizontal Force Magnetometer.										
	Noon.	1.	2.	3.	4.	5.	6.	7.	8.	9.	10.
1	—	—	—	—	At Fort Chipewyan.			—	—	—	—
2[b]	160·8	157·8	80·8	148·4	177·1	173·0	171·1	162·4	158·9	165·9	173·4
3	144·7	148·5[c]	156·4	160·9	158·5	159·6	153·5	156·1	159·5	155·9	155·7
4	159·4	160·6	159·1[d]	155·7	167·5	166·7	165·1	166·5	162·7	161·7	158·8
5	*107·6*	*71·6*	*165·8*	178·4	175·6	166·3	159·9	162·2	*133·5*	167·2	146·0
6	*89·8*	*108·1*	*196·4*	94·5	152·0	155·8	139·5	147·6	148·8	142·5	142·1
7	—	—	—	—	At Fort Chipewyan.			—	—	—	—
8	121·4	121·8	125·7	128·2	129·3	133·9	125·4	133·0	149·9	122·6	134·2
9[b]	125·7	133·9	129·8	138·9	145·9	140·1	128·8	143·1	143·9	140·5	142·5
10	128·5	127·9	126·9	130·9	135·1	140·4	140·2	138·1	136·5	143·0	138·1
11	141·7	134·1	112·4	144·0	146·4	144·5	141·5	136·8	138·1	144·9	146·1
12	146·9	148·9	152·4	150·4	150·0	156·2	144·0	142·4	142·8	144·2	143·0
13	142·7	145·0	143·8	146·1	146·3	144·7	143·7	141·5	154·2	142·6	144·4
14	—	—	—	—	At Fort Chipewyan.			—	—	—	—
15	143·6	142·5	142·9	141·5	142·3	141·6	140·0	140·5	138·2	140·4	141·5
16	156·3[a]	157·0[a]	156·8	156·4	155·5	155·3	153·4	152·5	148·8	148·6	151·2
17	*128·5*	135·3	147·8	150·4	147·7	145·7	144·3	142·4	139·3	139·7	144·4
18	137·3[a]	137·9	136·7	131·6	123·5	127·7	137·9	132·8	134·9	135·7	139·9
19	134·3	135·5	132·7	131·0	127·7	128·4	128·9	131·9	135·0	134·1	132·3
20	148·9[a]	141·2[a]	137·1	139·3	140·8	143·1	137·2	137·6	150·5	136·4	138·7
21	—	—	—	—	At Fort Chipewyan.			—	—	—	—
22	112·3	112·1	114·1	109·4	110·2	108·4	102·3	120·7	131·7	130·5	134·7
23	138·4	131·3	128·1	125·5	127·3	124·7	124·7	126·3	131·7	132·2	134·7
Term Day. 24	112·6	117·6	115·9	118·5	119·3	118·6	115·9	117·2	120·8	121·4	125·4
Term Day. 25	46·4[a]	24·5	66·9	129·3	140·4	141·4	134·5	130·8	139·0	142·3	145·1
26	127·4	130·0	124·0	129·3	125·5	125·3	134·4	127·8	127·4	125·7	128·4
27	[h]115·6[a]	111·3[a]	116·3	123·2	118·1	116·4	115·8	112·7	114·6	117·1	113·3
28	—	—	—	—	At Fort Chipewyan.			—	—	—	—
29	119·1	136·1	135·0	136·0	137·0	124·0	135·0	134·6	133·2	128·4	131·8
30	139·6	148·3	149·0	150·3	147·3	146·4	146·4	147·4	151·8	149·7	148·6
31	138·6[a]	140·1	138·4	139·2	137·6	133·2	133·0	139·7	138·5	133·8	135·0
Sums -	3016·6	3062·2	3210·1	3300·0	3360·9	3353·3	3301·5	3319·1	3361·4	3340·6	3353·4
Means -	125·69	127·59	133·75	137·50	140·04	139·72	137·56	138.30	140·06	139·19	139·72
Differences	8·41	10·31	16·47	20·22	22·76	22·44	20·28	21·02	22·78	21·91	22·44
$\frac{\Delta X}{X}$ }	·002986	·003660	·005847	·007178	·008080	·007966	·007199	·007462	·008087	·007778	00796
Jan. 31	—	—	[e] —	—	—	—	—	—	—	—	—
Feb. 1	*49·7*	*8·4*	*82·8*	*109·7*	120·5	110·1	132·4	141·5	139·5	125·0	135·6
2	133·7	130·0	132·1	132·5	117·3	112·4	*67·0*	*125·5*	139·2	149·3	152·0
3	145·0	133·8	137·8	138·2	137·5	138·6	132·0	118·2	144·4	139·3	151·0
4	—	—	—	—	At Fort Chipewyan.			—	—	—	—
5	*−18·0*	*69·0*	160·9	124·4	109·0	125·4	150·4	135·8	159·2	155·0	150·3
6	67·4	151·5	140·3	142·4	145·9	137·4	134·6	143·1	144·4	117·8	146·0
7	*145·0*	139·2	134·1	142·8	148·9	148·5	145·1	146·3	145·9	149·1	178·9
8	*74·5*	111·5[a]	123·8	133·1	122·2	*79·8*	136·9	154·7	153·9	155·3	165·8
9	162·1	164·3	162·3	162·7	161·4	161·3	161·8	161·0	161·9	163·3	163·(
10	162·5	155·8	153·6	153·9	152·6	153·1	153·3	155·6	151·1	148·4	162·4
11	—	—	—	—	At Fort Chipewyan.			—	—	—	—
12	140·8	142·5	143·7	141·4	142·8	139·4	140·8	144·7	141·4	146·4	143·
13	134·5	149·1	150·8	152·5	140·4	138·7	136·6	133·3	139·7	139·7	139·
14	150·2	149·7	148·1	147·2	144·7	145·2	146·3	147·2	147·8	148·3	151·
15	126·3	143·2	157·8	151·7	147·6	143·9	149·0	156·3	154·0	154·6	157·
16	174·6	172·4	162·8	164·2	180·4	174·2	173·5	163·6	176·4	178·3	170·
17	164·9	177·4	179·2	184·2	180·3	175·6	177·0	180·5	182·1	185·9	185·
18	—	—	—	—	At Fort Chipewyan.			—	—	—	—
19	179·0	180·2	183·3	183·2	183·6	182·7	182·4	181·6	182·6	184·9	185·
20	184·2	183·8	184·8	184·0	181·0	180·2	179·8	173·3	183·9	187·5	191·
21	158·2	166·4	169·4	175·6	181·0	179·1	178·2	176·2	177·0	176·9	179·
22	200·6	192·4	190·0	192·9	192·3	187·1	184·5	184·4	186·3	186·6	193·
Term Day. 23	197·3	194·9	196·6	197·6	200·2	199·6	197·3	197·6	197·9	199·4	198·
Term Day. 24	188·6	187·9	185·5	186·0	182·3	183·6	183·2	181·1	185·5	184·3	183·
25	—	—	—	—	At Fort Chipewyan.			—	—	—	—
26	176·7	162·6	157·3	179·9	181·9	181·8	179·6	177·6	181·4	178·9	187·
27	163·6	174·2[a]	175·1	180·4	176·0	174·4	174·2	171·4	173·6	173·2	176·
28	—	—	—	—	—	—	—	—	—	—	—
Sums -	3258·4	3428·4	3612·1	3660·5	3629·8	3551·6	3595·9	3655·5	3749·1	3727·4	3855
Means -	141·67	149·06	157·05	159·15	157·82	154·42	156·34	158·93	163·00	162·06	167·
Differences	0·00	7·39	15·38	17·48	16·15	12·75	14·67	17·26	21·33	20·61	25·
$\frac{\Delta X}{X}$ }	·000000	·002623	·005460	·006205	·005733	·004526	·005208	·006127	·007572	·007317	·009

[a] Visible aurora. [b] Omitted from the Means, as being incomplete. [c] Seve
[d] Fifteen minutes late. [e] At 21h 30m 133·7 [f] Seventeen minutes late. [g] Te
[h] Twenty minutes late. [k] Nine minutes late. [m] Included with February

Lake Athabasca—*continued*.

Abstract of Hourly Observations made during the months of January and February 1844.

Horizontal Force Magnetometer.

13.	14.	15.	16.	17.	18.	19.	20.	21.	22.	23.	Sums.	Means.	Fortnightly Means.
—	—	—	—	—	—	—	—	148·5	154·3	155·8	—	—	
173·1	—	174·7	168·5	171·8	165·3	164·2	158·6	159·3	159·7	156·4	—	162·01	
159·1	157·5	158·1	159·1	159·0	159·1	159·1	158·3	155·8	152·7	156·3	3756·8	156·53	
175·2	179·0	177·7	*212·2*	*205·0*	*196·5*	*224·0*	178·1	*54·7*	*117·0*	*−37·5*	3802·6	158·44	
161·0	158·1	159·3	158·7	154·6	156·2	182·7	151·9	157·8	156·6	*−49·0*	3506·1	146·09	
155·1	145·1	161·0	137·4	150·8	142·8	146·5	89·9		Sunday.		3152·5	131·35	
—	—	—	—	—	—	—	—	127·3	122·8	115·5			
152·2	144·1	*147·3*[a]	143·3	151·9	153·5	147·3	*139·3*[a]	*133·4*	154·1	137·9	3329·1	138·71	146·70
148·4	150·5	155·5	143·2[a]	145·5[a]	144·1	146·2	147·8	—[e]	131·3	125·7	—	140·71	
139·9	148·5	149·7	147·4	140·7	141·4	146·9	137·5	141·9	137·1	138·1	3368·6	140·36	
151·4	147·4	153·3	150·7	149·1	150·4	150·0	149·8	144·7	143·0	148·4	3461·9	144·25	
152·5	155·3	154·3	155·4	155·0	159·8	163·6	157·5	145·0	143·2	141·6	3607·2	150·30	
147·7	147·6	147·1	144·2	146·6	147·1	141·6	143·8		Sunday.		3468·9	144·54	
—	—	—	—	—	—	—	—	138·3	140·4	140·7			
144·6	147·6	150·3	158·2	153·8	151·0	157·3	159·2	158·4[f]	157·8	157·8	3531·7	147·15	
153·6	152·8	150·2	152·7	150·6	151·1	152·4	144·8[a]	154·0[a]	152·5	149·4	3661·4	152·56	
146·3	145·6	144·2	140·5	141·4	143·9	141·5	162·0	145·6[a]	142·5[a]	138·7	3440·5	143·35	
140·8	145·5	135·1	140·7	140·3	138·7[a]	135·8	138·5	124·9	132·5	137·5	3264·1	136·00	
141·2	147·8	155·4[a]	153·4[a]	152·4[a]	154·9[a]	155·8[a]	160·3[a]	104·5[a]	121·1[a]	143·5[a]	3304·5	137·69	
131·4	131·2	134·1	134·9	133·4	131·1	132·6[a]	128·5[a]		Sunday.		3223·9	134·33	
—	—	—	—	—	—	—	—	110·0[a]	118·1[a]	125·3[a]			
134·7	132·1	131·3	137·1	142·1[g]	147·1	142·8	156·8[a]	157·1[a]	145·8[a]	152·1[a]	3135·5	130·65	130·94
133·1	133·0	131·1	127·1	128·5	122·4	124·8	126·5	121·1	120·8	122·5	3083·1	128·46	
128·8	127·8	139·9[a]	147·3[a]	74·0[a]	126·5	112·0	52·6	51·7[a]	59·2[a]	35·6[a]	2609·8	108·74	
135·7	132·4	127·2	128·4	120·1	119·1	125·1	128·5	123·0	123·8	126·0	2903·3	120·97	
127·4	125·6	125·7	127·7	119·8[a]	131·8	124·9	122·8[a]	120·1	[a]*69·5*[h]	90·8[a]	2952·8	123·03	
118·1	122·7	121·1	125·5	126·0	127·7	135·5	108·3		Sunday.		2884·3	120·18	
—	—	—	—	—	—	—	—	134·0	128·4	124·7[a]			
138·0	137·8	141·0	140·8	141·0	140·5	145·3	145·5	143·6	143·8	106·3	3246·8	135·28	
147·4	151·5	150·3	140·6	151·1	159·4[a]	157·6	147·2	117·7[a]	137·3	138·6	3527·1	146·96	
140·4	139·0	137·7	139·2	143·9	157·4[a]	137·2	139·9	159·5[m]	144·3[m]	112·2[m]	3353·2	139·72	
3455·6	3455·0	3482·4	3506·5	3428·0	3511·4	3545·2	3320·5	3112·8	3187·1	2778·4	79636·5	—	
143·98	143·96	145·10	146·10	142·83	146·31	147·72	138·35	129·70	132·80	115·76	138·26	—	
26·70	26·68	27·82	28·82	25·55	29·03	30·44	21·07	12·42	15·52	0·00	—	—	
·009478	·009471	·009876	·010231	·009070	·010306	·010806	·007480	·004409	·005510	·000000	—	—	
—	—	—	—	—	—	—	—	159·5	144·3	112·2	—	—	
147·9	142·9	141·0	137·0	155·8[a]	163·0	*185·5*[c]	*171·0*	147·0	139·4	136·1	3085·9	128·58	
154·4	154·5	165·4	161·7	*155·7*	*170·4*	*204·8*	*201·6*	173·7	167·3	148·1	3556·3	148·18	
158·0	168·6	173·7	156·0	143·1	145·1	151·9	159·3		Sunday.		3482·2	145·09	
—	—	—	—	—	—	—	—	140·8	149·4	140·6			
176·3	178·4[a]	180·4[a]	*193·7*[a]	*214·8*[a]	*195·4*	178·4	153·0	*91·6*	*80·7*	133·4	3395·6	141·48	146·08
150·3	146·6	147·9	144·8	154·9	162·6	*198·2*	164·2	150·0	155·6	148·3	3428·8	142·87	
156·9	167·1	156·6	172·7	162·4	154·8	165·7	166·3	183·7	162·2	122·6	3703·3	154·30	
175·7	175·2	173·0	162·5	163·5	165·8	161·3	162·4	154·1	163·3	164·1	3567·9	148·66	
162·1	162·3	162·2	158·3	162·9	172·9	164·5	161·2	170·3	131·3	167·1	3886·1	161·92	
158·6	163·0	157·0	158·4	156·8	158·1	156·3	108·1		Sunday.		3641·6	151·73	
—	—	—	—	—	—	—	—	117·2[a]	133·8	128·1			
130·5	149·8	151·0	149·5	142·4	153·6[a]	155·4	158·4	135·9[a]	96·0[a]	140·0[a]	3442·2	143·42	
145·9	145·5	147·5	146·4	149·1	149·3	154·6[a]	159·3[a]	149·3	150·4	150·8	3493·9	145·58	
148·6	149·3	148·2	148·6	149·3	161·8	172·6	166·4	169·3	147·0	150·0	3633·9	151·41	
172·7	173·3	177·9	181·4	181·6	182·1	182·6	183·6[a]	179·1	177·0	174·5	3936·0	164·00	
144·8	187·5	189·2	185·6	193·5[a]	200·2[a]	184·1	178·5	*175·5*[a]	157·9[a]	171·7[a]	4260·2	177·51	
197·6	201·9	202·6[a]	204·6[a]	189·9	189·7	183·0	164·6		Sunday.		4459·6	185·82	
—	—	—	—	—	—	—	—	190·2	191·3	183·7			
190·7	189·3	189·3	184·6	187·1	187·6	186·6	186·8	186·8	175·8	179·2	4428·2	184·51	
196·6	196·9	197·0	192·0	192·6	192·6	191·4	188·7[a]	192·5[a]	181·7	172·6	4495·8	187·32	177·23
180·1	182·2	186·5	189·7	189·7	196·1	195·5	163·0[a]	186·7[a]	180·9[a]	178·6[a]	4309·7	179·57	
198·1	201·1	196·2	196·6	196·4	195·6	198·7	199·7	186·6	194·6	202·2	4643·9	193·50	
204·0	204·2	205·0	201·1	199·8	196·3	195·8	194·6	196·7	193·1	189·1	4756·8	198·20	
181·4	182·4	180·1	181·3	181·0	178·3	176·7	175·1		Sunday.		4361·8	181·74	
—	—	—	—	—	—	—	—	176·2	176·8	176·3			
190·4	190·5	192·3	191·9	195·9	193·9	198·3	202·9	191·7	175·0	*106·3*	4362·1	181·75	
175·8	173·4	174·1	170·9	167·0	168·5	172·2	169·9	—	—	—	—	172·91	
—	—	—	—[a]	—[a]	—	—[a]	—	—[a]	—[a]	—	—	—	
3957·4	3985·9	3994·1	3969·9	3985·2	4035·7	4044·1	3938·6	3734·4	3625·4	3578·2	90385·1	—	
172·06	173·30	173·66	172·60	173·27	175·47	175·83	171·24	162·37	157·63	155·57	3929·79	163·74	
30·39	31·63	31·99	30·93	31·60	33·80	34·16	29·57	20·70	15·96	13·90	—	—	
·010788	·011[illegible]	·011305	·010850	·011218	·011[illegible]	·012127	·010[illegible]	·007345	·00[illegible]	·00[illegible]	—	—	

[a] Twelve minutes late. [e] Three minutes late.

Increasing numbers denote increase of Horizontal Force.

FORT SIMPSON.

Abstract of Hourly Observations made during the months of April and May 184

Date. Gött. Mean Time.	Bifilar Magnetometer.										
	Noon.	1.	2.	3.	4.	5.	6.	7.	8.	9.	10.
	—	—	—	—	—	—	—	—	—	—	—
1	112·4	180·9	140·2	146·9	203·2	208·4	185·6	210·6	219·2	221·5	246·4
2	235·6	234·5	231·0	210·5	191·0	213·1	193·7	248·9	230·2	244·9	233·6
3	233·6	250·8	248·6	236·1	185·9	115·9	182·0	211·6	248·2	245·8	252·5
4	124·7	176·5	222·5	224·8	219·8	255·5	243·0	241·5	265·5	241·8	243·1
5	—	—	—	—	—	—	—	—	—	—	—
6	240·7	237·8	248·1	248·3	244·0	228·8	243·9	254·3	246·5	246·3	237·4
7	—	—	—	—	—	—	—	—	—	—	—
8	247·1	—	245·8	247·2	242·1	248·7	248·9	245·9	245·0	241·3	241·0
9	249·9	253·1	247·0	251·5	246·7	247·2	251·8	245·7	236·4	249·5	246·7
10	229·8	203·8	218·0	187·6	223·0	242·5	238·6	234·8	234·9	232·1	229·0
11	243·2[a]	221·3	188·9	229·7	243·6	237·1	239·2	233·7	229·9	228·4	226·6
12	221·3	225·5	238·2	236·8	237·4	233·8	228·3	228·8	236·8	220·8	220·0
13	225·8	233·5	230·0	233·5	232·0	228·0	237·5	224·0	218·2	216·8	214·3
	214·67	221·77	223·03	222·02	218·97	223·55	225·59	234·05	241·89	235·38	235·51
14	—	—	—	—	—	—	—	—	—	—	—
15	270·6	187·4	134·5	215·6	279·8	291·9	282·0	275·5	275·6	276·0	276·5
16	—	—	—	264·6	252·6	252·4	280·2	271·8	269·3	248·0	269·3
17	189·0	−60·8[b]	98·6	185·0	151·4	108·5	131·8	226·7	254·6	294·1	316·7
18	—	—	276·6	277·0	273·8	266·5	272·0	290·5	272·4	271·9	259·3
19	238·5	253·0	262·5	222·0	270·1	263·6	256·5	263·8	261·8	260·8	260·6
20	191·1	199·6	223·4	251·8	266·4	270·8	270·8	271·5	263·8	260·8	263·0
21	—	—	—	—	—	—	—	—	—	—	—
22	221·1	—	243·4	253·1	268·3	269·1	265·4	268·8	277·4	265·3	260·5
23	178·3	216·6	229·1	204·9	252·9	270·8	271·6	273·4	271·5	266·6	268·
24	267·3	—	272·8	279·8	275·6	275·5	271·4	269·2	269·8	268·6	269·
25	182·7	207·3	90·4	86·6	224·0	255·8	243·6	226·6	286·0	278·2	257·
26	96·8	98·0	215·3	—[c]	187·0	122·8	219·2	259·5	263·4	271·3	269·
27	187·8	247·8	238·3	250·1	291·8	270·0	232·8	250·2	263·5	268·5	263·
28	—	—	—	—	—	—	—	—	—	—	—
29	251·0	186·0	137·0	174·8	162·0	210·8	255·7	261·2	258·9	261·5	253·
30	148·0	178·6	250·6	260·0	256·8	253·9	258·8	237·8	250·1	251·5	249·
1	219·5	248·2	236·9	284·5	260·5	258·5	255·5	256·4	260·6	265·2	261·
2	—	207·8	259·2	267·2	277·6	250·6	268·8	270·5	266·0	260·8	251·
3	262·0	256·7	266·5	232·1	†43·0	191·6	220·4	268·6	259·8	261·3	260
4	263·9	256·7	258·3	249·7	266·2	265·8	267·4	260·4	257·4	253·4	256
5	—	—	—	—	—	—	—	—	—	—	—
6	249·6	229·6	262·6	269·0	243·7	235·3	243·2	262·0	262·9	260·9	264
7	237·8	248·9	225·6	246·9	234·3	240·9	260·0	263·7	261·1	255·6	256
8	247·7	—[d]	182·1	176·1	231·6	227·3	245·6	253·3	250·6	253·0	259
9	254·5	253·4	201·1	171·2	162·1	221·6	249·3	271·0	269·0	262·0	259
10	258·9	261·6	262·1	248·5	248·1	192·9	259·1	263·3	262·0	259·4	26
11	—	238·4	241·3	238·4	256·8	263·9	259·9	261·0	258·5	260·5	26
12	—	—	—	—	—	—	—	—	—	—	
13	264·9	263·4	243·0	242·1	269·6	267·0	260·5	259·5	259·8	260·8	26
14	223·4	246·1	243·4	257·1	286·9	278·2	264·5	267·3	264·2	264·8	26
15	268·3	268·1	252·1	250·8	254·8	256·7	243·0	259·8	251·3	261·4	26
16	212·1	215·8	219·3	228·1	236·3	250·1	247·9	254·6	254·4	258·5	25
17	247·5	247·9	256·1	259·3	259·3	257·3	257·4	250·4	249·4	242·9	24
18	252·1	242·4	249·5	258·6	254·7	251·9	266·6	248·9	252·6	249·3	25
19	—	—	—	—	—	—	—	—	—	—	
20	255·6	264·9	262·1	261·9	273·1	262·8	256·6	255·3	252·0	249·1	2
21	229·3	288·0	224·0	231·3	233·6	236·4	236·5	248·3	244·5	242·1	2
22	—	187·0	112·8	196·6	211·0	286·3	274·6	257·3	249·8	233·1	2
23	98·9	168·1	209·2	208·8	204·1	262·2	264·8	257·0	272·7	265·4	2
24	212·1	218·9[e]	160·8	215·3	237·2	237·4	259·5	251·4	250·9	248·9	2
25	—	—	—	—	—	—	—	—	—	—	
26	—	—	—	—	—	—	—	—	—	—	
27	—	—	—	—	—	—	—	—	—	—	
28	—	—	—	—	—	—	—	—	—	—	
29	—	—	—	—	—	—	—	—	—	—	
30	—	—	—	—	—	—	—	—	—	—	
31	—	—	—	—	—	—	—	—	—	—	
Mean 15th April to 24th May.	217·50	212·92	220·97	231·20	238·77	244·45	252·94	259·33	261·33	260·33	2

[a] Visible aurora. [b] Beyond the zero of the scale. [c] At 3h
[d] At 1h 30m 193·0. [e] Taken at 1h 20m.

FORT SIMPSON.

Abstract of Hourly Observations made during the months of April and May 1844.

Horizontal Force Magnetometer.

13.	14.	15.	16.	17.	18.	19.	20.	21.	22.	23.	Sums.	Means.	
—	—	—	—	—	—	—	—	217·7	163·5	233·2	614·4	—	
245·6	262·0	253·7	255·3	266·5[a]	253·6[a]	248·9	240·3	196·3	160·0	199·4	5120·6	213·36	
235·1	244·3	245·9	260·7	248·7[a]	276·3[a]	234·7	241·6[a]	238·5[a]	199·3[a]	79·3	5490·1	228·75	
242·0	254·3	263·6	276·8	261·3[a]	278·6[a]	262·8	251·2	250·5	229·1	214·3	5625·1	234·38	
254·7	249·3	257·3	253·5	300·8	294·9	281·9[a]	273·7	—	—	—	5837·9	243·25	
—	—	—	—	—	—[a]	—	—	237·7[a]	244·5	236·3			
522·2	263·8	233·0	286·1	291·4	261·1[a]	272·2	265·0[a]	—	—	—	6130·6	255·44	
—	—	—	—	—	—	—[a]	—	237·6	235·4	244·4[a]			
249·6	254·6	250.4	248·7	256·3	253·0	260·4	261·5	262·8	253·3	254·3	—	250·01	
246·0	249·5	251·3	256·9	250·1	249·9	248·0	245·5[a]	241·6[a]	195·6[a]	187·8	5897·9	245·75	
219·0	240·8	234·0	232·9	239·9	236·7[a]	228·0	247·6[a]	†16·6[a]	186·6	242·2	5356·4	223·18	
219·8	224·3	229·4	226·2	244·8	243·2	237·0[a]	227·0	220·2	231·3	230·9[a]	5512·0	229·67	
223·6	229·3	226·3	221·1	216·5	226·1	221·9	216·7	231·8	210·4	221·6	5417·1	225·71	
222·7	222·5	223·3	229·3	232·1	233·1	229·8[a]	223·5[a]	—	—	—	—	226·09	
237·30	244·97	247·03	249·86	255·31	256·95	252·33	244·87	220·95	209·91	213·06	—	—	
—	—	—	—	—	—	—	—	274·2[a]	279·8	287·6	841·6	—	
234·4	285·3	289·6	279·0	271·6	282·8	276·5[a]	216·6[a]	272·6	288·6	279·8	6287·3	261·97	
298·2	354·6	377·6	344·3	286·6[a]	193·5[a]	271·4[a]	208·7[a]	137·8[a]	139·0[a]	188·6[a]	—	255·88	
343·9	294·9	262·7	262·1	268·5	270·2	274·0	236·4	280·9	281·6[a]	285·5	5575·9	232·33	
269·8	281·7	262·7	264·9	262·4	268·8	272·2	269·4[a]	268·9[a]	265·3[a]	255·1	—	270·23	
265·6	259·1	260·6	259·6	265·6	269·3	272·5	269·1[a]	260·8[a]	244·7[a]	161·5	6129·6	255·40	
258·0	263·3	263·3	262·9	262·4	269·8	271·5	267·2[a]	—	—	—	6136·4	255·68	
—	—	—	—	—	—	—	—	264·5	251·0[a]	244·0[a]			
270·6	269·4	265·7	268·0	266·9	268·3	268·6	270·9	272·5	267·8	230·5	—	261·98	
265·9	270·5	273·3	275·5	278·0	282·4	274·1	266·1	246·9	278·5	254·5	6203·2	258·47	
273·0	275·1	270·9	276·1	286·2	292·7	304·2	288·6	284·3	266·6	264·9	—	275·84	
288·8	266·1	257·3	298·2	283·5	265·3	269·9	239·4	206·7	255·1[a]	244·9	5821·4	242·56	
296·3	297·1	280·3	265·0	277·4	200·9	196·3	290·9	232·1	255·8	228·3		230·65	
287·6	283·2	276·6	309·2	287·6	260·4	289·2	170·9	—	—	—	6154·6	256·44	
—	—	—	—	—	—	—	—	230·3	296·0	240·5			
259·3	260·3	285·1	273·6	262·5	281·5	268·3	281·3	261·6	201·1	256·3	5870·2	244·59	
280·4	326·4	324·9	306·7	306·9	278·6	265·3	246·1	80·0	118·2	169·7	5845·0	243·54	
288·0	280·8	287·3	278·6	288·2	265·3	242·5	294·7	279·7	263·7	274·3	6402·4	266·77	
303·6	296·0	336·4	297·6	309·2	298·6	230·6[a]	252·4	259·9	226·9	230·3	—	268·94	
280·3	264·1	264·2	271·2	266·4	261·6	267·6	269·4	267·6	267·3	264·6	5969·2	248·72	
268·7	267·0	267·4	269·2	278·6	285·7	286·7[a]	252·6	—	—	—	6247·6	260·32	
—	—	—	—	—	—	—	—	238·5	215·1[a]	240·3			
262·7	258·6	268·9	277·3	277·6	270·6	271·3	254·3	276·7	280·8	259·8	6264·8	261·03	
258·6	269·2	275·8	272·5	300·3	272·4	275·1	282·9	254·7	167·	257·3	6146·8	256·12	
306·7	298·2	294·3	294·4	284·4	298·3	286·5	238·1	226·3	241·	241·4	—	255·58	
270·3	259·5	276·1	263·9	264·4	262·6	255·2	252·7	262·1	254·	253·8	5998·0	249·92	
258·6	260·6	281·8	285·7	264·5	249·5	280·1	268·4	267·7	253·	259·4	6245·7	260·24	
264·3	262·9	267·9	261·3	272·1	262·3	266·3	267·9	—	—	—	—	260·79	
—	—	—	—	—	—	—	—	274·1	268·6	267·6			
269·9	269·7	271·3	268·2	292·5	325·6	305·1	290·0	273·7	247·8	270·6	6465·9	269·41	257·57
307·3	325·7	281·7	284·0	282·1	280·2	273·4	260·4	241·4	259·	270·1	6458·5	269·10	
291·8	266·1	268·3	268·4	257·8	263·2	262·4	256·7	262·8	240·	236·9	6242·1	260·09	
251·8	257·6	255·6	250·9	248·6	256·8	261·9	267·5	245·9	240·	253·9	5932·6	247·19	
258·4	248·4	245·2	250·9	248·1	252·4	233·0	262·4	246·6	248·	241·9	6043·0	251·79	
269·9	251·8	249·1	257·1	274·9	283·3	202·3	258·9	—	—	—	6151·5	256·31	
—	—	—	—	—	—	—	—	206·4	245·4	246·6			
250·3	251·6	250·6	250·9	251·1	254·8	259·7	259·8	238·1	250·1	238·7	6102·6	254·27	
278·9	267·9	266·1	264·7	259·4	251·4	249·9	254·9	251·7	259·2	254·8	6053·6	252·23	
158·9	325·1	293·5	267·5	225·6	233·4	222·7	282·4	246·2	233·2	140·5	—	235·38	
260·7	251·9	258·6	252·9	261·7	257·7	268·8	251·4	242·8	248·3	247·3	5761·0	240·04	
279·5	285·3	282·8	267·1	260·3	260·7	206·7	209·2	239·8	243·4	244·9	5916·2	246·51	
—	—	—	—	—	—	—	—	—	—	—	—	—	
—	—	—	—	—	—	—	—	—	—	—	—	—	
—	—	—	—	—	—	—	—	—	—	—	—	—	
—	—	—	—	—	—	—	—	—	—	—	—	—	
—	—	—	—	—	—	—	—	—	—	—	—	—	243·54
—	—	—	—	—	—	—	—	—	—	—	—	—	
—	—	—	—	—	—	—	—	—	—	—	—	—	
275·86	276·94	277·01	274·58	272·45	267·20	266·60	258·07	248·30	245·34	245·18	—	—	

Increasing numbers denote an increase of Horizontal Force.

LAKE ATHABASCA.

Abstract of Hourly Observations made during the month of October 1843.

Date. Gött. Mean Time.	Induction Inclinometer.										
	Noon.	1.	2.	3.	4.	5.	6.	7.	8.	9.	10.
1	—	—	—	—	—	—	—	—	—	—	—
2	—	—	—	—	—	—	—	—	—	—	—
3	—	—	—	—	—	—	—	—	—	—	—
4	—	—	—	—	—	—	—	—	—	—	—
5	—	—	—	—	—	—	—	—	—	—	—
6	—	—	—	—	—	—	—	—	—	—	—
7	—	—	—	—	—	—	—	—	—	—	—
8	—	—	—	—	—	—	—	—	—	—	—
9	—	—	—	—	—	—	—	—	—	—	—
10	—	—	—	—	—	—	—	—	—	—	—
11	—	—	—	—	—	—	—	—	—	—	—
12	—	—	—	—	—	—	—	—	—	—	—
13	—	—	—	—	—	—	—	—	—	—	—
14	—	—	—	—	—	—	—	—	—	—	—
15[b]	—	—	—	—	—	—	—	—	—	—	—
16	*335·9*[a]	*151·5*[a]	*146·9*[a]	*158·0*	*180·4*	123·9	86·9	118·8	120·9	133·8	135·7
17	150·8	*244·6*[a]	*227·7*	133·9	143·4	132·8	141·6	117·9	122·0	127·6	130·4
Term Day. 18	152·6	137·0	131·5	131·6	129·2	120 7	118·6	126·5	97·4	113·3	121·[illegible]
19	118·3	115·5	120·2	117·2	116·2	125·9	116·7	118·1	138·4	126·5	126·[illegible]
	—	—	—	—	—	—	—	—	—	—	—
20[b]	265·7	—	232·0	241·4	238·0	239·4	241·1	245·1	245·5	240·2	241·[illegible]
21	233·6	239·2	237·3	337·4	241·6	245·1	185·0	212·7	231·4	232·6	231·[illegible]
22	—	—	—	—	—	—	—	—	—	—	—
23	246·7	247·0	248·1	252·4	249·0	243·3	246·1	252·4	269·5	252·1	258·[illegible]
24	271·4	*333·2*	*230·7*	257·1	241·4	253·9	245·1	248·6	247·5	249·6	244·[illegible]
25	253·2	248·3	252·5	259·5	255·3	256·4	250·4	254·9	252·0	248·9	249·[illegible]
26	*391·5*[a]	*281·8*[a]	257·6	249·8	261·9	264·7	254·1	267·2	254·2	254·8	265·[illegible]
27	*354·5*[a]	*414·7*[a]	261·2	265·3	250·3	260·7	253·0	249·3	260·5	260·2	252·1
28	307·7	271·4	261·3	262·4	255·2	262·6	278·9	269·9	278·0	276·2	270·4
29	—	—	—	—	—	—	—	—	—	—	—
30	294·1	337·7	*306·8*	*273·9*	275·3	266·0	269·7	270·5	269·6	277·8	277·1
31	*337·6*	*341·0*	*276·2*	*255·7*	255·7	257·3	266·3	269·2	268·2	264·6	264·[illegible]
Sums -	3703·6	3352·9	3281·0	3095·6	3041·9	3052·7	2953·5	3021·1	3050·1	3058·0	3068·[illegible]
Means -	264·54	257·92	234·36	221·11	217·28	218·05	210·97	215·79	217·86	218·43	219·[illegible]

[a] Visible aurora.

[b] Not included in the Mean at the foot, being impe

LAKE ATHABASCA.

Abstract of Hourly Observations made during the month of October 1843.

* Means by the observations forming complete 8-hourly services.

Lake Athabasca—*continued.*

Abstract of Hourly Observations made during the months of November and Decemb

Date. Gött. Mean Time.	Induction Inclinometer.										
	Noon.	1.	2.	3.	4.	5.	6.	7.	8.	9.	10.
1	106·0	115·2	133·3	97·1	98·1	91·5	98·1	100·3	101·8	105·8	105·2
2	100·4	93·7	93·8	97·0	94·8	102·3	106·4	107·3	107·4	110·2	99·6
3	*222·0*	124·6	106·9	103·9[d]	99·9	104·5	106·7	110·8	114·3	113·1	109·2
4[b]	—	—	—	99·8	102·9	103·1	104·0	103·5	106·0	109·5	114·2
5	—	—	—	—	—	—	—	—	—	—	—
6	*223·4*	*127·9*	122·9	111·2	102·6	118·9	109·8	115·3	113·0	109·0	110·6
7	113·1	104·7	103·4	110·4	119·3	121·6	115·5	125·9	113·4	101·2	110·8
8	120·4	121·8	114·3	129·9	122·2	110·1	111·8	117·3	114·7	122·5	129·8
9	*135·9*	91·8	102·5	106·1	107·3	115·9	120·1	120·5	122·2	122·1	118·9
10	*113·9*	109·2	119·4	121·6	124·2	120·2	114·5	115·7	119·3	133·4	124·0
11	144·1	125·4	109·0	109·8	113·2	122·7	128·1	125·2	123·2	121·4	120·1
12	—	—	—	—	—	—	—	—	—	—	—
13	123·8	113·9	120·5	138·5	*243·9*	*156·4*	*138·7*	104·9	119·6	121·8	118·
14	117·9	119·9	134·5	118·2	124·0	112·5	119·1	121·5	119·4	124·4	123·
15	123·9	136·6	123·8	151·1	109·5	115·9	120·4	119·3	118·2	121·6	116·
16	118·4	121·1	117·0	117·2	116·1	119·7	127·1	132·2	123·9	132·4	134·
17	122·3	123·8	128·2	126·9	123·9	121·8	125·8	122·0	123·8	119·5	118·
18	· 124·7	122·7	121·7	121·8	121·5	122·0	131·3	131·4	125·3	124·2	124·
19	—	—	—	—	—	—	—	—	—	—	—
20	123·5	132·4	139·2	132·2	131·9	123·1	126·5	124·1	125·7	129·1	130·
21	130·9	126·4	124·7	127·1	123·1	126·7	136·4	137·7	137·9	142·7	133·
Term Day. 22	131·7	146·7	136·6	147·9	130·0	137·8	136·2	132·3	129·4	128·7	127·
23	127·5	124·1[a]	129·5[a]	126·6	126·8	127·2	136·5	137·6	135·2	134·7	132·
Term Day. { 24	139·9	*235·5*	144·1	139·0	126·0	130·9	132·6	127·5	135·3	135·6	136·
Term Day. { 25	136·0	136·3	135·8	132·7	133·8	133·6	135·8	138·3	142·0	138·8	134·
26	—	—	—	—	—	—	—	—	—	—	—
27	140·0	151·0	141·3	139·2	137·8	139·5	144·2	141·7	133·7	136·3	137
28	143·7	139·4	139·2	139·5	139·0	136·5	153·6	137·1	141·7	143·4	151
29	143·4	182·2[a]	165·4[a]	159·0	152·2	135·9	135·2	141·2	143·2	146·5	146
30	138·9	139·8	136·5	138·7	134·7	134·7	144·8	141·6	146·8	152·4	148
Sums -	3367·7	3269·1	3148·5	3142·6	3160·8	3086·6	3153·2	3128·7	3135·4	3171·1	3142
Means -	134·71	130·76	125·94	125·70	126·43	123·46	126·13	125·15	125·42	126·84	125·
1	149·7	163·8	145·3	148·9	139·4	139·8	132·0	143·7	135 2	124·3	14
2	*190·3*	*198·0*	*253·4*	*271·7*	*177·3*	138·5	134·1	134·4	150·1	149·6	14
3	—	—	—	—	—	—	—	—	—	—	-
4	157·0	159·0	158·2	158·0	155·5	155·4	155·8	157·1	155·3	155·7	15
5	158·1	157·4	157·0	143·5	149·4	131·9	147·6	150·3	150·1	152·7	15
6	*193·5*	*205·0*	*151·8*	141·9	141·9	140·7	153·9	156·7	159·2	153·2	14
7	161·2	155·6	146·6	151·7	150·1	149·6	154·8	156·7	154·8	154·5	15
8	155·9	151·0	154·3	156·3	158·3	155·5	178·5	178·9	194·0	161·8	11
9	150·8	154·6	181·4	185·5	158·7	146·9	159·9	151·9	155·1	151·0	15
10	—	—	—	—	—	—	—	—	—	—	
11	162·5	185·2	165·9	141·3	173·4	177·7	150·5	150·3	154·0	149·7	16
12	177·6	160·7	165·3	157·2	166·4	171·1	167·5	170·0	170·5	161·1	14
13	162·4	172·0	171·8	170·6	167·1	163·3	178·4	183·6	183·6	183·4	17
14	188·3[a]	188·5[a]	172·0	166·0	174·4	177·3	180·3	174·1	174·9	176·7	17
15	173·8	170·2	175·5	183·6	169·1	160·9	170·0	177·3	167·8	167·8	1
16	172·9	172·6	170·6	166·7	166·8	171·7	179·6	177·6	171·3	169·8	1
17	—	—	—	—	—	—	—	—	—	—	
18	179·3	166·0	166·9	159·9	167·0	153·1	163·0	166·0	170·8	171·0	1
19	164·8	171·2	166·1	167·0	162·7	159·6	150·9	170·6	173·3	170·8	1
Term Day. { 20	*161·8*	*189·4*	172·7	164·5[b]	160·0	158·5	160·9	161·3	157·7	166·7	1
Term Day. { 21	181·0	166·4	154·2	155·5	158·6	157·3	167·5	171·4	164·1	164·7	1
22	163·2	160·6	160·4	167·0	162·0	165·9	173·7	167·2	168·3	164·5	1
23	166·0	165·7	167·5	169·8	172·0	175·0	173·6	175·6	173·8	171·9	1
24	—	—	—	—	—	—	—	—	—	—	
25	· —	—	—	—	—	—	—	—	—	—	
26	182·8	171·7	156·5	163·0	164·8	161·3	163·2	173·0	173·2	171·7	1
27	*223·3*[a]	165·8	155·2	155·9	158·8	175·8	173·8	186·2	179·9	171·3	1
28	174·5	176·5	164·3	*229·5*	143·7	146·8	166·1	168·9	172·2	174·0	1
29	174·5	171·3	167·0	164·9	155·6	165·8	176·5	174·8	173·4	182·8	1
30	165·6	166·9	179·6	172·7	171·7	174·6	178·2	182·1	176·5	173·0	1
31	—	—	—	—	—	—	—	—	—	—	
Sums -	4309·8	4264·8	4179·2	4212·6	4034·7	4000·0	4099·3	4159·7	4159·1	4103·9	4
Means -	172·39	170·59	167·17	168·50	161·39	160·00	163·97	166·39	166·36	164·16	1

[a] Visible aurora. [b] This day is omitted from the Mean at the foot as

[c] Means by the observations furnishing complete 8-hourly series. [d] Taken eight minutes late. [e]

Lake Athabasca—*continued.*

Abstract of Hourly Observations made during the months of November and December 1843.

Induction Inclinometer.													
13.	14.	15.	16.	17.	18.	19.	20.	21.	22.	23.	Sums.	Means.	Fortnightly Means.
102·2	101·9	97·1	83·6	83·8	81·6	97·8	98·5	133·4[a]	117·6	102·8	2459·3	102·47	
81·1	96·9	101·6	95·8[a]	97·6[a]	109·8	134·4	173·9	186·9[a]	153·8[a]	255·8[a]	2796·6	116·52	
107·5	107·5	105·2	108·1	107·1	108·1	111·6	109·1	—	—	—	2918·1	117·45[c]	
112·4	107·5	102·2	102·5	108·0	105·5	111·0	117·1	—	—	—	—	119·90[c]	
—	—	—	—	—	—	—	—	159·4	171·2	183·5[a, f]	—		
112·6	118·0	108·4	107·1	105·7	110·2	114·7	116·1	110·9	115·3	131·7	2845·5	118·56	
107·2	104·5	108·3	104·0	102·8	105·9	114·8	115·5	111·3	112·3	113·4	2859·7	110·82	
105·3	99·5	90·8	117·2	101·5	70·1	104·4	106·3	139·4	187·2	217·2	2876·7	119·86	
121·0	113·5	122·7	116·1	119·9	115·7	117·3	107·7[a]	116·4	191·5	134·6	2877·4	119·89	
114·4	115·9	116·8	116·0	116·2	115·9	114·4	123·9	168·2	142·3	125·3	2919.3	121·64	
114·3	104·3	112·9	120·6	123·1	114·9	93·3	94·1[e]	—	—	—	2797·3	116·55	
—	—	—	—	—	—	—	—	113·9	113·6	113·2			
119·0	103·3	102·6[a]	100·5[a]	107·9	122·0	118·4[a]	170·0[a]	149·5	133·3	116·8	3102·2	129·26	120·12
121·1	116·5	116·1	110·3	117·8	114·5	108·9	113·6	112·6	117·9	115·8	2834·3	118·10	
112·0	120·6	110·5	117·3	118·5	120·1	112·9	120·1	131·4	119·0	133·4	2916·9	121·54	
118·4	118·0	115·8	120·2	114·8	84·2	97·8	121·2	119·8	118·3	123·4	2846·0	118·58	
123·2	121·0	123·4	121·9	121·2	123·4[a]	120·1[a]	115·8	124·0	124·9	123·8	2945·6	122·73	
120·2	122·1	122·4	120·0	126·4	125·0	125·0	131·5	—	—	—	2973·9	123·91	
—	—	—	—	—	—	—	—	122·8	128·2	125·4			
128·2	125·4	126·3	126·2	117·0	121·3	121·5	129·7	145·6	124·4	125·0	3073·1	128·05	
128·8	127·8	133·4	130·0	141·9	133·4	121·9	132·8	132·6	131·8	131·0	3155·5	131·48	
127·0	127·3	129·2	129·0	126·1	128·0	129·1[a]	126·2[a]	130·1[a]	138·4	142·0	3178·1	132·42	
129·9	129·5	126·5	126·6	130·7	128·8	126·2	131·5	127·9	131·1	129·1	3117·3	129·89	
133·1	128·5	126·1	140·8	131·5	118·8	136·9	132·9	135·9	140·0	135·7	3302·1	137·59	
131·2	129·4	134·8	133·5	133·0	135·6	136·0	133·1	—	—	—	3231·0	134·62	
—	—	—	—	—	—	—	—	133·7	133·7	135·5			
142·4	139·9	136·8	137·7	139·9	137·7	133·6	136·3[a]	142·3	141·8	137·2	3347·9	139·50	139·55
131·4	139·3	137·5	140·7	144·3	144·9	138·8[a]	146·6[a]	138·8[a]	138·3	141·6	3367·3	141·14	
134·6	135·4	131·7	137·3	135·9	132·1	144·3[a]	136·5[a]	167·4[a]	178·4	140·1[a]	3512·0	146·33	
134·7	145·1	142·0	140·7	150·3	133·1	137·1	134·9	137·5	140·8	141·9	3372·7	140·53	
3000·8	2991·1	2960·4	3000·7	3014·9	2935·1	3005·2	3159·8	3391·7	3445·1	3474·7	75445·8	—	
120·03	119·64	119·22	120·03	120·60	117·40	120·21	126·30	135·67	137·80	138·99	3017·83	125·74	
145·9	151·7	146·7	143·6	137·5	128·8	131·8	131·1	206·9	194·8	279·6	3661·6	152·57	
146·3	148·6	142·6	141·3	142·4	136·6	144·4	145·7	—	—	—	3850·5	160·44	
—	—	—	—	—	—	—	—	149·7	157·7	155·7			
149·5	148·5	155·9	155·1	151·7	150·8	150·6	152·6	146·4	150·9	153·1	3687·3	153·64	
156·6	149·5	148·3	149·3	147·6	149·5	151·3	151·5	186·9	237·9	225·0	3802·1	158·42	
142·9	150·0	145·8	142·9	124·5	117·9	147·6	150·3	155·6	159·1	160·6	3651·1	152·13	
151·1	152·8	157·8	154·4	151·1	154·0	151·0	149·7	150·8	159·8	155·4	3684·3	153·51	
153·3	152·2	152·6	139·9	155·2	189·5	183·4	152·2	151·1	158·4	156·0	3726·6	155·27	
156·7	151·9	146·4	129·2	146·0	151·9	150·1	151·7[a]	—	—	—	3716·4	154·85	
—	—	—	—	—	—	—	—	145·4	155·1	160·6			
141·2	132·4	152·6	147·0	143·0	144·0	147·1	147·4	160·3	165·4	160·5	3705·9	154·41	161·19
142·9	144·4	161·0	160·6	167·9	162·6	163·3	161·8	169·9	167·3	163·4	3875·6	161·48	
172·4	174·9	158·9	168·7	168·9	171·8	171·5	191·0	190·9	178·0	179·0[a]	4161·8	173·41	
171·9	172·1	176·8	165·8	169·9	171·0	169·1	173·9	167·6	167·1	169·0	4173·9	173·91	
164·5	168·8	170·3	167·2	166·0	166·1	169·6	169·8[a]	171·5	172·7	172·8	4086·5	170·27	
170·2	172·0	171·4	170·8	172·5	168·5	166·3	168·5	—	—	—	4124·4	171·85	
—	—	—	—	—	—	—	—	170·0	167·6	177·8			
163·7	161·2	163·0[k]	162·2	155·3	163·9	164·0	164·6	166·8[g]	165·9	169·4	3966·6	165·28	
165·4	163·2	162·8	161·8	166·8	166·8	164·0	184·5	172·0	241·3	255·8	4174·9	173·95	
162·7	167·7	171·7	164·7	165·3	165·4	170·4	172·3	176·4	172·6	170·9	4031·7	167·99	
167·3	165·9	159·0	160·7	162·9	159·5	161·5	162·5	179·8	177·2	168·3	3958·1	164·92	
165·0	163·7	163·1	161·8	165·1	170·5	172·7	172·7	176·7[a]	165·9	167·5	3997·8	166·57	
169·2	166·9	165·0	163·5	163·3	162·5	166·3	163·8	—	—	—	4068·3	166·89	
—	—	—	—	—	—	—	—	—	—	—			
—	—	—	—	—	—	—	—	178·8	177·9	186·9	—	—	169·80
169·3	167·9	166·2	173·5	169·9	163·6	173·3	182·7	197·7[a]	258·3[a]	209·2[a]	4311·7	179·65	
165·2	163·5	162·9	161·0	162·7	148·4	143·6	153·4	167·8	169·9	172·6	4031·4	167·98	
166·1	160·6	166·6	166·3	165·8	167·9	170·1	169·8	167·4	191·9	181·1	4098·9	170·79	
171·5	165·4	171·0	167·3	172·0	172·1	168·2	172·2	201·1	276·8	187·3	4252·6	177·19	
173·7	178·3	176·3	177·6	175·9	174·1	158·4	180·1	—	—	—	4233·5	176·40	
—	—	—	—	—	—	—	—	188·3	189·6	189·9			
4006·5	3996·1	4012·7	3956·2	3966·2	3947·7	3959·6	4075·8	4267·8	4572·6	4611·4	99033·5	—	
160·26	159·84	160·51	158·25	158·65	157·91	158·38	163·03	170·71	182·90	184·46	3961·33	165·06	

[f] Taken thirty minutes late. [g] Fifteen minutes late. [h] Five minutes late. [k] Twelve minutes late.

Lake Athabasca—*continued.*

Abstract of Hourly Observations made during the months of January and February

Date. Gött. Mean Time.	Induction Inclinometer.										
	Noon.	1.	2.	3.	4.	5.	6.	7.	8.	9.	10.
1	—	—	—	—	—	—	—	—	—	—	—
2[b]	190·1	192·3	231·6	204·2	149·1	175·3	180·2	189·0	199·7	188·7	183·7
3	198·1	186·8[d]	183·3	179·8	175·0	181·6	187·8	191·2	191·5	190·9	190·9
4	196·3	191·7	188·5[e]	190·5	175·5	175·5	182·9	180·9	188·7	185·7	187·5
5	*249·9*	*261·9*	*181·9*	167·8	173·5	196·5	180·1	185·3	213·6	189·1	204·1
6	*318·4*	*238·6*	*291·2*	215·1	169·3	183·5	183·1	189·3	182·3	190·3	199·1
7	—	—	—	—	—	—	—	—	—	—	—
8	209·9	209·5	205·6	196·9	194·8	192·9	206·5	198·6	175·8	176·0	196·1
9[b]	210·4	215·1	210·9	208·8	191·3	186·2	222·5	195·2	190·2	197·0	202·3
10	202·3	199·3	205·3	198·7	193·7	186·1	203·6	200·4	202·6	205·6	191·4
11	195·2	207·1	236·9	190·6	187·1	189·2	196·1	204·4	212·9	199·3	198·5
12	196·4	190·5	184·5	190·6	187·7	186·5	191·4	198·1	203·8	198·7	213·2
13	198·4	198·7	200·0	199·1	208·3	203·9	200·4	202·6	205·2	206·4	203·0
14	—	—	—	—	—	—	—	—	—	—	—
15	198·0	197·1	198·6	198·1	198·1	198·1	196·5	198·6	196·8	215·1	215·0
16	190·4[a]	196·5[a]	194·1	195·8	194·8	192·4	193·5	191·3	208·5	200·3	193·8
17	*233·5*	211·1	201·4	199·1	209·5	201·3	202·2	207·3	210·3	207·6	203·0
18	208·4[a]	213·9	206·1	214·3	207·4	204·3	201·5	207·5	201·7	204·2	212·0
19	206·1	202·9	207·6	199·6	205·3	206·8	206·2	211·7	207·3	208·2	206·5
20	201·5[a]	206·1[a]	201·4	200·1	201·3	207·2	216·8	205·5	202·5	199·4	207·4
21	—	—	—	—	—	—	—	—	—	—	—
22	209·0	212·8	215·6	218·8	225·5	217·8	261·3	229·5	212·2	204·3	203·7
23	194·6	207·2	206·1	211·1	211·3	212·5	211·1	211·2	203·5	198·9	205·7
Term Day. 24	208·5	212·7	209·9	200·5	205·8	214·3	219·8	215·1	209·9	212·7	202·3
Term Day. 25	343·6[a]	392·9	318·6	215·7	203·8	194·1	215·5	214·2	211·2	205·9	192·0
26	202·1	205·4	207·1	208·2	211·3	215·5	205·4	211·8	211·3	204·2	207·3
27	[a]212·6[a]	206·9[a]	205·2	207·0	206·5	212·4	215·6	209·5	208·0	209·9	205·3
28	—	—	—	—	—	—	—	—	—	—	—
29	228·2	205·2	200·9	200·0	207·2	199·7	203·9	198·7	204·7	204·9	203·5
30	203·0	196·5	193·6	191·1	198·9	199·5	203·3	203·4	199·1	201·8	201·3
31	208·1[a]	209·2	206·9	213·3	203·9	203·9	205·5	208·2	214·6	212·9	209·9
Sums -	5623·0	5561·9	5472·8	5214·8	5093·9	5135·0	5297·7	5258·5	5286·9	5218·0	5238·
Means -	216·27	213·92	210·49	200·57	195·92	197·50	203·76	202·25	203·57	200·69	201·4
1	*302·7*	*353·8*	*247·8*	*223·4*	207·1	231·9	213·4	201·3	208·2	222·5	212·
2	214·1	214·1	205·6	205·9	222·3	214·6	*276·7*	*219·6*	218·9	195·2	183·
3	196·1	194·7	193·5	195·2	196·7	197·9	203·4	224·8	200·5	192·3	192·
4	—	—	—	—	—	—	—	—	—	—	—
5	*474·6*	*291·8*	177·6	190·2	165·5	203·0	194·2	213·0	198·1	174·7	189·
6	285·5	182·3	195·9	189·6	183·0	202·7	199·2	189·0	185·2	184·2	191·
7	186·1	204·0	208·5	197·7	184·0	184·0	190·1	186·2	201·7	192·4	170·
8	*294·6*	239·4[a]	232·9	205·3	*226·8*	265·1	207·7	190·9	193·9	184·5	186·
9	185·3	188·6	187·6	184·6	189·5	185·1	193·1	188·7	188·1	186·8	187·
10	179·5	184·5	188·2	186·7	191·4	192·2	187·8	192·8	198·7	203·3	189·
11	—	—	—	—	—	—	—	—	—	—	—
12	225·7	194·2	196·5	198·1	196·8	206·4	202·4	194·4	197·6	196·9	196·
13	204·7	188·8	189·7	192·5	198·0	201·8	201·2	196·5	200·5	190·9	189·
14	192·9	190·4	195·3	—	204·1	209·1	210·5	204·3	207·0	205·6	206·
15	224·1	204·8	180·8	192·9	201·4	209·5	204·8	201·6	206·7	202·3	204·
16	195·9	194·8	203·2	199·4	193·7	202·1	199·0	218·0	197·9	196·1	197·
17	215·7	197·0	199·5	164·3	201·6	209·0	200·4	196·3	200·3	197·5	199·
18	—	—	—	—	—	—	—	—	—	—	—
19	205·5	201·7	200·3	200·7	200·6	196·7	199·7	200·7	202·7	198·2	200·
20	190·0	194·0	194·4	193·8	195·8	294·1	198·3	201·9	202·2	200·5	194·
21	232·6	213·0	206·1	202·5	199·8	196·0	188·2	199·5	190·6	199·3	193·
22	186·2	192·8	194·9	193·1	196·3	201·3	202·8	205·8	208·1	200·5	190·
Term Day. 23	190·9	196·8	193·1	192·8	192·7	178·1	192·2	199·0	195·6	196·6	196·
Term Day. 24	199·5	196·2	197·4	199·0	203·1	193·6	191·2	194·5	203·1	202·5	200·
25	—	—	—	—	—	—	—	—	—	—	—
26	199·9	216·4	210·0	194·1	193·9	188·5	196·2	199·0	197·3	197·3	190
27	208·0	199·3[a]	196·5	192·0	193·1	206·2	204·2	200·6	201·8	200·7	200
28	206·0	199·3	202·0	193·0	193·0	188·2	197·0	211·7	208·0	194·0	183
29	*346·1*[a]	225·2[a]	198·7[d]	208·2	193·4	195·5	211·0	215·2	170·9	197·2	201
	—	—	—	—	—	—	—	—	—	—	—
Sums -	5764·4	5259·9	5005·0	4724·7	4923·6	5062·6	5066·7	5045·3	4989·6	4921·0	4857
Means -	230·58	210·40	200·20	196·86	196·94	202·50	202·67	201·81	199·58	196·84	194·

[a] Visible aurora. [b] Day omitted as incomplete.
[c] This portion of a day is included under the same hours for December to make the numbe
[d] Taken seven minutes late. [e] Fifteen minutes late. [f] At 21h 3(

Lake Athabasca—*continued.*

Abstract of Hourly Observations made during the months of January and February 1844.

Induction Inclinometer.

13.	14.	15.	16.	17.	18.	19.	20.	21.	22.	23.	Sums.	Means.	Fortnightly Means.
—	—	—	—	—	—	—	—	188·3	189·6	189·6[a]	—	—	
185·9	—	168·4	180·0	172·4	180·4	178·1	167·5	185·6	182·6	185·1	4263·0	185·73	
186·1	185·4	183·4	186·7	187·7	182·3	185·1	189·6	193·5	196·4	192·6	4502·1	187·59	
180·5	173·0	170·8	*128·5*	*130·4*	*148·5*	*110·8*	164·2[h]	*338·9*	*280·7*	*477·9*	4650·1	193·75	
178·2	179·0	177·4	180·1	182·9	181·9	152·0	184·5	180·4	177·9	*330·8*	4716·8	196·53	
176·8	184·3	164·7	193·1	176·2	185·8	177·9	181·4	—	—	—	4781·2	199·22	
—	—	—	—	—	—	—	—	208·7	212·0	230·6			
191·5	189·5	*400·5*[a]	183·1	184·1	193·7	189·7	*205·4*[a]	*221·2*	185·9	189·5	4656·6	194·02	194·65
189·3	164·0	175·9	188·6[a]	195·3[a]	195·8	190·5	159·5	—	210·5	218·0	4490·5	195·93	
185·4	179·7	190·0	187·4	198·5	195·1	238·1	206·9	196·1	199·2	197·5	4732·0	197·17	
191·2	189·2	187·5	184·6	186·5	191·4	196·0	195·4	205·3	198·3	195·1	4723·2	197·61	
194·9	188·8	189·8	188·1	188·3	185·7	178·1	185·9	201·1	205·7	213·5	4599·8	191·66	
189·7	193·5	191·4	189·7	189·6	191·2	187·1	194·6	—	—	—	4766·6	198·61	
—	—	—	—	—	—	—	—	211·7	205·6	202·6			
196·5	195·4	199·7	196·8	193·8	197·8	193·8	196·9	201·4[s]	197·7	195·5	4766·7	198·61	
192·0	192·4	194·5	192·0	193·6	197·4	196·7	201·6[a]	191·1[a]	195·0	199·5	4688·5	195·35	
196·7	197·5	194·4	199·0	205·2	199·5	201·1	180·5	196·7[a]	207·5[a]	201·1	4875·0	203·13	
204·1	211·5	216·3	200·4	201·8	195·8[a]	210·5	208·9	220·2	218·1	206·1	4987·0	207·79	
209·7	208·5	211·8[a]	211·1[a]	213·0[a]	205·6[a]	204·0[a]	207·5[a]	262·8[a]	234·9[a]	208·5[a]	5062·6	210·93	
202·1	197·8	202·7	198·2	202·6	199·3	198·8[a]	196·7[a]	—	—	—	4896·4	204·02	
—	—	—	—	—	—	—	—	210·3[a]	207·0[a]	210·0[a]			
200·3	194·3	192·9	204·8	194·5[h]	194·2	190·4	173·6[a]	181·0[a]	181·3[a]	173·9	4894·7	203·93	
200·0	196·8	204·5	207·8	204·1	212·4	199·5	212·2	205·9	206·0	208·5	4944·9	206·04	
203·6	210·9	195·9[a]	203·3[a]	310·4[a]	218·3	247·8	381·2	358·1[a]	325·0[a]	375·5[a]	5755·0	239·79	209·74
205·8	210·5	210·3	207·8	204·8	206·5	221·1	214·1	212·4	201·9	199·0	5417·7	223·74	
210·1	206·7	201·7	198·8	202·8[a]	196·5	199·2	203·0[a]	203·8	*254·8*[k a]	247·9[a]	5042·7	210·11	
201·9	195·6	205·6	204·5	203·7	201·7	193·2	219·2	—	—	—	4960·6	206·69	
—	—	—	—	—	—	—	—	202·1	210·5	223·3			
204·5	202·9	189·1	190·7[h]	195·5	206·5	206·5	194·8	195·0	196·4	238·5	4878·4	203·27	
199·7	196·0	195·0	199·0	199·8	183·1[a]	187·8	178·4	229·7[a]	206·6	206·3	4777·2	199·05	
197·4	198·9	203·4	191·1	189·0	167·7[a]	191·9	201·2	175·0	191·3	219·4	4823·7	200·99	
5075·9	4842·1	5007·6	4989·2	3056·5	5014·1	5019·5	3204·2	5410·0	5484·2	5061·3	125659·0	—	
195·23	193·68	192·60	191·89	194·48	192·85	193·06	200·16	216·40	209·01	230·05	4848·80	202·03	

13.	14.	15.	16.	17.	18.	19.	20.	21.	22.	23.	Sums.	Means.	Fortnightly Means.
185·7	196·8	192·6	202·7	174·4[l]	163·9	*115·1*[m]	*156·5*	189·1	199·2	216·0	5029·3	209·55	
185·7	188·7	171·3	170·7	*164·4*	*174·7*	*153·3*	*164·9*	193·8	195·4	195·5	4714·0	196·42	
168·9	169·5	163·6	189·6	186·4	184·1	170·5	160·5	—	—	—	4519·1	188·30	
—	—	—	—	—	—	—	—	187·9	178·6	190·3			
157·9	151·6[a]	147·3[a]	*153·1*[a]	*121·6*[a]	*132·2*	183·2	220·6	*300·8*	*208·9*	213·0	5008·4	208·68	195·83
180·8	189·3	185·3	186·7	182·7	171·2	*213·4*	175·3	180·6	184·7	189·1	4623·2	192·63	
192·6	180·2	191·5	158·9	176·3	190·1	185·9	189·4	152·9	178·8	231·9	4506·7	187·78	
169·3	177·3	171·9	191·2	188·0	182·8	187·5	187·1	192·1	183·8	186·4	4799·5	199·98	
184·8	183·0	188·0	187·1	177·6	179·5	188·8	192·4	174·0	124·8	173·2	4390·6	182·94	
191·7	180·8	188·8	188·4	188·7	189·5	190·5	246·3	—	—	—	4673·4	194·72	
—	—	—	—	—	—	—	—	235·3[a]	205·8	215·7			
195·3	191·5	190·5	187·6	187·8	183·4[a]	183·1	179·5	211·4[a]	259·5[a]	205·9[a]	4775·5	198·98	
194·5	192·2	190·1	192·7	197·5	192·2	190·3[a]	184·9[a]	192·4	191·6	196·7	4660·4	194·18	
204·0	200·4	196·5	200·9	201·2	193·1	167·5	170·1	178·0	204·5	196·6	4540·2	189·18	
197·6	199·5	194·0	194·4	194·3	194·8	191·9	188·6[a]	194·5	197·8	194·0	4782·1	199·25	
190·0	180·9	184·7	181·1	174·7	181·0[a]	196·4[a]	192·1	*204·9*[a]	211·3[a]	207·7[a]	4700·1	195·84	
187·8	184·8	182·3	185·7[a]	185·2[a]	193·7	198·7	213·1	—	—	—	4715·8	196·49	
—	—	—	—	—	—	—	—	196·2	199·6	203·5			
195·6	195·7	192·5	191·4	196·5	193·9	192·5	196·1	197·1	205·1	196·0	4750·2	197·93	
191·4	189·2	190·8	193·3	192·9[a]	192·0	196·0	189·9[a]	191·1[a]	207·4	215·3	4701·9	195·91	197·14
199·5	194·3	192·5	191·5	194·9	185·1	187·7	199·3[a]	198·9[a]	203·6[a]	198·9[a]	4765·7	198·57	
194·4	186·7	193·9	189·3	191·7	195·1	193·4	192·5	204·5	193·4	188·8	4712·0	196·33	
189·2	189·7	191·2	187·9	190·1	197·3	194·1	193·6	191·2	191·5	192·9	4623·2	192·63	
196·2	197·1	200·3	194·2	196·8	196·0	199·4	197·8	—	—	—	4771·4	198·81	
—	—	—	—	—	—	—	—	205·7	202·6	201·5			
195·7	195·1	186·9	194·2	188·3	188·1	181·4	173·8	183·0	205·3	*277·9*	4751·4	197·97	
194·2	190·0	192·2	193·6	194·4	201·5	192·1	209·4	207·6	232·2	218·8	4807·3	200·30	
188·0	169·4	158·0	210·2[a]	144·8[a]	166·4	160·1[a]	186·8	*261·4[a]	277·0[a]	304·0	4796·4	199·81	
193·2	195·8	191·4	187·0	170·2	189·4	190·2	193·4	—	—	—	4279·4	204·62	
—	—	—	—	—	—	—	—	—	—	—	—	197·09	
4729·0	4671·5	4631·1	4693·4	4587·6	4640·0	4612·0	4757·9	4918·4	4932·4	3009·6	117390·2	—	
189·16	186·86	185·24	187·74	183·50	185·60	184·48	190·32	204·93	205·52	208·73	4728·45	197·02	

[s] Seventeen minutes late.
[l] Twelve minutes late.
[h] Ten minutes late.
[m] Three minutes late.
[k] Taken nine minutes late.
[a] Twenty minutes late.

FORT SIMPSON.

Abstract of Hourly Observations made during the months of April and May 18

Date. Gött. Mean Time.	Induction Inclinometer.										
	Noon.	1.	2.	3.	4.	5.	6.	7.	8.	9.	10.
1	394·1	251·4	268·1	201·6	255·6	137·6	177·9	132·0	100·0	100·0	84·4
2	144·0	139·8	138·1	183·3	219·5	170·0	193·7	101·1	153·7	108·7	136·3
3	235·9	120·2	123·1	159·7	368·2	337·1	245·4	181·2	121·9	127·8	120·2
4	368·1	274·4	180·9	170·3	142·6	123·4	135·7	137·8	127·0	137·5	144·3
5	—	—	—	—	—	—	—	—	—	—	—
6	145·9	148·5	138·3	132·4	136·9	167·1	140·0	120·5	131·1	112·1	150·1
7	—	—	—	—	—	—	—	—	—	—	—
8	148·2	—	152·2	153·0	141·2	150·7	150·5	194·7	163·2	153·7	158·9
9	137·6	136·6	143·5	138·5	146·1	145·9	128·1	133·2	138·9	140·6	144·6
10	191·6	216·1	220·0	272·2	250·5	112·2	120·0	124·0	120·3	127·7	125·0
11	486·4	142·2	189·8	130·6	123·4	127·4	125·3	126·6	125·3	126·4	132·7
12	133·5	134·7	132·2[b]	125·4	121·8	125·6	137·0	129·8	126·7	130·9	127·2
13	124·1	102·6	123·5	123·5	124·3	130·9	129·2	131·1	134·7	135·1	134·9
14	—	—	—	—	—	—	—	—	—	—	—
15	147·0	383·3	389·7	222·3	149·9	102·4	118·2	121·0	120·6	115·3	120·8
16	—	—	—	137·6	167·0	153·0	138·9	126·7	131·3	132·0	130·6
17	330·8	837·9	470·4	485·5	387·1	439·2	352·8	214·1	148·0	83·9	67·1
18	—	—	129·1	124·9	118·8	133·4	153·1	111·0	135·3	127·4	143·1
19	188·3	158·5	141·0	125·7	127·9	146·7	143·2	138·4	139·5	153·5	143·1
20	255·7	254·6	199·0	154·8	134·3	136·6	126·5	134·7	136·0	143·5	137·1
21	—	—	—	—	—	—	—	—	—	—	—
22	114·3	—	178·1	162·9	154·1	135·0	140·9	139·7	134·7	147·0	154·
23	304·8	242·9	246·0	277·3	184·5	146·7	147·1	148·4	140·8	159·8	157·
Term Day 24	159·8	—	152·0	139·7	147·1	142·1	146·3	146·3	153·0	153·7	161·7
Term Day 25	311·3	276·0	456·5	526·6	229·0	174·3	195·8	214·4	165·2	138·1	160·
26	645·9	460·1	240·3	—[c]	357·9	390·6	185·5	242·5	165·5	161·3	162·1
27	344·6	301·5	193·7	196·0	123·2	140·4	211·4	186·4	168·2	161·3	177·
28	—	—	—	—	—	—	—	—	—	—	—
29	200·1	307·0	483·3	301·8	318·1	251·8	157·4	154·9	163·4	161·1	175·
30	366·8	296·8	180·1	173·5	172·9	175·7	171·5	208·3	174·1	172·6	183·
May 1	256·9	210·7	231·7	144·0	176·3	179·1	185·9	180·6	171·5	180·3	171·
											Mea
Means -	240·61	215·18	217·34	199·87	191·47	175·69	163·74	153·05	142·27	138·13	142·4

Date.	Noon.	1.	2.	3.	4.	5.	6.	7.	8.	9.	10.
1	—	—	—	—	—	—	—	—	—	—	—
2	—	—	—	—	—	—	—	—	—	—	—
3	241·2	242·6	239·9	291·5	614·1	316·7	266·6	223·0	239·8	238·8	231·
4	233·2	242·0	244·4	250·4	228·8	231·6	227·0	236·9	238·8	247·5	263·
5	—	—	—	—	—	—	—	—	—	—	—
6	255·2	282·4	221·0	214·8	252·6	260·6	279·0	241·0	240·8	243·4	236·
7	276·2	264·8	300·4	267·0	290·8	276·0	248·0	233·0	244·0	249·6	247·
8	279·6	—[d]	400·8	387·0	295·2	289·8	265·0	253·8	252·0	261·0	240·
9	252·8	258·0	374·2	422·6	407·2	299·1	250·4	232·0	233·1	235·0	238·
10	258·4	253·4	260·8	281·6	279·6	257·5	250·0	252·0	255·4	262·0	260·
11	—	284·0	238·4	289·0	258·0	261·0	270·2	264·8	265·4	265·6	260·
12	—	—	—	—	—	—	—	—	—	—	—
13	231·0	240·0	286·2	276·6	233·0	234·0	245·0	246·0	245·0	244·5	242·
14	311·0	283·0	268·8	254·0	197·8	210·0	235·6	227·2	230·8	231·0	228·
15	205·6	235·0	257·0	258·8	231·0	247·2	270·0	240·0	252·0	233·0	223·
16	324·8	316·2	321·3	295·6	268·6	232·8	219·0	242·0	246·2	241·6	245·
17	246·0	251·0	249·8	240·6	238·0	243·0	249·0	251·0	253·0	246·6	254·
18	244·8	250·6	251·6	236·4	245·2	250·0	225·2	251·0	244·5	252·5	251·
19	—	—	—	—	—	—	—	—	—	—	—
20	257·0	246·0	258·0	240·8	236·0	237·2	248·0	248·0	259·8	265·0	262·
21	289·8	300·0	306·2	292·8	285·6	281·8	266·8	259·8	265·8	269·6	272·
22	—	484·0	501·6	360·6	315·4	246·0	241·5	275·8	266·4	303·0	306·
23	519·1	429·0	359·6	351·4	348·8	256·4	247·2	268·0	243·8	246·6	256·
24	341·2[e]	344·0	440·0	333·0	292·2	294·6	258·5	271·0	268·8	274·6	267·
25	301·9[f]	306·4[f]	236·2[f]	277·3[f]	269·1[f]	283·8[f]	271·6[f]	264·0[f]	266·0[f]	296·0[f]	266·
26	—	—	—	—	—	—	—	—	—	—	—
27	—	—	—	—	—	—	—	—	—	—	—
28	—	—	—	—	—	—	—	—	—	—	—
29	—	—	—	—	—	—	—	—	—	—	—
30	—	—	—	—	—	—	—	—	—	—	—
31	—	—	—	—	—	—	—	—	—	—	—
Mean 1st April to 25th May	280·41	294·15	306·84	292·13	291·47	259·23	250·63	248·23	249·76	253·21	251·9

[a] Visible aurora. [b] At 2^h 25^m. [c] At 3^h 30^m, 203·5. [d] At 1^h 30^m, 373·8, bo

[e] At 0^h 21^m. [f] Not included in the Mean at the foot.

FORT SIMPSON.

Abstract of Hourly Observations made during the months of April and May 1844.

13.	14.	15.	16.	17.	18.			21.	22.	23.	Sums.	Means.
88·1	65·8	78·3	71·9	61·2a	67·2a			279·1	296·9	209·5	3768·6	157·02
130·2	108·1	119·6	88·0	113·5a	62·5a			166·7a	231·1a	467·7	3597·2	149·88
144·1	124·6	116·0	86·6	107·4a	83·4a			135·4	151·5	177·4	3772·9	157·20
127·8	129·9	115·4	111·1	84·9	69·7			—	—	—	3428·4	142·85
—	—	—	—	—	—			152·4a	138·5	144·6		
129·1	105·9	81·6	63·0	95·9	86·4a			—	—	—	3033·1	126·38
—	—	—	—	—	—			151·5	158·9	152·9a		
145·4	148·2	136·7	136·7	132·5	145·6			133·5	129·3	138·6	3392·7	147·93
111·5	138·6	138·1	124·0	128·6	135·1			169·1a	240·0a	264·0	3575·8	148·99
139·0	113·6	124·6	125·1	126·6	196·6			280·1a	188·7a	130·7	3702·0	154·25
133·1	135·9	124·5	117·9	115·6	114·2			144·6	132·6	146·3a	3162·4	131·77
125·5	126·7	132·6	126·9	126·3	124·7			118·8	143·4	129·3	3093·5	128·90
129·6	124·3	126·9	133·9	122·0	115·9			—	—	—	3018·0	125·75
—	—	—	—	—	—			118·9a	126·7	117·7		
107·4	102·8	100·7	115·1	124·8	117·4			182·7	118·8	[illegible]	3574·1	148·92
99·4	44·0	89·8	89·2	112·8a	237·8a			335·2a	395·8a		3092·7	140·93
80·9	101·1	144·8	134·3	133·4	135·9			118·1	112·4a		5116·7	213·20
119·6	123·8	125·1	136·4	122·9	131·2			128·1a	133·2a		2805·0	127·15
132·6	131·6	140·3	128·4	120·6	128·5			139·5a	161·7a		3574·5	148·94
121·2	138·3	137·7	127·3	123·7	130·1			—	—		3597·9	149·92
—	—	—	—	—	—			141·3	156·1a			
137·3	140·5	145·7	131·4	137·7	142·5			140·4	146·8		3386·3	147·13
157·5	150·4	145·0	141·8	127·3	131·3			186·2	139·1		4109·8	171·24
153·0	166·1	160·9	155·9	133·6	127·2			137·7	160·7		3424·2	149·30
119·0	150·3	164·9	88·2	114·7	151·3			216·9	176·2a		5004·0	208·50
108·2	108·4	122·6	149·1	130·4	200·6			192·5	195·9		5380·7	224·20
115·8	138·0	169·5	85·2	92·1	168·6			—	—		4213·3	175·55
—	—	—	—	—	—			194·9a	252·3			
156·8	149·0	130·0	134·0	152·8	126·1			161·3	161·7		4596·4	191·52
143·7	80·7	81·3	117·3	148·8	147·5			499·1	415·8		4834·1	201·42
137·8	150·5	137·8	156·9	139·5	135·5			143·8	158·8		3908·7	165·36
to 1st May inclusive			—	—	—			—	—		—	159·01
125·33	121·51	121·97	116·37	120·60	132·04			181·84	183·80		160·55	—
—	—	—	—	—	—	—		—	—			—
—	—	—	—	—	158·3f	263·0f		206·7f	288·1f			—
237·0	232·3	217·9	204·0	227·1	222·4	214·3		217·1	231·3			252·27
220·6	212·2	216·0	211·6	202·4	194·2	184·0a		—	—			236·02
—	—	—	—	—	—	—		285·8	319·1a			
237·4	242·3	228·2	231·0	217·0	227·8	225·6		219·4	235·2			240·26
243·2	237·7	214·2	222·2	165·4	218·2	211·0		234·2	407·2			240·37
160·3	203·2	191·5	184·0	203·8	181·2	199·9		290·4	276·0			258·79
241·7	238·4	221·4	239·3	236·4	251·8	264·2		258·2	265·7			266·35
262·0	258·1	228·2	232·0	245·7	241·1	242·0		230·4	263·0			234·90
244·2	231·7	237·0	229·9	222·1	234·2	230·5		—	—			254·25
—	—	—	—	—	—	—		219·5	230·2			
229·0	224·1	219·9	225·8	196·2	134·0	170·2		208·0	270·2			228·43
176·5	138·4	215·8	218·7	202·0	207·8	219·9	[illegible]	274·0	244·8			228·76
198·0	228·3	222·4	211·8	229·0	235·9	226·0	240·2	231·1	261·8			236·46
242·4	223·1	224·0	235·6	240·9	240·0	228·0	221·1	260·3	257·3			252·48
227·4	237·1	238·0	237·8	241·3	245·3	245·0	229·2	252·2	243·2			243·90
235·4	253·1	250·0	247·8	218·0	219·6	241·6	237·0	—	—			245·22
—	—	—	—	—	—	—	—	248·2	264·0			
230·0	252·0	254·4	254·8	249·4	254·3	240·3	246·0	296·7	269·0			255·51
216·7	236·0	229·6	233·8	231·7	262·4	268·2	236·0	261·4	258·2			261·23
245·1	161·5	199·3	251·9	265·4	238·8	282·2	268·0	293·2	300·1			301·23
259·4	264·3	264·3	260·8	262·0	254·4	245·4	270·0	205·0	305·5			293·15
229·0	211·0	224·7	229·2	236·0	254·8	246·2	324·3	203·8	283·4			279·58
—	—	—	—	—	—	—	—	—	—			—
—	—	—	—	—	—	—	—	—	—			—
—	—	—	—	—	—	—	—	—	—			—
—	—	—	—	—	—	—	—	—	—			—
—	—	—	—	—	—	—	—	—	—			—
—	—	—	—	—	—	—	—	—	—			—
—	—	—	—	—	—	—	—	—	—			—
230·48	225·52	226·15	229·58	227·46	230·43	231·25	245·11	255·47	272·01			254·18

MAGNETICAL DISTURBANCES, LAKE ATHABASCA, 1843.

The Bifilar scale readings are reduced to a uniform temperature of 40° : the Inclinometer scale readings are reduced to a uniform reading of the Declinometer and Bifilar. The Declinometer was observed 1m before and the Inclinometer 2m after the time named.

October 15–17. | October 16–17—*continued.*

Gött. mean Time.	Declination.		Bifilar.		Inclinometer.		Approx. $\left(\frac{\Delta\phi}{\phi}\right)$	Gött. mean Time.	Declination.		Bifilar.		Inclinometer.		Approx. $\left(\frac{\Delta\phi}{\phi}\right)$
	Scale.	$(\Delta\psi)$	Scale corrected for Temp.	Approx. $\left(\frac{\Delta X}{X}\right)$	Scale corrected for Decl. and Bif.	Approx. $(\Delta\theta)$			Scale.	$(\Delta\psi)$	Scale corrected for Temp.	Approx. $\left(\frac{\Delta X}{X}\right)$	Scale corrected for Decl. and Bif.	Approx. $(\Delta\theta)$	
15 D. H. M.		′				′		16 D. H. M.		′				′	
21 0	390·0	−23·3	295·4	·0131	117·9	−12·8	·0076	2 35	419·5	6·2	260·8	·0013	129·5	−0·7	−·0001
22 0	401·6	−11·7	205·0	−·0178	202·6	12·0	·0059	40	416·1	2·8	261·1	·0014	147·2	2·3	·0059
5	—	—	215·4	−·0142	195·1	10·7	·0069	45	423·4	10·1	254·9	·0007	146·4	2·2	·0036
10	403·5	−9·8	235·3	−·0074	181·4	8·3	·0090	50	432·1	18·8	259·4	−·0008	158·6	4·3	·0077
15	400·0	−13·3	236·7	−·0070	182·0	8·4	·0096	55	424·3	11·0	228·7	·0097	157·8	4·2	−·0014
20	399·4	−13·9	248·8	−·0028	187·2	9·3	·0156	3 0	422·5	9·2	235·3	−·0074	158·0	4·2	·0009
25	399·5	−13·8	213·4	−·0149	228·2	16·5	·0157	5	426·3	13·0	268·8	−·0040	134·7	0·2	·0044
30	413·3	0·0	189·9	−·0229	222·6	13·5	·0033	10	420·2	6·9	275·6	·0063	138·4	0·8	·0079
35	411·0	−2·3	195·5	−·0210	232·6	17·3	·0131	15	420·4	7·1	276·5	·0066	132·7	−0·2	·0062
40	408·5	−4·8	179·7	−·0264	242·1	18·9	·0109	20	422·1	8·8	272·3	·0052	135·1	0·2	·0048
45	396·7	−16·6	164·1	−·0317	272·1	22·2	·0121	25	423·6	10·3	275·8	·0064	138·9	0·9	·0082
50	402·0	−11·3	147·3	−·0374	285·1	24·4	·0108	30	422·6	9·3	273·3	·0055	135·7	0·3	·0061
55	400·5	−12·8	122·0	−·0461	309·7	30·3	·0137	35	419·2	5·9	275·3	·0062	134·8	0·2	·0066
23 0	386·0	−27·3	109·3	−·0504	322·1	32·9	·0146	40	420·2	6·9	277·2	·0068	131·5	−0·4	·0076
5	352·4	−60·9	127·7	−·0441	—	—	—	45	420·1	6·8	276·5	·0066	138·5	0·8	·0082
10	383·5	−29·8	180·1	−·0263	233·3	17·4	·0081	50	419·4	6·1	277·6	·0070	135·2	0·2	·0074
15	379·1	−34·2	193·4	−·0217	220·8	13·2	·0044	55	416·8	3·5	276·2	·0065	133·5	−0·05	·0064
20	390·7	−22·6	181·1	−·0259	221·5	13·3	·0004	4 0	412·1	−1·2	278·9	·0074	130·4	−0·6	·0062
25	399·0	−14·3	196·8	−·0206	210·0	12·3	·0037	5	411·3	−2·0	281·6	·0084	124·1	−1·7	·0070
30	419·0	5·7	184·6	−·0247	220·5	13·1	·0012	10	409·6	−3·7	282·4	·0086	126·5	−1·3	·0060
35	423·0	9·7	160·9	−·0328	244·1	19·3	·0053	15	410·1	−3·2	282·7	·0087	124·7	−1·6	·0056
40	436·0	22·7	147·8	−·0373	262·3	22·4	·0069	20	412·5	−0·8	281·8	·0084	126·8	−1·2	·0060
45	458·0	44·7	129·5	−·0435	264·7	22·9	−·0001	25	410·0	−3·3	286·5	·0100	126·2	−1·2	·0076
50	507·9	94·6	112·3	−·0494	265·1	22·9	−·0042	30	410·5	−2·8	281·5	·0083	140·8	1·2	·0107
55	468·8	75·5	96·5	−·0551	317·4	32·2	·0085	35	416·6	3·3	274·7	·0060	156·8	4·0	·0139
								40	426·2	12·9	271·5	·0049	168·3	6·0	·0167

5	496·6	83·3	77·0	−·0614	341·6	36·2	·0101
10	457·0	43·7	128·2	−·0440	227·1	16·3	−·0119
15	471·5	58·2	141·1	−·0396	245·1	19·4	−·0013
20	444·6	31·3	175·9	−·0377	216·0	13·3	·0014
25	448·0	34·7	197·7	−·0203	211·4	12·5	·0044
30	443·5	30·2	193·4	−·0217	207·5	12·9	·0038
35	439·0	25·7	215·9	−·0140	184·5	8·8	·0034
40	431·6	18·3	235·6	−·C073	158·8	4·4	·0014
45	419·9	6·6	255·5	−·0005	61·3	4·8	· 09
50	422·7	9·4	264·9	·0027	161·9	4·9	·0124
55	418·7	5·4	260·8	·0012	162·0	4·9	·0109
1 0	417·0	3·7	244·9	−·0042	151·5	3·1	·0019
5	415·9	2·6	259·6	·0008	145·9	2·1	·0049
10	419·2	5·9	263·2	·0020	142·1	1·3	·0046
15	419·1	5·8	284·5	·C093	126·6	−1·2	· 00
20	409·5	−3·8	290·7	·0115	137·0	5·6	−· 04
25	418·0	4·7	268·1	·0037	143·8	11·7	· 00
30	422·0	8·7	253·3	−·0013	152·4	3·2	·0050
35	416·0	2·7	251·5	−·0019	161·2	4·8	·0076
40	425·4	12·1	258·6	·0005	160·2	4·6	·0096
45	418·6	4·3	274·7	· 60	139·3	1·0	·0079
50	415·9	2·6	281·3	·0082	139·0	0·9	·01C0
55	419·9	6·6	281·4	·0083	141·7	1·3	·0109
2 0	422·3	9·0	277·9	·0071	146·9	2·3	·0116
5	412·5	−0·8	261·5	·0015	136·7	0·5	·C025
10	421·4	8·1	248·6	−·0029	163·2	5·1	·0071
15	423·7	10·4	255·6	−·0012	159·0	4·4	·0075
20	419·5	6·2	263·2	·0021	142·5	1·3	·0047
25	418·4	5·1	272·6	·0053	137·6	0·7	·0C67
30	420·2	6·9	275·5	·0063	138·4	0·8	· 69

55	424·8	11·0	271·2	·0048	132·5	−0·2	· 64
5 0	420·0	6·7	273·0	·0054	123·9	−1·7	·0021
6 0	416·5	3·2	287·3	·0103	86·9	−8·2	−· 69
7 0	413·6	0·3	276·1	·0065	118·8	−2·6	·0014
18 0	409·0	−4·3	293·9	·0125	115·1	−3·3	·0060
19 0	407·3	−6·0	268·5	·0039	134·1	0·05	·0040
20 0	410·0	−3·3	211·4	−·0156	191·7	10·1	·0043
21 0	416·5	3·2	237·3	−·0067	155·9	3·9	·0010
22 0	420·0	6·7	228·0	−·C099	163·3	5·1	·0010
23 0	422·0	8·7	234·7	−·0076	155·6	3·8	−·0001
17 D.							
H. M.							
0 0	425·8	12·5	261·6	·C015	150·8	+3·0	·0074
1 0	—	—	—	—	—	—	—
5	446·0	32·7	148·1	−·0372	243·8	19·2	·0007
10	—	—	—	—	—	—	—
15	444·4	31·1	139·1	−·0403	236·7	18·0	−·0048
20	443·2	29·9	121·0	−·0464	269·8	23·8	·0006
25	459·0	45·7	129·8	−·0433	275·1	24·9	·0059
30	462·4	49·1	126·5	−·0445	261·2	22·2	−·0007
35	466·0	52·7	101·1	−·0532	329·9	34·4	·0147
40	465·0	51·7	74·6	−·0621	337·6	35·6	·0082
45	458·0	44·7	68·0	−·0645	362·8	40·0	·0145
50	445·0	31·7	74·3	−·0624	353·5	38·4	·0134
55	441·8	28·5	90·6	−·0566	324·9	33·5	·0096
2 0	413·0	−0·3	115·9	−·0481	287·7	26·9	·0050
5	422·1	8·8	163·3	−·0320	233·8	17·5	·0025
10	423·5	10·2	205·0	−·0178	192·0	9·2	·0004
15	423·0	9·7	216·9	−·0137	182·6	8·5	−·0010

The regular daily changes are included in these values throughout October. Co-ordinate mean values, declination, 413·3; bifilar, 257·1; inclinometer, 133·8.

October, 15ᵈ 21ʰ, Gött.—Faint aurora in bands from W. to E. 22ʰ. Faint diffused aurora both N. and S. of the zenith, in motion. 23ʰ. Aurora brighter, and gathered to a corona near the zenith at 16ᵈ 0ʰ. Two parallel arches of aurora in the N., brightest at the extremities E. and W., and striated. At 1ʰ, brightest portion of aurora to the S. of the zenith; faint auroral bands in the E. 2ʰ. Very faint aurora still visible in the E.

16ᵈ 18ʰ. Bright arch of aurora in the N. 19ʰ. A broad arch or band of aurora extending across the zenith; a brighter arch to the N. 20ʰ. A very faint arch of aurora in the north. 21ʰ. A faint auroral arch extending from N.E. to N.W., brightest at the extremities; a few faint detached patches in the N.W. 22ʰ. Arch less distinct; a broad band of aurora to the S. of the zenith. 23ʰ. Appearance of aurora nearly the same as before. 17ᵈ 0ʰ. Auroral arch much brighter than before at its N.W. extremity. 1ʰ. Faint arches or bands across the meridian in the zenith.

Magnetical Disturbances, Lake Athabasca, 1843—*continued*.

OCTOBER 17—*continued*.

Gött. mean Time.	Declination.		Bifilar.		Inclinometer.		Approx. $\left(\frac{\Delta\phi}{\phi}\right)$
	Scale.	(Δ ψ)	Scale corrected for Temp.	Approx. $\left(\frac{\Delta X}{X}\right)$	Scale corrected for Decl. and Bif.	Approx. (Δ θ)	
17 D. H. M.		′				′	
2 20	417·2	3·9	231·8	−·0086	184·5	8·8	−·0012
25	411·0	−2·3	228·6	−·0097	176·5	7·4	·0049
30	408·0	−5·3	241·8	−·0052	165·6	5·6	·0059
35	411·0	−2·3	239·7	−·0059	164·3	5·3	·0045
40	402·8	−10·5	251·5	−·0019	157·1	4·1	·0062
45	409·2	−4·1	256·2	−·0003	152·1	3·2	·0060
50	411·0	−2·3	263·2	·0021	147·2	2·3	·0066
55	414·1	0·8	269·3	·0042	147·0	2·3	·0087
3 0	410·0	−3·3	261·7	·0016	133·9	0·0	·0016
4 0	400·5	−12·8	253·7	−·0012	142·4	1·3	·0014
5 0	414·0	0·7	262·4	·0018	132·8	−0·2	·0022
15 0	409·0	−4·3	267·4	·0035	129·8	−0·7	·0027
16 0	411·0	−2·3	287·1	·0103	122·5	−2·0	·0064
17 0	410·2	−3·1	284·7	·0094	121·7	−2·1	·0053
30	440·6	27·3	264·4	·0025	119·1	−2·6	−·0026
35	448·0	34·7	259·9	·0009	97·1	−6·4	−·0135
40	445·3	32·0	253·8	−·0011	126·0	−1·3	−·0037
45	446·4	33·1	261·6	·0015	119·3	−2·5	−·0034
50	449·6	36·3	263·7	·0022	117·6	−2·8	−·0033
55	452·6	39·3	261·0	·0013	116·7	−3·0	−·0046
18 0	448·3	35·0	265·0	·0027	115·8	−3·1	−·0039
5	439·0	25·7	261·9	·0016	117·4	−2·9	−·0041
10	428·0	14·7	260·8	·0013	126·0	−1·3	−·0013
15	426·4	13·1	261·4	·0015	119·1	−2·6	−·0036
20	416·0	2·7	260·9	·0013	124·3	−1·6	·0010
25	411·8	−1·5	263·9	·0023	126·8	−1·2	·0021
30	413·6	0·3	263·3	·0021	131·1	−0·5	·0020
35	416·4	3·1	261·4	·0015	130·1	−0·6	·0014
[illegible]	[illegible]	[illegible]	[illegible]	[illegible]	[illegible]	[illegible]	·0022

OCTOBER 17—*continued*.

Gött. mean Time.	Declination.		Bifilar.		Inclinometer.		Approx. $\left(\frac{\Delta\phi}{\phi}\right)$
	Scale.	(Δ ψ)	Scale corrected for Temp.	Approx. $\left(\frac{\Delta X}{X}\right)$	Scale corrected for Decl. and Bif.	Approx. (Δ θ)	
17 D. H. M.		′			Div.	′	
45	416·2	2·9	267·7	·0036	127·5	−1·1	·0034
50	414·8	1·5	269·1	·0041	126·2	−1·2	·0039
55	414·0	0·7	274·5	·0059	124·2	−1·7	·0056
19 0	414·6	1·3	274·1	·0058	126·2	−1·2	·0056
5	414·8	1·5	271·3	·0048	128·2	−1·0	·0046
10	406·4	−6·9	269·8	·0043	142·7	1·3	·0045
15	400·0	−13·3	230·8	−·0090	146·2	2·2	−·0047
20	376·0	−37·3	214·2	−·0146	205·3	12·5	·0101
25	362·5	−50·8	174·5	−·0281	245·5	19·5	·0104
30	356·2	−57·1	171·9	−·0291	244·9	19·4	·0092
35	365·5	−47·8	151·3	−·0360	278·1	25·2	·0138
40	362·2	−51·1	149·1	−·0368	257·9	21·7	·0061
45	382·2	−31·1	157·3	−·0340	251·2	20·5	·0065
50	[illegible]0·5	−12·8	221·4	−·0121	252·1	20·0	·0274
55	389·6	−23·7	159·6	−·0336	246·0	19·6	·0051
20 0	392·3	−21·0	150·5	−·0362	245·5	19·5	·0023
5	390·3	−23·0	149·6	−·0367	251·9	20·6	·0040
10	400·4	−12·9	149·9	−·0366	253·9	21·0	·0049
15	404·4	−8·9	140·4	−·0398	250·8	20·4	·0005
20	—	—	—	—	—	—	—
25	410·5	−2·8	182·1	−·0255	211·1	12·5	−·0008
30	413·4	0·2	216·3	−·0105	184·3	8·8	·0069
35	420·5	7·2	215·7	−·0141	176·0	7·3	·0003
40	—	—	235·4	−·0074	161·2	4·8	·0021
45	406·7	−6·6	237·4	−·0067	161·4	4·8	·0028
50	405·0	−8·3	231·9	−·0086	160·2	4·6	·0005
55	409·0	−4·3	240·1	−·0058	150·1	2·8	−·0003
21 0	410·0	−3·3	255·2	−·0006	132·7	−0·2	−·0002

Magnetical Disturbances, Lake Athabasca, 1843—*continued.*

OCTOBER 18. Term Day.

Gött. mean Time.	Declination. Scale.	Declination. Δψ	Bifilar. Scale corrected for Temp.	Bifilar. Approx. ΔX/X	Inclinometer. Scale corrected for Decl. and Bif.	Inclinometer. $P_1 - P$.	Inclinometer. Approx. Δθ	Approx. Δφ/φ
18 D. H. M.		′				Div.	′	
10 0	408·0	+0·1	287·1	·0086	121·8	−10·3	−1·7	+·0051
5	405·6	−2·3	289·0	·0092	119·2	−12·7	−2·2	+·0049
10	406·7	−1·3	288·7	·0091	115·7	−16·1	−2·8	+·0036
15	408·0	−0·1	276·9	·0050	115·9	−15·7	−2·7	−·0004
20	406·8	−1·4	279·4	·0058	118·9	−12·6	−2·2	+·0015
25	410·6	+2·3	282·4	·0068	115·1	−16·2	−2·8	+·0012
30	409·4	+1·0	287·5	·0085	116·7	−14·5	−2·5	+·0035
35	404·8	−3·6	289·2	·0091	115·5	−15·5	−2·7	+·0038
40	404·0	−4·5	285·8	·0079	114·8	−16·0	−2·7	+·0024
45	405·1	−3·5	292·5	·0101	113·5	−17·2	−3·0	+·0042
50	409·0	+0·3	285·0	·0076	120·7	−9·8	−1·7	+·0042
55	409·0	+0·2	285·1	·0076	121·1	−9·3	−1·6	+·0044
11 0	409·8	+0·9	277·8	·0051	120·9	−8·3	−1·4	+·0022
5	411·0	+2·0	281·1	·0062	128·9	−1·2	−0·2	+·0060
10	408·1	−1·0	279·6	·0056	113·7	−16·3	−2·8	·0000
15	411·0	−1·9	276·5	·0046	127·3	−2·5	−0·4	+·0038
20	407·0	−2·2	274·5	·0039	124·1	−5·6	−1·0	+·0020
25	409·0	−0·3	278·5	·0052	122·4	−7·2	−1·2	+·0027
30	409·2	+0·4	278·0	·0051	120·4	−9·1	−1·5	+·0020
35	410·0	+0·5	273·6	·0035	114·7	−14·7	−2·6	−·0014
40	409·3	−0·2	272·4	·0031	115·3	−14·0	−2·4	−·0019
45	408·6	−1·0	270·2	·0023	123·5	−5·6	−1·0	+·0004
50	410·0	+0·4	279·6	·0055	122·3	−6·7	−1·1	+·0032
55	410·5	+0·8	278·1	·0050	118·7	−10·2	−1·7	+·0015
12 0	408·0	−1·8	275·4	·0041	122·3	−6·6	−1·1	+·0018
5	409·0	−0·9	273·6	·0034	120·1	−8·6	−1·5	+·0003
10	412·0	+2·1	281·3	·0060	117·9	−10·7	−1·8	+·0023
15	410·0	0·0	281·3	·0060	113·9	−14·6	−2·5	+·0030

OCTOBER 18—*continued.*

Gött. mean Time.	Declination. Sale.	Declination. Δψ	Bifilar. Scale corrected for Temp.	Bifilar. Approx. ΔX/X	Inclinometer. Scale corrected for Decl. and Bif.	Inclinometer. $P_1 - P$.	Inclinometer. Approx. Δθ	Approx. Δφ/φ
18 D. H. M.		′				Div.	′	
20	407·0	−3·0	273·5	·0033	115·3	−13·0	−2·2	−·0012
25	407·5	−2·6	268·5	·0016	125·8	−2·4	−0·4	+·0008
30	409·9	−0·2	271·6	·0026	117·8	−10·3	−1·7	−·0009
35	408·0	−2·1	268·9	·0017	118·1	−9·9	−1·7	−·0017
40	407·0	−3·2	269·9	·0020	116·7	−10·1	−1·7	−·0015
45	408·0	−2·2	270·7	·0022	117·0	−10·7	−1·9	−·0013
50	409·5	−0·7	267·5	·0011	120·2	−7·4	−1·3	−·0014
55	408·0	−2·3	268·0	·0012	118·0	−9·4	−1·6	−·0020
13 0	410·0	−0·3	267·0	·0009	117·5	−9·8	−1·7	−·0024
5	411·0	+0·7	265·0	·0002	118·2	−9·0	−1·6	−·0029
10	414·0	+3·6	259·6	−·0017	120·9	−6·3	−1·1	−·0039
15	413·0	+2·5	253·2	−·0039	121·6	−5·5	−0·9	−·0058
20	413·1	+2·6	249·0	−·0053	127·4	0·4	0·1	−·0052
25	414·0	+3·4	252·5	−·0041	125·0	−1·9	−0·3	−·0047
30	415·0	+4·4	253·2	−·0039	123·1	−3·8	−0·6	−·0052
35	413·0	+2·3	255·8	−·0030	120·8	−6·0	−1·0	−·0051
40	413·0	+2·2	258·6	−·0020	118·3	−8·4	−1·4	−·0049
45	413·5	+2·6	258·2	−·0022	121·4	−5·2	−0·9	−·0040
50	412·0	+1·0	259·7	−·0016	121·6	−5·0	−0·8	−·0033
55	412·0	+0·9	259·4	−·0017	128·2	1·6	0·3	−·0022
14 0	412·0	+0·8	260·2	−·0015	126·4	0·0	0·0	−·0015
5	416·2	+5·0	260·2	−·0015	123·8	−2·6	−0·4	−·0024
10	414·4	+3·2	260·6	−·0013	121·7	−4·7	−0·8	−·0029
15	412·8	+1·5	258·1	−·0022	120·7	−5·6	−0·9	−·0041
20	414·0	+2·7	257·7	−·0023	121·8	−4·5	−0·8	−·0038
25	412·4	+1·1	258·6	−·0020	123·3	−3·0	−0·5	−·0030
30	410·0	−1·4	267·7	−·0023	123·1	−3·2	−0·6	−·0034
35	411·0	−0·4	261·4	−·0010	120·4	−5·9	−1·0	−·0030

* Co-ordinate mean values for the term observations, declinometer, 413·35; bifilar, 257·06; inclinometer, 133·83. Impossible to see aurora.

Magnetical Disturbances, Lake Athabasca, 1843—*continued.*

October 18–19. Term Day.

Gött. mean Time.	Declination. Scale.	Declination. Δ	Bifilar. Scale corrected for Temp.	Bifilar. Approx. ΔX/X	Inclinometer. Scale corrected for Decl. and Bif.	Inclinometer. $P_1 - P$.	Inclinometer. Approx. Δθ	Approx. Δφ/φ
18 D. H. M.		′				Div.	′	
14 40	410·0	−1·5	255·0	−·0032	120·4	−5·8	−1·0	−·0042
45	409·2	−2·3	263·0	−·0004	121·7	−4·5	−0·8	−·0019
50	407·0	−4·6	264·2	·0000	123·3	−2·9	−0·5	−·0010
55	408·2	−3·5	266·8	·0009	120·8	−5·4	−0·9	−·0009
15 0	407·6	−4·2	269·4	·0018	122·3	−3·8	−0·6	+·0007
5	410·6	−1·2	273·3	·0031	120·2	−6·0	−1·0	+·0011
10	412·4	+0·6	274·0	·0033	121·1	−5·2	−0·9	+·0015
15	415·4	3·6	276·2	·0041	115·2	−11·4	−1·9	+·0002
20	420·4	8·6	275·5	·0039	115·0	−11·5	−2·0	−·0001
25	426·0	14·2	281·0	·0057	109·7	−16·9	−2·9	−·0001
30	428·0	16·2	277·4	·0045	110·9	−15·7	−2·7	−·0011
35	430·0	18·2	271·9	·0026	114·4	−12·3	−2·1	−·0023
40	432·0	20·1	271·8	·0026	114·6	−12·2	−2·1	−·0016
45	428·0	16·1	272·5	·0028	113·1	−13·8	−2·4	−·0020
50	426·0	14·1	273·0	·0030	111·8	−15·2	−2·6	−·0022
55	422·6	10·7	283·5	·0066	120·7	−6·3	−1·1	+·0044
16 0	409·6	−2·3	266·1	·0006	115·9	−11·1	−1·9	−·0032
5	414·4	2·5	273·4	·0031	115·4	−11·4	−2·0	−·0008
10	416·0	4·1	275·1	·0038	111·4	−15·2	−2·6	−·0014
15	415·4	3·5	276·6	·0043	114·1	−12·3	−2·2	·0000
20	415·0	3·0	272·8	·0030	110·7	−15·5	−2·7	−·0023
25	413·0	+1·0	268·1	·0014	127·3	1·4	0·2	+·0019
30	408·8	−3·2	257·1	−·0023	83·1	−42·6	−7·4	−·0123
35	369·6	−42·4	231·1	−·0112	156·4	30·8	5·4	+·0005
40	401·0	−11·0	247·7	−·0055	132·8	7·5	1·3	−·0029
45	387·0	−25·1	269·1	−·0018	130·5	5·4	0·9	+·0001

October 18–19—*continued.*

Gött. mean Time.	Declination. Scale.	Declination. Δψ	Bifilar. Scale corrected for Temp.	Bifilar. Approx. ΔX/X	Inclinometer. Scale corrected for Decl. and Bif.	Inclinometer. $P_1 - P$.	Inclinometer. Approx. Δθ	Approx. Δφ/φ
18 D. H. M.		′				Div.	′	
20 20	414·0	2·7	251·7	−·0010	127·4	−8·8	−1·5	−·0040
25	414·0	2·6	254·0	−·0001	124·7	−12·3	−2·1	−·0048
30	420·0	8·6	256·1	·0006	124·9	−13·0	−2·2	−·0037
35	422·0	10·6	258·4	·0017	119·8	−19·1	−3·3	−·0048
40	421·5	10·0	260·7	·0027	122·5	−17·3	−3·0	−·0032
45	419·0	7·5	260·3	·0027	124·7	−16·0	−2·8	−·0028
50	415·0	3·5	261·0	·0031	123·3	−18·3	−3·1	−·0038
55	414·0	2·5	264·6	·0044	123·2	−19·3	−3·3	−·0021
21 0	415·0	3·5	266·8	·0054	122·0	−21·4	−3·7	−·0020
5	415·8	4·0	269·0	·0062	121·5	−22·0	−3·8	−·0014
10	415·0	3·0	276·5	·0089	109·5	−34·0	−5·9	−·0028
15	412·0	−0·1	280·3	·0103	111·7	−31·9	−5·6	−·0009
20	408·0	−4·4	272·8	·0079	120·5	−23·2	−4·0	·0000
25	405·0	−7·6	267·0	·0060	122·2	−21·6	−3·8	−·0014
30	408·9	−3·9	264·4	·0058	125·6	−18·2	−3·1	−·0009
35	412·6	−0·4	265·3	·0057	122·9	−21·0	−3·6	−·0015
40	414·0	+0·8	264·8	·0057	125·0	−19·0	−3·3	−·0008
45	414·5	+1·1	266·2	·0063	122·0	−22·0	−3·8	−·0012
50	412·0	−1·6	263·7	·0055	125·8	−18·3	−3·1	−·0007
55	410·5	−3·3	265·1	·0061	124·9	−19·3	−3·3	−·0007
22 0	410·5	−3·5	260·4	·0046	125·3	−19·0	−3·3	−·0020
5	410·0	−4·0	257·2	·0037	128·4	−16·5	−2·8	−·0020
10	410·0	−4·1	262·8	·0057	124·2	−21·3	−3·6	−·0010
15	409·5	−4·6	259·1	·0045	129·1	−17·1	−3·0	−·0014
20	413·4	−0·8	256·7	·0039	127·8	−19·0	−3·3	−·0027
25	414·4	+0·1	259·8	·0050	[illegible]	[illegible]	[illegible]	[illegible]

17	0	409·0	−3·2	311·7	·0164	83·6	−40·8	−7·1	+·0020
	5	403·0	−4·2	313·8	·0172	74·8	−49·6	−8·6	+·0008
	10	410·0	−2·3	314·1	·0172	87·3	−37·2	−6·4	+·0046
	15	418·2	5·9	310·1	·0158	81·9	−42·6	−7·2	+·0012
	20	418·6	6·2	308·4	·0151	95·6	−29·0	−5·0	+·0050
	25	415·0	2·6	297·9	·0115	98·9	−35·7	−6·1	−·0037
	30	406·5	−6·0	291·6	·0093	111·6	−13·0	−2·2	+·0048
	35	406·8	−5·8	278·4	·0048	116·3	−8·4	−1·4	+·0019
	40	409·8	−2·8	279·4	·0051	117·1	−7·6	−1·3	+·0025
	45	408·0	−4·7	281·6	·0058	116·6	−8·1	−1·3	+·0030
	50	410·0	−2 7	282·1	·0059	114·0	−10·7	−1·9	+·0022
	55	408·4	−4·4	283·8	·0064	107 8	−17·0	−3·0	+·0006
[illegible]	0	411·6	−1·2	285·0	·0068	110·4	−14·5	−2·5	+·0018
	5	410·0	−2·9	283·9	·0064	109·8	−15·1	−2·6	+·0012
	10	409·6	−2·3	282·7	·0061	109·5	−15·5	−2·7	+·0008
	15	411·7	−1·2	279·1	·0049	115·5	−9·6	−1·6	+·0016
	20	411·0	−1·9	276·5	·0040	117·6	−7·6	−1·3	+·0014
	25	410·5	−2·4	272·0	·0025	119·1	−6·1	−1·0	+·0005
	30	413·0	+0·1	265·6	·0004	121·9	−3·4	−0·6	−·0016
	35	412·2	−0·8	259·8	−·0015	121·8	−3·6	−0·6	−·0027
	40	415·0	+2·0	256·8	−·0025	122·3	−3·2	−0·6	−·0036
	45	415·0	+2·0	253·5	−·0036	122·1	−3·5	−0·6	−·0048
	50	414·0	+1·0	251·0	−·0044	123·4	−2·2	−0·3	−·0051
	55	414·0	+1·0	249·4	−·0049	124·4	−1·3	−0·2	−·0094
[illegible]	0	412·5	−0·5	249·2	−·0049	126·4	0·6	0·1	−·0029
	5	413·0	+0·1	248·3	−·0051	126·5	0·1	0·0	−·0051
	10	413·0	+0·3	248·4	−·0048	126·2	−0·7	−0·1	−·0024
	15	414·0	+1·4	253·0	−·0029	123·0	−4·5	−0·8	−·0044
	20	414·0	+1·6	250·6	−·0035	122·8	−5·2	−0·9	−·0053
	25	413·0	+0·7	257·8	−·0009	123·4	−5·2	−0·9	−·0026
	30	414·0	+1·9	257·7	−·0007	125·0	−4·1	−0·7	−·0021
	35	414·0	+2·0	259·0	·0000	121·1	−8·6	−1·5	−·0030
	40	414·0	+2·2	260·6	·0008	123·6	−6·7	−1·1	−·0031
	45	413·0	+2·9	263·8	·0020	124·8	−6·5	−1·1	−·0003
	50	412·5	+1·0	257·0	·0000	126·1	−5·3	−0·9	−·0018
	55	414·0	2·6	253·1	−·0013	127·2	−4·7	−0·8	−·0029
[illegible]	0	414·5	3·3	250·7	−·0019	126·2	−6·3	−1·1	−·0041
	5	415 6	4·3	249·8	−·0021	127·5	−5·9	−1·0	−·0041
	10	417·0	5·7	253·1	−·0008	124·1	−10·3	−1·8	−·0043
	15	416·0	4·7	253·3	−·0006	125·6	−9·6	−1·7	−·0039

	40	415·8	−1·4	265·7	·0074	122·2	−27·3	−4·7	−·0020
	45	415·0	−0·5	267·4	·0081	121·0	−29·1	−5·0	−·0021
	50	415·8	−1·2	266·8	·0080	120·4	−30·4	−5·3	−·0025
	55	416·0	−1·4	274·1	·0106	115·7	−35·7	−6·2	−·0017
23	0	413·2	−1·5	279·0	·0124	108·0	−44·0	−7·7	−·0027
	5	410·0	−5·0	272·2	·0101	118·9	−33·0	−5·8	−·0012
	10	410·0	−5·4	270·4	·0095	121·9	−29·9	−5·2	−·0005
	15	412·0	−3·7	269·0	·0091	121·9	−29·7	−5·2	−·0008
	20	412·2	−3·9	267·2	·0085	124·5	−27·0	−4·7	−·0008
	25	415·4	−1·0	265·1	·0078	127·6	−23·8	−4·1	−·0004
	30	415·0	−1·8	261·4	·0066	128·8	−22·4	−3·7	−·0011
	35	417·8	+0·7	262·3	·0069	126·5	−24·6	−4·2	−·0016
	40	418·2	+0·7	261·3	·0066	126·9	−24·1	−4·2	−·0017
	45	418·2	+0·4	260·8	·0064	126·9	−24·0	−4·2	−·0018
	50	419·0	+0·8	265·3	·0080	124·0	−26·5	−4·6	−·0002
	55	420·2	+1·8	266·8	·0086	120·2	−30·4	−5·3	−·0019
19 D.									
H.	M.								
0	0	418·4	+0·5	269·4	·0095	118·3	−32·2	−5·6	−·0006
	5	416·6	−1·6	270·9	·0099	117·1	−32·8	−5·7	−·0014
	10	417·0	−1·5	272·9	·0105	115·4	−33·9	−5·9	−·0012
	15	416·8	−2·0	273·6	·0106	115·6	−33·2	−5·8	−·0008
	20	416·0	−3·1	274·4	·0108	115·1	−33·1	−5·7	−·0007
	25	416·4	−2·9	275·2	·0110	116·3	−31·3	−5·4	+·0002
	30	416·6	−3·0	274·8	·0108	115·6	−31·5	−5·4	−·0001
	35	416·8	−3·1	275·0	·0108	115·3	−31·2	−5·4	+·0001
	40	416·4	−3·8	274·4	·0105	114·8	−31·1	−5·4	−·0005
	45	417·0	−3·5	273·9	·0103	114·6	−31·8	−5·4	−·0001
	50	416·6	−4·2	273·7	·0101	114·5	−30·3	−5·3	−·0004
	55	416·8	−4·3	274·1	·0102	114·4	−29·8	−5·2	−·0001
1	0	416·6	−4·7	273·6	·0099	115·5	−28·2	−4·9	+·0002
	5	416·0	−5·1	273·1	·0095	116·0	−27·1	−4·7	+·0002
	10	416·4	−4·4	272·9	·0092	117·4	−25·2	−4·3	+·0005
	15	417·0	−3·5	272·7	·0090	118·1	−23·9	−4·1	+·0008
	20	416·6	−3·7	271·3	·0083	117·0	−24·5	−4·2	−·0002
	25	416·2	−3·8	270·6	·0078	117·7	−23·2	−4·0	+·0001
	30	415·8	−3·9	271·8	·0081	117·6	−22·8	−4·0	+·0003
	35	415·4	−4·0	272·5	·0081	118·4	−21·4	−3·7	+·0008
	40	416·4	−2·8	269·9	·0070	119·4	−19·9	−3·5	+·0002

Magnetical Disturbances, Lake Athabasca, 1843—*continued.*

OCTOBER 19—*continued.*

Gött. mean Time.	Declination. Scale	Declination. Δ φ	Bifilar. Scale corrected for Temp.	Bifilar. Approx $\frac{\Delta X}{X}$	Inclinometer. Scale corrected for Decl. and Bif.	Inclinometer. P,—P.	Inclinometer. Approx. Δ θ	Approx. $\frac{\Delta \phi}{\phi}$
19 D. H. M.								
1 45	416·8	−2·1	238·7	·0054	118·7	−20·0	−3·5	−·0005
50	417·6	−1·0	270·2	·0067	119·2	−19·0	−3·3	·0030
55	418·2	−0·1	270·6	·0067	118·8	−18·8	−3·3	−·0001
2 0	419·6	+1·6	270·4	·0074	120·2	−16·9	−3·0	+·0006
5	419·4	+1·3	270·2	·0062	120·5	−16·4	−2·9	+·0007
10	419·4	+1·3	268·7	·0056	121·8	−14·8	−2·6	+·0005
15	420·0	+1·8	268·1	·0054	122·5	−14·8	−2·6	+·0003
20	420·0	+1·8	263·1	·0036	122·7	−13·5	−2·3	−·0011
25	420·2	+1·9	262·4	·0032	122·3	−13·7	−2·4	−·0015
30	419·1	+0·8	261·3	·0031	121·0	−14·8	−2·5	−·0020
35	418·0	−0·3	274·4	·0072	119·9	−15·6	−2·6	+·0020
40	421·5	3·1	266·7	·0045	120·6	−14·7	−2·5	−·0006
45	421·9	3·5	267·3	·0046	114·9	−20·2	−3·5	−·0023
50	418·0	−0·4	270·4	·0056	119·8	−16·1	−2·8	·0030
55	417·0	−1·5	271·9	·0060	117·7	−17·0	−3·0	−·0001
3 0	422·0	3·4	265·6	·0038	117·2	−17·3	−3·0	−·0022
5	422·0	3·7	274·2	·0066	118·2	−16·1	−2·8	+·0010
10	420·0	+1·9	271·5	·0056	106·6	−25·5	−4·3	−·0032
15	413·4	−4·5	269·4	·0048	112·0	−21·9	−3·8	−·0027
20	419·2	+1·5	256·0	·0035	121·6	−12·2	−2·1	−·0008
25	422·5	5·0	262·0	·0021	121·1	−12·5	−2·2	−·0025
30	423·0	5·8	263·5	·0026	119·1	−14·3	−2·5	−·0023
35	424·0	6·0	255·6	·0032	120·0	−13·2	−2·3	−·0013
40	421·0	7·2	262·6	·0021	120·0	−13·1	−2·3	−·0023
45	422·0	5·4	262·2	·0019	119·7	−13·2	−2·3	−·0026
[illegible]	[illegible]	6·1	256·7	·0034	116·1	−16·6	−2·9	−·0023

OCTOBER 19–26—*continued.*

Gött. mean Time.	Declination. Scale.	Declination. Δ ψ	Bifilar. Scale corrected for Temp.	Bifilar. Approx. $\frac{\Delta X}{X}$	Inclinometer. Scale corrected for Decl. and Bif.	Inclinometer. Δ θ	Inclinometer. Approx. Δ θ	Approx. $\frac{\Delta \phi}{\phi}$
19. D. H. M.								
6 30	415·6	3·6	281·4	·0086	113·8	−21·7	−3·7	·0011
35	414·0	2·4	281·2	·0085	113·7	−21·8	−3·7	·0010
40	412·6	+1·3	280·2	·0082	115·9	−19·6	−3·2	·0015
45	411·4	+0·5	279·6	·0080	116·7	−18·9	−3·2	·0015
50	412·0	+1·5	278·9	·0077	117·5	−17·1	−3·0	·0018
55	412·4	2·2	279·7	·0079	116·9	−18·8	−3·2	·0014
7 0	411·8	+1·7	279·6	·0079	118·1	−17·6	−3·1	+·0018
5	412·0	+2·0	279·0	·0076	118·3	−17·3	−3·0	+·0016
10	410·0	+0·1	277·1	·0070	118·0	−17·5	−3·0	+·0010
15	411·2	+1·4	275·6	·0064	120·5	−15·0	−2·6	+·0012
20	410·6	+0·9	276·2	·0066	122·2	−13·2	−2·3	+·0021
25	413·1	3·5	275·7	·0063	122·1	−13·2	−2·3	+·0018
30	415·6	6·1	268·4	·0038	124·9	−10·4	−1·8	+·0002
35	418·0	8·6	263·6	·0021	—	—	—	—
40	417·8	8·5	259·1	·0005	126·6	−8·5	−1·5	−·0024
45	415·0	5·7	255·2	−·0005	125·4	−9·7	−1·6	−·0038
50	418·0	8·8	253·4	−·0015	128·5	−6·5	−1·1	−·0037
55	415·8	6·6	246·9	−·0038	129·4	−5·5	−0·9	−·0057
8 0	415·0	5·9	251·5	−·0023	133·4	−1·4	−0·2	−·0028
5	416·0	6·9	219·2	−·0031	134·5	−0·2	−0·0	−·0031
10	415·6	6·6	245·9	−·0042	136·6	2·0	+0·3	−·0035
15	416·0	7·0	243·9	−·0049	136·6	2·2	0·4	−·0042
20	414·0	5·1	243·0	−·0052	136·7	2·4	0·4	−·0044
25	414·5	5·7	243·9	−·0049	137·7	3·5	0·6	−·0037
30	415·8	7·0	247·1	−·0038	135·4	1·4	0·2	−·0031
35	416·0	7·3	251·5	−·0024	132·9	−1·0	−0·2	−·0028
40	416·0	7·4	253·9	−·0015	130·8	−3·0	−0·5	−·0025

5	425·0	9·0	269·6	·0044	117·7	−14·7	−2·6	−·0007
10	424·0	7·9	267·3	·0036	116·2	−16·3	−2·8	−·0020
15	423·2	7·0	268·3	·0040	117·4	−15·1	−2·6	−·0012
20	420·2	3·9	270·1	·0046	115·4	−17·2	−3·0	−·0013
25	420·4	4·1	270·2	·0046	116·8	−15·9	−2·8	−·0009
30	419·6	3·2	282	·0043	117·5	−15·2	−2·6	−·0010
35	422·1	5·6	267·9	·0039	120·1	−12·7	−2·2	−·0005
40	423·6	7·0	264·0	·0025	120·5	−12·3	−2·1	−·0017
45	424·0	7·4	262·5	·0020	120·1	−12·8	−2·2	−·0028
50	422·4	5·7	259·4	·0010	121·7	−11·3	−2·0	−·0029
55	422·2	5·4	260·2	·0013	123·4	−9·6	−1·7	−·0019
5 0	419·6	2·8	257·7	·0004	125·9	−7·2	−1·2	−·0021
5	422·0	5·4	253·6	·0010	132·1	−1·2	−0·2	−·0014
10	419·0	2·6	256·1	·0001	130·4	−3·1	−0·5	−·0012
15	418·0	+1·8	258·7	·0008	129·8	−3·8	−0·6	−·0005
20	417·5	+1·5	259·7	·0012	125·2	−8·6	−1·5	−·0017
25	418·6	2·8	262·7	·0022	12[illegible]·0	−12·0	−2·1	−·0020
30	419·6	3·0	265·3	·0031	122·8	−11·4	−2·0	−·0008
35	[illegible]	4·6	266·7	·0036	123·6	−10·8	−1·9	−·0001
40	421·4	6·2	271·6	·0053	119·9	−14·6	−2·5	+·0003
45	420·0	5·0	271·8	·0054	122·6	−12·1	−2·1	·0012
50	421·6	6·8	272·2	·0056	122·3	−12·6	−2·2	·0013
55	419·6	5·0	274·1	·0062	118·4	−16·7	−2·9	·0004
60	418·2	3·8	275·6	·0068	116·7	−18·6	−3·2	·0004
5	418·4	4·4	280·6	·0084	116·1	−19·2	−3·3	·0018
10	418·0	4·4	280·1	·0083	116·0	−19·3	−3·3	·0016
15	416·2	3·0	280·1	·0082	115·1	−20·3	−3·5	·0013
20	417·0	−4·2	283·3	·0093	115·7	−19·7	−3·4	·0025
25	417·2	4·8	282·9	·0091	114·3	−21·1	−3·6	·0019

50	416·0	7·6			130·0	−3·5		
55	416·5	8·1			129·4	−4·0		
9 0	410·8	2·5			126·5	−6·7		
5	415·8	7·6			136·1	−7·1		
10	416·5	8·3			139·4	−3·7		
15	413·1	4·9			127·5	−5·5		
20	412·0	3·9			128·6	−4·3		
25	410·2	2·1			128·3	−4·5		
30	410·6	2·5			133·6	0·9		
35	412·0	4·0			128·7	−3·8		
40	412·4	4·4			129·3	−3·2		
45	412·2	4·2			128·6	−3·8		
50	410·4	2·5			126·4	−5·8		
55	410·6	2·7			128·0	−4·1		
10 0	411·8	3·9			126·9	−5·2		
26 D.								
H. M.								
15 0	412·0	−4·3			246·7	−15·8		
16 0	416·4	0·1			246·4	−16·1		−·0077
17 0	420·0	3·7			250·4	−12·1		−·0061
18 0	414·0	−2·3			241·6	−20·9		−·0094
19 0	388·0	−28·3			262·2	−0·3		−·0143
20 0	411·2	−5·1			295·4	32·9		−·0173
5	424·0	7·7			315·7	53·2		−·0054
10	417·0	0·7			315·3	52·8		−·0045
15	421·0	4·7			309·7	47·2		−·0079

Mar, 26^{d} 15^{h}. A faint auroral [illegible] at elevation 15° from N.E. to N.W. 16. Arch rising gradually and [illegible] brighter, elevation 18°; a second arch much fainter at [illegible] on 13°. 17^{h}. A faint auroral arch, elevation 5°, to the East of North. 18^{h}. A faint [illegible] arch at an [illegible] of 37°, lightly overcast to the [illegible]. 19^{h}. A faint auroral arch, elevation 5°. 20^{h}. Aurora in heavy masses of moderate [illegible], in [illegible] motion. Long streamers extending to the [illegible]. 20^{h} 40^{m}. Two long beams nearly stationary, altitude 54° and [illegible], brightest to westward of the zenith; [illegible] masses and unconnected streamers in N.N.W. and N.E. 21^{h}. Aurora faint, [illegible] like thin [illegible], in the zenith, with moderate rapid [illegible] to the E., [illegible] somewhat brighter in the N., faint streamers and flexuous [illegible] to the E. and W. 22^{h}. Aurora very faint, streamers [illegible] per[illegible]. 22^{h} 30^{m}. No aurora visible. 27^{d} 0^{h}. A fresh display of aurora, rising rapidly from the northern horizon and [illegible] itself from an [illegible] of 20° N. to 40° S.; at 0^{h} 30^{m} considerable disturbance, bright aurora in flexuous [illegible] to S. and S.E. of the zenith, an arch at elevation 36° extending [illegible] N.E. to N.W. 27^{d} 1^{h}. [illegible] clouded to S. and E.; patches of aurora of [illegible] brightness; at 1^{h} 5^{m} a large [illegible] of aurora passing to the zenith, where it [illegible] faint, in flexuous streaks, without perceptible motion. 2^{h}. No aurora.

* Co-ordinate mean values derived from the means for seven days, 20^{d} to 27^{d}; declinometer, 416·5; bifilar, 236·2; inclinometer, 255·2.

Magnetical Disturbances, Lake Athabasca, 1843—*continued.*

OCTOBER 26–27.

Gött. mean Time.	Declination. Scale.	Declination. Δ ψ	Bifilar. Scale corrected for Temp.	Bifilar. Approx. Δ X / X	Inclinometer. Scale corrected for Decl. and Bif.	Inclinometer. P,—P.	Inclinometer. Approx. Δ θ	Approx. Δ φ / φ
26 D. H. M.		′					′	
20	414·0	−2·3	180·1	−·0189	302·4	39·9	7·0	−·0051
25	425·6	9·3	161·1	−·0254	306·8	44·3	7·7	−·0102
30	413·0	−3·3	174·3	−·0209	299·7	37·2	6·5	−·0081
35	419·8	3·5	186·4	−·0167	292·6	30·1	5·3	−·0062
40	412·0	4·3	168·3	−·0229	311·8	49·3	8·6	−·0059
45	404·0	−12·3	162·5	−·0249	323·2	60·7	10·6	−·0040
50	414·0	−2·3	168·7	−·0228	314·2	51·7	9·0	−·0050
55	416·6	0·3	170·7	−·0221	308·6	46·1	8·1	−·0061
21 0	414·0	−2·3	174·4	−·0208	307·9	45·4	7·9	−·0052
5	414·0	−2·3	174·6	−·0208	304·7	42·2	7·4	−·0062
10	416·0	−0·3	185·2	−·0172	299·1	36·6	6·4	−·0046
15	418·0	1·7	185·8	−·0173	291·5	29·0	5·1	−·0072
20	419·4	3·1	189·6	−·0157	286·2	23·7	4·1	−·0076
25	418·0	1·7	192·1	−·0148	288·2	25·7	4·5	−·0059
30	417·0	0·7	200·5	−·0119	274·0	11·5	2·0	−·0080
35	417·4	1·1	201·3	−·0117	274·8	12·3	2·1	−·0076
40	419·2	2·9	195·5	−·0136	281·4	18·9	3·3	−·0071
45	420·6	4·3	187·9	−·0162	287·8	25·3	4·4	−·0075
50	421·8	5·3	179·5	−·0191	299·2	37·7	6·6	−·0061
55	419·8	3·5	171·9	−·0217	308·1	45·6	8·0	−·0059
22 0	418·6	2·3	166·5	−·0235	315·6	53·1	9·3	−·0051
5	419·0	2·7	167·8	−·0231	313·1	50·6	8·8	−·0037
10	419·6	3·3	177·0	−·0198	[illegible]	38·8	6·8	−·0064
15	418·4	2·1	178·7	−·0194	298·7	36·2	6·3	−·0070
[illegible]	[illegible]	[illegible]	[illegible]	[illegible]	[illegible]	30·5	5·3	−·0085

NOVEMBER 13.

Gött. mean Time.	Declination. Scale.	Declination. Δ ψ	Bifilar. Scale corrected for Temp.	Bifilar. Approx. Δ X / X	Inclinometer. Scale corrected for Decl. and Bif.	Inclinometer. P,—P.	Inclinometer. Approx Δ θ	Approx. Δ φ / φ
27 D. H. M.		′					′	
1 45	414·0	−2·3	184·7	−·0173	279·7	17·2	3·0	−·0124
50	417·0	0·7	195·4	−·0137	269·7	7·2	1·2	−·0113
55	417·0	0·7	202·4	−·0113	268·4	5·9	0·9	−·0096
2 0	414·0	−2·3	209·7	−·0088	261·2	−1·3	−0·2	−·0092
3 0	423·0	6·7	205·0	−·0104	265·3	2·8	0·5	−·0094
Nov. 13 D. H. M.								
3 0	436·0	13·9	194·6	−·0036	138·5	85·1	14·5	·0250
4 0	438·0	17·6	109·8	−·0333	243·9	125·1	21·4	·0090
5	425·0	4·6	105·6	−·0347	237·5	118·6	20·2	·005[illegible]
8	425·5	5·1	105·6	−·0347	293·6	120·6	20·6	·0060
11	423·0	2·6	117·0	−·0308	222·1	103·1	17·6	·0040
17	—	—	138·1	−·0236	190·5	71·5	12·3	·0007
20	420·0	−0·4	134·2	−·0249	142·7	79·0	13·5	·0018
23	423·0	2·6	136·1	−·0243	196·2	77·0	13·1	·0016
26	424·0	3·6	133·0	−·0253	198·4	79·2	13·5	·0014
29	426·0	5·6	137·9	−· 206	190·1	70·9	12·1	·0003
32	418·0	−2·4	151·8	−· 189	171·8	52·6	9·0	−·0011
35	419·5	−0·9	156·5	−·0173	165·3	46·0	7·8	−·0019
38	420·0	−0·3	159·0	−·0164	166·4	47·1	8·1	−·0004
41	424·0	3·7	161·8	−·0144	161·0	41·7	7·1	−·0004
44	417·0	−3·3	163·1	−·0150	174·4	55·0	9·4	·0036
[illegible]	[illegible]	−3·1	[illegible]	[illegible]	[illegible]	[illegible]	7·0	−·0015

40	421·0	4·7	192·6	−·0146	283·4	20·9	3·5	−·0077
45	417·6	1·3	196·0	−·0135	282·0	19·5	3·4	−·0068
50	418·2	1·9	192·7	−·0146	283·1	20·6	3·6	−·00 5
55	419·6	3·3	192·3	−·0147	280·4	17·9	3·1	−·0086
[illegible]3 0	418·4	2·1	196·8	−·0132	276·9	14·4	2·5	−·0083
5	418·2	1·9	198·3	−·0127	274·7	12·2	2·1	−·0086
10	418·0	1·7	200·4	−·0120	272·0	9·5	1·6	−·0088
15	418·0	1·7	205·5	−·0102	269·9	7·4	1·3	−·0076
20	417·6	1·3	208·0	−·0093	260·4	−2·1	−0·4	−·0101
25	417·2	0·9	212·6	−·0078	259·8	−3·8	−0·7	−·0092
30	417·0	0·7	213·2	−· 006	261·2	−1·3	−0·2	−·0080
35	417·6	0·3	206·7	−·0098	265·0	2·5	0·4	−·0090
40	418·2	1·9	203·2	−·0110	272·6	10·1	1·8	−· 0 5
45	421·0	4·7	182·5	−·0181	285·9	23·4	4·1	−·0100
50	421·8	5·5	187·4	−·0164	287·5	25·0	4·4	−·0077
55	422·4	6·1	168·9	−·0227	319·3	56·8	9·9	−·0032
[illegible]7 D.								
H. M.								
0 0	436·7	20·4	143·5	−·0314	354·5	92·0	16·1	·0004
5	438·0	21·7	140·2	−·0325	355·8	93·3	16·3	·00 3
10	442·0	25·7	118·7	−·0398	354·5	92·0	16·1	−·008
15	440·0	23·7	114·7	−·0412	367·3	104·8	18·3	−·0051
20	444·8	28·5	105·1	−·0445	390·2	127·7	22·3	−· 0 5
25	450·0	33·7	75·9	−·0544	415·7	153·2	26·8	−·0015
30	452·4	36·1	60·7	−·0596	449·8	187·3	32·8	·0052
35	450·0	33·7	67·4	−· 503	427·4	164·9	28·9	·0002
40	448·0	31·7	71·0	−·0561	424·6	162·1	28·4	·0000
45	444·0	27·7	64·5	−·0583	436·6	174·1	30·5	·0019
50	432·0	15·7	58·9	−·0600	441·4	178·9	31·3	·0018
55	442·0	25·7	52·5	−·0624	439·4	176·9	31·0	−·0012
1 0	450·5	33·2	62·8	−·0555	414·7	152·2	26·6	−·0030
5	458·0	41·7	81·5	−· 526	423·9	161·4	28·2	·0032
10	423·5	6·2	102·8	−· 66	380·6	117·1	20·4	−·0050
15	426·0	9·7	124·8	−·0378	347·1	84·6	14·8	−·0086
20	435·5	9·2	140·5	−·0324	337·3	74·8	13·1	−·0065
25	414·0	−2·3	160·5	−·0256	311·2	48·7	8·5	−·0088
30	421·5	5·2	186·0	−·0169	288·5	26·0	4·5	−·0080
35	410·0	−6·3	195·4	−·0137	289·9	27·4	4·8	−·0042
40	414·0	−2·3	183·2	−·0178	286·7	24·2	4·2	−·0095

56	412·0	−8·3	167·0	−·0136	161·5	42·0	7·2	·0006
59	418·0	−2·3	166·9	−·0137	155·7	36·1	6·2	−·0015
5 2	414·2	−6·1	171·7	−·0120	156·1	36·5	6·2	·0002
5	416·0	−4·1	173·3	−·0115	154·6	34·8	5·9	·0001
8	420·2	0·3	177·9	−·0099	147·4	37·4	6·4	·0027
11	420·2	0·3	176·7	−·0103	144·9	24·9	4·2	−·0020
14	422·4	2·7	178·5	−·0097	142·7	22·6	3·8	−·0022
17	420·0	0·3	179·3	−·0094	140·2	20·1	3·4	−·0027
20	424·0	4·5	176·7	−· 100	142·5	22·2	3·8	−·0028
23	420·2	0·9	178·3	−· 09	147·1	26·6	4·5	−·0008
26	422·4	3·1	176·7	−·0102	149·5	29·0	4·9	−·0005
29	419·6	0·5	177·9	−·0098	148·7	28·0	4·8	−·0003
32	412·8	−6·3	174·9	−·0109	150·8	30·1	5·1	−·0008
35	413·2	−5·7	174·9	−·0108	151·7	30·8	5·3	−·0003
38	412·0	−6·7	175·1	−·0107	152·1	31·1	5·3	−·0002
41	413·0	−5·7	170·3	−·0124	156·3	36·3	6·2	−·0002
44	416·0	−2·5	182·1	−· 083	151·5	30·3	5·2	·0020
47	417·0	−1·5	182·0	−·0084	148·5	27·3	4·6	·0007
50	410·0	−8·3	188·2	−·0062	137·5	16·1	2·7	−·0009
53	410·0	−8·0	192·7	−·0047	134·6	13·0	2·2	−·0004
56	411·6	−6·4	193·7	−·0043	138·4	16·8	2·9	+·0014
59	410·0	−7·9	187·2	−· 65	141·8	20·0	3·4	·0002
6 2	410·0	−7·9	193·1	−·0045	138·7	16·9	2·9	·0012
16	409·0	−7·8	192·7	−·0047	137·9	16·0	2·7	· 06
31	408·1	−7·6	193·7	−·0044	135·2	13·2	2·2	−·0001
46	407·2	−7·4	194·5	−·0042	134·1	12·0	2·0	−·0003
7 2	408·0	−5·6	204·7	−·0001	104·9	−17·3	2·9	−·0061
19 0	416·0	−0·5	202·8	−·0039	118·4	6·1	1·0	−·0020
20 0	407·0	−7·7	182·5	−·0083	170·0	51·0	8·7	·0089
10	362·0	−52·7	181·0	−·0086	170·4	49·6	8·5	·0082
13	353·0	−61·7	138·2	−· 030	201·1	79·4	13·5	·0037
16	349·0	−65·7	121·8	−· 286	229·1	107·4	18·3	· 0 [illegible]
19	358·2	−56·6	114·0	−·0311	230·6	107·9	18·4	·0052
22	378·0	−36·8	120·2	−·0290	223·4	100·7	17·3	·0052
25	382·1	−32·7	125·6	−·0270	202·3	78·8	13·4	−·0005
28	384·0	−30·9	124·9	−·0271	214·5	90·1	15·4	·0033
31	402·0	−12·9	124·5	−·0272	204·7	80·3	13·7	−·0001
34	408·0	−6·9	130·4	−·0250	189·9	64·5	11·0	−·0033
37	419·8	4·9	136·3	−·0230	187·3	61·9	10·5	−·0042
40	416·2	1·3	137·1	−·0226	185·1	58·8	10·0	−·0029

Clouded, with snow. Co-ordinate mean values, declinometer, 416·1; bifilar, 207·6; inclinometer, 120·4. Correction to declinometer differences, +0'·50.

Magnetical Disturbances, Lake Athabasca, 1843—*continued.*

NOVEMBER 13—*continued.*

Gött. mean Time.	Declination. Scale.	Declination. Δψ	Bifilar. Scale corrected for Temp.	Bifilar. Approx. ΔX/X	Inclinometer. Scale corrected for Decl. and Bif.	Inclinometer. $P_1 - P$.	Inclinometer. Approx. Δθ	Approx. Δφ/φ
13 D. H. M.		′				Div.	′	
20 43	414·4	−0·6	139·2	−·0217	187·8	60·6	10·3	−·0014
46	412·2	−2·8	142·0	−·0208	179·3	52·1	8·9	−·0032
49	396·2	−18·8	159·0	−·0148	168·0	39·9	6·8	−·0014
52	398·0	−17·0	173·8	−·0098	152·1	24·0	4·1	−·0017
55	402·0	−13·0	176·2	−·0088	143·3	14·3	2·4	−·0041
58	404·0	−11·0	186·5	−·0052	138·5	8·6	1·5	−·0022
21 1	402·6	−12·4	182·4	−·0066	149·5	19·6	3·3	−·0001
4	404·4	−10·8	179·5	−·0074	145·6	15·6	2·7	−·0021
7	406·0	−9·2	182·2	−·0065	145·2	15·2	2·6	−·0014
10	406·4	−9·0	183·0	−·0061	139·1	9·1	1·6	−·0029
13	408·2	−7·4	185·2	−·0052	139·8	9·7	1·6	−·0020
16	410·0	−5·6	188·1	−·0042	138·9	8·8	1·5	−·0012
19	407·6	−8·2	187·8	−·0042	140·9	10·7	1·8	−·0007
22	404·4	−11·4	185·3	−·0050	136·5	6·3	1·1	−·0028
25	402·8	−13·2	185·1	−·0050	140·7	10·4	1·8	−·0015
28	410·4	−5·8	185·5	−·0047	139·1	8·8	1·5	−·0017
31	411·3	−4·9	183·9	−·0053	135·4	5·1	0·9	−·0035
34	410·6	−5·8	184·5	−·0050	133·0	2·6	0·4	−·0042
37	410·4	−6·0	185·7	−·0046	132·0	1·6	0·3	−·0040
40	411·6	−5·0	186·3	−·0043	132·1	1·6	0·3	−·0037
43	411·4	−5·4	184·5	−·0047	132·8	2·3	0·4	−·0039
46	411·6	−5·2	185·7	−·0043	131·3	1·8	0·3	−·0037
49	411·4	−5·6	186·8	−·0038	132·3	1·7	0·3	−·0032
52	411·8	−5·2	185·4	−·0043	134·1	3·5	0·6	−·0031
[illegible]	[illegible]	[illegible]	[illegible]	[illegible]	133·8	3·1	0·5	−·0021

DECEMBER 1–2.

Gött. mean Time.	Declination. Scale.	Declination. Δψ	Bifilar. Scale corrected for Temp.	Bifilar. Approx. ΔX/X	Inclinometer. Scale corrected for Decl. and Bif.	Inclinometer. $P_1 - P$.	Inclinometer. Approx. Δθ	Approx. Δφ/φ
1 D. H. M.		′				Div.	′	
21 46	417·6	−0·0	139·4	−·0091	188·9	27·8	4·7	·0002
49	415·8	−2·0	151·2	−·0050	178·3	17·1	2·9	·0007
52	406·	−11·8	166·6	−·0003	161·9	0·7	0·1	−·0008
55	386·0	−32·1	157·4	−·0027	178·3	17·0	2·9	·0030
58	396·0	−22·5	132·5	−·0111	173·3	11·9	2·0	−·0072
22 0	442·0	23·7	142·6	−·0077	194·3	32·9	5·6	·0034
3	435·5	17·1	146·6	−·0062	182·4	20·4	3·5	·0007
6	445·0	26·6	117·6	−·0161	224·4	62·4	10·6	·0048
9	414·0	−4·4	137·6	−·0091	210·3	47·7	8·1	·0069
12	331·0	−37·4	136·1	−·0096	254·6	92·0	15·7	·0224
15	324·0	−94·5	125·6	−·0131	216·4	53·1	9·1	·0049
18	366·0	−52·5	103·4	−·0205	245·6	81·7	13·9	·0069
21	380·8	−37·7	109·0	−·0186	233·9	70·0	11·9	·0049
24	386·0	−32·6	109·1	−·0185	243·4	78·8	13·5	·0082
27	403·9	−14·7	100·7	−·0214	244·8	80·2	13·7	·0068
30	412·0	−7·5	98·6	−·0218	245·4	80·2	13·7	·0058
33	419·0	0·8	97·7	−·0221	241·3	75·4	12·9	·0033
36	436·0	16·2	96·3	−·0226	237·7	71·8	12·4	·0017
39	449·0	29·1	85·8	−·0261	261·7	95·1	16·2	·0060
42	434·0	14·1	67·3	−·0324	288·4	121·8	20·8	·0087
45	434·6	14·7	62·9	−·0341	313·2	146·0	24·9	·0152
48	420·0	0·0	57·7	−·0354	298·1	130·2	22·3	·0096
51	401·0	−19·0	80·4	−·0277	269·2	101·3	17·2	·0065
54	413·6	−6·4	91·5	−·0238	249·3	80·8	13·8	·0036
57	418·0	−2·0	96·0	−·0222	262·7	94·2	16·0	·0096
[illegible] 0	407·8	−11·2	90·8	−·0239	279·6	110·5	18·9	·0134

7	416·6	−0·8	189·1	−·0027	135·7	4·3	0·7	−·0013
10	418·4	0·9	189·5	−·0024	133·2	1·2	0·2	−·0020
13	418·0	0·5	189·7	−·0023	131·6	−1·1	−0·2	−·0027
16	418·4	0·8	189·1	−·0028	131·7	−1·0	−0·2	−·0032
19	418·6	0·9	189·3	−·0023	129·8	−3·5	−0·6	−·0035
22	417·4	−0·3	185·9	−·0034	130·0	−3·3	−0·6	−·0046
25	418·2	0·5	186·2	−·0029	229·1	−4·9	−0·8	−·0045
28	418·0	0·2	187·6	−·0026	126·3	−8·3	−1·4	−·0054
31	417·6	−0·2	187·4	−·0027	123·3	−11·3	−1·9	−·0063
34	419·2	1·4	187·8	−·0024	121·1	−14·2	−2·4	−·0071
37	420·4	2·6	188·1	−·0023	118·8	−16·5	−2·8	−·0078
40	418·6	0·7	189·9	−·0016	119·1	−16·9	−2·9	−·0073
43	418·4	0·4	193·9	−·0001	118·2	−18·4	−3·1	−·0062
46	419·0	1·0	195·9	·0006	116·6	−20·0	−3·4	−·0061
49	416·2	−1·8	197·3	·0012	114·6	−22·7	−3·9	−·0065
52	418·4	0·4	201·7	·0027	116·3	−21·0	−3·6	−·0044
55	420·0	1·9	204·9	·0039	116·4	−21·5	−3·7	−·0034
58	[illegible]	2·4	205·3	·0042	160	−22·5	−3·8	−·0033
23 1	420·4	2·2	207·8	·0050	116·8	−21·7	−3·7	−·0023
0 0	416·0	−6·4	202·2	·0034	117·9	−19·1	−3·3	−·0031
Dec. 1 d. h. m.								
20 0	412·2	−3·3	192·5	·061	131·1	−18·5	−3·2	−·0002
21 0	390·0	−25·9	123·8	−·0155	208·9	+48·4	82	·0007
5	397·0	−19·0	106·5	−·0213	250·4	89·8	15·3	·0089
10	415·2	−3·0	110·5	−·0200	224·9	+64·3	11·0	·0017
13	413·5	−2·9	121·6	−·0160	214·2	53·5	9·1	·0020
16	416·5	−0·1	114·5	−·0183	227·9	67·2	11·5	·0043
19	388·0	−28·6	138·7	−·099	156·2	−4·6	−0·84	−·0083
22	396·0	−20·6	155·7	−·0041	182·8	22·0	3·7	+·0032
25	4032	−13·6	151·6	−·0054	168·7	7·8	13	−·0028
28	410·2	−6·3	146·8	−·0070	66·8	5·9	1·0	−·0050
31	419·0	+2·0	146·7	−·0070	174·3	1·8	0·3	−·0064
34	416·0	−1·2	142·9	−·082	178·5	17·5	3·0	−·0023
37	412·6	−4·8	141·4	−·0087	190·6	39·6	6·8	+·0047
40	416·0	−1·4	142·4	−·082	192·7	31·6	5·4	·0025
43	404·6	−13·0	142·3	−·081	189·3	18·2	4·8	·0014

9	398·0	−21·7	93·1	−·0231	250·4	81·5	13·9	·0044
12	400·4	−19·3	1 021	−·0200	248·4	79·5	13·6	·0069
15	405·0	−15·1	105·6	−·0188	239·3	60·6	10·3	·0023
18	414·0	−6·4	105·3	−·0189	236·8	68·2	11·6	·0042
21	126·0	+5·6	99·6	−·0208	237·0	68·4	11·7	·0025
24	416·0	−4·8	104·3	−·192	233·7	65·2	11·1	·0029
27	423·0	+2·2	115·5	−·0154	220·9	52·4	8·9	·0024
30	123·0	+1·9	119·5	−·0140	236·4	68·1	11·6	·0090
33	429·6	8·2	111·5	−·0166	227·0	58·8	10·0	·0032
36	436·0	14·6	118·6	−·0142	205·0	36·8	6·2	−·0018
39	450·8	29·0	122·5	−·0129	211·3	43·1	7·3	·0018
42	453·2	31·4	118·6	−·0141	217·6	49·5	8·5	·0026
45	446·0	23·8	119·5	−·0139	219·4	51·4	8·7	·0035
48	434·6	12·1	126·9	−·0113	213·8	46·2	7·9	·0043
51	432·0	9·5	134·9	−·0086	197·5	29·9	5·1	·0015
54	137·0	14·1	134·1	−·0088	201·4	33·7	5·7	·0026
57	434·0	11·1	136·9	−·0079	196·1	23·4	4·9	·0017
2 d. h. m.								
0 0	134·0	10·8	142·0	−·0061	190·3	22·7	3·9	·0020
3	434·6	11·2	142·0	−·0062	191·5	24·5	4·2	·0021
6	437·6	14·2	142·0	−·0062	188·5	21·5	3·6	·0011
9	435·0	11·4	146·6	−·0049	186·2	19·8	3·4	·0018
12	436·8	13·2	146·8	−·0046	185·4	19·0	3·2	·0017
15	438·6	14·8	144·4	−·0056	188·9	23·0	3·9	·0022
18	134·0	10·2	146·2	−·0050	186·2	20·9	3·6	·0020
21	430·4	5·4	147·1	−·0047	184·4	19·1	3·2	·0017
24	427·0	2·8	153·8	−·0025	172·6	7·9	1·3	·0001
27	430·0	6·8	155·9	−·0018	176·7	12·0	2·0	·0022
30	433·0	8·6	149·7	−·0040	177·3	13·1	2·2	·0004
33	434·4	9·8	145·7	−·0054	180·1	16·5	2·8	·0002
36	434·6	9·8	146·5	−·0052	186·4	23·4	4·0	·0028
39	132·8	8·0	146·8	−·0052	182·2	19·7	3·4	·0015
42	433·8	90	142·6	−·0066	185·9	22·9	3·9	·0011
45	432·0	7·0	136·7	−·0087	191·7	29·2	5·0	·0012
48	132·1	6·9	137·2	−·0085	192·9	31·0	5·3	·0019
51	432·0	6·8	137·4	−·0085	198·7	36·8	6·3	·0039
54	444·0	18·6	136·4	−·0089	192·3	31·0	5·3	·0015
57	444·4	19·0	132·9	−·0101	201·1	39·8	6·8	·0033
1 0	442·0	16·3	131·3	−·0107	198·0	37·3	6·5	·0019

Clouded. Co-ordinate mean values, declinometer, 417·4; bifilar, 175·2; inclinometer, 151·0. Clouded throughout. A violent N.W. gale from 2d 5h to 2d 5h.
Co-ordinate mean values, declinometer, 417·4; bifilar, 175·2; inclinometer, 151·0.

Magnetical Disturbances, Lake Athabasca, 1843—*continued.*

December 2—*continued.*

Gött. mean Time.	Declination. Scale.	Declination. Δ ψ	Bifilar. Scale corrected for Temp.	Bifilar. Approx. Δ X / X	Inclinometer. Scale corrected for Decl. and Bif.	Inclinometer. P,—P.	Inclinometer. Approx. Δ θ	Approx. Δ φ / φ
2 D. H. M.		′					′	
1 3	452·0	26·5	131·3	—·0110	198·0	37·8	6·6	·0017
6	454·4	28·9	125·7	—·0129	207·6	47·4	8·1	·0032
9	460·6	35·3	119·7	—·0151	215·6	55·9	9·5	·0038
12	458·0	32·7	129·6	—·0117	203·6	43·9	7·5	·0030
15	449·4	24·3	134·0	—·0104	198·3	39·2	6·7	·0028
18	446·2	21·3	137·6	—·0094	195·0	36·4	6·2	·0023
21	441·4	16·5	135·8	—·0100	198·9	40·3	6·9	·0035
24	440·2	15·5	135·0	—·0107	202·1	44·1	7·5	·0041
27	440·0	15·3	135·5	—·0103	203·4	45·4	7·7	·0050
30	447·6	23·1	125·7	—·0139	211·3	43·8	7·5	·0009
33	446·0	21·7	122·6	—·0151	211·0	54·1	9·2	·0031
36	444·6	20·3	116·6	—·0172	219·4	54·5	9·3	·0012
39	448·7	24·6	111·6	—·0191	231·6	75·2	12·9	·0063
42	446·0	21·9	107·3	—·0206	237·9	81·5	13·9	·0069
45	448·7	24·8	102·1	—·0225	250·0	94·2	16·1	·0092
48	451·2	27·5	97·9	—·0241	245·7	90·4	15·4	·0064
51	448·0	24·3	102·5	—·0226	237·9	82·6	14·1	·0053
54	449·8	26·3	107·2	—·0212	234·9	80·2	13·7	·0060
57	451·4	27·9	106·4	—·0215	243·0	88·3	15·1	·0084
2 0	450·1	26·9	88·1	—·0279	253·1	98·9	16·9	·0054
3	451·5	28·3	86·5	—·0285	254·1	100·1	17·1	·0052
6	453·7	30·5	85·9	—·0287	264·7	110·7	18·9	·0087
9	462·0	38·8	68·3	—·0348	275·5	123·8	21·2	·0070
12	462·6	39·4	63·6	—·0365	288·4	134·7	23·0	·0092
15	458·5	35·3	57·3	—·0387	302·7	149·2	25·6	·0018
[illegible]	[illegible]	[illegible]	[illegible]	[illegible]	[illegible]	[illegible]	[illegible]	·0113

Dec. 28. Jan. 4, 1844.

Gött. mean Time.	Declination. Scale.	Declination. Δ ψ	Bifilar. Scale corrected for Temp.	Bifilar. Approx. Δ X / X	Inclinometer. Scale corrected for Decl. and Bif.	Inclinometer. P,—P.	Inclinometer. Approx. Δ θ	Approx. Δ φ / φ
2 D. H. M.		′					′	
4 24	426·4	5·2	177·7	+·0010	145·0	—4·8	—0·8	—·0027
27	425·7	4·5	173·6	—·0004	153·3	3·5	0·6	·0008
30	428·4	7·2	167·2	—·0026	158·0	8·2	1·4	·0001
33	428·6	7·4	167·7	—·0024	156·6	6·7	1·1	—·0001
36	428·4	7·2	168·7	—·0021	154·1	4·2	0·7	—·0006
39	426·1	4·9	168·6	—·0021	150·9	1·0	0·2	—·0018
42	423·2	2·0	169·9	—·0017	147·3	—2·6	—0·4	—·0026
45	424·2	3·0	175·4	·0002	148·5	—1·9	—0·3	—·0005
48	423·8	2·6	177·9	·0011	146·9	—3·2	—0·5	—·0001
51	424·1	2·9	181·8	·0024	150·1	0·0	0·00	·0024
54	423·8	2·7	181·0	·0021	140·1	—10·0	—1·7	—·0012
57	423·4	2·3	179·7	·0017	138·9	—11·2	—1·9	—·0020
5 0	422·0	0·8	179·4	·0016	138·5	—11·7	—2·0	—·0023
6 0	416·0	—2·7	184·7	·0036	134·1	—18·3	—3·1	—·0026
28 D. H. M.								
1 0	421·4	—6·8	175·4	·0060	176·5	—	—1·3	·0033
2 0	400·4	—25·4	164·9	—·0000	164·3	—	—2·3	—·0056
3 0	477·8	49·5	112·9	—·0188	229·5	—	9·3	—·0004
6	465·0	36·9	120·9	—·0161	235·7	—	10·3	·0001
9	454·0	26·0	135·1	—·0113	219·5	—	7·6	·0037
12	422·0	—6·0	153·5	—·0051	201·4	—	4·5	·0039
15	415·8	—12·0	166·8	—·0006	169·6	—	—0·9	—·0023
[illegible]	[illegible]	[illegible]	[illegible]	[illegible]	[illegible]	[illegible]	[illegible]	—·0036

27	472·0	48·9	50·3	−·0412	324·3	171·2	29·2	·0166	23	439·0	11·3	189·7	·0071	147·0	—	−·7	−·0021
30	475·2	52·1	49·5	−·0440	316·5	163·6	27·9	·0113	27	434·8	7·2	188·1	· 66	153·9	—	−3·4	−·0004
33	474·4	51·3	41·7	−·0443	328·9	176·3	30·1	·0151	30	432·0	4·4	188·6	·0063	155·5	—	−3·2	−·0000
36	449·0	18·9	47·1	−·0425	309·9	145·3	24·8	·0066	33	429·0	1·7	187·6	·0063	158·3	—	−2·7	·0009
39	439·8	16·7	59·1	−·0385	301·3	148·9	25·4	·0112	36	429·6	2·3	188·6	· 66	154·9	—	−3·2	·0002
42	420·2	−2·9	74·5	−·0332	288·3	130·9	22·3	· 00	39	428·0	0·8	185·6	·0055	159·5	—	−2·4	·0007
45	422·0	−1·0	81·5	−·0309	259·9	107·7	18·4	·0055	42	428·0	0·8	188·1	·0064	156·5	—	−2·9	·0005
48	448·0	25·0	83·3	−·0304	265·5	113·4	19·4	·0075	45	425·0	−2·0	192·6	·0078	150·1	—	−4·0	−·0001
51	448·4	25·4	80·1	−·0315	255·7	103·7	17·7	· 66	48	425·8	−1·1	195·2	·0087	151·5	—	−3·7	−·0003
54	440·6	17·0	76·1	−·0329	282·6	130·8	22·3	·0113	51	421·5	−5·4	197·4	·0095	148·0	—	−4·3	·0008
57	452·1	29·1	72·1	−·0343	277·0	125·2	22·4	·0080	54	422·5	−4·2	198·4	·0097	147·3	—	−4·4	·0009
3 0	446·0	23·0	64·3	−·0370	271·7	120·1	20·5	·0085	57	424·5	−2·2	198·8	·0099	145·4	—	−4·8	·0005
3	456·4	33·6	62·1	−·0378	263·0	111·6	19·0	−·0008	4 0	424·0	−2·6	199·3	· 00	143·7	—	−5·0	−·0000
6	456·0	33·2	59·0	−·0389	271·3	119·9	20·5	·0016	5 0	426·0	−0·5	200·6	· 05	146·8	—	−4·6	·0014
9	458·2	34·5	61·1	−·0382	269·5	118·3	20·2	·0017									
12	460·4	37·7	63·3	−·0375	266·9	115·7	19·8	·0015	1844.								
15	470·0	47·5	65·0	−·0370	256·3	105·8	18·1	−·0011	Jan.								
18	476·5	54·0	68·2	−·0359	256·9	106·0	18·1	−·0002	4 d.								
21	481·8	60·4	68·8	−·0358	256·6	105·7	18·0	−·0001	h. m.								
24	482·2	59·9	68·5	−·0359	248·9	98·2	16·8	−·0007	15 0	426·2	1·3	177·6	·0074	170·7	—	−2·4	·0026
27	484·0	61·7	70·6	−·0351	247·9	97·2	16·6	−·0023	16 0	429·0	4·0	212·1	·0192	122·5	—	−10·8	−·0021
30	479·2	47·1	77·4	−·0329	240·9	90·4	15·4	−·0023	9	417·0	−8·1	211·4	·0190	138·2	—	−8·0	·0030
33	466·0	44·0	80·3	−·0320	239·7	89·4	15·3	−·0020	12	430·2	5·1	205·6	·0170	134·7	—	−8·6	−·0001
36	459·6	37·6	99·6	−·0254	209·8	58·5	10·0	−·0057	15	430·0	4·9	201·2	·0152	147·0	—	−6·5	·0022
39	456·0	34·1	108·9	−·0223	204·2	54·0	9·2	−·0041	18	423·0	−2·1	203·4	·0197	138·0	—	−8·0	·0038
42	455·8	33·9	112·6	−·0211	204·8	54·6	9·3	−·0026	21	423·8	−1·3	204·6	· 67	134·9	—	−8·6	−·0017
45	452·4	30·7	114·9	−·0203	200·7	50·7	8·7	−·0032	24	420·0	−5·1	216·6	·0208	120·9	—	−10·9	−·0007
48	446·8	25·2	109·9	−·0221	204·3	65·0	11·1	−·0001	27	423·8	−1·3	216·2	· 97	123·0	—	−10·5	−·0002
51	449·0	27·4	106·9	−·0153	201·2	51·4	8·6	−·0021	30	420·0	−5·2	218·6	·0215	118·0	—	−11·3	−·0010
54	446·0	24·6	132·6	−·0144	181·7	32·1	5·5	−·0036	33	415·1	−10·1	214·9	·0203	126·9	—	−9·8	·0008
57	441·7	20·3	133·2	−·0142	182·0	32·4	5·5	−·0032	36	424·2	−1·0	26·6	·0208	123·7	—	−10·3	·0004
4 0	442·2	20·9	137·6	−·0128	177·3	27·9	4·8	−·0033	39	425·0	−0·2	213·4	·0198	146·8	—	−6·4	·0071
3	440·0	18·7	142·2	−·0112	168·5	19·0	3·2	−·0048	42	138·2	13·0	209·4	·0184	131·3	—	−9·0	·0006
6	439·3	18·5	149·0	−·0089	164·2	14·7	2·5	−·0039	45	444·8	19·6	213·6	·0199	114·1	—	−11·9	−·0037
9	435·9	14·6	157·3	−·0060	159·6	10·0	1·7	−·0026	48	449·6	24·3	221·8	·0227	104·6	—	−13·3	−·0041
12	433·8	17·5	158·6	−·0056	156·8	7·2	1·2	−·0031	51	448·2	22·9	221·3	·0225	111·7	—	−12·3	−·0018
15	431·0	12·7	173·0	−·0006	145·4	−4·2	−0·7	−·0020	54	448·4	23·1	219·6	·0220	112·6	—	−12·1	−·0020
18	430·6	9·3	175·4	·0002	143·1	−6·6	−1·1	−·0021	57	439·7	14·4	218·3	·0215	117·2	—	−11·3	−·0009
21	428·0	6·7	174·5	−·0001	147·5	−2·2	−0·4	−·0009	17 0	444·0	18·7	205·8	·0173	130·4	—	−9·0	−·0005

Declinometer, 422·8 ; bifilar, 170·3 ; inclinometer, 174·6.

Magnetical Disturbances, Lake Athabasca, 1844—*continued.*

January 4—*continued.*

Gött. mean Time.	Declination. Scale.	Declination. Δ ψ	Bifilar. Scale corrected for Temp.	Bifilar. Approx. $\frac{\Delta X}{X}$	Inclinometer. Scale corrected for Decl. and Bif.	Inclinometer. P, − P.	Inclinometer. Approx. Δ θ	Approx. $\frac{\Delta \phi}{\phi}$
4 D. H. M.		′				Div.	′	
17 3	436·0	10·7	191·0	·0122	146·9	—	−6·2	−·0001
6	442·0	16·7	193·9	·0132	139·6	—	−7·1	−·0009
9	444·0	18·6	189·1	·0115	147·9	—	−6·0	−·0004
12	444·2	18·8	188·2	·0112	147·7	—	−6·1	−·0008
15	440·6	15·1	185·0	·0100	143·0	—	−6·9	−·0036
18	438·2	12·7	187·5	·0105	152·5	—	−5·2	·0001
21	440·2	14·7	194·5	·0132	144·2	—	−6·7	·0003
24	440·0	14·4	198·3	·0145	144·5	—	−6·6	·0013
27	436·8	11·2	203·5	·0163	135·5	—	−8·1	·0030
30	437·6	12·0	208·2	·0178	140·0	—	−7·4	·0013
33	434·2	8·5	207·8	·0176	134·0	—	−8·4	·0009
36	440·0	14·3	210·4	·0185	133·8	—	−8·5	·0018
39	436·0	10·2	206·9	·0173	148·2	—	−6·0	·0053
42	435·2	9·4	202·4	·0157	146·1	—	−6·4	·0031
45	432·0	6·2	200·4	·0150	144·6	—	−6·6	·0018
48	432·4	6·6	194·6	·0130	147·8	—	−6·1	·0009
51	426·0	0·1	196·4	·0136	148·6	—	−6·0	·0018
54	426·8	0·9	198·5	·0143	144·3	—	−6·7	·0009
57	423·0	−2·9	198·3	·0142	144·8	—	−6·6	·0010
18 0	422·0	−4·0	198·1	·0141	148·5	—	−6·0	·0021
3	421·6	−4·4	198·5	·0143	146·7	—	−6·3	·0018
6	421·0	−5·0	203·1	·0158	141·8	—	−7·1	·0017
9	420·6	−5·4	207·9	·0175	137·2	—	−7·9	·0019
12	420·0	−6·0	213·1	·0193	128·3	—	−9·4	−·0018
15	418·2	−7·8	206·7	·0171	132·0	—	−8·8	−·0026

January 4–5.

Gött. mean Time.	Declination. Scale.	Declination. Δ ψ	Bifilar. Scale corrected for Temp.	Bifilar. Approx. $\frac{\Delta X}{X}$	Inclinometer. Scale corrected for Decl. and Bif.	Inclinometer. P, − P.	Inclinometer. Approx. Δ θ	Approx. $\frac{\Delta \phi}{\phi}$
4 D. H. M.		′				Div.	′	
21 27	412·0	−13·7	78·2	−·0214	289·7	—	14·9	·0081
30	419·0	−6·9	82·8	−·0197	303·3	—	17·2	·0143
33	445·0	18·9	95·5	−·0153	307·9	—	18·0	·0203
36	418·0	−8·1	83·8	−·0192	286·3	—	14·3	·0090
39	415·0	−11·3	93·5	−·0158	254·8	—	8·9	·0015
42	445·0	18·7	84·5	−·0189	286·3	—	14·3	·0094
45	420·0	−6·5	89·0	−·0172	270·1	—	11·5	·0055
48	438·0	11·3	91·1	−·0164	248·2	—	7·7	−·0011
51	442·0	15·3	107·0	−·0110	219·1	—	2·7	−·0054
54	470·0	43·1	116·8	−·0075	240·7	—	6·4	·0052
57	448·0	21·1	120·0	−·0064	245·3	—	7·2	·0071
22 0	448·0	20·9	116·9	−·0074	226·7	—	4·0	·0006
3	448·4	21·3	120·6	−·0060	216·0	—	2·1	−·0020
6	463·8	36·7	135·6	−·0009	195·8	—	−1·5	−·0042
9	467·8	40·6	136·5	−·0004	216·3	—	2·0	·0036
12	451·2	24·0	132·0	−·0019	223·1	—	3·2	·0044
15	438·0	10·7	138·3	·0008	220·5	—	2·6	·0055
18	430·4	3·1	136·5	−·0003	224·3	—	3·2	·0061
21	435·0	7·6	140·3	·0011	205·7	—	0·0	·0012
24	446·4	19·0	144·9	·0028	195·5	—	−1·8	−·0009
27	450·0	22·6	152·8	·0055	191·1	—	−2·6	·0004
30	452·0	24·5	157·0	·0070	190·0	—	−2·9	·0014
33	451·2	23·6	150·7	·0053	203·8	—	−0·6	·0037
36	449·2	21·6	135·0	−·0003	233·7	—	4·5	·0084
39	432·0	4·3	116·3	−·0032	257·6	—	8·4	·0135
42	424·0	−3·7	69·0	−·0227	330·2	—	20·8	·0185

H.	M.								
	27	417·3	−8·8	209·4	·0182	135·7	—	−8·2	·0019
	30	411·0	−15·1	208·0	·0178	133·4	—	−8·6	·0007
	33	414·0	−12·1	202·8	· 61	141·1	—	−7·4	·0016
	36	413·8	−12·3	202·4	·0159	145·9	—	−6·5	·0029
	39	416·0	−10·1	206·4	·0174	141·5	—	−7·3	·0029
	42	418·2	−7·9	208·4	· 160	134·0	—	−8·6	·0011
	45	416·4	−9·7	205·0	·0169	137·5	—	−8·0	· 010
	48	404·6	−21·5	198·2	· 166	155·4	—	−4·9	·0048
	51	410·4	−15·7	180·2	·0085	171·8	—	−2·3	·0039
	54	432·2	8·1	202·9	·0162	144·5	—	−6·6	·0031
	57	438·4	12·3	209·5	·0185	115·2	—	−11·8	−·0049
19	0	422·0	−4·2	224·7	·0237	110·3	—	−12·7	−·0014
	3	419·7	−6·5	225·8	·0243	108·8	—	−13·0	−· 016
	6	406·2	−19·8	235·7	·0277	85·5	—	−17·0	−·0087
	9	396·2	−29·7	239·0	·0287	105·3	—	−12·0	·0048
	12	396·7	−29·2	228·5	·0254	112·8	—	−12·4	·0008
	15	405·4	−20·3	220·1	·0227	125·6	—	−8·6	·0056
	18	400·0	−25·6	211·3	·0200	146·9	—	−6·8	·0072
	21	426·0	0·4	210·5	· 107	138·4	—	−8·3	·0033
	24	416·0	−9·4	211·7	·0203	132·6	—	−9·4	·0018
	27	411·0	−14·4	206·9	· 67	144·3	—	−7·3	·0040
	30	404·1	−21·2	202·6	· 104	147·5	—	−6·9	·0038
	33	401·8	−23·3	201·2	· 071	148·1	—	−6·9	·0037
	36	[illegible]62	−18·9	199·8	·0167	150·0	—	−6·6	·0036
	39	18·1	−11·9	204·4	·0184	142·0	—	−8·0	·0025
	42	422·3	−2·7	201·4	·0174	149·4	—	−6·8	·0040
	45	419·6	−5·2	194·4	· 102	154·2	—	−6·0	·0033
	48	416·7	−8·1	190·7	· 102	155·6	—	−5·9	·0025
	51	419·5	−5·2	181·9	·0146	162·6	—	−4·7	·0052
	54	414·6	−9·9	180·5	·0109	163·8	—	−4·6	·0018
	57	414·4	−10·1	178·3	· 01	164·2	—	−4·5	·0012
20	0	414·0	−10·4	177·7	·0101	164·2	—	−4·6	·0010
21	0	384·0	−40·7	54·7	· 600	358·9	—	26·7	·0229
	6	—	—	57·7	−·0289	357·1	—	26·4	·0245
	9	422·8	−2·3	53·8	−·0302	300·5	—	16·8	·0036
	12	430·0	4·9	86·1	−· 101	302·1	—	17·0	· 106
	15	384·6	−40·7	71·4	−·0240	290·8	—	15·1	·0059
	18	415·0	−10·3	84·0	−· 105	312·5	—	18·8	·0176
	21	436·0	10·5	83·5	−· 107	307·8	—	18·0	· 109
	24	414·6	−11·1	83·1	−· 137	304·0	—	17·3	·0145

H.	M.								
	51	160·0	32·3	−17·9	−·0522	414·5	—	35·0	·0170
	54	475·0	47·2	−3·4	−·0471	426·3	—	36·9	·0259
	57	464·0	36·2	−12·2	−·0501	426·3	—	36·9	·0229
23	0	460·0	32·2	−37·4	−·0585	472·9	—	44·8	·0299
	3	456·0	27·8	−51·9	−·0635	477·0	—	45·5	·0268
	6	468·0	39·8	−33·7	−·0573	445·1	—	40·0	·0219
	9	458·0	29·5	−16·4	−·0513	441·6	—	39·4	·0267
	12	[illegible]5·0	26·5	−21·4	−·0531	455·3	—	41·8	·0296
	15	432·0	13·1	−13·3	−·0503	137·6	—	38·8	·0264
	18	417·0	−12·2	19·4	−·0392	377·9	—	28·6	·0176
	21	427·2	−2·0	15·6	−·0404	393·6	—	31·13	·0215
	24	423·2	−6·4	11·1	−·0419	403·5	—	33·0	·0234
	27	418·0	−11·6	24·3	−·0374	352·9	—	24·4	·0109
	30	433·5	3·6	38·3	−·0322	364·0	—	26·3	·0198
	33	437·0	6·7	30·9	−·0351	376·1	—	28·4	·0216
	36	440·5	10·2	34·4	−·0339	352·8	—	24·4	·0144
	39	444·0	13·4	51·6	−·0280	357·9	—	25·3	·0122
	42	455·0	24·4	58·7	−·0256	325·6	—	19·8	·0136
	45	434·0	3·0	72·0	−·[illegible]20	307·1	—	16·7	·0121
	48	444·8	13·5	80·3	−·0182	293·0	—	14·3	·0101
	51	445·0	13·7	89·6	−· 160	284·4	—	12·8	·0104
	54	446·0	14·4	85·6	−·0163	274·3	—	11·1	·0046
	57	448·0	16·4	93·5	−· 166	265·5	—	9·6	·0054
5 p.									
H. M.									
0	0	449·8	17·8	107·6	−·0088	249·9	—	7·0	· 051
	3	451·2	18·9	112·5	−·0072	243·6	—	5·9	·0045
	6	442·2	9·9	119·6	−·0048	226·3	—	2·9	· 010
	9	444·6	12·1	131·9	−·0006	226·8	—	3·2	−·0002
	12	444·0	11·5	134·6	·0003	214·7	—	1·1	·0025
	15	440·0	7·3	142·6	·0029	213·9	—	1·1	·0051
	18	441·5	8·6	143·6	·0032	209·4	—	0·4	·0040
	21	441·0	8·1	139·6	· 018	204·5	—	−0·4	·0009
	24	436·0	3·0	149·6	· 051	211·6	—	0·9	·0031
	27	440·0	7·0	146·0	·0039	196·5	—	−1·7	·0005
	30	438·0	4·7	140·3	· 09	193·0	—	−2·4	−·0028
	33	436·0	2·5	116·5	−·0063	230·8	—	4·4	·0023
	36	448·5	15·0	99·6	−·0121	256·9	—	8·8	·0053
	39	446·0	12·3	88·5	−· 160	273·8	—	11·8	·0077
	42	458·0	24·3	35·0	−·0342	249·2	—	7·4	−·0196

Magnetical Disturbances, Lake Athabasca, 1844—*continued.*

JANUARY 5—*continued.*

Gött. mean Time.	Declination. Scale.	Declination. Δψ	Bifilar. Scale corrected for Temp.	Bifilar. Approx. ΔX/X	Inclinometer. Scale corrected for Decl. and Bif.	Inclinometer. $P_1 - P$.	Inclinometer. Approx. Δθ	Approx. Δφ/φ
5 D. H. M.		′					′	
0 45	436·5	2·6	44·5	−·0310	329·5	—	21·4	·0118
48	485·0	50·9	44·4	−·0312	314·0	—	18·8	·0061
51	468·0	33·9	71·6	−·0219	291·6	—	15·0	·0059
54	483·0	48·7	63·9	−·0246	290·1	—	14·9	·0048
57	500·0	65·7	56·5	−·0271	309·8	—	18·2	·0089
1 0	497·2	62·7	71·4	−·0221	261·9	—	10·1	−·0022
3	470·2	35·9	82·6	−·0185	276·6	—	12·7	·0066
6	474·0	39·7	71·2	−·0224	274·5	—	12·4	·0021
9	473·0	38·9	74·4	−·0215	274·1	—	12·4	·0030
12	462·8	28·7	82·8	−·0186	280·6	—	13·5	·0081
15	463·1	29·2	84·9	−·0181	272·7	—	12·2	·0061
18	459·2	25·3	92·6	−·0157	269·9	—	11·9	·0079
21	456·0	22·3	111·2	−·0098	267·1	—	11·4	·0103
24	461·4	27·9	92·6	−·0159	259·4	—	10·2	·0042
27	450·9	17·4	111·6	−·0094	250·0	—	8·61	·0075
30	449·2	15·9	103·4	−·0124	245·7	—	7·9	·0030
33	446·4	13·3	124·4	−·0055	218·9	—	3·5	·0014
36	439·2	6·1	144·3	·0013	198·2	—	−0·1	·0000
39	436·0	4·1	163·1	·0088	190·4	—	−1·3	·0062
42	429·8	−3·1	169·3	·0097	175·1	—	−3·9	·0019
45	421·0	−11·7	173·5	·0109	179·0	—	−3·2	·0047
48	423·2	−9·3	170·0	·0095	179·6	—	−3·0	·0037
51	426·0	−6·5	169·8	·0095	174·1	—	−3·9	·0017
54	422·4	−9·9	171·6	·0099	181·9	—	−2·5	·C049
57	424·3	−8·0	166·1	·0080	183·6	—	−2·2	·0037
[illegible]	[illegible]	[illegible]	165·9	·0075	181·9	—	−2·4	·0027
								·0029

JANUARY 5–6.

Gött. mean Time.	Declination. Scale.	Declination. Δψ	Bifilar. Scale corrected for Temp.	Bifilar. Approx. ΔX/X	Inclinometer. Scale corrected for Decl. and Bif.	Inclinometer. $P_1 - P$.	Inclinometer. Approx. Δθ	Approx. Δφ/φ
5 D. H. M.		′					′	
8 57	422·0	0·7	161·9	·0038	183·6	—	−1·5	·0009
9 0	422·6	1·3	167·1	·0055	189·1	—	−0·5	·0045
10 0	420·0	−1·0	145·7	−·0027	204·1	—	2·3	·0018
11 0	419·0	−3·0	164·2	·0033	177·4	—	−2·0	·0011
22 0	420·0	−7·1	156·3	·0061	177·9	—	4·3	·0025
23 0	408·4	−19·4	−49·5	−·0626	380·3	—	28·9	−·0054
3	428·0	0·2	−29·5	−·0558	372·0	—	27·5	−·0014
6	441·4	13·2	−51·5	−·0633	355·7	—	24·8	−·0143
9	438·0	9·5	−31·7	−·0566	356·2	—	24·9	−·0075
12	432·0	3·5	−21·5	−·0497	367·1	—	26·7	·0031
15	429·4	0·5	−4·1	−·0471	376·6	—	28·4	·0090
18	430·2	1·3	−13·5	−·0504	376·8	—	28·4	·0059
21	443·0	13·8	13·7	−·0410	343·0	—	22·6	·0028
24	453·9	24·3	20·5	−·0387	347·2	—	23·6	·0081
27	458·0	28·4	16·5	−·0400	347·9	—	23·5	·0065
30	458·4	28·5	17·7	−·0396	341·3	—	22·4	·0048
33	462·4	32·1	13·3	−·0410	340·8	—	22·3	·0032
36	456·8	26·5	31·9	−·0347	315·6	—	18·0	·0009
39	462·0	31·4	31·5	−·0349	319·0	—	18·6	·0018
42	462·8	32·2	39·2	−·0322	305·1	—	16·3	·0000
45	467·7	36·7	40·5	−·0318	312·7	—	17·6	·0020
48	472·4	41·4	42·5	−·0311	300·4	—	15·6	−·0003
51	472·0	40·7	39·6	−·0320	304·9	—	16·3	−·0032
54	458·4	26·8	49·5	−·0286	292·9	—	14·2	−·0006
57	450·9	19·3	52·9	−·0275	301·5	—	15·7	·0035

9	432·4	0·4	156·3	·0043	174·1	—	−3·6	−·0028
12	428·0	−4·0	154·7	·0037	192·1	—	−0·6	·0027
15	424·0	−8·0	155·9	·0041	196·7	—	0·2	·0046
18	425·7	−6·3	155·3	·0038	193·8	—	−0·2	−·0006
21	426·2	−5·8	160·1	·0054	186·4	—	−1·5	·0025
24	426·4	−5·5	164·2	·0067	180·6	—	−2·4	·0020
27	426·0	−5·9	164·9	·0070	180·4	—	−2·5	·0021
30	426·8	−5·1	167·2	·0076	179·4	—	−2·6	·0028
33	425·0	−6·9	171·6	·0091	175·5	—	−3·2	·0027
36	424·4	−7·5	172·0	·0092	173·4	—	−3·5	·0025
39	422·1	−9·8	172·2	·0092	169·6	—	−4·2	·0008
42	418·0	−13·9	178·0	·0112	168·9	—	−4·3	·0026
45	417·8	−14·1	178·2	·0112	168·2	—	−4·4	·0024
48	418·0	−13·8	179·8	·0116	165·9	—	−4·7	·0022
51	418·0	−13·8	182·1	·0124	165·0	—	−4·9	·0027
54	419·6	−12·1	177·7	·0108	167·6	—	−4·4	·0021
57	420·0	−11·7	176·1	·0103	170·4	—	−3·9	·0025
3 0	419·0	−12·3	178·1	·0109	167·8	—	−4·3	·0022
4 0	424·2	−5·9	175·5	·0092	173·5	—	−3·0	·0033
5 0	432·0	−2·0	166·2	·0062	196·5	—	0·8	·0044
6 0	420·0	−7·5	159·5	·0041	180·1	—	−2·4	−·0007
7 0	426·0	−2·8	162·1	·0047	185·3	—	−1·6	·0012
8 0	419·0	−3·3	133·1	−·0057	213·6	—	3·4	·0020
3	—	—	149·6	·0000	203·8	—	1·7	·0037
6	414·2	−8·0	141·3	−·0029	199·6	—	1·0	·0007
9	413·8	−8·3	146·8	−·0011	203·4	—	1·7	·0023
12	414·2	−7·9	144·3	−·0019	207·8	—	2·4	·0029
15	417·0	−5·0	156·3	·0022	192·3	—	−0·1	·0018
18	417·8	−4·2	154·7	·0015	193·0	—	−0·1	·0016
21	422·8	0·9	160·9	·0037	184·3	—	−1·5	·0048
24	423·5	1·6	153·7	·0012	190·0	—	−0·5	·0022
27	429·0	7·1	165·2	·0051	199·4	—	1·1	·0073
30	427·0	5·2	164·6	·0049	189·4	—	−0·6	·0038
33	423·0	1·3	166·3	·0054	194·0	—	0·2	·0050
36	424·0	2·3	157·9	·0060	187·0	—	−1·0	·0040
39	428·0	6·4	170·6	·0069	184·6	—	−1·3	·0042
42	422·0	0·4	173·8	·0080	177·0	—	−2·7	·0026
45	422·2	0·6	174·0	·0080	175·5	—	−2·9	·0022
48	425·0	3·4	172·3	·0074	167·8	—	−4·2	−·0039
51	425·0	3·5	166·0	·0053	180·7	—	−2·0	·0010
54	426·4	5·1	161·6	·0037	194·5	—	0·4	·0045

6 D.								
H. M.								
0 0	452·0	20·0	29·5	−·0354	328·4	—	20·3	·0047
3	446·0	14·0	29·7	−·0354	317·7	—	18·6	·0013
6	436·2	3·9	41·5	−·0314	318·5	—	18·8	·0057
9	437·2	4·7	47·5	−·0294	317·8	—	18·8	·0076
12	430·2	−2·3	52·8	−·0276	302·2	—	16·1	·0042
15	434·0	1·3	70·1	−·0215	274·1	—	11·4	·0010
18	432·6	−0·3	80·5	−·0184	271·2	—	11·0	·0033
21	429·0	−3·9	87·3	−·0157	264·3	—	9·8	·0037
24	427·6	−5·4	92·0	−·0145	254·5	—	8·2	·0017
27	426·8	−6·2	95·8	−·0132	249·7	—	7·4	·0014
30	430·4	−2·9	97·6	−·0127	248·8	—	7·3	·0019
33	428·0	−5·5	93·6	−·0141	240·9	—	6·1	−·0020
36	427·4	−6·1	106·3	−·0098	242·4	—	6·3	·0027
39	434·4	0·7	108·0	−·0110	243·4	—	6·6	·0021
42	436·6	2·7	103·4	−·0108	244·4	—	6·8	·0026
45	438·8	4·9	102·8	−·0112	241·3	—	6·3	·0014
48	440·0	5·9	102·8	−·0113	235·7	—	5·5	−·0004
51	432·0	−2·1	95·5	−·0138	250·7	—	8·0	·0022
54	441·0	6·7	91·7	−·0151	267·1	—	11·0	·0065
57	446·2	11·9	101·0	−·0119	237·3	—	5·8	−·0003
1 0	445·2	10·7	102·4	−·0115	232·6	—	5·1	·0001
3	440·0	5·7	108·5	−·0096	228·6	—	4·5	−·0006
6	437·0	2·7	113·1	−·0081	237·0	—	6·0	·0038
9	436·4	2·3	113·7	−·0081	227·6	—	4·5	·0008
12	410·4	6·3	105·7	−·0108	239·9	—	6·6	·0022
15	444·0	10·1	98·6	−·0131	—	—	—	—
18	—	—	87·8	−·0 73	—	—	—	—
21	452·0	18·3	86·6	−·0177	258·0	—	9·8	·0017
24	450·4	16·9	87·6	−·0176	266·1	—	11·3	·0043
27	456·0	22·5	83·8	−·0189	254·1	—	9·3	−·0006
30	457·5	24·2	72·5	−·0231	261·7	—	10·7	−·0020
33	458·0	24·9	83·0	−·0196	243·7	—	7·7	−·0043
36	456·0	22·9	103·8	−·0121	229·4	—	5·3	−·0021
39	452·5	19·6	105·5	−·0121	235·3	—	6·3	·0021
42	455·5	22·6	112·8	−·0096	213·0	—	2·5	·0040
45	449·0	16·3	128·2	−·0045	206·9	—	1·6	·0010
48	440·8	8·3	135·4	·0011	194·5	—	−0·4	·0002
51	457·8	5·3	131·2	−·0037	196·9	—	0·0	·0038
54	437·0	4·7	128·4	−·0049	196·5	—	0·0	·0048

Magnetical Disturbances, Lake Athabasca, 1844—*continued*.

JANUARY 6—*continued*.

Gött. mean Time.	Declination. Scale.	Declination. Δψ	Bifilar. Scale corrected for Temp.	Bifilar. Approx. ΔX/X	Inclinometer. Scale corrected for Decl. and Bif.	Inclinometer. P_1-P.	Inclinometer. Approx. Δθ	Approx. Δφ/φ
6 D. H. M.		′				Div.	′	
1 57	435·5	3·2	134·3	—·0029	205·4	—	1·5	·0002
2 0	434·0	1·9	125·9	—·0059	221·2	—	4·3	·0027
3	—	—	119·3	—·0083	225·4	—	5·1	·0018
6	439·2	7·1	120·6	—·0078	211·5	—	2·7	—·0024
9	440·0	8·0	120·3	—·0080	—	—	—	—
12	442·5	10·5	121·3	—·0077	226·8	—	1·9	—·0037
2 15	439·4	7·4	137·6	—·0022	215·6	—	3·5	·0048
18	434·0	2·0	122·7	—·0073	194·5	—	—0·1	·0075
21	440·0	8·0	139·2	—·0017	204·3	—	1·6	·0014
24	435·0	3·1	135·3	—·0031	195·2	—	0·1	·0030
27	440·0	8·1	139·3	—·0018	207·4	—	2·2	·0026
30	436·8	4·9	132·6	—·0042	210·0	—	2·6	·0011

JANUARY 6—*continued*.

Gött. mean Time.	Declination. Scale.	Declination. Δψ	Bifilar. Scale corrected for Temp.	Bifilar. Approx. ΔX/X	Inclinometer. Scale corrected for Decl. and Bif.	Inclinometer. P_1-P.	Inclinometer. Approx. Δθ	Approx. Δφ/ψ
6 D. H. M.		′				Div.	′	
2 33	436·0	4·1	125·4	—·0067	209·1	—	0·8	·0048
36	440·0	8·1	122·4	—·0077	225·7	—	5·3	·0029
39	437·5	5·9	124·3	—·0072	216·2	—	3·8	·0008
42	440·0	8·1	128·7	—·0057	208·9	—	2·5	·0000
45	438·6	6·7	131·4	—·0048	201·8	—	1·3	·0018
48	440·0	8·2	133·5	—·0042	208·4	—	2·5	·0007
51	443·6	11·8	134·8	—·0038	219·0	—	4·3	·0049
54	441·0	9·3	131·4	—·0050	202·5	—	1·5	·0015
57	442·6	10·9	129·2	—·0057	215·3	—	3·7	·0016
3 0	444·0	12·2	123·6	—·0077	215·1	—	3·7	—·0004
4 0	434·6	4·5	151·5	·0011	169·3	—	3·7	·0063
5 0	436·0	6·0	156·6	·0029	183·5	—	—1·4	·0001

Magnetical Disturbances, Lake Athabasca, 1844—*continued.*

JANUARY 24. Term Day.

Gött. mean Time.	Declination. Scale.	Declination. Δψ	Bifilar. Scale corrected for Temp.	Bifilar. B,—B.	Bifilar. Approx. ΔX/X	Inclinometer. Scale corrected for Decl. and Bif.	Inclinometer. P,—P.	Inclinometer. Approx. Δθ	Approx. Δφ/φ
24 D. H. M.		′						′	
10 0	415·8	1·1	125·1	−5·1	−·0017	202·3	−8·9	−1·5	−·0049
5	415·3	0·6	126·7	−3·3	−·0011	205·2	−5·8	−1·0	−·0032
10	415·0	0·2	—	—	—	204·5	−6·4	−1·1	—
15	416·0	1·1	123·3	−7·1	−·0024	207·9	−2·8	−0·5	−·0034
20	422·0	7·1	123·4	−7·1	−·0024	208·2	−2·4	−0·4	−·0033
25	420·6	5·6	125·2	−5·4	−·0018	207·8	−2·6	−0·4	−·0027
30	421·0	5·9	125·3	−5·5	−·0019	209·1	−1·2	−0·2	−·0023
35	423·2	8·0	126·9	−4·9	−·0017	209·0	−1·1	−0·2	−·0021
40	425·0	9·7	125·0	−6·0	−·0020	211·0	1·1	0·2	−·0016
45	426·0	10·6	124·6	−6·5	−·0022	210·4	0·6	0·1	−·0021
50	426·0	10·5	124·7	−6·4	−·0022	208·2	−1·4	−0·2	−·0027
55	424·8	9·2	125·1	−6·1	−·0021	209·2	−0·3	0·0	−·0022
11 0	428·4	12·7	125·0	−6·2	−·0021	209·8	0·5	0·0	−·0023
5	428·0	12·3	124·2	−7·1	−·0024	209·3	0·1	0·0	−·0024
10	427·0	11·2	126·2	−5·1	−·0017	206·4	−2·7	−0·6	·0030
15	426·0	10·7	125·8	−5·6	−·0019	209·1	0·2	0·0	−·0018
20	431·6	15·6	126·5	−4·9	−·0017	210·9	2·1	0·3	−·0024
25	431·0	14·9	123·7	−7·7	−·0026	214·0	5·3	0·9	−·0007
30	437·6	21·4	124·8	−6·7	−·0023	212·8	4·2	0·7	−·0008
35	426·0	9·7	123·7	−7·9	−·0027	206·1	−2·4	−0·4	−·0036
40	430·0	13·7	123·7	−7·9	−·0027	208·7	0·3	0·0	−·0026
45	429·6	12·2	124·0	−7·7	−·0026	206·4	−1·8	−0·3	−·0032
50	426·0	8·5	123·1	−7·6	−·0026	205·3	−2·8	−0·5	−·0036
55	428·0	10·4	125·6	−6·1	−·0021	207·0	−1·0	−0·2	−·0024
12 0	430·0	13·3	125·2	−6·5	−·0022	204·7	−3·3	−0·6	−·0033
5	427·8	11·1	125·3	−6·4	−·0022	201·6	−6·2	−1·1	−·0044
10	427·0	10·3	124·0	−7·9	−·0027	203·0	−4·7	−0·8	−·0044
15	425·0	8·2	121·5	−10·4	−·0035	205·3	−2·3	−0·4	−·0043

JANUARY 24—*continued.*

Gött. mean Time.	Declination. Scale.	Declination. Δψ	Bifilar. Scale corrected for Temp.	Bifilar. B,—B.	Bifilar. Approx. ΔX/X	Inclinometer. Scale corrected for Decl. and Bif.	Inclinometer. P,—P.	Inclinometer. Approx. Δθ	Approx. Δφ/φ
24 D. H. M.		′							
12 20	424·0	7·2	123·5	−8·5	−·0029	201·1	−6·3	−1·1	−·0052
25	424·8	7·9	124·9	−7·2	−·0025	203·4	−3·9	−0·7	−·0039
30	427·5	10·6	124·6	−7·6	−·0026	198·7	−8·5	−1·4	−·0057
35	426·0	9·1	129·3	−2·9	−·0010	199·0	−8·1	−1·4	−·0039
40	429·5	12·5	135·4	3·1	−·0011	193·9	−13·0	−2·2	−·0036
45	424·0	7·0	131·8	−0·6	−·0002	199·0	−7·8	−1·3	−·0050
50	424·0	7·0	127·3	−5·2	−·0018	199·7	−7·0	−1·2	−·0043
55	425·0	8·0	127·3	−5·3	−·0018	203·5	−3·0	−0·5	−·0029
13 0	425·6	8·6	128·0	−4·8	−·0016	203·6	−2·8	−0·5	−·0026
5	424·1	7·0	128·0	−4·7	−·0016	202·5	−3·8	−0·6	−·0030
10	422·6	5·4	128·6	−4·1	−·0014	202·5	−3·8	−0·6	−·0028
15	421·6	4·3	127·5	−5·2	−·0018	201·5	−4·7	−0·8	−·0035
20	420·0	2·6	127·5	−5·2	−·0018	199·3	−6·8	−1·2	−·0032
25	419·9	2·4	127·0	−5·7	−·0019	199·2	−6·8	−1·2	−·0043
30	422·4	4·8	127·7	−5·0	−·0017	203·2	−2·8	−0·5	−·0027
35	421·6	3·9	126·2	−6·5	−·0022	198·6	−7·3	−1·3	−·0048
40	429·8	12·1	126·0	−6·7	−·0023	199·4	−6·4	−1·1	−·0046
45	420·0	2·2	126·9	−5·8	−·0020	204·9	−0·8	−0·1	−·0023
50	419·6	1·7	127·3	−5·4	−·0018	203·4	−2·3	−0·4	−·0026
55	429·9	11·9	127·8	−4·9	−·0017	203·1	−2·5	−0·4	−·0026
14 0	430·0	11·9	127·3	−5·4	−·0018	210·9	5·4	1·0	·0001
5	430·4	12·3	127·3	−5·4	−·0018	210·3	4·8	0·8	−·0001
10	432·0	13·8	130·2	−2·5	−·0009	209·5	4·0	0·7	·0005
15	432·4	14·2	127·0	−5·6	−·0019	214·1	8·7	1·5	·0012
20	430·8	12·5	127·5	−5·1	−·0017	206·4	1·0	0·2	·0014
25	431·7	13·4	134·0	1·4	·0005	211·3	5·9	1·0	·0026
30	428·0	9·7	132·8	0·3	·0001	203·0	−2·4	−0·4	·0009
35	424·6	6·2	132·8	0·3	·0001	206·0	0·6	0·1	·0003

Magnetical Disturbances, Lake Athabasca, 1844—*continued.*

January 14—*continued.*

Gott. mean Time.	Declination. Scale.	Δ ψ	Bifilar. Scale corrected for Temp.	P − P.	Approx. Δ X / X	Inclinometer. Scale corrected for Decl. and Bif.	P, − P.	Approx Δ X / X	Approx. Δ φ / φ
24 D. H. M.								′	
14 40	424·4	6·0	135·6	3·1	·0011	201·2	−4·1	−0·7	−·0008
45	422·8	4·3	140·6	8·1	·0028	197·2	−8·1	−1·4	−·0001
50	423·8	5·3	139·1	6·6	·0023	194·7	−10·6	−1·9	−·0013
55	424·0	5·5	140·0	7·6	·0026	194·7	−10·6	−1·9	−·0012
15 0	424·1	5·5	139·6	7·2	·0025	195·9	−9·3	−1·6	−·0008
5	423·6	5·0	139·4	7·0	·0024	191·0	−14·5	−2·4	−·0027
10	423·2	4·6	141·0	8·6	·0029	189·3	−16·1	−2·7	−·0029
15	424·0	5·4	140·8	8·4	·0029	180·0	−25·1	−4·3	−·0062
20	426·2	7·6	149·9	17·5	·0060	186·2	−19·4	−3·3	−·0010
25	431·0	12·4	150·6	18·2	·0062	185·4	−20·3	−3·5	−·0011
30	435·0	16·3	147·2	14·8	·0050	185·4	−20·3	−3·5	−·0023
35	435·2	16·5	151·7	19·3	·0066	176·1	−28·7	−4·9	−·0037
40	432·0	13·3	153·0	20·6	·0070	185·8	−20·1	−3·4	−·0002
45	435·0	16·3	151·4	19·0	·0065	189·5	−16·5	−2·8	·0036
50	429·7	11·0	141·7	9·3	·0032	197·3	−8·8	−1·5	·0000
55	429·0	10·3	132·7	0·3	·0001	213·9	7·8	1·3	·0029
16 0	418·0	−0·7	146·9	13·5	·0046	203·3	−2·8	−0·5	·0036
5	428·0	9·3	150·8	18·4	·0063	203·5	−2·4	−0·4	·0055
10	434·0	15·3	137·0	4·7	·0016	201·9	−3·8	−0·6	·0002
15	432·4	13·6	128·5	−3·8	−·0013	205·0	−0·5	0·8	·0019
20	430·2	11·4	124·7	−7·5	−·0026	223·7	18·4	3·1	·0040
25	425·2	6·4	125·3	−6·9	−·0024	219·0	14·0	2·4	·0026
30	420·0	1·2	116·0	−16·1	−·0055	239·1	34·3	5·8	·0068
35	430·0	11·1	137·1	5·0	·0017	210·3	5·9	1·0	·0037
40	432·0	13·1	142·8	10·7	·0037	195·6	−8·8	−1·5	·0005
[illegible]	[illegible]	[illegible]	[illegible]	9·1	·0011	207·7	3·5	0·6	·0024
[illegible]	[illegible]	[illegible]	[illegible]	[illegible]	[illegible]	[illegible]	[illegible]	[illegible]	·0009

January 14.

Gott. mean Time.	Declination. Scale.	Δ ψ	Bifilar. Scale corrected for Temp.	B, − B.	Approx. Δ X / X	Inclinometer. Scale corrected for Decl. and Bif.	P, − P.	Approx. Δ θ	Approx. Δ φ / φ
24 D. H. M.			′					′	
20 15	428·0	9·9	56·7	−66·2	−·0226	333·8	121·5	20·7	·0210
20	417·5	−0·6	54·2	−68·7	−·0235	341·3	126·0	21·5	·0218
25	379·2	−38·9	26·9	−95·5	−·0326	372·1	156·0	26·6	·0235
30	396·2	−22·1	15·2	−106·8	−·0365	401·9	184·9	31·6	·0259
35	406·0	−12·2	19·5	−102·1	−·0348	377·9	159·9	27·3	·0225
40	401·8	−16·5	32·2	−88·8	−·0303	327·8	108·9	18·6	·0088
45	430·4	12·1	51·2	−69·4	−·0237	342·8	123·0	21·0	·0205
50	439·6	21·3	42·9	−77·3	−·0264	354·6	133·9	22·9	·0217
55	447·0	28·7	47·8	−71·9	−·0245	331·6	110·0	18·8	·0150
21 0	471·0	52·6	51·3	−68·0	−·0232	358·1	135·6	23·1	·0252
5	450·8	32·3	41·1	−77·8	−·0265	360·1	137·5	23·5	·0229
10	437·0	18·3	53·9	−64·3	−·0219	342·3	119·7	20·4	·0211
15	419·8	0·9	63·7	−54·5	−·0186	339·5	116·8	19·9	·0234
20	401·8	−17·3	75·2	−42·6	−·0145	333·9	111·1	18·9	·0254
25	419·8	0·5	66·0	−49·5	−·0169	309·2	86·3	14·6	·0141
30	453·4	33·9	66·1	−51·1	−·0174	313·8	90·9	15·5	·0151
35	463·0	43·3	56·9	−59·9	−·0204	334·2	111·2	19·9	·0195
40	463·2	43·3	50·4	−66·1	−·0226	319·6	96·5	16·4	·0121
45	451·3	81·2	63·0	−58·1	−·0181	297·1	74·0	12·6	·0085
50	500·0	79·7	70·2	−45·6	−·0156	296·7	73·5	13·5	·0108
55	490·4	69·8	67·2	−48·2	−·0164	307·3	84·0	14·3	·0138
22 0	480·4	59·6	58·8	−55·9	−·0191	325·0	101·6	17·3	·0174
5	484·0	63·1	53·4	−60·9	−·0208	343·3	118·3	20·1	·0217
10	476·0	55·1	46·7	−67·6	−·0231	348·5	123·9	21·1	·0233
15	450·0	29·0	63·0	−50·0	−·0171	308·9	83·6	14·2	·0131
20	452·8	31·8	51·8	−61·8	−·0211	343·9	118·0	20·0	·0213
25	441·4	20·3	56·9	−66·4	−·0227	337·9	111·3	19·0	·0174
[illegible]	[illegible]	[illegible]	[illegible]	[illegible]	·0196	354·3	127·1	21·7	·0260

17 0	413·8	−5·2	73·4	−57·3	−·0195	310·4	106·9	18·3	·0363	35	422·4	0·7	−37·6	−150·2	−·0513	373·1	145·2	24·8	·0008
5	509·5	90·5	109·9	−22·0	−·0075	130·4	−73·1	−12·5	−·0337	40	422·4	0·0	−29·8	−141·0	−·0478	394·4	165·8	28·3	·0116
10	515·0	95·9	175·5	43·5	·0148	198·4	−5·2	−0·9	·0129	45	438·2	15·8	−11·5	−123·4	−·0421	397·7	168·6	28·7	·0185
15	471·6	52·5	116·4	−15·8	−·0054	202·3	−1·3	−0·2	−·0059	50	430·0	7·5	5·2	−106·0	−·0362	390·8	160·9	27·4	·0216
20	136·0	16·8	116·9	−15·4	−·0053	226·6	22·9	3·9	+·0029	55	426·2	3·7	14·8	−96·4	−·0329	383·8	153·3	26·2	·0322
25	452·0	32·7	101·7	−30·7	−·0105	220·8	17·1	2·9	−·0044	23 0	416·0	−5·5	35·3	−75·6	−·0258	372·5	141·4	24·2	·0249
30	421·0	1·7	61·4	−71·2	−·0243	282·9	79·2	13·5	·0043	5	417·2	−4·7	35·4	−75·2	−·0257	369·1	138·1	23·6	·0238
35	414·0	−5·4	65·0	−67·7	−·0231	336·9	133·1	22·7	·0253	10	422·0	−0·2	46·7	−64·0	−·0219	371·6	140·7	24·0	·0286
40	419·3	0·3	77·2	−55·7	−·0190	293·6	89·8	15·3	·0133	15	430·2	7·6	59·1	−51·5	−·0176	345·6	114·9	20·0	·0237
45	428·0	8·5	90·1	−42·9	−·0146	289·9	86·1	14·7	·0163	20	466·7	43·8	47·5	−63·1	−·0215	329·8	99·2	16·9	·0141
50	429·8	10·2	102·0	−31·2	−·0106	236·3	32·5	5·5	·0010	25	499·8	76·5	51·8	−58·7	−·0200	360·5	130·0	22·2	·0267
55	417·5	−2·1	111·1	−22·2	−·0076	232·4	28·5	4·9	·0026	30	470·4	46·8	59·4	−51·0	−·0174	357·1	126·8	21·7	·0282
18 0	421·0	1·3	125·6	−6·9	−·0024	218·3	13·4	2·3	·0024	35	460·6	36·7	61·5	−48·8	−·0167	342·6	112·4	19·1	·0236
5	422·0	2·3	123·8	−7·5	−·0026	218·0	14·0	2·4	·0024	40	429·9	5·6	65·4	−44·9	−·0153	345·1	115·0	19·6	·0260
10	428·0	8·3	114·2	−18·9	−·064	237·9	33·8	5·8	·0057	45	442·0	17·3	5303	−56·8	−·0194	346·0	116·0	19·8	·0223
15	434·0	14·3	119·3	−13·8	−·0047	235·1	30·9	5·2	·0075	50	444·2	19·2	63·4	−46·7	−·0193	327·4	98·8	16·7	·0158
20	435·5	15·8	117·4	−15·5	−·0053	240·3	36·0	6·1	·0076	55	467·6	42·2	29·7	−80·3	−·0274	328·7	99·0	16·9	·0081
25	432·0	12·2	122·2	−10·4	−·0035	232·2	27·9	4·8	·0065										
30	439·0	19·2	113·3	−19·1	−·065	233·9	29·5	4·0	·0041										
35	156·0	36·2	101·5	−30·8	−·0105	258·3	53·8	9·2	·0088	25 D.									
40	442·2	22·4	100·8	−31·3	−·0107	248·1	43·5	7·4	·0049	H. M.									
45	128·0	8·2	80·6	−51·4	−·0175	272·8	68·1	11·6	·0070	0 0	484·2	58·5	46·1	−63·9	−·0218	343·6	114·0	19·5	·0192
50	439·5	19·7	100·4	−31·6	−·0108	250·8	56·1	9·6	·0093	5	478·0	52·1	51·3	−58·9	−·0201	343·6	114·6	19·6	·0211
55	427·5	7·7	96·5	−35·4	−·0121	294·2	89·4	15·3	·0200	10	475·6	49·5	48·4	−62·0	−·0212	404·4	176·0	30·1	·0420
19 0	128·2	8·3	111·2	−20·7	−·0071	247·8	42·9	7·3	·0083	15	564·0	137·7	64·1	−46·6	−·0159	390·0	162·1	27·7	·042[illegible]
5	427·0	7·2	74·6	−56·6	−·0193	293·1	87·6	14·9	·0122	20	541·0	114·5	58·0	−52·9	−·0180	385·1	157·8	27·0	·0387
10	421·0	1·3	73·5	−57·1	−·0195	299·2	93·2	15·9	·0140	25	476·0	49·3	66·0	−45·2	−·0154	352·8	126·1	21·5	·0298
15	418·0	−1·5	85·0	−45·0	−·0153	279·6	73·0	12·4	·0110	30	490·0	63·1	66·8	−44·6	−·0152	343·7	117·5	20·0	·0271
20	432·0	12·7	75·5	−53·9	−·0184	284·6	77·5	13·2	·0094	35	463·0	35·9	63·9	−47·7	−·0163	363·8	138·2	23·6	·0335
25	442·0	22·9	82·2	−46·6	−·0159	293·1	85·4	14·6	·0148	40	494·0	66·7	57·8	−54·1	−·0185	367·3	142·3	24·2	·0326
30	424·0	5·1	81·6	−46·6	−·0159	268·1	59·9	10·2	·0056	45	506·0	78·5	45·7	−66·4	−·0227	354·0	129·5	22·1	·0239
35	436·0	17·2	71·6	−56·6	−·0193	297·3	87·9	15·0	·0123	50	475·0	47·3	55·8	−56·6	−·0224	382·1	158·2	27·0	·0245
40	444·6	25·9	101·9	−25·1	−·0086	252·5	42·6	7·2	·0067	55	515·2	87·3	14·7	−97·9	−·0334	340·6	117·3	20·0	·0087
45	421·0	2·5	82·8	−43·6	−·0149	293·2	77·7	13·4	·0130	1 0	551·0	122·8	23·7	−89·1	−·0304	392·9	170·2	29·0	·0308
50	421·0	2·7	63·8	−62·0	−·0212	326·1	115·1	19·6	·0202	5	524·0	96·0	23·2	−90·2	−·0308	375·0	152·8	26·1	·0241
55	416·0	−2·2	55·7	−73·5	−·0268	331·4	120·4	20·5	·0165	10	514·0	86·2	24·2	−89·8	−·0306	358·6	136·9	23·4	·0186
20 0	437·5	19·5	52·4	−72·3	−·0247	381·2	168·6	28·7	·0180	15	499·0	71·4	35·0	−79·6	−·0272	385·9	164·8	28·1	·0320
5	399·4	−18·6	44·1	−79·8	−·0272	362·6	150·1	25·6	·0267	20	481·4	54·0	34·0	−81·2	−·0277	371·7	150·9	25·8	·0265
10	449·0	30·9	38·5	−84·8	−·0289	347·1	133·7	22·9	·0192	25	506·4	79·2	26·2	−89·6	−·0306	388·8	168·8	28·9	·0300

Declination, a = 1′·0; bifilar, k = ·0003412; inclinometer, k = 0′·171.

Magnetical Disturbances, Lake Athabasca, 1844—*continued.*

JANUARY 25—*continued.*

Gött. mean Time.	Declination. Scale.	Δψ	Bifilar. Scale corrected for Temp.	$B_1 - B$	Approx. $\frac{\Delta X}{X}$	Inclinometer. Scale corrected for Decl. and Bif.	$P_1 - P$	Approx. $\Delta\theta$	Approx. $\frac{\Delta\phi}{\phi}$
25 D. H. M.		′				Div.		′	
1 30	488·5	61·5	40·7	−75·7	−·0258	362·5	143·0	24·4	·0255
35	481·8	55·0	66·7	−50·3	−·0172	320·0	101·1	17·3	·0191
40	miss[d]	—	66·4	−51·2	−·0175	325·1	106·7	18·2	·0208
45	476·0	49·6	67·0	−51·1	−·0174	307·7	89·9	15·5	·0154
50	504·0	77·8	68·8	−49·9	−·0170	287·6	70·3	12·0	·0083
55	491·2	65·2	57·9	−61·4	−·0209	346·0	129·3	22·0	·0255
2 0	484·0	58·2	66·2	−53·7	−·0183	318·6	92·4	15·7	·0149
5	469·6	43·9	63·2	−56·9	−·0194	316·1	100·1	17·0	·0165
10	460·0	34·3	79·1	−41·3	−·0141	283·4	67·7	11·5	·0102
15	478·0	52·3	71·5	−49·1	−·0168	298·2	82·7	14·0	·0098
20	482·0	56·3	78·7	−42·0	−·0143	294·4	79·1	13·5	·0141
25	484·0	58·4	88·4	−32·7	−·0112	266·5	51·4	8·8	·0073
30	462·5	36·9	101·5	−19·9	−·0068	260·3	45·4	7·7	·0095
35	458·8	33·2	113·1	−8·5	+·0029	255·7	41·1	7·0	·0119
40	466·0	40·4	106·0	−15·9	−·0054	257·6	43·2	7·3	·0101
45	469·4	43·9	114·4	−7·7	−·0026	249·4	29·2	5·0	·0079
50	464·6	39·1	109·5	−12·9	−·0044	237·4	24·4	4·1	·0043
55	452·0	36·5	122·9	0·3	−·0001	229·9	16·1	2·7	·0059
3 0	454·0	28·5	128·8	6·0	−·0020	215·7	1·1	0·2	·0024
5	455·0	29·7	127·9	4·9	−·0017	220·0	6·6	1·1	·0040
10	455·0	29·8	121·3	−1·9	−·0006	243·1	29·9	5·1	·0058
15	447·6	22·6	119·1	−4·3	−·0015	228·8	15·8	2·6	·0042
20	453·0	28·1	114·6	−9·0	−·0031	217·7	4·8	0·8	·0014
25	460·0	35·2	118·3	−5·4	−·0014	225·4	12·7	2·1	·0031
30	454·0	29·6	109·5	−14·4	−·0049	245·3	32·8	5·6	·0069
[illegible]	[illegible]	[illegible]	119·7	−10·4	−·0035	246·4	33·1	5·6	·0084
[illegible]	[illegible]	[illegible]	[illegible]	[illegible]	[illegible]	[illegible]	19·1	2·2	·0037

JANUARY 25—*continued.*

Gött. mean Time.	Declination. Scale	Δψ	Bifilar. Scale corrected for Temp.	$B_1 - B$	Approx. $\frac{\Delta X}{X}$	Inclinometer. Scale corrected for Decl. and Bif.	$P_1 - P$	Approx. $\Delta\theta$	Approx. $\frac{\Delta\phi}{\phi}$
25 D. H. M.		′						′	
5 50	425·6	4·0	134·6	10·4	·0035	209·8	−4·2	−0·7	·0020
55	426·8	5·4	134·2	10·1	·0034	214·1	−0·1	0·0	·0034
6 0	428·0	6·8	133·6	9·5	·0032	215·5	1·1	−0·2	·0036
5	426·0	5·2	134·7	10·6	·0036	214·2	−0·2	0·0	·0028
10	425·2	4·7	135·5	11·3	·0039	211·4	−3·0	−0·5	·0028
15	422·4	2·3	136·0	11·8	·0040	208·5	−6·0	−1·0	·0019
20	423·2	3·4	135·1	10·8	·0037	211·2	−3·3	−0·6	·0025
25	424·0	4·6	134·2	9·8	·0033	213·6	−0·9	−0·1	·0030
30	428·0	8·9	134·1	9·7	·0033	212·8	−1·8	−0·3	·0027
35	426·4	7·7	135·5	11·0	·0038	210·9	−3·7	−0·6	·0025
40	424·8	5·9	135·6	11·1	·0038	211·5	−3·1	−0·5	·0027
45	424·0	6·0	135·7	11·1	·0038	211·3	−3·4	−0·6	·0026
50	425·2	7·6	136·0	11·4	·0039	210·4	−4·3	−0·7	·0024
55	424·6	7·3	136·9	12·3	·0042	210·3	−4·4	−0·7	·0026
7 0	424·2	7·3	136·8	12·0	·0041	214·2	−0·6	−0·1	·0039
5	426·0	9·2	135·5	10·6	·0036	213·3	−1·4	−0·2	·0031
10	425·4	8·7	135·7	10·7	·0037	213·2	−1·4	−0·2	·0032
15	428·0	11·4	137·6	12·4	·0042	208·4	−6·2	−1·0	·0020
20	425·2	8·7	143·3	18·0	·0064	210·7	−3·8	−0·6	·0050
25	422·0	5·6	139·4	13·9	·0047	209·0	−5·4	−0·9	·0028
30	420·4	4·0	138·0	12·4	·0042	210·0	−4·4	−0·7	·0026
35	420·0	3·6	136·7	11·0	·0038	212·5	−1·8	−0·3	·0032
40	418·2	1·9	139·5	13·6	·0040	211·2	−3·0	−0·5	·0029
45	418·0	1·8	138·2	12·2	·0042	213·7	−0·5	−0·1	·0040
50	418·2	2·1	135·2	9·0	·0031	205·6	−8·5	−1·5	·0000
55	426·2	10·2	139·5	13·2	·0045	214·0	0·0	0·0	·0045
8 0	422·0	6·0	138·3	11·9	·0041	211·2	−2·7	−0·4	·0031
5	[illegible]	7·2	145·7	19·2	·0066	216·2	2·4	0·4	·0057

−4·6	−0·8
0·2	0·0
−3·2	−0·5
−5·5	−0·9
−6·2	−1·1
−5·1	−0·9
−6·6	−1·1
−6·4	−1·1
−6·2	−1·1
−6·5	−1·1
−6·4	−1·1
−8·3	−1·4
−8·1	−1·4
−8·9	−1·5
−10·0	−1·7
−8·8	−1·5
—	—
−11·6	−2·0
−14·5	−2·5
−23·9	−4·1
−19·1	−3·2
−17·7	−3·0
−19·2	−3·3

−·0026	25	
·0011	30	
−·0002	35	
·0022	40	
−·0010	45	
−·0023	50	
−·0043	55	419·
−·0017	9 0	422·0
−·0020	5	418·0
−·0017	10	418·2
−·0031	15	420·2
−·0039	20	419·6
−·0007	25	418·0
·0002	30	421·0
·0006	35	420·0
−·0001	40	418·
·0014	45	416·0
·0012	50	416·2
·0006	55	419·2
·0013	10 0	420·0
·0014		

[illegible]	184·1
·0074	183·0
·0056	194·1
·0057	194·6
·0059	196·6
·0055	199·1
·0050	198·7
·0044	204·8
·0037	206·3
·0035	205·4
·0035	207·4
·0034	208·2

Magnetical Disturbances, Lake Athabasca, 1844—*continued.*

JANUARY 31. FEB. 1.

Gött. mean Time.	Declination. Scale.	Declination. Δ ψ	Bifilar. Scale corrected for Temp.	Bifilar. Approx. ΔX/X	Inclinometer. Scale corrected for Decl. and Bif.	Inclinometer. Approx. Δ θ	Approx. Δ φ/φ
31 D. H. M.		′					
17 0	416·8	-6·1	143·6	-·0012	189·0	-0·2	-·0065
18 0	420·8	2·8	157·4	·0029	167·7	-4·0	-·0050
18 30	410·4	-13·3	145·2	-·0009	170·5	-3·5	-·0078
19 0	408·4	-15·4	136·9	-·0036	191·9	0·0	-·0035
20 0	394·4	27·5	139·5	-·0002	201·2	0·4	·0007
21 0	411·0	-11·3	159·1	·0088	173·9	-6·1	-·0037
22 0	417·0	-7·7	143·9	·0045	191·1	-3·3	-·0020
23 0	424·0	-1·4	111·6	-·0050	221·0	0·5	·0040
Feb. 1 D. H. M.							
0 0	430·2	2·6	49·3	-·0260	302·7	14·7	·0030
3	422·0	-5·8	33·8	-·0314	336·3	20·5	·0101
6	419·0	-8·8	26·3	-·0339	335·9	20·5	·0066
9	414·0	-14·0	22·3	-·0353	328·1	19·2	·0026
12	430·0	2·8	28·6	-·0332	329·5	19·5	·0053
15	442·4	14·0	29·1	-·0331	334·9	20·5	·0074
18	440·4	12·0	23·3	-·0351	340·4	21·5	·0074
21	430·0	1·6	8·6	-·0402	390·4	30·1	·0192
24	442·0	13·4	-1·7	-·0438	381·3	28·6	·0127
27	449·6	21·0	-9·7	-·0465	387·6	29·7	·0122
30	443·0	14·2	3·9	-·0417	357·6	24·6	·0069
[illegible]	[illegible]	[illegible]	[illegible]	·0429	370·5	27·0	·0104

FEBRUARY 1—*continued.*

Gött. mean Time.	Declination. Scale.	Declination. Δ ψ	Bifilar. Scale corrected for Temp.	Bifilar. Approx. ΔX/X	Inclinometer. Scale corrected for Decl. and Bif.	Inclinometer. Approx. Δ θ	Approx. Δ φ/φ
1 D. H. M.		′				′	
2 42	464·6	37·2	68·1	-·0236	271·3	11·9	-·0001
45	465·0	37·6	57·6	-·0273	292·6	15·6	·0035
48	468·2	40·8	69·2	-·0234	265·2	11·0	-·0017
51	450·7	23·3	92·3	-·0156	237·2	6·2	-·0034
54	442·0	14·7	101·1	-·0126	229·1	4·8	-·0031
57	443·1	15·8	106·8	-·0107	224·9	4·1	-·0026
3 0	444·8	16·4	109·4	-·0099	223·1	3·8	-·0024
3	442·0	14·8	118·4	-·0069	211·8	1·9	-·0032
6	444·0	16·8	124·4	-·0048	209·6	1·6	-·0016
9	442·8	15·7	124·2	-·0049	209·5	1·6	-·0017
12	442·0	14·9	122·5	-·0055	207·0	1·2	-·0031
15	438·2	11·2	121·3	-·0060	214·3	2·4	-·0013
18	440·0	13·2	120·4	-·0064	221·8	3·7	·0010
21	446·2	19·4	117·8	-·0073	220·5	3·5	-·0004
24	449·6	12·9	116·9	-·0076	223·7	4·1	·0005
27	457·2	30·5	109·0	-·0104	219·5	3·4	-·0037
30	458·4	31·9	121·0	-·0063	201·0	0·2	-·0057
33	450·0	23·6	115·0	-·0084	209·6	1·8	-·0049
36	449·0	22·6	121·1	-·0063	197·1	-0·4	-·0055
39	447·6	21·3	109·1	-·0105	227·6	4·8	-·0010
42	451·0	24·7	107·8	-·0109	224·2	4·1	-·0028
45	445·4	19·3	107·4	-·0112	221·3	3·8	-·0037
48	445·2	19·2	106·7	-·0114	227·1	4·8	-·0019
51	444·7	18·7	108·1	-·0109	227·8	3·2	-·0046
54	439·0	13·2	113·6	-·0091	213·8	2·6	-·0040
57	434·7	8·9	118·0	-·0076	208·2	1·6	-·0044
[illegible]	[illegible]	[illegible]	[illegible]	[illegible]	[illegible]	[illegible]	-·0040

45	460·0	30·6	4·7	−·0419	337·1	21·4	·0015		18 0*	421·8	0·4	162·7	·0047	168·9	−3·8	−·0028
48	460·0	30·4	11·0	−·0399	317·5	18·2	−·0039		19 0	—	—	—	—	—	—	—
51	460·0	30·4	21·5	−·0363	317·6	18·2	−·0008		3	444·0	22·6	185·4	·0132	115·4	−13·2	−·0129
54	465·0	35·2	9·5	−·0405	338·9	22·0	·0031		6	—	—	—	—	—	—	—
57	470·0	40·0	−0·9	−·0440	356·3	24·9	·0053		9	457·0	35·7	160·9	·0051	145·6	−8·1	−·0109
1 0	476·0	45·9	−3·5	−·0450	353·8	24·6	·0037		12	468·2	46·9	143·5	·0025	153·1	−6·8	−·0108
3	478·4	48·6	13·6	−·0393	355·8	25·0	·0101		15	460·0	38·9	147·8	·0008	165·0	−4·9	−·0089
6	—	—	26·1	−·0351	332·4	21·0	·0065		18	415·0	−6·0	150·9	·0020	177·4	−2·8	−·0035
9	454·0	24·4	39·6	−·0307	303·1	16·1	·0013		21	380·0	−41·0	142·1	−·0010	184·1	−1·7	·0043
12	458·0	28·4	42·6	−·0297	306·8	16·8	·0046		24	382·6	−38·2	121·5	−·0078	222·6	4·7	·0015
15	457·0	27·6	48·7	−·0277	295·7	15·0	·0018		27	341·0	−79·8	110·6	−·0115	247·7	9·0	·0063
18	453·2	24·0	49·9	−·0276	316·1	18·5	·0091		30	352·4	−68·3	102·3	−·0141	—	—	—
21	476·1	46·9	52·8	−·0266	282·0	12·7	−·0015		33	372·4	−48·1	87·3	−·0190	272·7	13·1	·0059
24	460·2	31·2	51·7	−·0272	288·8	14·0	·0004		36	374·4	−46·1	73·6	−·0237	289·9	16·1	·0071
27	471·8	42·8	58·4	−·0249	283·5	13·1	·0010		39	378·0	−42·3	72·5	−·0239	271·0	12·7	·0022
30	470·0	41·2	64·1	−·0231	273·9	11·5	−·0004		42	380·0	−40·3	93·6	−·0167	258·9	10·7	·0044
33	478·8	50·2	63·2	−·0236	280·1	12·7	·0014		45	385·5	−34·7	106·6	−·0124	238·9	7·2	·0018
36	486·7	58·1	50·7	−·0279	277·1	12·2	−·0038		48	385·0	−35·0	104·0	−·0127	247·4	8·5	·0041
39	484·2	55·8	50·0	−·0284	270·3	11·1	−·0065		51	391·0	−29·0	111·8	−·0101	212·3	2·7	−·0048
42	455·0	26·6	79·9	−·0182	252·7	8·1	−·0022		54	401·0	−18·9	144·9	·0014	194·9	−0·5	·0004
45	453·2	25·0	78·8	−·0187	260·4	9·5	−·0001		57	401·0	−18·9	148·9	·0028	194·9	−0·5	·0018
48	450·0	21·8	74·4	−·0204	267·6	10·8	−·0009		20 0	412·0	−7·7	170·6	·0104	156·5	−7·2	−·0038
51	448·5	20·3	92·3	−·0143	235·7	5·3	−·0038		3	423·0	3·0	163·0	·0079	193·6	−1·0	·0058
54	452·1	24·3	100·2	−·0118	235·4	5·4	−·0011		6	412·0	−7·3	159·3	·0067	184·0	−2·6	·0016
57	460·0	32·2	81·0	−·0184	253·9	8·6	−·0014		9	392·0	−27·8	137·6	−·0002	182·7	−3·0	−·0061
2 0	462·8	35·1	82·4	−·0177	247·8	7·6	−·0027		12	401·4	−18·4	146·9	·0026	165·1	−2·6	−·0025
3	463·8	36·2	75·0	−·0207	256·6	9·1	−·0026		15	409·8	−10·0	146·3	·0026	190·3	−1·9	−·0011
6	462·4	34·8	61·5	−·0253	261·9	10·0	−·0056		18	410·3	−9·5	140·4	·0007	194·5	−1·3	−·0019
9	461·2	33·6	61·9	−·0252	274·5	12·4	−·0007		21	415·0	−4·8	140·8	·0008	197·5	−0·8	−·0008
12	457·0	29·4	50·7	−·0291	311·4	18·5	·0074		24	416·2	−3·7	138·6	·0002	198·3	−0·8	−·0014
15	466·2	38·6	46·9	−·0304	291·6	15·2	−·0004		27	416·8	−3·1	144·9	·0024	186·1	−2·9	−·0033
18	460·0	32·5	36·6	−·0340	301·7	17·0	−·0004		30	423·4	3·3	147·6	·0034	185·5	−3·2	−·0029
21	454·2	26·7	39·6	−·0330	314·6	20·9	·0083		33	419·9	−0·1	144·3	·0025	191·1	−2·4	−·0022
24	459·0	31·5	43·6	−·0 17	294·5	15·8	−·0005		36	418·2	−1·8	147·0	·0034	183·7	−3·6	−·0037
27	457·8	30·3	52·6	−·0287	282·7	13·1	−·0028		39	417·2	−2·8	149·6	·0045	184·8	−3·6	−·0026
30	461·4	34·0	53·1	−·0252	276·1	12·7	−·0001		42	419·2	−0·8	149·8	·0045	181·9	−4·1	−·0036
33	456·2	28·7	70·3	−·0228	255·7	9·2	−·0046		45	419·0	−1·0	148·8	·0047	182·9	−4·1	−·0034
36	465·2	37·7	71·6	−·0223	262·6	8·7	−·0051		48	418·0	−2·0	147·6	·0041	187·7	−3·5	−·0028
39	468·0	40·6	81·6	−·0190	254·0	9·0	−·0012		51	417·2	−2·8	146·0	·0035	194·1	−2·3	−·0010

* Clouded throughout. Co-ordinate mean values: declinometer, 421·8; bifilar, 140·8; inclinometer, 201·3.

Magnetical Disturbances, Lake Athabasca, 1844—*continued.*

February 1—*continued.*

Gott. mean Time.	Declination. Scale.	Declination. Δ ψ	Bifilar. Scale corrected for Temp.	Bifilar. Approx. ΔX/X	Inclinometer. Scale corrected for Decl. and Bif.	Inclinometer. Approx. ΔX/X	Approx. Δ φ
1 D. H. M.		′				′	
20 54	414·7	−5·3	148·8	·0046	193·7	−2·5	−·0003
57	415·9	−4·1	148·5	·0045	190·7	−3·1	−·0016
21 0	417·0	−3·1	147·0	·0042	189·1	−3·5	−·0027
22 0	420·2	−2·3	139·4	·0030	199·2	−1·9	−·0007
2 D. H. M.							
5 0*	435·8	10·1	112·6	−·0101	214·6	2·8	−·0046
6 0	425·0	−1·8	67·0	−·0255	276·7	13·1	·0004
3	—	—	—	—	279·2	13·5	—
6	419·0	−3·9	67·0	−·0255	289·7	15·3	·0047
9	417·0	−5·5	66·2	−·0258	375·1	12·8	−·0005
12	414·1	−8·4	71·0	−·0242	276·3	13·0	·0015
15	422·4	0·2	73·8	−·0232	276·8	13·1	·0017
18	419·0	−2·8	82·0	−·0204	274·0	12·6	·0045
21	408·4	−13·4	86·7	−·0188	268·5	11·7	·0043
24	411·0	−10·4	84·2	−·0197	268·6	11·7	·0034
27	413·2	−3·2	87·9	−·0185	266·9	11·4	·0041
30	410·8	−10·3	89·7	−·0178	263·5	10·8	·0035
33	408·2	−12·5	99·7	−·0145	246·6	7·9	+·0011
36	399·8	−20·9	118·8	−·0080	220·0	3·7	−·0007
39	392·6	−27·8	130·2	−·0041	211·4	3·6	·0030
42	390·0	−30·4	132·8	−·0032	212·1	2·0	·0007
[illegible]	[illegible]	[illegible]	134·8	−·0025	215·4	2·6	·0026
[illegible]	[illegible]	[illegible]	[illegible]	[illegible]	[illegible]	4·5	·0019

February 2–5.

Gott. mean Time.	Declination. Scale.	Declination. Δ ψ	Bifilar. Scale corrected for Temp.	Bifilar. Approx. ΔX/X	Inclinometer. Scale corrected for Decl. and Bif.	Inclinometer. Approx. Δ θ	Approx. Δ φ/φ
2 D. H. M.		′				′	
18 12	412·4	−9·2	172·6	·0075	168·3	−3·7	·0002
15	416·8	−4·9	173·2	·0077	167·1	−3·9	·0000
18	422·0	0·3	172·5	·0075	186·9	−0·5	·0064
21	416·4	−5·3	170·7	·0069	179·5	−1·8	·0034
24	424·6	2·9	180·7	·0104	160·2	−5·1	·0003
27	425·0	3·3	182·6	·0111	165·2	−4·2	·0018
30	420·8	−1·1	180·0	·0102	164·1	−4·4	·0015
33	424·0	2·2	184·7	·0119	165·2	−4·3	·0034
36	422·4	0·2	186·9	·0126	168·5	−3·7	·0053
39	426·0	4·6	189·9	·0137	149·5	−7·0	−·0001
42	423·2	1·4	191·0	·0141	146·9	−7·4	−·0005
45	433·0	11·2	191·6	·0143	149·3	−7·0	·0005
48	436·0	14·2	189·8	·0137	163·7	−4·6	·0046
51	432·4	10·6	196·7	·0161	145·5	−7·6	·0011
54	420·2	−1·6	226·9	·0264	114·0	−13·1	·0005
57	422·2	0·4	210·8	·0209	124·4	−11·3	−·0014
19 0	420·0	−1·9	205·0	·0189	143·3	−8·3	·0025
3	416·7	−5·0	206·2	·0196	135·7	−9·5	·0008
6	419·6	−2·1	205·6	·0194	132·8	−10·0	−·0003
9	410·8	−10·8	206·2	·0198	138·6	−9·1	·0018
12	401·6	−20·0	205·8	·0197	144·9	−8·0	·0039
15	401·4	−20·0	202·2	·0186	144·3	−8·2	·0024
18	413·2	−8·1	200·8	·0184	150·7	−7·2	·0042
21	402·0	−19·3	198·2	·0175	146·0	−8·0	·0017
24	403·6	−17·5	199·4	·0181	152·2	−7·0	·0043
27	410·4	−10·7	201·7	·0189	149·6	−7·5	·0041
[illegible]	[illegible]	−11·0	197·9	·0178	161·2	−5·6	·0067
[illegible]	[illegible]	[illegible]	[illegible]	[illegible]	[illegible]	−7·3	·0095

57	392·0	−27·3	110·9	−·0107	256·5	6·2	+·0015
7 0	373·0	−45·9	125·9	−·0056	219·6	3·3	+·0009
3	380·0	−38·8	119·3	−·0079	230·6	5·1	+·0022
6	386·0	−32·8	114·0	−·0097	229·0	4·9	+·0000
9	387·2	−31·5	119·8	−·0078	227·0	4·6	+·0013
12	389·6	−29·1	122·2	−·0104	234·0	5·7	+·0008
15	390·4	−28·2	117·8	−·0085	224·5	4·1	−·0001
18	394·0	−24·5	119·7	−·0079	223·7	5·0	+·0020
21	400·0	−18·5	116·7	−·0089	229·6	5·0	+·0018
24	403·8	−17·6	117·1	−·0085	228·1	4·8	+·0010
27	402·0	−16·4	119·9	−·0079	224·6	4·2	+·0004
30	402·2	−16·2	121·8	−·0073	229·4	5·0	+·0026
8 0	411·0	−7·0	139·3	−·0016	218·9	3·3	+·0049
9 0	414·0	−3·0	149·4	·0015	195·2	−0·5	+·0005
16 0†	409·6	−11·1	161·7	·0040	170·7	−3·6	−·0031
17 0	425·6	−4·6	155·6	·0022	184·4	−0·8	·0006
6	406·0	−15·0	198·6	·0168	138·6	−8·6	−·0002
9	414·0	7·1	198·5	·0167	125·6	−10·9	−·0048
12	424·0	2·9	206·6	·0194	118·9	−12·0	−·0043
15	422·4	1·2	208·2	·0199	119·0	−12·0	−·9038
18	420·0	−1·2	195·3	·0155	121·1	−11·7	−·0076
21	422·2	1·0	195·8	·0157	139·8	−8·5	−·0011
24	422·4	1·1	196·1	·0157	131·1	−9·9	·0038
27	402·4	−18·9	196·0	·0157	129·3	−10·3	−·0046
30	406·0	−15·3	194·7	·0148	109·1	−13·7	−·0123
33	404·4	17·0	196·3	·0157	146·6	−7·3	·0013
36	422·6	1·2	188·2	·0129	152·5	−6·8	+·0005
39	406·0	−15·4	185·3	·0119	138·6	−8·7	−·0053
42	386·4	−35·0	195·5	·0154	152·2	−6·4	·0028
45	388·0	−33·5	192·4	·0143	162·1	−4·7	·0050
48	403·0	−18·5	182·7	·0109	167·5	−3·7	·0036
51	423·0	1·5	171·0	·0069	185·1	−0·7	·0055
54	414·4	−7·2	165·0	·0052	188·6	0·2	·0055
57	412·8	−8·8	159·9	·0031	178·4	−1·9	−·0006
18 0	410·0	−11·7	170·4	· 66	174·7	−2·6	·0015
3	412·2	−9·4	169·8	·0064	176·7	−2·2	·0021
6	414·0	−7·6	172·3	·0073	175·0	−2·5	·0024
9	410·2	−11·4	171·5	·0071	176·9	−2·2	·0028

36	419·6	−1·2	198·0	·0180	156·2	−6·5	·0052
39	425·2	0·0	208·3	·0217	151·1	−7·5	·0069
42	422·0	1·4	199·1	·0186	149·4	−7·8	·0032
45	426·4	5·9	207·0	·0215	157·5	−6·5	·0087
48	410·8	−9·5	204·0	·0207	154·0	−7·2	·0065
51	410·2	−10·1	192·8	·0169	165·0	−5·3	·0065
54	402·0	−18·2	198·6	·0190	167·0	−5·1	·0089
57	408·0	−12·2	195·8	·0181	166·1	−5·2	·0078
20 0	412·0	−8·0	201·6	·0202	164·9	−5·5	·0093
3	419·0	−1·1	195·6	·0184	165·0	−5·7	·0071
6	420·8	0·7	190·6	·0167	179·1	−3·3	·0102
9	419·8	−0·3	186·7	·0155	182·0	−2·9	·0097
12	419·7	−0·4	186·4	·0154	180·5	−3·2	·0091
15	418·4	−1·7	188·5	·0162	175·8	−4·1	·0081
18	422·0	1·9	185·8	·0155	180·3	−3·5	·0085
21	422·0	1·9	191·0	·0168	172·9	−4·8	·0072
24	420·3	0·1	188·8	·0169	174·6	−4·6	·0077
27	420·2	0·0	187·3	·0161	183·2	−3·2	·0098
30	418·0	−2·2	182·0	·0145	187·3	−2·6	·0092
33	424·0	3·7	184·3	·0154	185·6	−2·9	·0096
36	426·0	5·7	182·3	·0147	193·3	−1·8	·0112
39	426·0	5·7	180·4	·0143	194·3	−1·8	·0107
42	424·0	3·7	182·1	·0148	191·8	−2·2	·0105
45	423·0	2·7	181·1	·0146	193·6	−2·0	·0106
48	421·8	1·5	180·0	·0144	193·8	−2·2	·0102
51	422·4	2·1	180·7	·0146	194·6	−2·0	·0106
54	422·0	1·7	179·2	·0143	—	—	—
57	423·5	3·2	177·3	·0136	—	—	—
21 0	421·2	0·9	173·6	·0125	193·8	−2·4	·0077
22 0	420·2	−2·6	167·3	·0118	195·4	−2·3	·0072
23 0	420·0	−3·5	148·1	·0067	195·5	−3·6	−·0004
4 D.							
H. M.							
23 0	‡397·6	−27·0	140·7	·0042	190·3	−4·2	−·0043
5 D.							
H. M.							
0 0	399·1	−29·7	−17·9	−·0497	494·8	48·0	·0451
3	459·0	30·0	−71·0	−·0679	495·3	48·2	·0272
6	496·0	67·0	−50·9	−·0609	436·6	37·2	·0126

* Clouded throughout. Co-ordinate mean values: declinometer, 421·9; bifilar, 141·8; inclinometer, 199·8. † Clouded throughout.
‡ Clouded throughout. Co-ordinate mean values: declinometer, 423·0; bifilar, 144·2; inclinometer, 198·2.

Magnetical Disturbances, Lake Athabasca, 1844—*continued.*

FEBRUARY 5—*continued.*

Gött. mean Time.	Declination.		Bifilar.		Inclinometer.		Approx. $\frac{\Delta \phi}{\phi}$
	Scale.	$\Delta \psi$	Scale corrected for Temp.	Approx. $\frac{\Delta X}{X}$	Scale corrected for Decl. and Bif.	Approx. $\Delta \theta$	
5 D. H. M.		′				′	
18 9	503·2	74·0	−31·5	−·0544	413·7	34·3	·0133
12	466·2	17·0	−19·8	−·0505	532·1	54·6	·0573
15	439·6	10·2	−55·3	−·0626	564·0	60·2	·0563
18	457·0	27·4	−37·2	−·0565	477·4	45·4	·0332
21	504·4	74·8	−27·8	−·0533	409·5	33·8	·0134
24	470·0	40·2	21·0	−·0368	371·1	27·4	·0173
27	386·0	−43·8	47·1	−·0279	361·2	25·7	·0229
30	348·0	−82·0	53·2	−·0259	396·5	31·8	·0369
33	360·0	−70·2	−22·7	−·0518	433·4	39·9	·0270
36	399·8	−30·4	−28·0	−·0537	442·6	39·8	·0249
39	391·2	−39·2	− 4·0	−·0459	419·9	36·0	·0252
42	398·7	−31·7	10·3	−·0414	378·2	28·9	·0157
45	429·0	−1·6	9·0	−·0411	366·1	26·9	·0120
50	454·4	23·6	29·2	−·0344	339·5	22·4	·0078
55	441·1	10·1	43·1	−·0297	326·9	20·4	·0106
1 0	436·0	4·7	69·4	−·0208	291·8	15·3	·0094
5	440·0	9·0	80·2	−·0173	281·6	12·8	·0080
10	442·5	11·7	91·0	−·0139	370·0	10·9	·0076
15	443·1	12·5	92·9	−·0134	356·9	7·8	·0020
20	450·4	20·0	107·0	−·0088	232·0	4·6	·0008
25	449·6	19·4	116·1	−·0059	220·5	2·8	−·0004
30	454·0	24·0	138·7	·0016	194·3	−1·6	−·0016
35	447·0	17·2	144·7	·0034	198·2	−0·8	·0018
40	444·0	14·4	143·1	·0027	196·3	−1·1	·0005
45	437·4	8·0	153·8	·0062	195·6	−1·1	·0040

FEBRUARY 5–6.

Gött. mean Time.	Declination.		Bifilar.		Inclinometer.		Approx. $\frac{\Delta \phi}{\phi}$
	Scale.	$\Delta \psi$	Scale corrected for Temp.	Approx. $\frac{\Delta X}{X}$	Scale corrected for Decl. and Bif.	Approx. $\Delta \theta$	
5 D. H. M.		′				′	
17 57	416·5	−6·2	193·8	·0146	140·1	−8·2	−·0016
18 0	419·6	−3·2	195·6	·0152	152·2	−6·1	·0032
3	424·6	1·9	202·0	·0174	147·5	−6·9	·0038
6	423·7	−1·0	195·6	·0153	151·7	−6·2	·0031
9	423·9	1·2	190·7	·0136	154·4	−5·8	·0022
12	423·3	1·6	187·3	·0125	158·7	−5·0	·0026
15	420·0	−2·8	186·8	·0123	162·5	−4·4	·0036
18	420·0	−2·8	185·8	·0120	166·2	−3·8	·0045
21	422·3	−0·5	184·1	·0115	171·3	−2·9	·0059
24	420·6	−2·2	186·2	·0123	167·4	−3·6	·0052
27	421·2	−1·6	182·9	·0112	167·9	−3·5	·0043
30	418·0	−4·9	182·8	·0112	167·5	−3·6	·0041
33	416·5	−6·4	170·2	·0069	180·4	−1·4	·0041
36	417·5	−5·4	178·7	·0098	172·9	−2·7	·0045
39	417·0	−5·9	182·7	·0113	171·3	−3·0	·0054
42	417·0	−5·9	182·7	·0113	172·8	−2·7	·0060
45	420·0	−2·9	185·9	·0124	164·2	−4·2	·0041
48	418·7	−4·2	185·4	·0124	165·0	−4·1	·0043
51	414·5	−8·4	184·4	·0119	168·6	−3·5	·0050
54	414·0	−8·9	183·7	·0117	171·8	−2·9	·0060
57	417·2	−5·7	182·5	·0113	175·9	−2·2	·0070
19 0	424·0	−1·0	178·6	·0099	183·2	−1·0	·0079
20 0	424·8	3·7	152·9	·0036	220·6	−4·2	−·0047
21 0	—	—	—	—	—	—	—
9	435·0	13·2	21·4	−·0392	399·8	31·2	·0224
21	416·0	−6·5	107·1	−·0096	279·1	12·3	·0147
[illegible]	[illegible]	−0·5	106·7	−·0096	282·9	12·9	·0159

h	m
3	0
4	0
14	0*
15	0
16	0
	18
	21
	24
	27
	30
	33
	36
	39
	42
	45
	48
	51
	54
	57
17	0
	3
	6
	9
	12
	15
	18
	21
	24
	27
	30
	33
	36
	39
	42
	45
	48
	51
	54

* Aurora visible from 14^h to 17^h. See p.

† Overcast. Co-ordinate mean values; declination, 422·6; bifilar, 148·9; inclinometer, 194·1. Correction to declination differences, + 1′·7.

Magnetical Disturbances, Lake Athabasca, 1844—*continued.*

FEBRUARY 7–8. FEBRUARY 8.

Gött. mean Time.	Declination. Scale.		Temp.	Gött.	Bifilar. Scale corrected for Temp.	Bifilar. Approx. $\frac{\Delta X}{X}$	Inclinometer. Scale corrected for Decl. and Bif.	Inclinometer. Approx. $\Delta \theta$	Approx. $\frac{\Delta \phi}{\phi}$
7 D. H. M.								′	
22 0*	421·6	3°	62°	4 2	111·8	−·0085	239·4	6·0	·0033
23 0	433·2	7°	22°	449 2	124·3	−·0067	232·9	6·0	·0051
				436 0	133·4	−·0042	205·3	1·7	0009
				446 2	122·6	−·0090	226·8	5·7	·0022
				446 4	79·7	−·0235	265·1	2·1	·0004
8 D.				447 0	85·7	−·0215	278·8	4·4	·0069
H. M.				445 4	86·0	−·0214	280·7	4·8	·0078
0 0	450·6			147 2	85·9	−·0214	278·4	4·4	·0070
6	—			438 0	88·3	−·0206	282·9	5·1	·0092
9	439·0			442 4	90·2	−·0199	287·2	5·8	·0113
12	446·0			438 0	93·1	−·0189	275·4	3·8	·0088
15	454·5			43 4	101·2	−·0161	260·7	1·3	·0062
18	459·0			[illegible] 0	108·7	−·0136	242·2	8·1	·0024
21	452·0				113·7	−·0119	234·3	6·7	·0013
24	452·0				116·4	−·0110	235·6	6·9	·0026
27	457·5				118·7	−·0102	223·9	4·9	·0005
30	448·0				124·5	−·0082	225·4	5·1	·0019
33	439·6				128·8	−·0067	218·6	4·0	·0012
36	444·2				132·0	−·0056	215·4	3·4	·0011
39	444·0				131·8	−·0056	214·8	3·3	·0009
42	445·8				131·6	−·0057	218·3	3·8	·0018
45	447·0				130·7	−·0060	211·5	2·7	·0007
48	450·0				135·4	−·0043	209·6	2·3	·0002
51	448·0				134·5	−·0046	209·9	2·4	·0001
54	446·0				137·5	−·0036	207·7	2·0	·0003
57	446·0				155·1	·0022	190·3	1·0	·0002

APRIL 2.

Gött. mean Time.	Declination. Scale.	Declination. Δ ψ	Bifilar. Scale corrected for Temp.	Bifilar. Approx. $\frac{\Delta X}{X}$	Inclinometer. Scale corrected for Decl. and Bif.	Inclinometer. Approx. Δ θ	Approx. $\frac{\Delta \phi}{\phi}$
2 D. H. M.		′				′	
16 0	338·2	*−14·8	260·7	*·0112	88·0	*−9·1	−·0074
17 0	334·2	−18·8	248·6	·0073	113·5	−4·5	−·0019
18 0	361·4	8·4	276·3	·0162	62·5	−12·4	−·0091
19 0	345·6	−7·4	285·0	·0189	58·0	−13·0	−·0103
20 0	362·8	9·8	241·5	·0051	117·5	−5·3	−·0071
21 0	354·4	1·4	238·5	·0041	166·7	1·1	·0063
22 0	331·8	−21·2	199·6	−·0082	231·1	9·4	·0129
15	374·0	21·0	Beyond the scale, $\frac{\Delta X}{X}$ exceeds	−·048	695·0	69·4	Not determinable.
18	364·0	11·0			742·3	75·5	
21	360·4	7·4			743·2	75·6	
24	338·2	−14·8			633·9	61·4	
27	310·0	−43·0			481·0	41·7	
30	332·4	−20·6			634·9	61·6	
35	312·4	−40·6			596·8	43·8	
40	508·4	155·4			623·6	60·1	
45	354·0	1·0			617·7	59·4	
50	301·4	−51·6			633·6	61·4	
55	384·2	31·2			623·8	61·4	
23 0	442·1	89·1			467·7	40·0	
5	444·0	91·0	98·9	−·0402	387·7	29·7	·0213
10	409·6	56·6	153·7	−·0228	274·8	15·0	·0078
15	382.0	29·0	172·9	−·0167	275·4	15·1	·0140
20	384·4	31·4	178·7	−·0149	265·5	13·8	·0132

APRIL 2–3.

Gött. mean Time.	Declination. Scale.	Declination. Δ ψ	Bifilar. Scale corrected for Temp.	Bifilar. Approx. $\frac{\Delta X}{X}$	Inclinometer. Scale corrected for Decl. and Bif.	Inclinometer. Approx. Δ θ	Approx. $\frac{\Delta \phi}{\phi}$
2 D. H. M.		′				′	
23 25	350·8	−2·2	196·1	−·0098	227·9	8·9	·0090
30	356·0	3·0	217·7	−·0025	184·1	3·3	·0043
3 D. H. M.							
0 0	354·6	1·6	233·5	·0025	235·9	10·0	·0178
1 0	344·0	−9·0	250·7	·0080	120·2	−4·9	−·0020
2 0	352·2	−0·8	248·5	·0073	123·1	−4·3	−·0010
3 0	360·0	7·0	225·0	−·0001	159·7	0·2	·0002
4 0	450·0	97·0	124·6	−·0320	368·2	27·1	·0232
3	427·0	74·0	118·6	−·0340	346·4	24·3	·0156
6	422·0	69·0	116·6	−·0346	402·6	31·6	·0297
9	423·5	70·5	104·4	−·0385	393·8	30·4	·0236
12	434·9	81·9	94·1	−·0418	417·5	33·6	·0264
15	433·0	80·0	69·6	−·0196	464·2	39·5	·0309
18	444·6	91·6	75·6	−·0477	—	—	—
21	459·0	6·0	73·2	−·0483	456·2	38·5	·0301
24	472·0	119·0	70·7	−·0492	474·0	40·8	·0345
27	462·3	109·3	66·7	−·0505	451·6	38·0	·0268
30	458·4	105·4	66·7	−·0505	484·5	42·2	·0355
33	430·4	77·4	73·7	−·0483	429·6	35·1	·0203
36	448·0	95·0	89·7	−·0432	474·1	40·7	·0401

* Co-ordinate mean values: declination, 353·0; bifilar, 225·5; inclinometer, 158·4. The quantities given in the columns Δ ψ, $\frac{\Delta X}{X}$, and Δ θ, for the observations at Fort Simpson, are referred to the true means for the 24 H., the regular diurnal change not having been estimated. These values are found by making each day of observation the centre of a group of days, usually from five to eight in number, and taking the mean of all the numbers in that group. Where the disturbance includes two days, these two are taken as the centre of such a group.

Magnetical Disturbances, Fort Simpson, 1844.

April 3–16.

Gött. mean Time.	Declination. Scale.	Declination. Δψ	Bifilar. Scale corrected for Temp.	Bifilar. Approx. ΔX/X	Inclinometer. Scale corrected for Decl. and Bif.	Inclinometer. Approx. Δθ	Approx. Δφ/φ
3 D.							
H. M.		′				′	
4 39	478·2	125·2	69·7	−·0496	442·5	36·7	·0250
42	419·0	66·0	103·8	−·0387	362·6	26·4	·0151
45	410·0	57·0	111·6	−·0363	367·0	26·9	·0187
48	419·0	66·0	96·8	−·0409	395·9	30·7	·0216
51	410·4	57·4	87·8	−·0435	380·8	28·7	·0148
54	400·2	47·2	103·8	−·0387	433·9	35·7	·0340
57	416·4	63·4	78·3	−·0468	419·1	33·7	·0219
5 0	376·6	23·6	114·8	−·0352	357·1	23·1	·0119
15	392·2	39·2	121·0	−·0333	314·6	20·2	·0080
30	390·4	37·4	126·8	−·0314	320·3	20·9	·0114
45	393·0	40·0	159·4	−·0210	275·7	15·2	·0099
6 0	382·0	29·0	180·7	−·0142	245·4	11·2	·0085
7 0	378·0	25·0	211·5	−·0045	181·2	2·9	·0019
16 D.							
H. M.							
13 0	*360·1	−19·5	298·5	·0123	99·4	−8·8	−·0057
14 0	347·0	−32·6	354·7	·0282	11·0	−20·1	−·0131
3	342·4	−37·2	348·5	·0264	16·2	−19·6	−·0135
6	344·0	−35·6	351·3	·0272	19·7	−19·3	−·0117
9	348·0	−31·6	347·3	·0261	29·5	−17·9	−·0102
12	356·1	−23·5	339·7	·0239	26·4	−18·3	−·0133
15	349·8	−29·8	349·3	·0267	6·6	−20·8	−·0158
[illegible]	[illegible]	[illegible]	354·3	·0281	1·3	−21·5	−·0158
					[illegible]	−20·9	−·0139

April 16—*continued.*

Gött. mean Time.	Declination. Scale.	Declination. Δψ	Bifilar. Scale corrected for Temp.	Bifilar. Approx. ΔX/X	Inclinometer. Scale corrected for Decl. and Bif.	Inclinometer. Approx. Δθ	Approx. Δφ/φ
16 D.							
H. M.		′				′	
16 27	350·0	−29·6	330·4	·0213	45·8	−15·8	−·0108
30	351·0	−28·6	328·4	·0207	47·0	−15·6	−·0110
33	348·2	−31·4	331·9	·0217	46·8	−15·6	−·0101
36	344·4	−35·2	330·7	·0214	48·5	−15·4	−·0100
39	345·6	−34·0	328·2	·0207	59·0	−14·0	−·0079
42	346·8	−32·8	324·9	·0198	58·3	−14·1	−·0091
45	344·0	−35·6	320·7	·0186	57·7	−14·2	−·0104
48	345·8	−33·8	319·7	·0183	66·5	−13·1	−·0084
51	346·2	−33·4	310·9	·0158	75·5	−11·9	−·0085
54	348·8	−30·8	299·7	·0126	93·6	−9·6	−·0069
57	346·2	−33·4	287·5	·0092	105·7	−8·0	−·0071
17 0	346·2	−33·4	286·9	·0090	112·8	−7·1	−·0054
15	344·4	−35·2	281·5	·0075	116·7	−6·6	−·0059
30	346·0	−33·6	249·5	−·0016	130·0	−4·9	−.0115
39	391·0	11·4	152·9	−·0289	331·7	21·2	·0129
42	382·4	2·8	139·4	−·0327	327·3	20·6	·0094
45	304·2	−75·4	190·4	−·0381	325·4	20·4	0034
48	342·0	−37·6	106·8	−·0420	385·8	28·2	·0155
51	321·4	−58·2	120·6	−·0381	348·7	23·4	·0098
54	337·8	−41·8	144·7	−·0312	278·0	14·3	−·0021
57	334·4	−45·2	171·5	−·0237	257·7	11·6	·0001
18 0	330·2	−49·4	193·9	−·0173	237·8	9·1	·0014
3	327·2	−52·4	191·3	−·0181	225·4	7·5	−·0028
6	306·2	−73·4	201·5	−·0152	275·2	14·0	·0185
9	312·3	−67·3	195·2	−·0170	232·3	8·4	·0001
12	316·4	−63·2	201·6	−·0151	252·6	11·0	·0072
15	296·6	−83·0	219·4	−·0101	137·5	−3·9	−·0180
			265·2	·0029	126·4	−5·3	−·0080
					[illegible]	−5·8	−·0086

33	351·7	−27·9	349·5	·0267	12·5	−20·0	−·0141	24	316·0	−63·6	244·3	−·0031	168·5	0·1	−·0028
36	352·6	−27·0	341·6	·0245	33·4	−17·3	−·0108	27	317·8	−61·8	253·1	−·0006	145·4	−2·9	−·0064
39	344·2	−35·4	346·2	·0258	39·7	−16·4	−·0079	30	322·8	−56·8	261·1	·0017	111·0	−7·3	−·0132
42	347·2	−32·4	354·6	·0282	3·1	−21·3	−·0152	33	329·8	−49·8	285·2	·0085	104·5	−8·1	−·0081
45	354·0	−25·6	355·4	·0284	6·8	−20·8	−·0141	36	344·8	−34·8	261·6	·0018	132·5	−4·5	−·0174
48	346·0	−33·6	365·7	·0313	−2·3	−22·0	−·0130	39	339·8	−39·8	283·5	·0080	114·1	−6·9	−·0061
51	348·8	−30·8	371·7	·0330	−23·7	−24·7	−·0173	42	329·2	−50·4	271·5	·0046	134·8	−4·2	−·0040
54	352·4	−27·2	376·7	·0345	−26·1	−25·0	−·0166	45	320·2	−49·4	265·9	·0031	118·6	−6·3	−·0098
57	338·2	−41·4	374·5	·0338	−16·6	−24·8	−·0148	48	329·8	−49·8	267·2	·0034	146·3	−2·7	−·0020
15 0	338·1	−41·5	377·5	·0346	−29·5	−25·5	−·0183	51	335·0	−44·6	249·2	−·0017	189·1	2·8	·0040
3	336·2	−43·4	378·1	·0348	−27·6	−25·2	−·0179	54	338·2	−41·4	201·2	−·0153	144·2	−3·0	−·0214
6	349·2	−30·4	374·7	·0338	−20·3	−24·3	−·0157	57	340·0	−39·6	208·9	−·0131	201·3	4·3	−·0042
9	348·0	−31·6	369·4	·0323	−11·0	−23·1	−·0147	19 0	335·9	−43·7	271·7	·0047	62·5	−13·6	−·0230
12	355·2	−24·4	366·9	·0316	−20·0	−24·3	−·0179	5	422·0	42·4	248·1	−·0020	172·6	0·6	−·0005
15	342·1	−37·5	365·6	·0313	−0·2	−21·7	−·0129	10	406·0	26·4	247·6	−·0021	186·5	2·4	·0029
18	350·2	−29·4	363·8	·0308	6·5	−20·8	−·0117	15	380·2	0·6	224·4	−·0087	200·1	4·2	−·0002
21	343·7	−35·9	353·9	·0280	1·9	−21·4	−·0158	20	412·4	32·8	212·4	−·0121	222·1	7·0	·0023
24	339·8	−39·8	354·0	·0280	3·8	−21·2	−·0152	25	390·4	10·8	200·9	−·0153	227·1	7·7	·0004
27	403·4	−39·2	356·1	·0286	5·8	−20·9	−·0141	30	382·0	2·4	190·6	−·0183	240·9	9·5	·0011
30	343·6	−36·0	360·6	·0298	2·9	−21·3	−·0135	35	333·2	−46·4	206·8	−·0137	225·6	7·5	·0016
33	334·2	−45·4	360·7	·0299	6·8	−20·8	−·0126	40	360·4	−19·2	200·3	−·0155	252·6	11·0	·0069
36	346·4	−33·2	363·7	·0318	1·3	−21·5	−·0121	45	382·2	2·6	60·6	−·0550	332·7	21·3	−·0115
39	332·2	−47·4	362·9	·0305	7·6	−20·7	−·0114	50	290·2	−89·4	110·2	−·0410	291·6	16·0	−·0084
42	324·2	−55·4	362·8	·0305	1·6	−21·5	−·0133	55	342·4	−37·2	220·5	−·0098	213·0	4·6	−·0004
45	330·8	−48·8	371·6	·0330	−1·7	−21·9	−·0116	20 0	348·4	−31·2	203·2	−·0147	251·1	10·8	·0073
48	353·2	−26·4	375·8	·0342	−28·5	−25·2	−·0173	5	348·2	−31·4	203·0	−·0147	245·1	10·0	·0057
51	345·6	−34·0	368·8	·0322	3·3	−21·1	−·0111	10	331·6	−48·0	225·6	−·0083	209·5	5·4	·0027
54	341·0	−38·6	361·8	·0302	10·0	−20·4	−·0114	15	334·4	−45·2	197·6	−·0163	220·6	6·8	−·0023
57	363·2	−16·4	353·6	·0279	24·6	−18·5	−·0097	20	392·2	12·6	201·1	−·0152	278·3	14·3	·0140
16 0	358·2	−21·4	344·6	·0253	39·2	−16·5	−·0108	25	363·7	−15·9	225·5	−·0084	210·4	5·5	·0030
3	348·2	−31·4	341·3	·0244	38·4	−16·6	−·0096	30	360·2	−19·4	211·2	−·0124	239·4	9·3	·0060
6	357·2	−22·4	337·4	·0233	47·5	−15·6	−·0084	35	354·0	−25·6	225·0	−·0085	232·6	8·4	·0086
9	362·1	−17·5	333·6	·0222	54·4	−14·6	−·0076	40	367·8	−11·8	208·7	−·0145	257·9	11·7	·0092
12	254·6	−25·0	330·7	·0214	58·9	−14·1	−·0073	45	376·1	−3·5	93·9	−·0456	467·8	38·8	·0334
15	353·2	−26·4	330·3	·0213	54·4	−14·6	·0086	50	492·0	112·4	112·7	−·0408	405·7	30·8	·0224
18	355·0	−24·6	328·4	·0207	60·7	−13·8	−·0074	55	480·4	100·8	117·2	−·0391	427·0	33·6	·0293
21	352·0	−27·6	320·7	·0186	57·2	−14·3	−·0106	21 0	394·6	15·0	137·8	−·0332	355·2	24·3	·0162
24	351·0	−28·6	328·9	·0209	39·7	−16·4	−·0128	3	378·0	−1·6	143·8	−·0315	320·6	19·8	·0088

* Brilliant aurora visible from 16 D. 17 H. 38 M. to 17 D. 0 H. See p. 182–3. Co-ordinate mean values: declination, 379·6; bifilar, 255·1; inclinometer, 167·6.

Magnetical Disturbances, Fort Simpson, 1844—*continued.*

April 16–17—*continued.*

Gött. mean Time.	Declination. Scale.	Declination. Δψ	Bifilar. Scale corrected for Temp.	Bifilar. Approx. ΔX/X	Inclinometer. Scale corrected for Decl. and Bif.	Inclinometer. Approx. Δθ	Approx. Δφ/φ
16 D. H. M.		′				′	
21 6	392·0	12·4	126·6	−·0364	424·5	33·3	·0314
9	428·2	48·6	102·2	−·0433	443·2	35·6	·0294
12	442·0	62·4	105·0	−·0425	432·5	34·3	·0274
15	411·4	31·8	119·9	−·0383	382·6	27·7	·0184
18	388·0	8·4	112·0	−·0405	380·8	27·6	·0155
21	387·2	7·6	115·7	−·0394	282·7	27·8	·0175
24	443·0	63·4	148·7	−·0302	367·7	25·9	·0126
27	413·8	34·2	160·7	−·0267	322·2	23·0	·0140
30	370·2	−9·4	90·6	−·0465	464·5	38·5	·0320
33	434·0	54·4	90·8	−·0465	469·8	39·1	·0342
36	486·0	106·4	91·3	−·0464	447·5	36·2	·0271
39	500·3	120·7	93·9	−·0456	438·8	35·1	·0287
42	489·8	110·2	145·7	−·0310	361·2	25·0	·0197
45	475·0	95·4	151·6	−·0293	340·8	22·4	·0164
48	392·1	12·5	156·5	−·0279	340·1	22·3	·0076
51	385·0	5·4	125·4	−·0367	377·6	27·1	·0186
54	370·8	−8·8	125·3	−·0367	368·6	26·0	·0162
57	382·0	2·4	149·7	−·0298	344·8	22·9	·0169
22 0	393·2	13·6	159·0	−·0329	325·8	20·4	·0088
3	430·0	50·4	156·9	−·0278	356·6	24·4	·0223
6	402·0	22·4	—	—	365·1	26·6	—
9	401·8	22·2	147·8	−·0304	356·1	24·4	·0193
12	411·4	31·8	165·7	−·0253	325·9	20·5	·0164
15	410·6	31·0	165·2	−·0254	324·9	20·3	·0161
18	401·8	22·2	177·2	−·0220	311·1	18·6	·0158
[illegible]	[illegible]	[illegible]	[illegible]	−·0238	320·5	19·6	·0166
[illegible]	[illegible]	[illegible]	[illegible]	[illegible]	[illegible]	[illegible]	·0193

April 17—*continued.*

Gött. mean Time.	Declination. Scale.	Declination. Δψ	Bifilar. Scale corrected for Temp.	Bifilar. Approx. ΔX/X	Inclinometer. Scale corrected for Decl. and Bif.	Inclinometer. Approx. Δθ	Approx. Δφ/φ
17 D. H. M.		′				′	
0 12	444·0	64·4	207·8	−·0134	284·6	15·1	·0175
15	417·5	37·9	191·9	−·0179	280·6	14·6	·0119
18	407·0	27·4	193·1	−·0176	272·3	13·9	·0100
21	392·0	12·4	187·1	−·0192	304·6	17·7	·0169
24	409·0	29·4	172·3	−·0234	331·8	21·2	·0201
27	414·0	34·4	149·6	−·0300	350·2	23·6	·0184
30	418·0	38·4	162·4	−·0262	353·2	24·0	·0227
33	499·0	119·4	160·5	−·0268	325·3	20·4	·0184
36	478·0	98·4	205·6	−·0140	220·2	6·8	−·0002
39	377·0	−2·6	209·7	−·0131	265·9	12·7	·0128
42	401·0	21·4	180·8	−·0210	313·3	18·8	·0173
45	436·0	56·4	176·3	−·0223	317·9	19·4	·0173
48	421·2	41·6	203·3	−·0147	271·5	13·4	·0127
51	420·0	40·4	240·1	−·0043	427·6	33·6	·0642
54	371·0	−8·6	−0·8	−·0724	620·9	58·6	·0470
57	485·0	105·4	3·2	−·0710	807·8	82·9	·0979
1 0	424·0	44·4	*−60·	−·0889	837·9	86·7	·0959
3	484·2	104·6	−30·	−·0807	764·6	76·4	·0749
6	604·0	224·4	−20·	−·0778	732·3	73·2	·0713
9	580·0	200·4	−10·	−·0747	737·0	73·7	·0754
12	520·0	140·4	−20·	−·0778	734·3	73·4	·0718
15	428·0	48·4	−40·	−·0835	814·6	82·4	·0843
18	545·0	165·4	−100·	−·1005	768·7	77·7	·0578
21	677·0	297·4	−65·	−·0906	752·9	75·8	·0638
24	780·0	400·4	0·0	−·0722	720·4	71·4	·0732
27	746·0	366·4	−200·0	−·1288	814·0	83·6	·0415
30	574·0	194·4	0·1	−·0721	602·4	56·4	·0428
[illegible]	[illegible]	16·4	60·2	−·0552	478·4	40·2	·0259

Time			
		51·8	63·2
		52·4	73·1
		84·4	68·0
		98·4	93·5
		31·4	105·6
		21·0	142·7
		30·4	167·1
17 D.			
H. M.			
0 0	449·0	69·4	159·0
3	436·5	56·9	200·5
6	418·4	38·8	205·1
9	470·0	90·4	219·7

* The negative readings from hence to 27 M. are approximations only. The point reflected being beyond the limits of the scale, was measured but roughly in the hurry of the moment.

Magnetical Disturbances, Fort Simpson, 1844—*continued.*

APRIL 17—*continued.*

Gött. mean Time.	Declination. Scale.	Declination. Δψ	Bifilar. Scale corrected for Temp.	Bifilar. Approx. $\frac{\Delta X}{X}$	Inclinometer. Scale corrected for Decl. and Bif.	Inclinometer. Approx. Δθ	Approx. $\frac{\Delta\phi}{\phi}$
17 D. H. M.		′				′	
3 30	453·0	73·4	87·5	—·0474	491·9	41·8	·0380
33	410·0	30·4	74·6	—·0510	478·4	40·2	·0309
36	504·6	125·0	111·7	—·0406	580·3	53·3	·0680
39	508·4	128·8	89·6	—·0469	488·9	41·5	·0379
42	471·9	92·3	61·7	—·0547	458·1	37·6	·0218
45	598·0	218·4	86·1	—·0478	442·3	34·2	·0221
48	514·0	134·4	96·1	—·0450	—	—	—
51	540·0	160·4	146·5	—·0308	417·8	32·3	·0352
54	459·0	79·4	167·5	—·0248	—	—	—
57	359·0	—20·6	176·8	—·0222	343·9	22·8	·0244
4 0	429·0	49·4	151·7	—·0292	387·1	28·4	·0285
15	396·0	16·4	154·0	—·0285	348·2	23·4	·0186
30	489·6	110·0	113·5	—·0401	444·3	35·8	·0326
45	494·0	114·4	98·6	—·0443	435·1	34·6	·0273
5 0	483·5	103·9	108·5	—·0415	432·2	34·2	·0282
15	—	—	—	—	—	—	—
30	521·0	141·4	138·6	—·0329	349·9	23·6	·0152
45	486·0	106·4	67·0	—·0532	461·6	38·0	·0242
48	—	—	—	—	445·8	36·0	—
51	459·5	79·9	127·1	—·0362	397·5	29·7	·0244
54	453·0	73·4	122·4	—·0376	375·4	26·9	·0172
57	427·0	47·4	128·2	—·0360	351·0	23·8	·0127

APRIL 17—*continued.*

Gött. mean Time.	Declination. Scale.	Declination. Δψ	Bifilar. Scale corrected for Temp.	Bifilar. Approx. $\frac{\Delta X}{X}$	Inclinometer. Scale corrected for Decl. and Bif.	Inclinometer. Approx. Δθ	Approx. $\frac{\Delta\phi}{\psi}$
17 D. H. M.		′				′	
6 0	442·5	62·9	132·0	—·0348	352·8	23·9	·0140
3	462·0	82·4	144·7	—·0312	311·7	18·6	·0067
6	482·0	102·4	140·5	·0325	310·8	18·4	·0053
9	460·8	81·2	173·0	—·0232	273·4	13·7	·0046
12	463·0	83·4	176·8	—·0222	277·4	14·2	·0069
15	465·2	85·6	183·8	—·0202	268·7	13·1	·0065
30	434·0	54·4	215·7	—·0112	200·8	4·3	—·0024
45	416·0	36·4	902·9	—·0148	224·2	7·3	·0001
7 0	442·5	62·9	226·7	—·0080	214·1	6·0	·0042
8 0	375·0	—4·6	255·0	·0000	153·8	—1·8	—·0037
9 0	362·0	—20·4	294·2	·0097	83·9	—9·5	—·0096
10 0	368·0	—14·4	317·1	·0162	67·1	—13·0	—·0103
11 0	333·0	—49·4	325·8	·0187	50·8	—15·1	—·0119
12 0	348·2	—31·4	333·5	·0222	45·7	—15·8	—·0100
13 0	350·6	—29·0	344·1	·0252	20·9	—19·0	—·0135
9	344·2	—35·4	338·5	·0236	33·5	—17·3	—·0116
18	358·0	—21·6	311·6	·0159	75·9	—11·9	—·0063
27	353·4	—26·2	309·4	·0153	81·9	—11·1	—·0073
36	356·2	—23·4	300·5	·0128	85·6	—10·6	—·0088
45	358·2	—21·4	290·6	·0100	109·3	—7·5	—·0053
14 0	350·2	—29·4	295·2	·0113	109·1	—7·6	—·0042
15 0	369·8	—9·8	263·0	·0023	144·8	—2·9	—·0036

Magnetical Disturbances, Fort Simpson, 1844—*continued.*

April 24. Term Day.

Gött. mean Time.	Declination.		Bifilar.		Inclinometer.		Approx. $\frac{\Delta\phi}{\phi}$
	Scale.	$\Delta\psi$	Scale corrected for Temp.	Approx. $\frac{\Delta X}{X}$	Scale corrected for Decl. and Bif.	Approx. $\Delta\theta$	
24 D. H. M.		′				′	
10 0	378·0	−8·6	269·5	·0043	161·7	−2·8	−·0017
5	380·5	−6·1	270·5	·0046	157·4	−3·3	−·0027
10	378·6	−8·0	272·2	·0051	156·0	−3·5	−·0026
15	379·0	−7·6	271·8	·0050	155·5	−3·6	−·0028
20	380·0	−6·6	272·7	·0052	153·1	−3·9	−·0030
25	380·0	−6·6	272·3	·0051	153·5	−3·8	−·0030
30	381·0	−5·6	272·9	·0053	154·3	−3·8	−·0028
35	380·4	−6·2	273·4	·0054	152·7	−3·9	−·0032
40	380·4	−6·2	272·8	·0053	154·7	−3·7	−·0012
45	380·0	−6·6	272·8	·0053	154·9	−3·7	−·0027
50	380·4	−6·2	270·7	·0046	154·7	−3·7	−·0033
55	380·0	−6·6	272·6	·0052	155·2	−3·6	−·0027
11 0	380·0	−6·6	272·0	·0050	158·5	−3·2	−·0019
5	380·6	−6·0	272·7	·0052	158·4	−3·2	−·0017
10	379·4	−7·2	272·5	·0052	156·5	−3·4	−·0023
15	377·2	−9·4	272·5	·0052	154·2	−3·7	−·0030
20	378·0	−8·6	272·3	·0051	154·2	−3·7	−·0031
25	377·6	−9·0	272·1	·0050	155·8	−3·5	−·0027
30	376·7	−9·9	270·5	·0046	155·8	−3·5	−·0031
35	377·8	−8·8	271·3	·0048	156·4	−3·4	−·0027
40	377·2	−9·4	271·9	·0050	155·2	−3·6	−·0029
45	378·0	−8·6	272·4	·0052	154·8	−3·7	−·0029
50	377·7	−8·9	273·7	·0055	154·2	−3·7	−·0027
55	375·2	−11·4	271·7	·0049	152·4	−4·0	−·0037
12 0	374·0	−12·6	273·1	·0053	151·1	−4·1	−·0037
5	374·0	−12·6	273·2	·0053	151·1	−4·1	−·0037
10	374·0	−12·6	272·2	·0051	157·3	−3·3	−·0032

April 24—*continued.*

Gött. mean Time.	Declination.		Bifilar.		Inclinometer.		Approx. $\frac{\Delta\phi}{\phi}$
	Scale.	$\Delta\psi$	Scale corrected for Temp.	Approx. $\frac{\Delta X}{X}$	Scale corrected for Decl. and Bif.	Approx. $\Delta\theta$	
24 D. H. M.		′					
12 15	375·4	−11·2	271·1	·0048	154·3	−3·7	−·0034
20	374·0	−12·6	272·1	·0053	153·4	−3·8	−·0033
25	374·0	−12·6	271·5	·0049	155·0	−3·6	−·0030
30	376·0	−10·6	270·1	·0045	154·7	−3·7	−·0035
35	376·0	−10·6	270·5	·0046	157·4	−3·3	−·0027
40	375·0	−11·6	271·8	·0050	151·7	−4·1	−·0039
45	374·0	−12·6	274·5	·0057	150·9	−4·2	− 0033
50	374·6	−12·0	273·7	·0055	152·4	−4·0	−·0032
55	374·8	−11·8	274·2	·0056	152·5	−3·9	−·0030
13 0	376·0	−10·6	272·9	·0047	153·0	−3·9	−·0032
5	374·6	−12·0	270·3	·0045	155·6	−3·5	−·0032
10	376·0	−10·6	269·8	·0044	155·7	−3·5	−·0033
15	374·0	−12·6	270·0	·0044	158·7	−3·1	−·0024
20	374·0	−12·6	270·2	·0045	154·3	−3·6	−·0036
25	376·0	−10·6	270·0	·0045	150·1	−4·3	−·0049
30	376·0	−10·6	271·1	·0048	155·4	−3·6	−·0031
35	376·2	−10·4	271·5	·0049	158·9	−3·1	−·0021
40	376·0	−10·6	271·2	·0048	159·8	−3·0	−·0018
45	374·0	−12·6	271·7	·0049	159·0	−3·1	−·0019
50	374·2	−12·4	271·8	·0050	159·3	−3·0	−·0018
55	375·5	−11·1	271·5	·0049	161·2	−2·8	−·0013
14 0	376·0	−10·6	275·0	·0059	166·1	−2·2	−·0010
5	378·2	−8·4	270·4	·0046	162·9	−2·6	−·0011
10	377·8	−8·8	271·1	·0050	157·3	−3·3	−·0022
15	378·0	−8·6	272·2	·0050	155·4	−3·5	−·0027
20	377·8	−8·8	271·9	·0050	155·0	−3·5	−·0027
25	378·2	−8·4	272·5	·0052	158·0	−3·2	−·0019

Overcast throughout. Co-ordinate mean values: declination, 386·6; bifilar, 254·3; inclinometer, 183·2.

Magnetical Disturbances, Fort Simpson, 1844—*continued.*

April 24—*continued.*

Gött. mean Time.	Declination.		Bifilar.		Inclinometer.		Approx. $\frac{\Delta\phi}{\phi}$
	Scale.	$\Delta\psi$	Scale corrected for Temp.	Approx. $\frac{\Delta X}{X}$	Scale corrected for Decl. and Bif.	Approx. $\Delta\theta$	
24 D. H. M.		′				′	
14 30	379·3	−7·3	271·1	·0048	160·3	−2·9	−·0017
35	378·7	−7·9	271·3	·0048	162·0	−2·7	−·0012
40	379·3	−7·3	270·6	·0046	162·6	−2·6	−·0012
45	381·4	−5·2	269·5	·0043	163·3	−2·6	−·0012
50	382·3	−4·3	269·2	·0042	164·4	−2·4	−·0011
55	381·4	−5·2	269·8	·0044	161·2	−2·8	−·0018
15 0	381·2	−5·4	270·9	·0047	160·9	−2·9	−·0015
5	381·0	−5·6	270·9	·0047	162·3	−2·7	−·0012
10	382·7	−3·9	269·7	·0044	164·3	−2·4	−·0009
15	382·6	−4·0	268·5	·0041	163·2	−2·6	−·0016
20	382·1	−4·5	267·9	·0039	164·9	−2·3	−·0013
25	382·4	−4·2	268·5	·0040	162·6	−2·7	−·0018
30	382·1	−4·5	271·3	·0048	159·2	−3·1	−·0019
35	382·0	−4·6	274·5	·0057	153·1	−3·9	−·0027
40	380·2	−6·4	277·5	·0066	151·2	−4·1	−·0025
45	380·0	−6·6	278·2	·0067	149·8	−4·3	−·0026
50	379·6	−7·0	277·6	·0066	150·9	−4·1	−·0025
55	379·2	−7·4	276·5	·0063	152·0	−4·0	−·0025
16 0	379·4	−7·2	276·0	·0062	155·9	−3·5	−·0017
5	380·0	−6·6	278·6	·0069	151·4	−4·1	·0020
10	379·8	−6·8	278·9	·0070	151·2	−4·1	−·0020
15	380·0	−6·6	278·9	·0070	151·5	−4·1	−·0019
20	379·8	−6·8	278·3	·0068	150·3	−4·2	−·0024
25	380·0	−6·6	277·9	·0067	159·4	−3·1	−·0028
	[illegible]	6·6	279·0	·0070	157·4	−3·3	−·0030
					145·4	−4·9	−·0036

April 24–25.

Gött. mean Time.	Declination.		Bifilar.		Inclinometer.		Approx. $\frac{\Delta\phi}{\phi}$
	Scale.	$\Delta\psi$	Scale corrected for Temp.	Approx. $\frac{\Delta X}{X}$	Scale corrected for Decl. and Bif.	Approx. $\Delta\theta$	
24 D. H. M.		′				′	
20 10	384·0	−2·6	283·1	·0082	139·5	−5·6	−·0041
15	384·0	−2·6	278·1	·0067	153·9	−3·8	−·0024
20	385·0	−1·6	279·9	·0073	155·3	−3·6	−·0007
25	385·5	−1·1	280·3	·0074	140·5	−5·5	−·0046
30	392·5	5·9	284·8	·0086	135·1	−6·2	−·0049
35	390·0	3·4	288·3	·0096	132·4	−6·6	−·0047
40	388·0	1·4	292·3	·0108	127·7	−7·2	−·0051
45	381·0	−5·6	292·2	·0107	127·9	−7·1	−·0048
50	380·0	−6·6	287·9	·0095	131·4	−6·7	−·0050
55	379·0	−7·6	286·5	·0091	134·7	−6·3	−·0042
21 0	380·0	−6·6	284·2	·0085	137·7	−5·9	−·0043
5	384·0	−2·6	282·2	·0079	141·0	−5·4	−·0040
10	381·6	−5·0	281·3	·0076	145·5	−4·9	−·0033
15	384·5	−2·1	280·1	·0073	144·9	−4·9	−·0035
20	385·2	−1·4	279·9	·0073	143·7	−5·1	−·0043
25	384·0	−2·6	278·4	·0068	145·1	−4·9	−·0044
30	384·6	−2·0	276·5	·0063	148·7	−4·4	−·0041
35	381·2	−5·4	274·0	·0056	149·3	−4·3	−·0044
40	379·2	−7·4	272·6	·0052	148·9	−4·4	−·0054
45	377·7	−8·9	269·3	·0042	154·6	−3·7	−·0055
50	378·5	−8·1	263·4	·0025	162·9	−2·6	−·0054
55	382·7	−3·9	265·4	·0031	162·2	−2·7	−·0026
22 0	382·8	−3·8	266·1	·0033	160·7	−2·9	−·0030
5	382·6	−4·0	270·9	·0047	158·2	−3·2	−·0023
10	382·6	−4·0	269·4	·0043	159·9	−3·0	−·0023
15	379·0	−7·6	269·2	·0042	160·6	−2·9	−·0021
20	373·7	−12·9	272·5	·0052	156·5	−3·4	−·0023
	[illegible]	−2·6	274·4	·0057	149·5	−4·3	−·0038

H. M.							
50	378·2	−8·4	282·7	·0080	143·0	−5·2	−·0033
55	377·0	−9·6	283·6	·0083	139·7	−5·6	−·0039
17 0	381·0	−5·6	286·2	·0090	133·6	−6·4	−·0069
5	381·0	−5·6	287·0	·0093	133·3	−6·4	−·0068
10	382·2	−4·4	289·7	·0100	127·9	−7·1	−·0055
15	384·0	−2·6	290·8	·0103	126·6	−7·3	−·0054
20	382·2	−4·4	291·6	·0106	126·6	−7·3	−·0052
25	381·6	−5·0	290·2	·0101	126·1	−7·4	−·0059
30	383·2	−3·4	290·7	·0100	127·4	−7·2	−·0058
35	383·6	−3·0	290·3	·0102	129·7	−6·9	−·0049
40	383·6	−3·0	289·4	·0099	132·5	−6·5	−·0044
45	383·0	−3·6	293·2	·0110	123·3	−7·7	−·0038
50	384·0	−2·6	294·7	·0114	114·9	−8·8	−·0078
55	386·4	−0·2	301·1	·0132	119·9	−8·2	−·0045
18 0	394·8	8·2	294·2	·0113	127·2	−7·2	−·0044
5	391·4	4·8	291·2	·0104	135·5	−6·2	−·0050
10	400·6	14·0	288·1	·0096	134·6	−6·3	−·0041
15	394·0	7·4	285·6	·0089	133·9	−6·4	−·0050
20	393·8	7·2	290·9	·0104	117·9	−8·4	−·0080
25	386·6	0·0	298·5	·0125	118·7	−8·3	−·0056
30	394·0	7·4	311·5	·0162	97·8	−11·0	−·0078
35	387·0	0·4	311·5	·0162	98·7	−10·9	−·0075
40	374·2	−12·4	316·2	·0175	88·8	−12·3	−·0093
45	373·0	−13·6	314·4	·0170	94·9	−11·4	−·0077
50	376·5	−10·1	315·4	·0173	87·3	−12·4	−·0097
55	371·5	−15·1	310·8	·0160	101·3	−10·6	−·0071
19 0	376·5	−10·1	305·4	·0145	109·5	−9·5	−·0153
5	378·0	−8·6	300·8	·0132	119·0	−8·3	−·0049
10	379·0	−7·6	296·9	·0121	118·7	−8·3	−·0060
15	378·6	−8·0	293·8	·0112	122·6	−7·8	−·0058
20	376·4	−10·2	298·1	·0124	119·1	−8·3	−·0056
25	378·4	−8·2	299·1	·0098	117·3	−8·5	−·0087
30	374·0	−12·6	294·3	·0113	120·1	−8·2	−·0064
35	378·4	−8·2	296·0	·0118	121·7	−7·9	−·0055
40	383·0	−3·6	296·3	·0119	120·9	−8·1	−·0056
45	382·0	−4·6	295·3	·0116	122·6	−7·8	−·0053
50	381·0	−5·6	293·6	·0111	125·7	−7·4	−·0050
55	384·0	−2·6	290·1	·0101	129·1	−7·0	−·0051
20 0	386·0	−0·6	288·1	·0096	134·7	−6·3	−·0041
5	384·5	−2·1	284·5	·0085	137·5	−5·9	−·0043

H. M.							
30	383·2	−3·4	273·3	·0054	157·1	−3·4	−·0020
35	384·3	−2·3	274·5	·0057	151·4	−4·0	−·0032
40	384·0	−2·6	276·4	·0060	145·8	−4·8	−·0043
45	387·1	0·5	276·4	·0062	149·5	−4·3	−·0032
50	386·0	−0·6	277·3	·0065	152·9	−3·9	−·0020
55	394·0	7·4	249·2	−·0015	179·9	−0·4	−·0024
23 0	402·3	15·7	264·8	·0030	173·6	−1·2	·0302
5	400·3	13·7	267·1	·0036	182·2	−0·1	·0033
10	396·7	10·1	251·3	−·0008	228·7	5·9	·0119
15	394·0	7·4	207·3	−·0133	296·4	14·6	·0185
20	416·6	30·0	206·2	−·0136	275·8	11·9	·0124
25	402·4	15·8	197·9	−·0160	286·3	13·3	·0130
30	403·0	16·4	197·0	−·0162	284·0	13·0	·0121
35	409·6	23·0	204·6	−·0141	288·8	13·6	·0156
40	409·7	23·1	187·8	−·0188	332·4	19·3	·0231
45	418·0	31·4	184·0	−·0199	315·5	17·1	·0173
50	412·9	26·3	180·7	−·0208	329·9	18·9	·0204
55	408·0	21·4	171·3	−·0235	304·2	15·5	·0105
25 D.							
H. M.							
0 0	414·2	27·6	182·6	−·0203	311·3	16·4	·0157
5	405·6	19·0	197·3	−·0161	281·8	12·7	·0116
10	412·2	25·6	192·6	−·0175	282·6	12·9	·0105
15	416·0	29·4	189·9	−·0182	297·1	14·7	·0138
20	419·6	33·0	202·7	−·0146	274·2	11·8	·0110
25	402·0	15·4	206·4	−·0136	282·0	12·8	·0142
30	420·6	34·0	209·0	−·0128	274·9	11·8	·0130
35	404·4	17·8	215·8	−·0109	261·5	10·1	·0111
40	404·3	17·7	209·9	−·0126	268·3	11·0	·0102
45	418·4	31·8	214·4	−·0113	256·9	9·5	·0094
50	419·6	33·0	215·0	−·0111	253·6	9·1	·0087
55	410·3	23·7	210·4	−·0124	270·7	11·3	·0120
1 0	403·7	17·1	207·2	−·0133	276·0	12·0	·0127
5	413·0	26·4	231·2	−·0065	238·6	7·1	·0090
10	411·7	25·1	235·3	−·0054	223·3	5·2	·0059
15	408·9	22·3	239·5	−·0042	223·2	5·2	·0070
20	403·4	16·8	235·4	−·0053	228·9	5·9	·0075
25	420·3	33·7	224·0	−·0086	276·2	12·0	·0175

Magnetical Disturbances, Fort Simpson, 1844—*continued.*

April 25—*continued.*

Gott. mean Time.	Declination. Scale.	Declination. Δ ψ	Bifilar. Scale corrected for Temp.	Bifilar. Approx. $\frac{\Delta X}{X}$	Inclinometer. Scale corrected for Decl. and Bif.	Inclinometer. Approx. Δ θ	Approx. $\frac{\Delta \phi}{\phi}$
25 D. H. M.		′				′	
30	440·0	53·4	229·2	−·0071	239·9	7·3	·0088
35	441·8	55·2	229·0	−·0072	224·3	5·3	·0034
40	408·0	21·4	219·1	−·0100	254·6	9·2	·0099
45	420·8	34·2	241·7	−·0036	210·5	3·5	·0042
50	440·0	53·4	196·6	−·0163	332·9	19·3	·0247
55	471·2	84·6	132·3	−·0340	430·5	32·0	·0355
2 0	446·0	59·4	90·3	−·0464	456·5	35·3	·0304
5	518·2	131·6	100·4	−·0432	477·2	38·0	·0391
10	494·4	107·8	104·9	−·0423	481·6	38·6	·0416
15	466·3	79·7	88·8	−·0472	505·6	41·8	·0434
20	470·4	83·8	60·0	−·0556	508·8	42·0	·0359
25	510·6	124·0	59·1	−·0562	511·4	42·5	·0360
30	555·0	168·4	53·3	−·0567	561·0	48·9	·0496
35	538·0	151·4	64·2	−·0538	580·0	51·3	·0577
40	505·0	118·4	85·1	−·0478	547·4	47·1	·0545
45	515·0	128·4	76·3	−·0504	543·4	46·6	·0427
50	535·3	148·7	68·9	−·0530	579·4	51·0	·0581
55	584·0	197·4	47·4	−·0585	493·4	40·1	·0293
3 0	526·2	159·6	86·5	−·0473	526·6	44·4	·0492
5	567·4	180·8	64·0	−·0539	554·9	49·0	·0504
10	513·0	126·4	113·3	−·0399	397·1	27·6	·0202
15	513·8	127·2	131·5	−·0346	433·1	32·3	·0357
20	528·0	141·4	126·3	−·0362	395·5	27·3	·0235
[illegible]	[illegible]	99·4	147·8	−·0302	342·7	20·6	·0146
[illegible]	[illegible]	[illegible]	[illegible]	[illegible]	361·7	23·1	·0201
[illegible]	[illegible]	[illegible]	[illegible]	[illegible]	[illegible]	[illegible]	·0237

April 25—*continued.*

Gott. mean Time.	Declination. Scale.	Declination. Δ ψ	Bifilar. Scale corrected for Temp.	Bifilar. Approx. $\frac{\Delta X}{X}$	Inclinometer. Scale corrected for Decl. add Bif.	Inclinometer. Approx. Δ θ	Approx. $\frac{\Delta \phi}{\phi}$
25 D. H. M.		′					
50	406·0	19·4	248·5	−·0016	183·2	0·0	−·0016
55	399·0	12·4	253·0	−·0004	175·3	−1·0	−·0026
6 0	407·0	20·4	243·5	−·0031	195·8	1·6	−·0005
5	408·3	21·7	243·3	−·0031	194·8	1·5	−·0001
10	409·8	23·2	242·2	−·0031	193·7	1·3	−·0002
15	415·3	28·7	248·1	−·0018	188·4	0·7	−·0003
20	411·4	24·8	248·1	−·0018	184·0	0·1	−·0015
25	416·0	29·4	246·5	−·0022	181·3	−0·2	−·0017
30	416·8	30·2	242·6	−·0033	203·8	2·6	·0025
35	422·6	36·0	238·5	−·0045	208·0	3·2	·0025
40	418·0	31·4	227·5	−·0076	222·4	5·1	·0044
45	410·3	23·7	235·3	−·0054	205·8	2·9	·0010
50	400·0	13·4	236·7	−·0050	201·2	2·3	·0009
55	400·2	13·6	230·7	−·0067	214·3	4·1	·0021
7 0	409·8	23·2	226·5	−·0079	214·4	4·0	·0009
5	410·6	24·0	230·4	−·0068	223·3	5·2	·0045
10	412·0	25·4	226·0	−·0080	228·4	5·0	·0047
15	412·8	26·2	228·1	−·0074	225·4	4·6	·0044
20	409·8	23·2	234·3	−·0057	213·2	3·9	·0028
25	410·0	23·4	230·9	−·0066	220·6	4·8	·0039
30	406·2	19·6	233·0	−·0060	218·1	4·5	·0038
35	392·1	5·5	242·6	−·0033	190·5	0·9	−·0013
40	379·3	−7·3	259·5	·0015	165·0	−2·3	−·0036
45	371·8	−14·8	261·2	·0019	160·9	−2·9	−·0043
50	368·0	−18·6	274·5	·0057	148·4	−4·5	−·0041
55	374·2	−12·4	282·3	·0079	131·6	−6·6	−·0064
[illegible]	[illegible]	−18·4	285·9	[illegible]	165·2	[illegible]	[illegible]
[illegible]	[illegible]	[illegible]	[illegible]	[illegible]	118·5	−8·3	−·0079

·0083	10	359·2	−27·4	288·3	·0096	124·9	−7·5	−·0068
·0055	15	358·0	−28·6	295·1	·0115	130·5	−6·8	−·0033
·0041	20	358·0	−28·6	308·7	·0097	141·6	−5·4	−·0020
·0031	25	358·6	−28·0	301·1	·0132	141·1	−5·4	−·0086
−·0022	30	362·4	−24·2	292·3	·0108	133·7	−6·4	−·0052
−·0020	35	374·2	−12·4	277·5	·0066	136·4	−6·0	−·0066
·0001	40	372·0	−14·6	271·3	·0048	152·1	−4·0	−·0139
−·0034	45	374·2	−12·4	269·5	·0043	162·2	−2·7	−·0016
·0025	50	377·2	−9·4	269·4	·0043	162·2	−2·7	— 0016
−·0026	55	372·6	−14·0	265·7	·0032	166·8	−2·1	−·0014
·0023	9 0	383·0	−3·6	278·1	·0067	133·1	—	—
−·0011	5	372·2	−14·4	276·8	·0064	146·5	−4·9	−·0039
·0006	10	375·5	−11·1	270·6	·0046	158·1	−3·2	−·0024
−·0085	15	378·2	−8·4	266·6	·0035	158·0	−3·2	−·0036
−·0023	20	375·6	−11·0	254·5	·0001	177·5	−0·7	−·0015
−·0055	25	379·5	−7·1	257·0	·0008	173·0	−1·3	−·0021
−·0088	30	379·0	−7·6	256·4	·0006	169·7	−1·7	−·0032
−·0063	35	381·5	−5·1	253·4	−·0002	169·9	−1·8	−·0040
−·0026	40	378·5	−8·1	255·8	·0004	172·0	−1·4	−·0027
−·0067	45	381·8	−4·8	257·3	·0008	167·1	−2·1	−·0036
−·0012	50	378·0	−8·6	257·9	·0010	165·2	−2·3	−·0040
·0024	55	384·0	−2·6	258·1	·0011	165·8	−2·2	−·0038
·0015	10 0	378·0	−8·6	257·0	·0008	160·2	−3·0	−·0057
·0004								

Magnetical Disturbances, Fort Simpson, 1844—*continued.*

April 25–26.

Gott. mean Time.	Declination. Scale.	Declination. Δ ψ	Bifilar. Scale corrected for Temp.	Bifilar. Approx. $\frac{\Delta X}{X}$	Inclinometer. Scale corrected for Decl. and Bif.	Inclinometer. Approx. Δ θ	Approx. $\frac{\Delta \phi}{\phi}$
25* D. H. M.		′				′	
19 0	372·2	−20·6	270·4	·0052	140·8	−5·9	−·0069
20 0	374·0	−18·8	239·5	−·0035	346·3	20·7	·0386
3	376·0	−16·8	65·9	−·0526	545·6	46·4	·0410
6	392·4	−0·4	60·8	−·0541	515·3	42·5	·0318
9	414·0	21·2	113·7	−·0392	386·2	25·8	·0125
12	442·2	49·4	125·5	−·0358	356·7	22·1	·0080
15	422·0	29·2	169·6	−·0233	281·5	12·3	·0017
18	410·2	17·4	170·4	−·0231	277·0	11·7	·0007
21	388·0	−4·8	180·5	−·0202	266·3	10·2	·0006
24	416·0	23·2	169·7	−·0233	316·2	16·8	·0109
27	408·2	15·4	174·4	−·0220	292·7	13·7	·0059
30	390·0	−2·8	189·4	−·0177	256·3	9·0	·0006
33	364·3	−28·5	140·4	−·0316	319·9	17·2	·0034
36	375·2	−17·6	131·1	−·0343	314·4	16·5	−·0007
39	362·3	−30·5	118·2	−·0379	335·2	19·2	·0012
42	358·8	−34·0	137·7	−·0323	318·6	17·1	·0025
45	335·7	−57·1	158·0	−·0266	301·0	14·8	·0035
48	347·8	−45·0	156·7	−·0270	305·8	15·4	·0044
51	358·8	−34·0	177·8	−·0210	285·5	12·8	·0051
54	370·2	−22·6	170·4	−·0231	288·0	13·1	·0036
57	388·4	−4·4	165·0	−·0246	291·8	13·6	·0041
21 0	388·0	−4·8	209·1	−·0122	216·9	3·9	−·0043
15	358·0	−34·8	238·6	−·0038	179·1	−1·0	−·0017
[illegible]	[illegible]	[illegible]	210·0	−·0119	227·5	5·3	−·0011
					[illegible]	1·4	−·0021

April 26–27.

Gott. mean Time.	Declination. Scale.	Declination. Δ ψ	Bifilar. Scale corrected for Temp.	Bifilar. Approx. $\frac{\Delta X}{X}$	Inclinometer. Scale corrected for Decl. and Bif.	Inclinometer. Approx. Δ θ	Approx. $\frac{\Delta \phi}{\phi}$
26 D. H. M.		′				′	
1 45	446·0	53·2	201·4	−·0143	291·2	13·5	·0129
48	453·0	60·2	188·9	−·0178	—	—	—
51	457·0	64·2	188·7	−·0179	270·7	10·9	·0042
54	432·6	39·8	198·4	−·0152	288·1	13·1	·0116
57	423·4	30·6	212·2	−·0112	248·4	8·0	·0051
2 0	423·6	30·8	215·2	−·0104	240·3	7·0	·0038
5 0	470·0	77·2	122·3	−·0367	390·6	26·4	·0171
6 0	432·0	39·2	219·2	−·0093	185·5	−0·1	−·0096
17 0†	390·0	−2·8	277·1	·0071	130·4	7·3	·0216
18 0	308·4	−84·4	201·0	−·0144	290·6	13·4	·0130
15	344·6	−48·2	283·4	·0089	145·4	−5·3	−·0019
30	369·0	−23·8	281·4	·0083	167·2	−2·5	·0032
45	332·0	−60·8	247·4	−·0296	308·0	15·7	·0024
50	358·0	−34·8	219·3	−·0092	190·6	0·5	−·0082
55	357·1	−35·7	292·9	·0116	121·1	−8·5	−·0057
19 0	326·0	−66·8	186·5	−·0185	286·3	12·9	·0077
5	350·0	−42·8	153·4	−·0279	343·4	20·3	·0134
10	358·3	−34·5	145·8	−·0300	311·6	16·2	·0030
15	338·0	−54·8	156·6	−·0270	307·2	15·6	·0027
20	356·8	−36·0	153·6	−·0278	348·5	20·9	·0147
25	381·0	−11·8	150·7	−·0287	353·3	21·6	·0153
30	398·0	5·2	148·3	−·0293	350·7	21·2	·0139
35	374·2	−18·6	186·7	−·0184	297·4	14·2	·0109
40	384·3	−8·5	176·7	−·0213	314·6	16·5	·0124
45	386·0	−6·8	228·9	−·0065	233·6	6·1	·0056
50	383·9	−8·9	257·1	·0014	133·4	−6·9	−·0126
[illegible]			[illegible]	·0128	122·8	−8·3	−·0041

26 D.							
H. M.							
0 0	422·0	29·2	26·9	—·0638	645·9	59·4	·0572
3	380·0	—12·8	—0·7	—·0715	781·5	76·7	·0347
6	407·8	15·0	—46·7	—·0845	[illegible]	74·9	·0681
9	462·0	69·2	—18·3	—·0765	719	68·8	·0636
12	412·0	19·2	—1·9	—·0719	658·7	69·9	·0521
15	415·4	22·6	—2·0	—·0719	628·9	57·1	·0446
18	340·8	—52·0	7·9	—·0691	605·5	54·2	·0413
21	341·8	—51·0	16·8	—·0666	602·3	53·6	·0430
24	321·6	—71·2	38·4	—·0605	546·3	46·5	·0344
27	325·5	—67·3	19·9	—·0657	525·1	43·8	·0235
30	361·0	—31·8	5·4	—·0698	659·1	61·2	·0547
33	373·0	—19·8	17·3	—·0664	582·6	51·2	·0380
36	392·0	—0·8	21·5	—·0652	590·6	52·2	·0412
39	379·2	—13·6	30·7	—·0626	566·0	49·1	·0376
42	402·2	9·4	68·1	—·0521	500·2	40·6	·0307
45	400·4	7·6	91·3	—·0455	471·4	36·8	·0296
48	402·4	9·6	102·6	—·0423	432·1	31·7	·0225
51	406·4	13·6	110·0	—·0402	447·7	33·8	·0286
54	420·0	27·2	104·5	—·0418	432·8	31·9	·0232
57	403·0	10·2	94·4	—·0446	451·0	34·2	·0251
1 0	429·0	36·2	93·2	—·0449	460·1	35·3	·0271
3	443·6	50·8	85·3	—·0472	—	—	—
6	416·2	23·4	99·9	—·0430	437·2	32·4	·0230
9	378·0	—14·8	132·7	—·0338	364·3	23·0	·0131
12	363·8	—29·0	160·8	—·0258	328·8	18·4	·0116
15	366·8	—26·0	171·0	—·0229	308·3	15·7	·0092
18	388·6	—4·2	178·8	—·0207	302·3	15·0	·0098
21	415·6	22·8	185·9	—·0187	288·5	13·2	·0082
24	423·8	31·0	182·1	—·0198	291·2	13·5	·0078
27	409·8	17·0	194·9	—·0161	271·4	11·0	·0062
30	412·4	19·6	211·4	—·0114	240·9	7·0	·0028
33	411·0	18·2	215·7	—·0103	256·4	8·9	·0078
36	434·0	41·2	197·8	—·0153	256·9	9·1	·0032
39	450·8	58·0	205·1	—·0133	257·5	9·2	·0044
42	442·4	49·6	198·0	—·0152	296·2	14·2	·0137

21 0	347·2	—45·6	232·1	—·0057	192·5	0·7	—·0043
22 0	384·5	—8·3	255·8	·0011	195·9	1·2	·0035
23 0	400·0	7·2	228·3	—·0067	198·6	1·6	—·0034
36	422·1	29·3	135·3	—·0330	368·6	23·5	·0150
39	407·8	15·0	145·8	—·0300	366·8	23·3	·0175
42	384·6	—8·2	121·7	—·0369	388·9	26·2	·0165
45	383·0	—9·8	124·5	—·0361	352·7	20·0	·0077
48	391·5	—1·3	150·3	—·0285	357·2	22·1	·0164
51	398·4	5·6	138·7	—·0321	354·5	21·7	·0121
54	403·0	10·2	145·8	—·0299	363·4	22·7	·0166
57	382·0	—10·8	140·6	—·0315	347·6	20·8	·0108
27 D.							
H. M.							
0 0	372·4	—20·4	139·0	—·0323	344·6	20·4	·0098
3	379·4	—13·4	136·6	—·0326	377·4	24·7	·0177
6	—	—	—	—	363·2	22·8	—
9	394·1	1·3	139·8	—·0318	357·2	22·1	·0132
12	410·6	17·8	143·2	—·0308	369·6	23·7	·0175
15	413·0	20·2	133·2	—·0336	415·2	29·6	·0266
18	426·9	34·1	136·1	—·0328	357·6	22·1	·0122
21	390·9	—1·9	137·8	—·0324	341·1	20·0	·0084
24	377·2	—15·6	146·6	—·0298	343·0	20·1	·0114
27	397·4	4·6	154·0	—·0277	324·8	17·9	·0086
30	415·7	22·9	165·8	—·0244	295·3	14·1	·0042
33	424·0	31·2	199·7	—·0148	302·9	15·0	·0158
36	404·3	11·5	194·5	—·0162	264·0	10·0	·0031
39	414·0	21·2	213·8	—·0108	257·5	9·2	·0079
42	413·0	20·2	224·4	—·0078	245·4	7·6	·0077
45	409·0	16·2	240·3	—·0033	204·7	2·3	·0015
48	396·3	3·5	247·0	—·0014	216·5	3·9	·0065
51	390·6	—2·2	244·0	—·0023	223·3	4·8	·0075
54	401·8	9·0	237·4	—·0041	226·8	5·2	·0065
57	399·8	7·0	235·9	—·0045	215·6	3·7	·0031
1 0	396·4	3·6	247·0	—·0014	201·5	1·9	·0025
3	396·0	3·2	253·6	·0005	195·6	1·2	·0020

* Generally clouded, but faint aurora visible at 25ᵈ 22ʰ, p. 185. Co-ordinate mean values, declination, 392·8; horizontal force, 252·0; inclinometer, 186·6.

† 26ᵈ 17ʰ overcast, with rain, during this disturbance. Co-ordinate mean values, declination, 392·8; horizontal force, 252·0; inclinometer, 186·6.

Magnetical Disturbances, Fort Simpson, 1844—*continued.*

April 27. May 2. | May 2–3.

Gott. mean Time.	Declination.		Bifilar.		Inclinometer.		Approx. $\frac{\Delta\phi}{\phi}$	Gott. mean Time.	Declination.		Bifilar.		Inclinometer.		Approx. $\frac{\Delta\phi}{\phi}$
	Scale.	Δψ	Scale corrected for Temp.	Approx. $\frac{\Delta X}{X}$	Scale corrected for Decl. and Bif.	Approx. Δθ			Scale.	Δψ	Scale corrected for Temp.	Approx. $\frac{\Delta X}{X}$	Scale corrected for Decl. and Bif.	Approx. Δθ	
27 D. H. M.		′				′		27 D. H. M.		′				′	
1 6	395·0	2·2	255·8	·0011	185·2	−0·2	−·0007	21 24	400·0	3·9	251·3	−·0018	242·7	0·2	−·0012
9	395·0	2·2	265·1	·0037	173·2	−1·7	·0002	30	399·8	3·7	246·3	−·0032	232·2	−1·4	−·0060
12	396·0	3·2	264·8	·0036	180·4	−0·8	·0020	36	419·8	23·7	226·6	−·0088	289·1	7·4	·0062
15	398·6	5·8	258·7	·0019	193·0	0·8	·0035	39	422·0	25·9	222·4	−·0100	283·3	6·5	·0032
18	401·0	8·2	255·0	·0008	194·0	1·0	·0028	42	403·4	7·3	230·9	−·0076	279·7	6·0	·0046
21	400·6	7·8	261·3	·0026	185·9	−0·1	·0025	45	404·7	8·6	227·4	−·0086	282·6	6·4	·0044
24	399·0	6·2	255·7	·0010	188·2	0·2	·0015	48	398·0	1·9	245·2	−·0035	256·0	2·3	·0012
27	401·0	8·2	263·4	·0032	180·7	−0·8	·0016	51	406·0	9·9	240·5	−·0049	246·6	0·9	−·0030
30	400·8	8·0	265·6	·0038	172·6	−1·8	·0001	54	403·2	7·1	249·4	−·0023	241·4	0·1	−·0022
33	403·2	10·4	269·0	·0048	174·6	−1·5	·0018	57	390·0	−6·1	249·9	−·0022	249·4	1·3	·0006
36	406·0	13·2	260·5	·0024	189·9	0·4	·0032	22 0	385·8	−10·3	226·9	−·0087	288·1	7·3	·0062
39	408·0	15·2	263·7	·0033	173·9	−1·6	·0001	3	394·0	−2·1	224·4	−·0094	285·6	6·9	·0046
42	406·4	13·6	262·5	·0029	179·6	−0·9	·0010	6	393·7	−2·4	229·8	−·0079	277·4	5·6	·0035
45	404·0	11·2	269·3	·0049	173·0	−1·7	·0014	9	392·4	−3·7	247·0	−·0030	254·8	2·1	·0013
48	404·0	11·2	262·7	·0030	178·8	−1·0	·0010	12	403·2	7·1	246·8	−·0031	250·3	1·4	−·0003
51	404·0	11·2	263·5	·0032	181·3	−0·7	·0017	15	412·2	16·1	236·9	−·0059	264·8	3·7	·0016
54	409·0	16·2	250·3	−·0005	200·4	1·8	·0032	18	410·6	14·5	242·9	−·0042	259·2	2·8	·0015
57	406·0	13·2	256·0	·0011	177·7	1·1	·0033	21	412·2	16·1	246·9	−·0031	258·1	2·6	·0022
2 0	399·0	6·2	258·3	·0018	193·7	0·9	·0036	24	407·2	11·1	243·4	−·0041	267·2	4·0	·0040
3 0	417·8	25·0	250·4	·0005	196·0	1·2	·0019	27	404·0	7·9	—	—	—	—	—
								30	—	—	—	—	260·0	2·9	—
								33	396·0	−0·1	248·6	−·0026	261·1	3·1	·0037
								36	396·4	0·3	251·6	−·0017	258·0	2·6	·0036
								39	396·4	0·3	250·4	−·0021	254·4	2·1	·0022
								42	402·8	6·7	252·2	−·0016	254·6	2·1	·0027
								45	404·0	7·9	251·6	−·0017	256·4	2·4	·0032
								48	400·0	3·9	252·0	−·0016	252·9	1·8	·0021
									[illegible]	0·1	252·1	−·0016	238·0	0·5	−·0006

MAY,

		·0078	181·9	−9·1	−·0108
−6·1	276·0	·0052	208·1	−5·1	−·0053
−9·9	251·7	−·0017	192·8	−7·5	−·0169
−0·1	274·9	·0049	213·4	−4·3	−·0037
−18·1	261·6	·0011	211·4	−4·6	−·0082
−12·1	241·9	−·0045	259·9	2·9	−·0015
−15·9	244·0	−·0039	278·5	5·8	−·0080
−12·8	247·8	−·0028	244·3	0·5	−·0017
−4·1	244·0	−·0039	258·1	2·6	·0014
−0·9	234·7	−·0065	264·4	3·6	·0018
−7·1	241·5	−·0046	264·7	3·8	·0031
−6·9	228·1	−·0084	273·6	5·0	·0018
−15·1	235·4	−·0063	254·8	2·1	−·0020
−34·1	231·2	−·0075	272·7	4·9	·0025
−29·9	240·6	−·0048	247·0	0·9	−·0030
−23·5	239·0	−·0053	241·3	0·0	−·0052
−10·1	260·4	·0008	208·7	−5·0	−·0093
−3·7	256·5	−·0003	230·7	−1·6	−·0036
−8·1	263·1	·0015	219·1	−3·4	−·0054
−2·3	268·9	·0032	201·1	−6·2	−·0086

		−4·1	−·0095
		−6·0	−·0120
		−6·6	−·0129
		−5·5	−·0087
		−2·9	−·0039
		0·3	·0006
		0·0	·0013
		0·2	·0003
		0·2	·0029
		7·8	·0087
		57·9	·0572
		55·2	·0542
		59·8	·0599
·0604	624·2	59·3	·0604
−·0604	608·9	55·9	·0535
−·0573	595·0	54·8	·0543
−·0559	589·6	52·8	·0516
−·0590	626·9	59·8	·0628
−·0587	600·6	55·4	·0541
−·0490	551·5	48·1	·0490
−·0428	486·6	38·0	·0346
−·0408	488·3	38·3	·0372
−·0386	465·9	34·8	·0323
−·0352	441·1	31·0	·0279
−·0325	430·7	29·4	·0274
−·0322	422·1	28·0	·0248
−·0305	400·3	24·7	·0198
−·0257	369·4	18·3	·0115
−·0212	348·3	16·6	·0126
−·0216	355·2	17·7	·0145
−·0186	316·7	11·7	·0052
−·0106	266·6	4·0	−·0026

* The inclinometer was re-adjusted on the 2d May, before the commencement of these observations. Aurora visible at 2[d] 19[h], p. 187. Co-ordinate mean values; declination, 396·1; horizontal force, 257·7; inclinometer, 241·0.

Magnetical Disturbances, Fort Simpson, 1844—*continued.*

May 22.

Gött. mean Time.	Declination. Scale.	Declination. Δ ψ	Bifilar. Scale corrected for Temp.	Bifilar. Approx. ΔX/X	Inclinometer. Scale corrected for Decl. and Bif.	Inclinometer. Approx. Δ θ	Approx. Δφ/φ
22 D. H. M.		′				′	
0 0*	No observation.		—	—	—	—	—
1 0	496·0	73·3	130·7	−·0319	484·0	31·0	·0312
3	496·0	73·3	101·5	−·0402	519·2	36·4	·0339
6	490·0	67·3	104·3	−·0394	528·6	37·9	·0378
9	530·8	78·1	133·9	−·0310	478·7	30·2	·0305
12	512·2	89·5	125·7	−·0333	496·4	32·9	·0337
15	535·0	112·3	129·6	−·0323	504·5	31·8	·0325
18	516·0	93·3	127·1	−·0329	497·0	33·0	·0343
21	482·0	59·3	125·4	−·0334	506·8	34·5	·0369
24	492·4	69·7	122·6	−·0343	502·9	33·9	·0347
27	503·2	85·5	112·7	−·0370	526·1	37·6	·0396
30	538·0	115·3	112·3	−·0372	538·0	39·3	·0428
33	554·0	131·3	102·3	−·0400	542·0	40·1	·0417
36	546·0	123·3	99·5	−·0408	558·8	42·6	·0458
39	528·0	105·3	91·9	−·0429	574·2	44·9	·0486
42	522·0	99·3	87·8	−·0442	553·0	41·7	·0407
45	511·0	88·3	98·1	−·0411	548·5	41·0	·0424
48	512·0	89·3	110·2	−·0377	526·0	37·5	·0387
51	528·0	105·3	104·0	−·0395	538·0	44·0	·0501
54	496·0	73·3	103·2	−·0397	539·2	39·5	·0416
57	513·6	90·9	107·9	−·0385	534·8	38·8	·0405
2 0	500·8	78·1	113·3	−·0369	501·6	33·7	·0317
3	480·0	51·3	119·6	−·0351	492·6	32·3	·0307
6	467·2	44·5	130·9	−·0319	491·6	32·2	·0337
[illegible]	[illegible]	[illegible]	[illegible]	−·0338	506·1	34·4	·0362
[illegible]	[illegible]	[illegible]	[illegible]	[illegible]	[illegible]	[illegible]	·0279

May 22—*continued.*

Gött. mean Time.	Declination. Scale.	Declination. Δ ψ	Bifilar. Scale corrected for Temp.	Bifilar. Approx. ΔX/X	Inclinometer. Scale corrected for Decl. and Bif.	Inclinometer. Approx. Δ θ	Approx. Δφ/φ
22 D. H. M.		′				′	
12 18	379·2	−43·5	324·9	·0230	138·7	−22·5	−·0228
21	391·9	−30·8	342·5	·0280	125·5	−24·5	−·0219
24	370·0	−52·7	337·0	·0263	147·2	−21·0	−·0165
27	372·0	−50·7	343·6	·0283	128·7	−24·0	−·0206
30	370·6	−52·1	344·3	·0285	121·9	−25·1	−·0226
33	378·0	−44·7	334·0	·0256	125·5	−24·5	−·0243
36	399·4	−23·3	331·3	·0248	126·5	−24·4	−·0249
39	407·6	−15·1	329·0	·0242	132·0	−23·5	−·0237
42	389·2	−33·5	318·5	·0212	137·7	−22·7	−·0250
45	382·0	−40·7	320·2	·0217	138·9	−22·3	−·0237
48	380·2	−42·5	310·8	·0191	143·6	−21·7	−·0251
51	389·4	−33·3	310·4	·0190	145·7	−21·2	−·0242
54	380·0	−42·7	281·2	·0107	213·9	−10·8	−·0113
57	384·2	−38·5	263·3	·0056	239·0	−7·0	−·0087
13 0	411·8	−10·9	253·3	·0028	246·1	−5·9	−·0098
3	412·0	−10·7	226·2	−·0049	295·1	1·7	−·0014
6	425·8	3·1	223·8	−·0056	307·3	3·6	·0017
9	424·0	1·3	231·3	−·0035	284·0	0·0	−·0035
12	408·0	−14·7	239·3	−·0012	265·0	−2·9	−·0071
15	402·0	−20·7	247·3	·0011	248·8	−5·4	−·0101
18	416·7	−6·7	249·6	·0017	226·9	−8·8	−·0162
21	414·0	−8·7	278·4	·0099	200·5	−12·9	−·0164
24	405·9	−16·8	290·0	·0132	176·5	−16·6	−·0206
27	393·2	−29·5	290·1	·0132	180·1	−16·0	−·0194
30	386·4	−36·3	304·6	·0173	171·3	−17·4	−·0181
33	372·6	−50·1	306·8	·0179	159·1	−19·4	−·0216
[illegible]	392·0	−30·7	317·7	·0210	151·0	−20·4	−·0205
[illegible]	[illegible]	[illegible]	[illegible]	·0224	139·1	−22·4	−·0232

h	m		
	21		
	24		
	27		
	30		
	33		
	36		
	39		
	42		
	45		
	48		
	51		
	54		
	57		
3	0		
	3		
	6		
	9		
	12		
	15		
	18		
	21		
	24		
	27		
	30		
	33		
	36		
	39		
	42		
	45		
	48		
	51	9	
	54	35·3	209·0
	57	35·1	204·5
4	0	21·9	211·1
5	0	34·3	261·5
11	0	−19·7	253·7
12	0	−26·5	309·0
[illegible]	15	−40·7	336·8

* Strong twilight. Co-ordinate mean values: declination, 422·7; horizontal force, 243·5; inclinometer, 284·0.

Magnetical Disturbances, Fort Simpson, 1844—*continued.*

May 22–23. | May 23—*continued.*

Gött. mean Time.	Declination. Scale.	Declination. Δψ	Bifilar. Scale corrected for Temp.	Bifilar. Approx. ΔX/X	Inclinometer. Scale corrected for Decl. and Bif.	Inclinometer. Approx. Δθ	Approx. Δφ/φ	Gött. mean Time.	Declination. Scale.	Declination. Δψ	Bifilar. Scale corrected for Temp.	Bifilar. Approx. ΔX/X	Inclinometer. Scale corrected for Decl. and Bif.	Inclinometer. Approx. Δθ	Approx. Δφ/φ
22 D. H. M.		′				′		23 D. H. M.		′				′	
15 36	403·7	−19·0	284·4	·0116	222·5	−9·5	−·0077	0 24	458·0	35·3	168·2	−·0213	432·7	23·0	·0255
39	402·0	−20·7	284·3	·0115	222·4	−9·5	−·0078	27	455·0	32·3	177·5	−·0187	429·3	22·5	·0271
42	401·6	−21·1	278·2	·0099	228·1	−8·7	−·0080	30	474·0	51·3	168·0	−·0213	435·4	23·4	·0264
45	406·2	−16·5	272·6	·0082	240·8	−6·7	−·0054	33	483·2	60·5	182·3	−·0173	401·9	18·3	·0200
48	404·2	−18·5	267·5	·0068	252·7	−5·0	−·0034	36	453·0	30·3	195·3	−·0135	379·3	14·8	·0166
51	404·4	−18·3	263·7	·0057	254·2	−4·6	−·0037	39	428·0	5·3	194·9	−·0138	395·1	17·2	·0203
54	403·4	−19·3	266·9	·0066	238·9	−7·0	−·0077	42	409·8	−12·9	195·4	−·0136	396·8	13·3	·0135
57	407·2	−15·5	276·7	·0094	235·3	−7·5	−·0059	45	401·0	−21·7	200·6	−·0129	378·5	14·6	·0179
16 0	410·2	−12·5	269·8	·0075	251·9	−5·0	−·0027	48	398·0	−24·7	184·8	−·0166	395·3	17·2	·0187
15	400·4	−22·3	278·5	·0099	227·2	−8·8	−·0080	51	388·2	−34·5	177·3	−·0187	410·3	19·6	·0212
30	390·0	−32·7	243·9	·0001	260·5	−3·6	−·0072	54	406·8	−15·9	176·4	−·0190	411·5	19·7	·0211
45	390·2	−32·5	240·6	−·0008	264·8	−3·0	−·0053	57	416·2	−6·5	168·7	−·0212	426·6	22·0	·0236
17 0	409·0	−13·7	225·8	−·0050	255·4	−4·4	−·0040	1 0	440·0	17·3	162·0	−·0230	429·0	22·5	·0228
15	424·7	2·0	264·0	·0058	278·1	−0·9	·0040	3	430·0	7·3	172·5	−·0201	423·1	21·3	·0233
30	403·2	−19·5	252·3	·0025	233·7	−7·8	−·0134	6	427·8	5·1	164·4	−·0224	426·3	22·0	·0228
45	412·0	−10·7	237·0	−·0018	287·0	0·5	−·0008	9	420·0	−2·7	167·1	−·0216	421·5	21·1	·0224
18 0	394·4	−28·3	233·6	−·0028	288·8	0·7	−·0012	12	441·0	18·3	164·9	−·0222	439·5	24·1	·0269
30	387·6	−35·1	238·3	−·0015	275·3	−1·3	−·0044	15	436·0	13·3	160·0	−·0236	437·0	23·7	·0247
19 0	390·2	−32·5	233·1	−·0029	282·2	−0·3	−·0035	18	436·0	13·3	169·2	−·0210	418·1	19·2	·0182
20 0	402·1	−10·7	282·6	·0111	268·0	−2·5	·0060	21	423·8	1·1	175·9	−·0191	396·3	17·4	·0165
21 0	410·0	−12·7	246·0	·0007	293·2	1·4	·0035	24	420·0	−2·7	177·6	−·0187	392·2	16·7	·0156
22 0	406·6	−16·1	233·3	−·0029	300·1	2·5	·0022	27	421·6	−1·1	176·6	−·0190	404·2	18·6	·0190
23 0	534·0	111·3	140·9	−·0290	519·2	36·4	·0451	30	442·0	19·3	177·0	−·0188	402·2	18·3	·0285
3	543·6	120·9	104·5	−·0394	553·1	41·7	·0455	33	447·6	24·9	176·5	−·0190	400·3	18·0	·0177
[illegible]	[illegible]	100·6	97·9	−·0412	552·1	41·4	·0431	36	461·8	39·1	177·8	−·0186	402·1	18·3	·0187
					615·3	51·3	·0567	39	486·0	63·3	190·6	−·0150	420·9	21·0	·0278
							·0559	42	512·2	89·5	156·3	−·0247	461·1	27·5	·0313
								[illegible]	[illegible]	118·9	154·3	−·0252	466·1	28·3	·0324

Magnetical Disturbances, Fort Simpson, 1844—*continued.*

May 24. Term Days.

Gött. mean Time.	Declination. Scale.	Declination. $\Delta\psi$	Bifilar. Scale corrected for Temp.	Bifilar. Approx. $\frac{\Delta X}{X}$	Inclinometer. Scale corrected for Decl. and Bif.	Inclinometer. Approx. $\Delta\theta$	Approx. $\frac{\Delta\phi}{\phi}$
24 D. H. M.		′				′	
10 0	421·8	−2·5	255·0	·0044	267·3	−3·7	−·0038
5	424·6	0·3	263·4	·0065	254·7	−5·7	−·0059
10	426·0	1·7	267·0	·0075	247·5	−6·8	−·0073
15	423·0	−1·3	268·2	·0078	240·2	−8·1	−·0094
20	422·2	−2·1	266·1	·0072	252·9	−6·0	−·0066
25	420·0	−4·3	265·4	·0070	251·3	−6·2	−·0065
30	421·0	−3·3	270·2	·0084	241·6	−7·7	−·0084
35	421·8	−2·5	272·6	·0091	238·8	−8·3	−·0087
40	422·2	−2·1	266·6	·0074	248·0	−6·7	−·0073
45	421·0	−3·3	266·0	·0072	252·6	−6·0	−·0059
50	417·8	−6·5	265·1	·0069	253·8	−5·8	−·0057
55	416·2	−8·1	961·4	·0059	252·6	−6·0	−·0072
11 0	414·2	−10·1	267·2	·0075	248·7	−6·6	−·0068
5	418·0	−6·3	265·4	·0070	243·0	−7·5	−·0093
10	418·0	−6·3	267·6	·0076	242·7	−7·5	−·0098
15	420·2	−4·1	272·4	·0090	236·1	−8·7	−·0096
20	439·7	15·4	273·0	·0092	237·9	−8·5	−·0096
25	420·1	−4·2	278·4	·0107	248·9	−6·6	−·0036
30	421·0	−3·3	278·4	·0107	243·6	−7·4	−·0054
35	416·3	−8·0	264·1	·0067	257·9	−5·2	−·0046
40	419·2	−5·1	258·4	·0050	261·0	−4·7	−·0052
45	417·6	−6·7	263·4	·0065	255·0	−5·6	−·0048
50	416·4	−7·9	265·4	·0070	254·5	−5·7	−·0054
55	417·8	−6·5	262·6	·0062	251·1	−6·2	−·0073
[illegible]	[illegible]	[illegible]	[illegible]	[illegible]	251·0	−6·3	−·0078

May 24—*continued.*

Gött. mean Time.	Declination. Scale.	Declination. $\Delta\psi$	Bifilar. Scale corrected for Temp.	Bifilar. Approx. $\frac{\Delta X}{X}$	Inclinometer. Scale corrected for Decl. and Bif.	Inclinometer. Approx. $\Delta\theta$	Approx. $\frac{\Delta\phi}{\phi}$
24 D. H. M.		′				′	
15 30	420·4	−3·9	279·7	·0111	219·8	−11·1	−·0130
35	426·0	1·7	284·3	·0124	220·8	−10·9	−·0114
40	419·6	−4·7	282·6	·0119	217·2	−11·5	−·0131
45	426·0	1·7	272·1	·0089	224·8	−10·3	−·0135
50	420·1	−4·2	268·0	·0078	229·0	−10·0	−·0132
55	418·0	−6·3	262·9	·0063	224·4	−10·4	−·0162
16 0	413·2	−11·1	267·1	·0075	229·2	−9·6	−·0134
5	414·4	−9·9	262·9	·0063	233·1	−9·0	−·0133
10	412·2	−12·1	262·3	·0061	234·9	−8·9	−·0129
15	412·0	−12·3	261·0	·0058	234·3	−9·0	−·0134
20	412·2	−12·1	274·3	·0095	215·9	−11·7	−·0159
25	415·3	−9·0	272·1	·0089	228·4	−9·8	−·0123
30	418·0	−6·3	263·3	·0064	240·7	−8·0	−·0106
35	418·6	−5·7	260·7	·0057	267·7	−3·8	−·0023
40	414·0	−10·3	258·3	·0050	265·0	−4·1	−·0039
45	424·1	−0·2	258·0	·0049	261·1	−4·7	−·0053
50	422·6	−1·7	258·4	·0050	262·6	−4·5	−·0047
55	420·2	−4·1	258·1	·0050	253·3	−5·9	−·0079
17 0	418·0	−6·3	260·3	·0056	256·0	−5·5	−·0064
5	418·0	−6·3	261·9	·0060	248·2	−6·7	−·0085
10	418·0	−6·3	266·0	·0072	250·4	−6·4	−·0066
15	417·6	−6·7	266·3	·0073	247·4	−6·8	−·0076
20	422·0	−2·3	271·9	·0089	238·9	−8·3	−·0088
25	418·6	−5·7	295·0	·0154	239·7	−8·0	−·0020
30	420·0	−4·3	266·9	·0074	249·0	−6·6	−·0068
35	422·0	−2·3	264·9	·0068	252·0	−6·1	−·0064

h. m.							
20	416·2	−8·1	283·8	·0122	220·8	−10·9	−·0115
25	418·2	−6·1	293·9	·0151	211·9	−12·3	−·0114
30	425·0	0·7	289·7	·0139	211·6	−12·3	−·0119
35	423·8	0·5	287·8	·0134	215·6	−11·7	−·0121
40	425·8	1·5	281·1	·0115	212·7	−12·2	−·0160
45	414·0	−10·3	286·1	·0129	211·0	−12·4	−·0141
50	414·0	−10·3	287·9	·0134	213·0	−12·2	−·0130
55	416·0	−8·3	285·5	·0127	221·0	−10·9	−·0110
13 0	418·0	−6·3	279·5	·0110	229·0	−9·7	−·0100
5	416·0	−8·3	272·3	·0090	231·9	−9·2	−·0110
10	414·6	−9·7	268·8	·0080	242·7	−7·5	−·0084
15	416·0	−8·3	269·3	·0081	238·1	−8·4	−·0119
20	415·8	−8·5	266·4	·0073	245·1	−7·2	−·0083
25	414·0	−10·3	271·5	·0087	238·2	−8·4	−·0092
30	414·0	−10·3	276·3	·0101	229·9	−9·5	−·0106
35	414·2	−10·1	279·1	·0109	234·1	−9·0	−·0084
40	417·6	−6·7	273·6	·0093	230·8	−9·4	−·0111
45	412·8	−11·5	275·1	·0098	231·2	−9·3	−·0105
50	410·0	−14·3	286·7	·0130	226·8	−10·0	−·0087
55	408·0	−16·3	281·6	·0106	218·0	−11·4	−·0131
14 00	404·0	−20·3	285·3	·0127	211·0	−12·5	−·0144
5	406·0	−18·3	287·5	·0133	212·2	−12·3	−·0143
10	406·4	−17·9	284·5	·0124	213·8	−12·0	−·0137
15	406·0	−18·3	290·9	·0142	210·2	−12·6	−·0130
20	406·0	−18·3	292·7	·0147	206·0	−13·2	−·0140
25	406·0	−18·3	293·4	·0149	208·1	−12·9	−·0129
30	406·3	−18·0	286·8	·0131	210·6	−12·5	−·0141
35	412·4	−11·9	281·4	·0115	222·1	−10·7	−·0118
40	415·2	−9·1	275·9	·0100	229·4	−9·6	−·0109
45	414·0	−10·3	276·7	·0102	229·0	−9·7	−·0108
50	408·1	−16·2	287·6	·0133	231·2	−9·3	−·0061
55	408·3	−16·0	284·9	·0125	228·4	−9·8	−·0087
15 00	409·4	−14·9	282·8	·0119	224·7	−10·3	−·0105
5	410·0	−14·3	295·0	·0154	215·1	−11·8	−·0103
10	409·8	−14·5	295·0	·0154	200·8	−14·0	−·0151
15	409·0	−15·3	296·0	·0157	195·6	−14·8	−·0166
20	413·2	−11·1	308·5	·0192	184·4	−16·6	−·0170
25	418·0	−6·3	286·5	·0130	215·2	−11·7	−·0123

h. m.							
50	428·0	3·7	270·6	·0085	243·6	−7·4	−·0076
55	—	—	266·1	·0072	—	—	—
18 0	426·0	1·7	260·7	·0057	254·8	−5·7	−·0067
5	420·4	−3·9	257·4	·0048	264·5	−4·2	−·0043
10	421·8	−2·5	251·3	·0030	268·6	−3·5	−·0047
15	423·4	−0·9	254·9	·0040	264·6	−4·2	−·0050
20	423·0	−1·3	255·3	·0042	259·0	−5·0	−·0068
25	423·0	−1·3	258·1	·0050	260·4	−4·8	−·0055
30	417·8	−6·5	259·3	·0053	256·3	−5·4	−·0065
35	417·6	−6·7	260·7	·0057	258·1	−5·1	−·0055
40	419·0	−5·3	261·2	·0058	253·1	−5·9	−·0071
45	418·4	−5·9	268·0	·0078	245·3	−7·1	−·0078
50	419·8	−4·5	265·3	·0070	248·4	−6·7	−·0075
55	423·4	−0·9	267·4	·0076	248·3	−6·7	−·0069
19 0	420·4	−3·9	266·7	·0074	246·2	−7·0	−·0078
5	418·4	−5·9	268·3	·0078	242·8	−7·5	−·0085
10	416·8	−7·5	265·2	·0070	244·7	−7·2	−·0088
15	420·6	−3·7	265·2	·0073	248·8	−6·6	−·0071
20	423·8	−0·5	252·8	·0035	262·6	−4·5	−·0063
25	431·0	6·7	261·1	·0058	248·8	−6·6	−·0086
30	419·4	−4·9	248·6	·0023	266·1	−3·9	−·0063
35	420·4	−3·9	240·1	−·0001	288·3	−0·4	−·0008
40	402·0	−22·3	237·1	−·0010	303·2	1·8	·0029
45	404·0	−20·3	219·8	−·0059	294·0	0·4	−·0051
50	430·0	5·7	263·7	·0065	251·3	−6·2	−·0070
55	415·0	−9·3	230·6	−·0028	305·9	2·2	·0020
20 0	405·6	−13·7	209·2	−·0089	324·3	5·1	·0021
5	407·8	−16·5	212·1	−·0080	329·5	5·9	·0047
10	418·0	−6·3	212·4	−·0080	329·7	5·9	·0048
15	415·8	−8·5	225·0	−·0044	323·9	5·0	·0065
20	427·0	2·7	225·1	−·0044	321·3	4·6	·0056
25	422·4	−1·9	243·0	·0007	269·3	−3·4	−·0068
30	415·0	−9·3	254·7	·0040	264·0	−4·3	−·0052
35	413·0	−11·3	251·4	·0031	264·2	−4·2	−·0061
40	410·6	−13·7	250·6	·0028	263·6	−4·3	−·0065
45	406·2	−18·1	242·2	·0005	300·5	1·4	−·0026
50	412·8	−11·5	250·3	·0027	263·3	−4·4	−·0067
55	408·6	−15·7	246·5	·0017	253·1	−5·3	−·0112

Co-ordinate mean values for the term day: declination, 424·3; bifilar, 240·6; inclinometer, 291·5.

Magnetical Disturbances, Fort Simpson, 1844—*continued.*

May 24-25.

Gött. mean Time.	Declination. Scale.	Declination. Δ ψ	Bifilar. Scale corrected for Temp.	Bifilar. Approx. $\frac{\Delta X}{X}$	Inclinometer. Scale corrected for Decl. and Bif.	Inclinometer. Approx. Δ θ	Approx. $\frac{\Delta \phi}{\phi}$
24 D. H. M.		′				′	
21 00	410·0	−14·3	239·8	−·0002	295·8	0.7	−·0012
5	406·4	−17·9	238·5	−·0006	291·0	0·0	−·0007
10	412·0	−12·3	233·3	−·0020	287·9	−0·6	−·0033
15	413·0	−11·3	231·9	−·0025	303·2	1·8	·0015
20	410·7	−13·6	234·2	−·0018	300·0	1·3	·0010
25	418·7	−5·6	233·8	−·0019	295·3	0·6	−·0006
30	413·1	−11·2	229·1	−·0033	305·1	2·1	·0013
35	432·9	8·6	239·1	−·0004	304·9	2·1	·0041
40	416·0	−8·3	228·3	−·0035	307·9	2·5	·0020
45	416·3	−8·0	230·9	−·0027	303·2	1·8	·0012
50	418·2	−6·1	235·9	−·0013	294·2	−0·4	−·0005
55	418·0	−6·3	237·0	−·0010	288·8	−0·4	−·0019
22 00	418·4	−5·9	243·0	·0007	283·4	−1·2	−·0020
5	420·0	−4·3	245·2	·0013	277·2	−2·2	−·0035
10	422·0	−2·3	245·0	·0012	278·7	−2·0	−·0030
15	421·9	−2·4	246·1	·0016	276·0	−2·4	−·0037
20	418·1	−6·2	245·1	·0013	278·2	−2·0	−·0032
25	421·6	−2·7	247·6	·0020	272·2	−3·0	−·0040
30	418·4	−5·7	253·5	·0037	259·2	−5·0	−·0069
35	423·2	−1·1	256·4	·0045	263·8	−4·3	−·0050
40	425·2	0·9	255·0	·0041	265·1	−4·1	−·0048
45	427·5	3·2	249·3	·0025	281·9	−1·5	−·0007
50	423·2	−1·1	246·3	·0016	267·2	−3·8	−·0065
[illegible]	[illegible]	[illegible]	246·7	·0017	275·8	−2·4	−·0036
[illegible]	[illegible]	[illegible]	[illegible]	[illegible]	[illegible]	−3·2	−·0039

May 25—*continued.*

Gött. mean Time.	Declination. Scale.	Declination. Δ ψ	Bifilar. Scale corrected for Temp.	Bifilar. Approx. $\frac{\Delta X}{X}$	Inclinometer. Scale corrected for Decl. and Bif.	Inclinometer. Approx. Δ θ	Approx. $\frac{\Delta \phi}{\phi}$
25 D. H. M.		′				′	
2 15	435·9	11·6	241·5	·0003	295·3	0·6	·0010
20	437·6	13·3	241·6	·0003	281·3	−1·6	−·0031
25	440·2	15·9	242·9	·0007	284·5	−1·0	−·0017
30	440·0	15·7	243·3	·0008	284·0	−1·2	−·0017
35	440·0	15·7	242·3	·0005	286·0	−0·9	−·0014
40	436·0	11·7	242·6	·0006	280·7	−1·7	−·0031
45	439·0	14·7	243·3	·0008	279·5	−1·9	−·0033
50	438·0	13·7	240·4	−·0001	285.0	−1·0	−·0031
55	440·0	15·7	245·0	·0012	272·2	−3·0	−·0052
3 0	442·0	17·7	242·7	·0006	277·3	−2·2	−·0041
5	441·0	16·7	247·1	·0018	275·5	−2·5	−·0035
10	440·0	15·7	244·5	·0011	278·4	−2·0	−·0033
15	439·2	14·9	242·7	·0006	281·9	−1·5	−·0026
20	440·6	16·3	238·9	−·0005	284·9	−1·2	−·0030
25	441·6	17·3	257·8	−·0008	288·1	−0·5	−·0019
30	441·8	17·5	240·6	·0000	280·0	−1·8	−·0039
35	441·8	17·5	242·1	·0004	285·7	−0·9	−·0015
40	444·0	19·7	237·9	−·0008	284·5	−1·1	−·0031
45	447·0	22·7	237·9	−·0008	287·7	−0·6	−·0021
50	443·6	19·3	238·5	−·0006	284·7	−1·0	−·0029
55	446·2	21·9	242·9	·0007	275·7	−2·4	−·0047
4 0	444·0	19·7	248·8	·0023	269·1	−3·5	−·0052
5	441·4	17·1	249·6	·0025	269·9	−3·3	−·0047
10	440·4	16·1	247·6	·0020	272·0	−3·0	−·0046
15	440·4	16·1	243·3	·0008	284·2	−1·1	−·0017
20	440·0	15·7	243·6	·0008	275·2	−2·5	−·0046
[illegible]	[illegible]	[illegible]	[illegible]	·0014	266·6	−3·8	−·0070

D.	H.	M.
		20
		25
		30
		35
		40
		45
		50
		55
25	0	0
		5
		10
		15
		20
		25
		30
		35
		40
		45
		50
		55
	1	0
		5
		10
		15
		20
		25
		30
		35
		40
		45
		50
		55
	2	0
		5
		10

Magnetical Disturbances, Fort Simpson, 1844—*continued.*

May 25—*continued.*

Gött. mean Time.	Declination. Scale.	Declination. Δψ	Bifilar. Scale corrected for Temp.	Bifilar. Approx. ΔX/X	Inclinometer. Scale corrected for Decl. and Bif.	Inclinometer. Approx. Δθ	Approx. Δφ/φ
25 D. H. M.		′				′	
7 55	436·0	11·9	246·0	·0015	268·0	−3·6	−·0054
8 0	432·0	7·7	246·5	·0017	266·0	−3·8	−·0069
5	436·0	11·7	248·2	·0022	277·0	−2·2	−·0027
10	430·0	5·7	248·2	·0022	265·5	−4·0	−·0066
15	432·0	7·7	248·4	·0022	271·0	−3·1	−·0047
20	430·0	5·7	246·9	·0018	273·0	−2·9	−·0044
25	430·0	5·7	246·8	·0018	269·0	−3·5	−·0058
30	432·0	7·7	249·7	·0026	266·0	−3·9	−·0060
35	430·0	5·7	247·7	·0020	271·5	−3·1	−·0047
40	433·0	8·7	247·9	·0021	269·5	−3·4	−·0053
45	432·0	7·7	247·0	·0018	273·0	−2·9	−·0044
50	430·0	5·7	243·5	·0008	281·5	−1·5	−·0025
55	431·0	6·7	241·2	·0002	284·5	−1·1	−·0022

May 25—*continued.*

Gött. mean Time.	Declination. Scale.	Declination. Δψ	Bifilar. Scale corrected for Temp.	Bifilar. Approx. ΔX/X	Inclinometer. Scale corrected for Decl. and Bif.	Inclinometer. Approx. Δθ	Approx. Δφ/φ
25 D. H. M.		′				′	
9 0	420·0	−4·3	232·8	−·0022	296·0	0·8	−·0039
5	420·0	−4·3	232·8	−·0022	287·9	−0·6	−·0035
10	416·2	−8·1	238·6	−·0006	289·4	−0·3	−·0002
15	417·0	−7·3	239·9	−·0002	285·5	−0·9	−·0019
20	422·0	−2·3	246·0	·0015	281·2	−1·6	−·0019
25	421·6	−2·7	251·6	·0031	266·7	−3·8	−·0052
30	421·0	−3·3	249·6	·0025	263·6	−4·3	−·0068
35	419·8	−4·5	252·3	·0033	264·6	−4·2	−·0059
40	419·0	−5·3	251·3	·0030	267·7	−3·7	−·0050
45	419·6	−4·7	252·4	·0033	262·2	−4·5	−·0065
50	418·0	−6·3	254·6	·0040	262·0	−4·5	−·0060
55	418·0	−6·3	253·6	·0037	262·3	−4·5	−·0061
10 0	417·5	−6·8	251·6	·0031	266·5	−3·9	−·0053

SIR JOHN RICHARDSON'S MAGNETICAL OBSERVATIONS

REDUCED AND DISCUSSED

BY CAPTAIN YOUNGHUSBAND, R.A.

THE Magnetic Instruments supplied to Sir John Richardson for observation in North America were—

1. An Azimuth Compass.
2. A Declinometer for observing Changes of Declination.
3. An Inclinometer, fitted with Deflection Apparatus. This instrument is constructed for observing the Magnetic Inclination or Dip in the usual manner, and then, by deflecting the dipping needle by another magnet, results are obtained from which (combined with an observation made with the Unifilar) the total force can be obtained in absolute measure.
4. A Unifilar Magnetometer for determining the absolute Horizontal Force, and to be used as a Declinometer in observing Changes of Declination.

Instructions for the use of these instruments in the observations recommended to be made with them were contained in the following letter and memorandum addressed by Colonel Sabine to Sir John Richardson previous to his starting on the expedition:—

From Colonel SABINE to Sir JOHN RICHARDSON.

My dear Sir, Woolwich, 22d March 1848.

I hope that you will find the subjoined directions sufficient for the use of the magnetic instruments with which you are supplied; and it remains only that I should indicate to you the points which it appears to me are most deserving of your attention.

1. Azimuths everywhere; we cannot have too many determinations in the quarter to which you are going, on account of the convergence of the lines, and the importance of ascertaining the point or points towards which they converge.

2. I hope that you will find the Declinometer of service in enabling you to record some of the principal disturbances of the Declination during your winter residence. It will in particular enable you to observe the movements of the magnet accompanying momentary auroral phenomena.

3. The diurnal variation of the Declination, both in amount and in turning hours, would be an important determination, especially if

you should find it convenient to determine them for each of the months of winter and of spring.

4. The determination of the Horizontal and Total Forces in absolute measure at your winter residence, is of an importance which I will venture to say will recompense all the time you bestow upon it. It would probably be referred to for centuries to come in connexion with the secular changes of the magnetic elements.

5. If you should be able, without too great a sacrifice of time or convenience, to obtain a second good determination of the absolute total force in a longitude more to the east than your winter resi dence, say between 90° and 100° west longitude, it will be extremely valuable. You will see by the directions that this may be don without carrying about the Unifilar Magnetometer.

Most cordially wishing you and Doctor Rae health for your nobl enterprise, and a safe return to England, either with Sir John Franklin, or to be welcomed by him on your route,

I remain,

Sincerely yours,

EDWARD SABINE.

MAGNETIC DIRECTIONS for Sir JOHN RICHARDSON.

Woolwich, 22d March 1848

The magnetic instruments furnished to Sir John Richardson four in number; viz.

An Azimuth Compass.
A Declinometer.
A Dipping Needle, with Deflection Apparatus.
A Unifilar Magnetometer.

1. Azimuth Compass.—No directions respecting this instr are considered necessary, except the caution that the magnet always be lowered very gently on its support, for fear of i either the cup or the point, which are very carefully worked point is of steel, and a spare one is also sent which is of i When azimuths are observed, the wooden stand on whi instrument is placed should be correctly levelled.

2. Declinometer.—This instrument is intended to serve the purpose of observing the diurnal variation of the Declination fluctuations of the Declination in times of magnetic distu The magnet may be used either on a point or suspended b thread. The point support is preferable, except it should that in very high magnetic latitudes the friction on the poin the free movement of the magnet, and thus prevents it

up its true direction. In such case the silk suspension must be resorted to.

The instrument being placed on its support, the bottom plate levelled, and the suspension tube in its place, screw in the steel point, place the magnet upon it, and fit on the top cover; then raise the point support (and with it of course the magnet) until the opposite portions of the graduated ring are seen in good focus in the two microscopes. The instrument is then ready for use.

If it is found that the magnet when resting on the point does not return after vibration in small arcs to (nearly) the same division in successive trials, it may be necessary to employ the silk suspension. In such case, the magnet resting on the point support, lower the support as far as it will go, lift the magnet off it, lower the screw attached to the thread, so that there may be no danger of breaking the thread whilst the screw is fastening; having then attached the magnet by the screw to the suspension thread, replace the magnet on the point support and raise the support as high as it will go, then shorten the thread until the magnet is relieved from the support and lower the support; now examine if the magnet must be either raised or lowered (by the thread), in order that the graduation on the ring may be in focus. When in focus, examine whether the suspension be truly central as regards the microscopes; it is so when the ring is seen in each microscope in about the same part of the field, and when precisely opposite divisions of the ring are cut by the two microscope wires. If this adjustment be not correct, make it so by moving the suspension either way by means of the adjusting screws which act on the suspension tube; fasten on the top cover, and the instrument is ready for use.

The magnet and ring are correctly balanced at present for the Dip at Woolwich; in any other Dip the balance may require to be adjusted afresh, by means of the cross of wires attached to the magnet. This adjustment is proved by the graduation being in distinct focus in both microscopes at the same time: 1°, when the magnet is in its natural position; and 2°, when it is deflected by another magnet 90° from its natural position. The instrument must be correctly levelled when this adjustment is made.

When the magnet is suspended by a silk thread the influence of torsion should be ascertained in the usual manner, *i.e.* by turning the torsion circle through an angle of 90°, first in one direction and then the other, and noting the difference of the readings in the microscopes in the three positions of the torsion circle, viz., before the torsion circle is turned, and when it is turned 90° on either side.

The agreement or otherwise of the diurnal variation observed by this instrument and by the Unifilar Magnetometer will assist in judging whether the friction on the point support operates unfavourably or not.

3. *A Dip Circle, with Apparatus for Deflection.*—This instrument is to serve the double purpose of observing the Dip, and of determining the ratio of the magnetic moments of each of the two 3·67 inch magnets of the Unifilar Magnetometer to the total force of the earth's magnetism. As the latter determination is novel, and as circumstances have prevented Sir John Richardson's practice with the instrument, full directions may be required.

Two dipping needles are supplied, A 1 and A 2; both are to be used and the same observations made with each. I shall therefore describe the process with A 1, only premising that a precisely similar process is to be pursued with A 2.*

4. If Sir John Richardson should have leisure to make a determination of the Total Force at any other than the winter station, deflections of the dipping needle will suffice, and the Unifilar Magnetometer need not be employed on that occasion; but the values of m X and $\frac{m}{X}$ must be ascertained by experiments with the Unifilar Magnetometer, either at the same or some other station, as soon after as leisure and circumstances will permit.

Be very careful at all times to pack the magnets S and C with their *opposite* poles adjacent to each other, and attend to the same precaution with the dipping needles A 1 and A 2. The deflecting tube and its counterpoise are only to be screwed on the vernier plate during the experiments of deflection, as if left on during travelling they might strain the instrument. The temperature should be noted at each determination of u or u'. The thermometer should be near the deflecting magnet.

Unifilar Magnetometer.—This instrument is supplied for the purpose of determining the absolute value of the horizontal component of the magnetic force, and (in conjunction with the deflection apparatus accompanying the dipping needle) the absolute value of the *total* magnetic force. Sir John Richardson is already furnished with printed instructions, in a paper entitled "On Magnetic Observations, by Lieutenant-Colonel Sabine," containing directions for the use of the Unifilar; and for his further guidance he is referred to the observations of the absolute Horizontal Force, which have been made at Woolwich by Captain Younghusband with the instrument supplied to Sir John Richardson, and have been entered in the

* The process is the same which has been subsequently printed in the "Admiralty Manual," pp. 34 to 39, and 44 to 48; and is therefore omitted here.

register books furnished for the record of the observations in America.

There are two deflecting magnets, C and S, each 3·67 inches in length, which will require to have their respective moments of inertia carefully determined. . For this purpose three inertia rings are supplied, and each of the three will have to be used with each of the magnets C and S. The observations for this purpose may be made at any time in the winter when most convenient.

Instead of the distances between the magnets named in Lieutenant-Colonel Sabine's paper (above noticed), Sir John Richardson had better employ those adopted for this occasion by Captain Younghusband, viz., 1·1 foot and 1·4 foot. Should it be convenient to make one determination of the values of *m* and X early in the winter, and a second towards its close, the two determinations may suffice. Six repetitions of the experiments at the two distances (*i.e.*, six at each distance) may be considered to constitute a determination of the values of *m* and X with the Unifilar.

When not employed in experiments on the magnetic force, the Unifilar Magnetometer may be used for determining the diurnal variation of the Declination, and should give results in accordance with those of the Declinometer. The hourly observations for this purpose need not be carried on through the twenty-four hours. From 6 or 7 A.M. to 8 or 9 P.M., on days on which it may be otherwise inconvenient to observe *hourly*, will probably suffice.

EDWARD SABINE.

The instruments were adjusted at Fort Confidence on Great Bear Lake, situated in 66° 54′ north latitude, and 118° 49′ longitude west of Greenwich; this being the first opportunity afforded to Sir John Richardson for employing the instruments since arriving in America, owing to the rapid rate he found it necessary to travel, so as to be able to explore the coast lying between the Mackenzie and Coppermine rivers, and reach Fort Confidence before the close of the season.

DECLINATION.

Absolute Value.—The absolute value of the Declination at Fort Confidence was determined in March and April 1849, with the Azimuth Compass. The observations are as follow, and show the reading of the Declinometer corresponding to each absolute observation.

TABLE I.

Date.	Absolute Declination.	Corresponding Reading of the Declination Magnetometer.	Date.	Absolute Declination.	Corresponding Reading of the Declination Magnetometer.
1849:	° ′	° ′	1849:	° ′	° ′
March 31st -	50 26 E.	4 21	April 21st -	50 34 E.	4 12
" 31st -	49 52 "	4 22	" 21st -	49 05 "	4 50
" 31st -	50 26 "	4 19	" 21st -	50 33 "	4 08
" 31st -	50 16 "	4 22	" 21st -	49 32 "	4 50
" 31st -	50 30 "	4 18	" 21st -	50 12 "	4 08
" 31st -	50 15 "	4 22	" 21st -	49 30 "	4 50
" 31st -	50 26 "	4 15	May 7th -	47 27 "	5 20
" 31st -	50 04 "	4 22	" 8th -	49 32 "	4 35
April 4th -	53 37 "	1 28	" 12th -	49 17 "	5 24
" 4th -	53 54 "	1 58	" 14th -	52 53 "	4 45
" 4th -	53 56 "	2 06	" 19th -	49 53 "	5 08
" 16th -	53 36 "	3 18	" 19th -	49 29 "	5 27
" 16th -	48 53 "	4 18	" 21st -	48 51 "	5 50
" 16th -	55 21 "	3 13	" 21st -	52 40 "	5 21
" 16th -	49 16 "	4 41	General Mean -	50 42 "	4 17

From the mean of these observations it appears that the Declination was 50° 42′ E. at the period referred to, and that this Declination corresponded to the reading 4° 17′ of the Declinometer scale; whence, having the mean reading of the Declinometer for each month of observation, we may obtain the mean absolute values of the Declination for the same periods. Table II. contains these values.

TABLE II.

Date.	Mean Reading of the Declinometer.	Differences from the Zero 4° 17′.	Absolute Declination 50° 42′ + Diff.
	° ′	′	° ′
October 1848. - -	4 49·5	−32·5	50 09·5
November " - -	4 27·5	−10·5	50 31·5
December " - -	3 58·7	+18·3	51 00·3
January 1849. - -	3 57·0	+20·0	51 02·0
February " - -	4 01·3	+15·7	50 57·7
March " - -	4 12·7	+04·3	50 46·3
April " - -	4 22·6	−05·6	50 36·4
		Mean -	50 43·4

Diurnal Variation.—The Declinometer with which the principal series of observations was made has been fully described in Colonel Sabine's memorandum above; the instrument was used as directed by the instructions, and nothing further seems necessary to be stated with reference to it, except that of the two modes of adjusting the needle of which the instrument is capable, that was chosen in which the magnet is made to traverse upon the steel point instead of being suspended by a silk thread; this is now to be regretted, as the observations show that the friction was so great as to impede the free movement of the needle. The observations made with another magnet, suspended by a silk thread in the Unifilar, prove that the range of movement of the Declinometer Magnet was limited by the friction on the point, but that the *direction* of the movement was recorded faithfully. The additional observations with the Unifilar extend over a portion of three months, and are valuable as confirmatory of the general accuracy in direction of the movements of the Declinometer Magnet, and as affording a truer value of the extent of the diurnal change in those months than can be obtained from the impeded action of the Declinometer Magnet.

Observations were made with the Declinometer during the months of October, November, and December, 1848, January, February, March, and April, 1849, commencing at 6 A.M. and continued hourly until 9 P.M. Occasionally, observations were taken at 4 and 5 A.M., and on two days generally in every month an observation was made at the night hours omitted on ordinary occasions. These observations are given in full, pp. 28 to 35, Table VIII.

The mean monthly diurnal variation appears in Table III.

TABLE III.

Mean Diurnal Variation in the several Months of Observation.

Mean Time at Fort Confidence. Astronomical Reckoning.	Noon.	1.	2.	3.	4.	5.	6.	7.	8.	9.	10.	11.
	′	′	′	′	′	′	′	′	′	′	′	′
1848:												
October	23·2	20·3	29·5	38·8	48·7	53·4	53·3	53·6	56·8	61·2	—	—
November	1·0	1·6	7·4	10·7	15·6	19·9	19·5	23·6	26·3	27·1	—	—
December	8·0	9·1	10·4	11·6	13·8	14·9	15·8	16·3	16·8	18·6	—	—
1849:												
January	2·3	5·4	4·5	7·7	9·5	11·9	13·2	14·8	15·8	17·7	—	—
February	0·0	0·9	3·7	4·4	8·3	11·5	13·3	16·2	17·8	18·1	20·8	—
March	6·6	7·7	13·1	18·9	20·9	23·6	25·9	27·1	29·2	29·7	30·6	—
April	10·8	18·5	26·0	33·0	37·2	44·3	45·8	49·4	52·2	52·5	55·3	—
Means	7·4	9·1	13·5	17·9	22·0	25·7	26·7	28·7	30·7	32·1	—	—
Means reduced	4·6	6·3	10·7	15·1	19·2	22·9	23·9	25·9	27·9	29·3	—	—

Mean Time at Fort Confidence. Astronomical Reckoning.	Mid-night.	13.	14.	15.	16.	17.	18.	19.	20.	21.	22.	23.
	′	′	′	′	′	′	′	′	′	′	′	′
1848:												
October	—	—	—	—	—	—	44·0	38·2	13·9	0·0	4·1	17·7
November	—	—	—	—	—	26·9	15·9	12·2	14·4	7·7	7·3	0·0
December	—	—	—	—	—	—	13·6	11·9	0·0	0·6	2·1	5·5
1849:												
January	—	—	—	—	—	12·8	4·3	5·3	3·8	2·5	0·0	0·3
February	—	—	—	—	—	—	20·4	20·9	19·4	9·9	6·0	2·5
March	—	—	—	—	—	—	27·8	26·6	22·0	8·7	0·0	3·9
April	—	—	—	—	—	—	49·0	45·3	30·8	3·4	0·0	2·5
Means	—	—	—	—	—	—	25·0	22·9	14·9	4·7	2·8	4·6
Means reduced	—	—	—	—	—	—	22·2	20·1	12·1	1·9	0·0	1·8

From these observations, it appears that on taking the first observation in the morning, the north end of the needle was found to be proceeding eastward, that the easterly extreme was attained by a rather rapid movement about 22^h (the time varied in the different months between 20^h and 0^h), that the north end then moved westward and continued a tolerably uniform movement until 9^h, after which no observation was made. The westerly extreme was attained some time in the course of the night, but from no observation having been made later than 9 P.M. the exact time is unknown, the north end of the magnet being found moving eastward, as has been already stated, on taking the first observation in the morning.

If we compare these movements with the diurnal variation in middle latitudes in the Northern Hemisphere, we find a very striking dissimilarity, and the difference is worthy of attention, because it appears probable that the diurnal varintion of the Declination needle in high latitudes follows a law differing materially from that of the diurnal variation in middle latitudes, now well known and established. The principal feature of the diurnal variation in middle latitudes of the Northern Hemisphere is the attainment by the north end of the magnet of its extreme westerly position about 1 o'clock P.M. daily, whereas, as stated above, at Fort Confidence the extreme easterly position occurs at 11 A.M., from whence a movement westward takes place, continuing until the latest hour at night at which any observation is made, the extreme being attained between 9^h and 18^h.

So few observations of the diurnal variation of the Declination in high latitudes are up to this time at command, that not even an approach can be made towards indicating a general law of the phenomena in such localities; we can only present those facts that have been already obtained, and direct attention to points of similarity and of discordance from movements in middle latitudes. In the accompanying plate is drawn the curve of the diurnal variation of the Declination at a number of places, for the purpose of showing the diurnal movement at Fort Confidence and at other stations in high latitudes in comparison with each other and with the movement in middle latitudes. It is thus shown that at Reikiavik in Iceland, the extreme easterly position is attained at 2 P.M., which is about the hour that the westerly extreme is attained at every other station from which we have observations, with the exception of Fort Confidence; thus presenting, perhaps, as strong an instance as could be ound of the widely differing phenomena of the diurnal variation in high from middle latitudes. The curve at Reikiavik differs also from that at Fort Confidence in the westerly extreme, which occurs in Iceland at 22^h, the same hour that an easterly extreme is attained at Fort Confidence. It is true that the observations at Reikiavik

Plate II Showing the Diurnal Variation of the Declination at

Fort Confidence, *from Oct.r 1848 to March 1849*
Iceland, *from 21st to 28th August 1836*
Lake Athabasca, *from Oct.r 1843 to Feb.y 1844*
Christiania, *from June to Nov.r 1842*
Bossekop *from Oct.r 1838 to April 1839*
St. Petersburg *from Oct.r to April; 1841 to 1845.*

Catherinenbourg, *from Oct.r to April 1841 to 18..*
Barnaoul, *from Oct.r to April, 1841 to 18..*
Nertchinsk, *from Oct.r to April, 1841 to 18..*
Sitka, *from October to April; 1842 to 18..*
Toronto *from Oct.r to April 1843 to 18..*
Greenwich, *from Oct.r to April, 1841 to 1846*

Scale 0·1 Inch = 1′ of arc; ↑ North end moving toward the W.; ↓ towards the E.

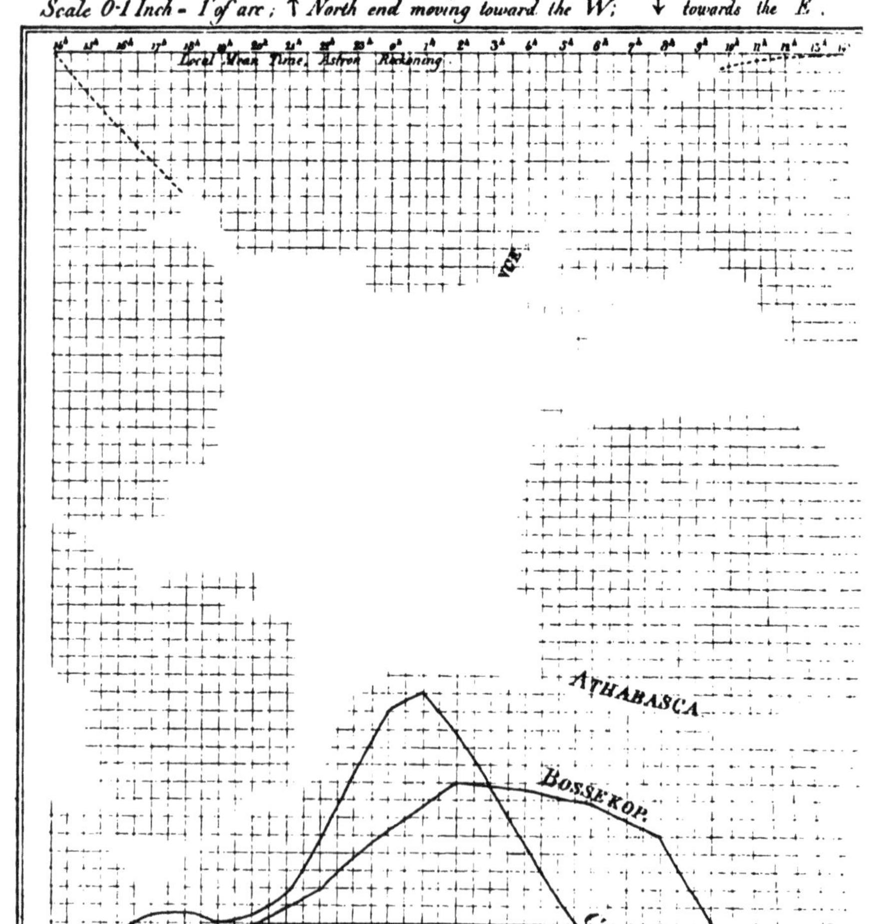

include only one week's series, in the month of August 1836, taken hourly; they are, however, very accordant, and the curve drawn from the observations proceeds from a maximum to a minimum and back again in a regular continuous progression.

Plate II. contains the curves of the diurnal variation at several stations in the Northern Hemisphere; it illustrates better than any verbal description the great change that takes place in the law of the diurnal variation when advancing into high latitudes. The stations are :—

	Latitude.	Longitude.
Fort Confidence	66° 54′ N.	118° 49′ W.
Reikiavik, Iceland	64° 08′ N.	21° 55′ W.
Athabasca	58° 41′ N.	111° 18′ W.
Bossekop	69° 58′ N.	21° 10′ E.
Christiania	59° 55′ N.	10° 34′ E.
St. Petersburg	59° 57′ N.	30° 19′ E.
Catherinenburg	56° 50′ N.	60° 34′ E.
Barnaoul	53° 20′ N.	83° 27′ E.
Nertchinsk	51° 18′ N.	119° 21′ E.
Sitka	57° 03′ N.	135° 18′ W.
Toronto	43° 39′ N.	77° 05′ W.
Greenwich	51° 29′ N.	00° 00′

By this plate we perceive that at Christiania, St. Petersburg, Catherinenburg, Sitka, and Toronto the extreme westerly position of the north end of the magnet was attained at 1^h for the period included in the observations; at Athabasca, Bossekop, and Greenwich at 2^h, and at Barnaoul at 3^h, while at Reikiavik a westerly maximum was reached at 22^h, and again at 8^h, and at Fort Confidence some time in the course of the night between 9^h and 18^h; thus at once pointing out the irregularity in the law of diurnal movement in high latitudes, and the marked contrast from the persistent general law of movement which everywhere obtains in middle latitudes. The curves are all drawn to the same scale, and show the relative amounts of the daily excursions at each place.

We may now examine in more detail the diurnal movement at the several stations. At Reikiavik we find the north end of the needle at its extreme observed westerly position at 9^h, from whence it proceeds uniformly to a secondary east at 17^h, thence to west at 22^h, again to east (the maximum) at 2^h, then back again to west at 9^h; the double curve is very fully exemplified by the two excursions (west to east and back again), being very nearly the same in extent.

The observations at Athabasca are those of Captain Lefroy, printed and discussed in the second part of this volume. They comprise hourly observations made in the months of October, November, and December, 1843, January and February 1844; the several months' observations accord with each other, and having been made regularly at each hour in the twenty-four, very valuable evidence is afforded of the diurnal movement at that station; the results may perhaps with advantage be again stated in this place for the sake of making the account of the comparison more complete. At Athabasca we find the easterly extreme occurring at 17^h, viz., the same hour at which one of the easterly extremes was attained at Reikiavik; a fact specially worthy of notice, because at no other station is this period of the day marked by a similar position of the magnet.

From 17^h the north end proceeds pretty uniformly to extreme west at 2^h, showing an accordance in this respect with the general law of the diurnal variation; thence again to east at 17^h. An interruption occurs from 10^h to 13^h, when the north end turns and moves west, but the retrograde movement is insignificant compared with the whole diurnal excursion.

At Bossekop a single curve is formed. Extreme west at 2^h; extreme east at 14^h.

At Christiania a double curve. Extreme west at 1^h; extreme east at 10^h; a secondary west at 17^h; a secondary east at 19^h. This curve is similar in form to that at St. Petersburg and several other places, but the amount of the excursion is much greater, as, for example, at Christiania, 20′; at St. Petersburg, 6′. It must, however, be remarked, that the months of observation are not the same at the two stations; they are June to November at Christiania, and October to April at St. Petersburg.

At Catherinenburg the curve is nearly the same as at St. Petersburg, showing at these two places the evening easterly extreme greater than the morning easterly extreme. At Barnaoul, Nertchinsk, Sitka, and Toronto, the morning easterly extreme exceeds that of the evening, while at Greenwich we have the morning easterly movement nearly obliterated, and the extreme easterly position at 10^h.

It will now be seen that there are no *general* characteristics of the diurnal variation of the Declination in very high latitudes; also, that in middle latitudes there is a consistent law, the most prominent feature of which is the occurrence in the Northern Hemisphere of the maximum westerly Declination during the day at 1^h—2^h. This is invariable. There is, also, a distinction between the character of the movement on the Siberian from that on the American continent, particularly if we study the law of the variation during the several seasons of the year; but such considerations are not relevant to the present discussion, and need not be here further noticed.

It is necessary to mention that every observation made has been included in forming the mean results upon which the discussion of the phenomena of the diurnal variation at Fort Confidence is founded, consequently the diurnal variation spoken of includes the modification to which it is subject, caused by the disturbed observations remaining, these disturbances having themselves a distinct and different law. Now it may be reasonably assumed that the effect due to disturbance varies considerably at different stations, and it seemed not improbable that, from the position of Fort Confidence, the effect there might be greatly magnified, even so much so as to cause the diurnal variation, without the disturbances, to present a very different aspect from the curve drawn from all the observations without omission. It may therefore be satisfactory to state that the process of eliminating the disturbances was undertaken, and that the diurnal curve of the residual observations was found only modified in the extent of range, but not altered in general character.

Absolute Horizontal Force.

The instrument employed by Sir John Richardson in observing the Absolute Horizontal Force was a portable Unifilar Magnetometer of the usual construction, viz., one in which the deflecting magnet is kept at right angles to the suspended magnet, and the angle of deflection read on the horizontal circle. Two magnets, 3·00 inches in length were supplied for suspension in the deflection apparatus. The deflecting magnets were C 1 and S 1, each 3·67 inches in length. These magnets were also employed as deflecting magnets in Dr. Lloyd's Inclinometer for determining the Total Force in absolute measure. The distances from the suspended magnet at which the deflectors were placed were in the Unifilar 1·1 and 1·4 feet, and these two distances were employed whenever experiments of deflection were made. The nearer distance, 1·1 feet, was chosen as being just beyond the limits of the quantity expressed by three times and a half the length of the longer magnet, and the second distance is in proportion to the first as 1·3 to 1 nearly.

For the experiments of vibration the magnets were suspended in the wooden box allotted to this purpose; the same stirrup was used during the whole series, and the moments of inertia of the magnets and stirrup determined by means of Dr. Lamont's Inertia Rings. The temperature coefficients of the deflecting magnets were determined after the instruments had been returned to Woolwich; and as the range of temperature to which they had been exposed during the winter had been very great, involving consequently very large corrections, the experiments for determining the coefficients were conducted with particular care, and it is believed that the value of the coefficients at different parts of the thermometric scale is known with sufficient accuracy. The observations made at Fort Confidence included temperatures varying from $-36°$ to $+70°$ Fahrenheit.

The value of the coefficient P, depending upon the distribution of magnetism in the suspended and deflecting magnets, was found to be inappreciably small when the suspended magnet was one of the 3·0 inch Unifilar magnets, but to have a sensible value when the dipping needle was suspended; the corrections on this account have been applied in calculating the value of the Total Force.

Table I. contains the data from which the values of K, the moment of inertia of the deflecting magnets, C 1 and S 1, were calculated. Three rings were employed in their determination, of which the weights and dimensions are as follows:

	Outer Diameter. Inches.	Inner Diameter. Inches.	Weight. Grains.
Ring 5 -	3·586	2·951	1493·14
„ 6 -	3·026	2·472	960·14
„ 8 -	3·002	2·477	638·64

The value of K′ for each ring was calculated by the formula $K' = \frac{1}{2}(r^2 + r_{,}^2)\, w$, when r and $r_{,}$ denote respectively the outer and inner radii of the rings in decimals of a foot, and w the weight in grains; whence we have

For Ring 5; K′=27·949 - - - Log. =1·44636
„ 6; K′=12·724 - - - Log. =1·10461
„ 8; K′= 8·397 - - - Log. =0·92412

Table IV.

Observations for the Moment of Inertia of the Magnet and Stirrup.

Magnet C 1.				Magnet S 1.			
Date.	No. of Ring.	Vibrations *with* Ring. Logs. of T'^2.	Vibrations *without* Ring. Logs. of T^2.	Date.	No. of Ring.	Vibrations *with* Ring. Logs. of T'^2.	Vibrations *without* Ring. Logs. of T^2.
1848: Nov. 16	—	—	1·83537 (10°·0)	1848: Oct. 16	—	—	1·92848 (30°·0)
20	—	—	1·83822 (10·0)	17	—	—	1·92943 (30·0)
20	—	—	1·83541 (10·0)	17	—	—	1·92371 (30·0)
21	5	2·71178 (10°·0)	—	Nov. 9	—	—	1·95449 (0·0)
21	5	2·71552 (10·0)	—	9	5	2·82535 (0°·0)	—
22	5	2·70639 (10·0)	—	13	5	2·83219 (0·0)	—
23	6	2·42341 (10·0)	—	13	6	2·55273 (0·0)	—
23	6	2·42876 (10·0)	—	13	6	2·55232 (0·0)	—

The degrees following the logs of the squares of the vibrations, with and without rings, signify the temperatures corresponding to the vibrations.

TABLE IV.—*continued.*

MAGNET C 1.

Date.	No. of Ring.	Vibrations *with* Ring. Logs. of T'^2.	Vibrations *without* Ring. Logs. of T^2.
1848: Nov. 27	8	2·29678 (10°·0)	— °
27	8	2·29515 (10·0)	—
Dec. 23	—	—	1·83550 (−20·0)
23	—	—	1·83608 (−20·0)
1849: Mar. 23	—	—	1·84473 (−5·0)
23	—	—	1·84645 (−5·0)
24	—	—	1·85101 (−5·0)
24	—	—	1·85019 (−5·0)
April 3	—	—	1·84173 (5·0)
3	—	—	1·84093 (5·0)
3	—	—	1·84565 (5·0)
4	8	2·33157 (5·0)	—
4	8	2·32559 (5·0)	—
5	8	2·32532 (5·0)	—
5	6	2·45582 (5·0)	—
5	6	2·45681 (5·0)	—
6	6	2·46098 (5·0)	—
6	5	2·73246 (5·0)	—
7	5	2·73330 (5·0)	—
7	5	2·73375 (5·0)	—
9	—	—	1·85028 (5·0)
9	—	—	1·85498 (5·0)
9	—	—	1·84705 (5·0)
26	5	2·72626 (20·0)	—
26	6	2·46194 (20·0)	—
26	8	2·32597 (20·0)	—
26	—	—	1·84970 (20·0)

MAGNET S 1.

Date.	No. of Ring.	Vibrations *with* Ring. Logs. of T'^2.	Vibrations *without* Ring. Logs. of T^2.
1848: Nov. 14	8	2·43614 (0°·0)	— °
14	8	2·42177 (0·0)	—
15	—	—	1·95991 (0·0)
Dec. 18	—	—	1·95809 (−30·0)
18	—	—	1·96117 (−30·0)
1849: Mar. 20	—	—	2·00031 (−5·0)
20	—	—	2·00023 (−5·0)
20	—	—	1·98974 (−5·0)
21	—	—	1·99532 (−5·0)
21	—	—	1·99431 (−5·0)
April 10	—	—	1·98720 (5·0)
10	—	—	1·99406 (5·0)
10	—	—	1·99559 (5·0)
11	8	2·45641 (5·0)	—
11	8	2·45904 (5·0)	—
11	8	2·45763 (5·0)	—
13	—	—	1·99113 (5·0
14	6	2·58786 (5·0)	—
14	6	2·58680 (5·0)	—
16	6	2·59764 (5·0)	—
17	5	2·91789 (5·0)	—
17	5	2·91068 (5·0)	—
18	5	2·92048 (5·0)	—
18	—	—	1·99055 (5·0)
25	—	—	1·99079 (20·0)
25	8	2·46344 (20·0)	—
25	6	2·59175 (20·0)	—
25	5	2·93349 (20·0)	—

The deduced values of K for each magnet, and from the observations with each ring, are as follows:—

MAGNET C 1.

With Ring 5.	With Ring 6.	With Ring 8.
4·2941	4·4053	4·4626
4·1890	4·1444	4·1682

The mean of all these is 4·2772; = Log. 0·63116, which is the value employed in the calculations.

MAGNET S 1.

With Ring 5.	With Ring 6.	With Ring 8.
4·2774	4·3399	4·3300
4·3582	[3·7693]	4·2751

Some undiscovered source of error existed in the second series of experiments with Ring 6 and Magnet S 1; the result has therefore been omitted in taking the mean. The mean of all the others is 4·3161; = Log. 0·63509, which is the value of K employed in the calculations with this magnet.

Temperature Corrections.—The experiments for ascertaining the temperature coefficients of magnets C 1 and S 1 were conducted according to the method of deflection. The suspended magnet employed was 3·00 inches in length, and the deflecting magnet was placed at right angles to it at a distance of 9 inches; the mean deflection produced was 25°. The angle of deflection, by magnet C 1, was ascertained at the following temperatures, viz.:—

At 36°·57, 55°·42, 73°·62, and 90°·79; and by a second series of experiments, at 32°·71, 53°·39, 72°·47, and 88°·99; and the coefficient q determined as follows:—

q = ·000412 at a mean temperature 44°·5
= ·000488 .. 63°·7
= ·000496 „ 81°·5

For magnet S 1, the angle of deflection produced by it was ascertained at the temperatures 32°·61, 52°·89, 71°·34, and 88°·91; and by a second series of experiments, at 32°·04, 50°·82, 70°·45, and 88°·69; from whence the coefficient q for this magnet was determined, viz.:—

q = ·000309 at temperature 42°·0
= ·000375 „ 61°·3
= ·000382 „ 79°·8

The rapid decrease in the value of the coefficient between 64° and 44° in the case of one magnet, and between 61° and 42° in the case of the other, rendered it desirable that the coefficient should be ascertained at lower temperatures than those just stated. Accordingly an attempt was made, by subjecting the deflecting magnet to temperatures varying between 0° and 32°, to ascertain the angles of deflection at low temperatures, which should be known with tolerable exactness. A mixture of pounded ice and salt was employed to surround the magnet, and the temperature reduced to −6° (as indicated by the thermometer employed, whose index error at that temperature was not, however, precisely known). Deflections were observed at −6°, 14°, and 32°.

The observations are subject to some degree of error, dependent upon the rapid changing of the temperature of the freezing mixture at a degree so much below the natural temperature; but great pains were taken to sustain a constant circulation.

The final results obtained were for Magnet C 1—

$q =$ ·000361 at temperature 4°·6
and $q =$ ·000366 „ 23°·8

For magnet S 1, the results were—

$q =$ ·000298 at temperature 14°·3
and $q =$ ·000328 „ 21°·8

After allowing for the probable amount of error occasioned in these values from the cause already mentioned, it appeared evident that the coefficient did not diminish in value in the same rapid ratio below 42° as it was proved to do between 42° and 62°, and it was considered that the temperature coefficients were sufficiently well obtained for every purpose of correction in the observations under calculation. The experiments were continued in the hope of ascertaining more precisely the law of change with the change of temperature; but it is sufficient here to mention that the conclusions previously arrived at were substantially confirmed.

The magnetic moment (m) of the deflecting magnets was found by combining together the values of $\frac{m}{X} = \frac{1}{2} r^3 \sin. u$ obtained from the experiments of deflection, and of $m X = \frac{\pi^2 K}{T^2}$ from the experiments of vibration; here some difficulty occurred on account of the experiments of deflection and of vibration having been conducted in separate series on different days. A mode of grouping the results of each experiment was eventually adopted, which gave, it is believed, the most satisfactory value of the magnetic moments of the bars that could be obtained from the observations, as well as the rate of the loss of magnetism, which it was found had occurred largely in the case of each magnet. A value of m has accordingly been calculated for every day on which observations were made. The Horizontal Force (X) was then found by the usual formula.

TABLE V.

FORT. CONFIDENCE.

Horizontal Force.—5th October 1848 to 26th April 1849.

Date.	Magnets employed.		Experiments of Deflection.			
	Suspended in Unifilar.	Deflecting and Vibrating Magnet.	Observed Temp. of Magnet.	Distance.	Angle.	Log. Values of $\frac{m}{X}$.
1848:			°		° ' "	
Oct. 5	N° 1	C. 1	37·5	1·4	18 55 47	9·64844
6	1	S. 1	35·4	1·4	14 05 25	9·52375
16	—	S. 1	—	—	—	—
17	—	S. 1	—	—	—	—
17	—	S. 1	—	—	—	—
20	1	C. 1	30·2	1·1	41 24 10	9·64358
20	1	C. 1	30·1	1·4	18 12 19	9·63209
20	1	S. 1	31·0	1·1	29 45 16	9·51888
	1	S. 1	31·5	1·4	13 49 06	9·51546
24	1	S. 1	29·3	1·1	30 23 32	9·52722
24	1	S. 1	30·0	1·4	14 06 23	9·52425
27	1	C. 1	29·5	1·1	39 41 07	9·62836
27	1	C. 1	30·5	1·4	17 53 26	9·62477
27	1	C. 1	32·8	1·1	39 25 54	9·62603
27	1	C. 1	33·3	1·4	18 14 51	9·63304
27	1	S. 1	33·3	1·1	28 05 33	9·49596
27	1	S. 1	33·3	1·4	13 31 49	9·50648
28	N° 2	S. 1	30·7	1·1	28 41 38	9·50450
28	2	S. 1	32·0	1·4	13 28 35	9·50480
Nov. 2	2	C. 1	18·5	1·1	38 07 36	9·61372
2	2	C. 1	17·5	1·4	17 24 49	9·61340
2	2	S. 1	20·0	1·1	28 22 23	9·50006
2	2	S. 1	19·5	1·4	13 22 26	9·50152
9	—	S. 1	—	—	—	—
9	—	S. 1	—	—	—	—
13	—	S. 1	—	—	—	—
13	—	S. 1	—	—	—	—
13	—	S. 1	—	—	—	—
14	—	S. 1	—	—	—	—
14	—	S. 1	—	—	—	—
15	—	S. 1	—	—	—	—
16	—	C. 1	—	—	—	—
20	—	C. 1	—	—	—	—
20	—	C. 1	—	—	—	—
21	—	C. 1	—	—	—	—
21	—	C. 1	—	—	—	—
22	—	C. 1	—	—	—	—
23	—	C. 1	—	—	—	—
23	—	C. 1	—	—	—	—
27	—	C. 1	—	—	—	—
27	—	C. 1	—	—	—	—
Dec. 18	—	S. 1	—	—	—	—
18	—	S. 1	—	—	—	—
22	—	C. 1	—	—	—	—
22	—	C. 1	—	—	—	—
1849:						
March 1	N° 2	S. 1	−3·2	1·1	26 14 27	9·46
2	2	S. 1	−2·0	1·4	12 49 45	9·4
3	2	C. 1	1·3	1·1	36 55 06	9·6
3	2	C. 1	1·3	1·4	17 01 32	9·6
5	2	C. 1	4·3	1·1	36 49 27	9·6
5	2	C. 1	4·4	1·4	16 56 24	9·6

Table V.

FORT CONFIDENCE.

Horizontal Force.—5th October 1848 to 26th April 1849.

Experiments of Vibration.								
Observed Temp. of Magnet.	Observed Time of One Vibration.	Log. Values of T² corrected for Torsion of Thread and Rate of Chronometer.	Temp. to which the Values of T² are reduced.	Log. Values of *m* X.	Log. Values of *m* employed at Temp. 20°.	Resulting Values of X.	Monthly Means.	Remarks.
°	s		°					
—	—	—	—	—	9·71533	1·158		
—	—	—	—	—	9·62753	1·264		
28·5	9·194	1·92842	30·0	9·70097	9·62413	1·197		
28·0	9·205	1·92943	30·0	9·69996	9·62379	1·195		
28·0	9·143	1·92371	30·0	9·70568	9·62379	1·211		
—	—	—	—	—	9·71308	1·169		
—	—	—	—	—	9·71308	1·199		
—	—	—	—	—	9·62277	1·266		
—	—	—	—	—	9·62277	1·266		
—	—	—	—	—	9·62141	1·239	1·235	
—	—	—	—	—	9·62141	1·247		
—	—	—	—	—	9·71203	1·208		
—	—	—	—	—	9·71203	1·217		
—	—	—	—	—	9·71203	1·213		
—	—	—	—	—	9·71203	1·193		
—	—	—	—	—	9·62039	1·326		
—	—	—	—	—	9·62039	1·295		
—	—	—	—	—	9·62005	1·301		
—	—	—	—	—	9·62005	1·299		
—	—	—	—	—	9·71113	1·252		
—	—	—	—	—	9·71113	1·254		
—	—	—	—	—	9·61835	1·313		
—	—	—	—	—	9·61835	1·309		Vibrations
15·0	9·487	1·95449	0·0	9·67490	9·61597	1·138		with
15·5	25·772	2·82535	0·0	9·67768	9·61597	1·146	- -	Ring
2·8	25·898	2·83219	0·0	9·67084	9·61461	1·131	- -	Ring
3·2	18·805	2·55273	0·0	9·67304	9·61461	1·137	- -	Ring 6
3·8	18·796	2·55232	0·0	9·67345	9·61461	1·138	- -	Ring 6
1·1	16·444	2·43614	0·0	9·66241	9·61427	1·111	- -	Ring 8
1·3	16·203	2·42177	0·0	9·67678	9·61427	1·148	- -	Ring 8
−2·8	9·525	1·95991	0·0	9·66948	9·61393	1·130	1·198	
8·5	8·245	1·83537	10·0	9·79009	9·70903	1·202		
13·2	8·282	1·83822	10·0	9·78724	9·70843	1·195		
14·1	8·265	1·83541	10·0	9·79005	9·70843	1·202		
14·6	22·692	2·71178	10·0	9·79078	9·70828	1·205	- -	Ring 5
15·5	22·721	2·71552	10·0	9·78699	9·70828	1·194	- -	Ring 5
15·1	22·529	2·70839	10·0	9·79412	9·70813	1·215	- -	Ring 5
16·3	16·272	2·42341	10·0	9·80137	9·70798	1·235	- -	Ring 6
16·2	16·366	2·42876	10·0	9·79602	9·70798	1·220	- -	Ring 6
0·3	14·027	2·29678	10·0	9·80044	9·70738	1·234	- -	Ring 8
1·4	14·007	2·29515	10·0	9·80207	9·70738	1·239	- -	Ring 8
−36·6	9·498	1·95809	−30·0	9·67130	9·60271	1·154		
−36·0	9·533	1·96117	−30·0	9·66822	9·60271	1·146	1·177	
−22·4	8·265	1·83550	−20·0	9·78996	9·70363	1·201		
−22·3	8·265	1·83603	−20·0	9·78938	9·70363	1·199		
—	—	—	—	—	9·57789	1·295		
—	—	—	—	—	9·57755	1·249		
—	—	—	—	—	9·69298	1·242	1·210	of the
—	—	—	—	—	9·69298	1·236		
—	—	—	—	—	9·69268	1·240		
—	—	—	—	—	9·69268	1·240		

Fort Confidence—*continued.*

Horizontal Force.—5th October 1848 to 26th April 1849.

Date.	Magnets employed.		Experiments of Deflection.					
	Suspended in Unifilar.	Deflecting and Vibrating Magnet.	Observed Temp. of Magnet.	Distance.	Angle.			Log. Values of $\frac{m}{X}$
1849:			°		°	′	″	
March 6	C. 2	S. 1	6·4	1·1	27	10	50	9·48286
6	C. 2	S. 1	7·0	1·4	12	46	51	9·48216
7	C. 2	S. 1	−2·9	1·1	27	18	25	9·48473
7	C. 2	S. 1	−2·3	1·4	12	53	42	9·48598
7	C. 2	C. 1	−0·7	1·1	37	02	26	9·60302
7	C. 2	C. 1	−0·5	1·4	16	59	08	9·60291
8	C. 1	S. 1	−7·7	1·1	27	37	28	9·48937
8	C. 1	C. 1	−5·0	1·4	18	13	22	9·63251
8	C. 1	S. 1	−7·1	1·4	13	02	38	9·49086
8	C. 1	C. 1	−5·1	1·1	38	07	27	9·61369
9	C. 1	S. 1	−10·3	1·1	27	56	41	9·49397
9	C. 1	S. 1	−9·2	1·4	13	02	43	9·49091
9	C. 1	C. 1	−5·7	1·1	39	04	18	9·62269
9	C. 1	C. 1	−6·2	1·4	18	21	59	9·63579
12	C. 1	C. 1	−11·9	1·1	40	10	23	9·63278
12	C. 1	C. 1	−11·5	1·4	18	02	11	9·62818
13	C. 1	S. 1	−8·7	1·1	27	34	05	9·48854
13	C. 1	S. 1	−8·5	1·4	12	47	04	9·48231
20	—	S. 1	—	—		—		—
20	—	S. 1	—	—		—		—
20	—	S. 1	—	—		—		—
21	—	S. 1	—	—		—		—
21	—	S. 1	—	—		—		—
23	—	C. 1	—	—		—		—
23	—	C. 1	—	—		—		—
24	—	C. 1	—	—		—		—
24	—	C. 1	—	—		—		—
April 3	—	C. 1	—	—		—		—
3	—	C. 1	—	—		—		—
3	—	C. 1	—	—		—		—
4	—	C. 1	—	—		—		—
4	—	C. 1	—	—		—		—
5	—	C. 1	—	—		—		—
5	—	C. 1	—	—		—		—
5	—	C. 1	—	—		—		—
6	—	C. 1	—	—		—		—
6	—	C. 1	—	—		—		—
7	—	C. 1	—	—		—		—
7	—	C. 1	—	—		—		—
9	—	C. 1	—	—		—		—
9	—	C. 1	—	—		—		—
9	—	C. 1	—	—		—		—
10	—	S. 1	—	—		—		—
10	—	S. 1	—	—		—		—
10	—	S. 1	—	—		—		—
11	—	S. 1	—	—		—		—
11	—	S. 1	—	—		—		—
11	—	S. 1	—	—		—		—
13	—	S. 1	—	—		—		—
14	—	S. 1	—	—		—		—
14	—	S. 1	—	—		—		—
16	—	S. 1	—	—		—		—
17	—	S. 1	—	—		—		—

Fort Confidence—*continued.*

Horizontal Force.—5th October 1848 to 26th April 1849.

Experiments of Vibration.					Log. Values of m employed at Temp. 20°.	Resulting Values of X.	Monthly Means.	Remarks.
Observed Temp. of Magnet.	Observed Time of One Vibration.	Log. Values of T² corrected for Torsion of Thread and Rate of Chronometer.	Temp. to which the Values of T² are reduced.	Log. Values of m X.				
°	s		°					
—	—	—	—	—	9·57619	1·245		
—	—	—	—	—	9·57619	1·247		
—	—	—	—	—	9·57585	1·242		
—	—	—	—	—	9·57585	1·238		
—	—	—	—	—	9·69238	1·239		
—	—	—	—	—	9·69238	1·239		
—	—	—	—	—	9·57551	1·229		
—	—	—	—	—	9·69223	1·159		
—	—	—	—	—	9·57551	1·228		
—	—	—	—	—	9·69223	1·210		
—	—	—	—	—	9·57517	1·217		
—	—	—	—	—	9·57517	1·225		
—	—	—	—	—	9·69208	1·185		
—	—	—	—	—	9·69208	1·150	1·210	
—	—	—	—	—	9·69163	1·160		
—	—	—	—	—	9·69163	1·172		
—	—	—	—	—	9·57381	1·227		
—	—	—	—	—	9·57381	1·245		
−6·6	9·967	2·00021	−5·0	9·62918	9·57143	1·134		
−4·8	9·980	2·00023	−5·0	9·62916	9·57143	1·134		
−1·5	9·864	1·98974	−5·0	9·63965	9·57148	1·161		
−10·9	9·916	1·99532	−5·0	9·63407	9·57109	1·147		
−7·8	9·911	1·99431	−5·0	9·63508	9·57109	1·150		
−5·4	8·348	1·84473	−5·0	9·78466	9·68998	1·231		
−3·7	8·369	1·84645	−5·0	9·77901	9·68998	1·215		
−11·1	8·402	1·85101	−5·0	9·77445	9·68983	1·203		
−7·8	8·397	1·85019	−5·0	9·77527	9·68983	1·205		
4·7	8·320	1·84173	5·0	9·78373	9·68833	1·239		
6·3	8·407	1·84993	5·0	9·77553	9·68838	1·216		Vibrations
8·9	8·364	1·84565	5·0	9·77981	9·68833	1·228		with
4·5	14·616	2·33157	5·0	9·76565	9·68818	1·189	- -	Ring 8
8·3	14·525	2·32559	5·0	9·77163	9·68818	1·206	- -	Ring 8
4·8	14·515	2·32532	5·0	9·77190	9·68808	1·207	- -	Ring 8
13·5	16·896	2·45582	5·0	9·76896	9·68803	1·198	- -	Ring 6
14·6	16·916	2·45681	5·0	9·76797	9·68803	1·196	- -	Ring 6
5·1	16·963	2·46098	5·0	9·76380	9·68788	1·185	- -	Ring 6
12·7	23·050	2·73246	5·0	9·77005	9·68788	1·202	- -	Ring 5
10·2	23·122	2·73330	5·0	9·76921	9·68773	1·200	- -	Ring 5
8·2	23·124	2·73375	5·0	9·76876	9·68773	1·199	- -	Ring 5
5·2	8·387	1·85028	5·0	9·77518	9·68743	1·217		
1·4	8·420	1·85498	5·0	9·77048	9·68743	1·205	1·198	
3·5	8·347	1·84705	5·0	9·77841	9·68743	1·227		
−4·2	9·795	1·98720	5·0	9·64219	9·56429	1·191		
2·9	9·879	1·99406	5·0	9·63533	9·56429	1·172		
4·1	9·875	1·99359	5·0	9·63580	9·56429	1·174		
−0·2	16·808	2·45641	5·0	9·64214	9·56395	1·192	- -	Ring 8
3·6	16·844	2·45904	5·0	9·63951	9·56395	1·185	- -	Ring 8
5·5	16·828	2·45763	5·0	9·64092	9·56395	1·189	- -	Ring 8
0·7	9·840	1·99113	5·0	9·63826	9·56327	1·183		
1·0	19·512	2·58786	5·0	9·63791	9·56293	1·183	- -	Ring 6
9·1	19·521	2·58680	5·0	9·63897	9·56293	1·186	- -	Ring 6
8·1	19·755	2·59764	5·0	9·62813	9·56225	1·159	- -	Ring 6
6·0	28·294	2·91789	5·0	9·58514	9·56191	1·050	- -	Ring 5

Fort Confidence—*continued.*

Horizontal Force.—5th October 1848 to 26th April 1849.

Date.	Magnets employed.		Experiments of Deflection.					
	Suspended in Unifilar.	Deflecting and Vibrating Magnet.	Observed Temp. of Magnet.	Distances.	Angles.			Log. Values of $\frac{m}{X}$
1849:			°		°	′	″	
April 17	—	S. 1	—	—		—		—
18	—	S. 1	—	—		—		—
18	—	S. 1	—	—		—		—
19	C. 1	S. 1	4·3	1·1	29	37	12	9·51711
19	C. 1	S. 1	5·2	1·4	14	20	42	9·53130
20	C. 1	S. 1	6·2	1·1	28	11	38	9·49750
20	C. 1	S. 1	6·2	1·4	13	01	43	9·49037
20	C. 1	S. 1	9·7	1·1	27	40	37	9·49012
20	C. 1	S. 1	11·9	1·4	12	39	49	9·47823
20	C. 1	C. 1	14·1	1·1	36	50	10	9·60096
20	C. 1	C. 1	14·4	1·4	17	18	52	9·61094
21	C. 1	C. 1	13·0	1·1	38	58	38	9·62180
21	C. 1	C. 1	13·1	1·4	15	10	46	9·55541
21	C. 1	C. 1	18·0	1·1	35	10	14	9·58358
21	C. 1	C. 1	18·4	1·4	17	43	46	9·62098
23	C. 2	S. 1	18·2	1·1	26	02	59	9·46577
23	C. 2	S. 1	18·7	1·4	12	12	02	9·46230
23	C. 2	S. 1	19·7	1·1	26	02	50	9·46572
23	C. 2	S. 1	20·0	1·4	12	18	43	9·46620
23	C. 2	S. 1	18·5	1·1	25	46	08	9·46137
23	C. 2	S. 1	18·9	1·4	12	05	11	9·45830
24	C. 2	C. 1	9·5	1·1	36	32	34	9·59796
24	C. 2	C. 1	9·0	1·4	17	03	43	9·60481
24	C. 2	C. 1	12·0	1·1	37	18	16	9·60566
24	C. 2	C. 1	11·9	1·4	16	58	17	9·60253
24	C. 2	C. 1	14·3	1·1	37	07	27	9·60385
24	C. 2	C. 1	14·8	1·4	16	48	55	9·59867
25	—	S. 1	—	—		—		—
25	—	S. 1	—	—		—		—
25	—	S. 1	—	—		—		—
25	—	S. 1	—	—		—		—
26	—	C. 1	—	—		—		—
26	—	C. 1	—	—		—		—
26	—	C. 1	—	—		—		—
26	—	C. 1	—	—		—		—

Fort Confidence—*continued.*

Horizontal Force.—5th October 1848 to 26th April 1849.

Experiments of Vibration.								
Observed Temp. of Magnet.	Observed Time of One Vibration.	Log. Values of T² corrected for Torsion of Thread and Rate of Chronometer.	Temp. to which the Values of T² are reduced.	Log. Values of *m* X.	Log. Values of *m* employed at Temp. 20°.	Resulting Values of X.	Monthly Means.	Remarks.
°	s		°				Vibrations with	
10·2	28·096	2·91068	5·0	9·59235	9·56191	1·068	- -	Ring 5
−1·1	28·307	2·92048	5·0	9·58255	9·56157	1·045	- -	Ring 5
0·0	9·823	1·99055	5·0	9·63884	9·56157	1·189		
—	—	—	—	—	9·56123	1·112		
—	—	—	—	—	9·56123	1·076		
—	—	—	—	—	9·56089	1·162		
—	—	—	—	—	9·56089	1·181		
—	—	—	—	—	9·56089	1·181		
—	—	—	—	—	9·56089	1·213		
—	—	—	—	—	9·68578	1·218		
—	—	—	—	—	9·68578	1·190		
—	—	—	—	—	9·68563	1·161		
—	—	—	—	—	9·68563	1·355		
—	—	—	—	—	9·68563	1·266		
—	—	—	—	—	9·68563	1·161		
—	—	—	—	—	9·55987	1·243		
—	—	—	—	—	9·55987	1·252		
—	—	—	—	—	9·55987	1·242	1·98	
—	—	—	—	—	9·55987	1·241		
—	—	—	—	—	9·55987	1·255		
—	—	—	—	—	9·55987	1·264		
—	—	—	—	—	9·68518	1·227		
—	—	—	—	—	9·68518	1·209		
—	—	—	—	—	9·68518	1·204		
—	—	—	—	—	9·68518	1·213		
—	—	—	—	—	9·68518	1·209		
—	—	—	—	—	9·68518	1·223		
14·0	9·877	1·99079	20·0	9·63860	9·55919	1·201		
17·7	17·017	2·46344	20·0	9·63511	9·55919	1·191	- -	Ring 8
20·2	19·740	2·59175	20·0	9·63402	9·55919	1·188	- -	Ring 6
22·5	28·988	2·93349	20·0	9·56954	9·55919	1·289	- -	Ring 5
16·7	22·882	2·72626	20·0	9·77625	9·68488	1·234	- -	Ring 5
17·8	16·984	2·46194	20·0	9·76284	9·68488	1·197	- -	Ring 6
18·6	14·515	2·32597	20·0	9·77125	9·68488	1·220	- -	Ring 8
19·9	8·396	1·84970	20·0	9·77576	9·68488	1·233		

General Mean, X = 1·205

Total Force.—The values of $\frac{m}{\phi}$ were calculated by the formula

$$\frac{m}{\phi} = \frac{1}{2} r^3 \sin u \left(\frac{1}{1 + \frac{P}{r^2}} \right)$$

$r_{\prime}$ and $r_{\prime\prime}$, $u_{\prime}$ and $u_{\prime\prime}$, &c. being substituted in the formula for the value of $\frac{m}{\phi}$ at the second and third distances.

Three distances were employed in the experiments of deflection, viz., 8·0, 9·2, and 10·5 inches. It was found, in making the calculation for the value of P, that the formula for two distances was preferable to that which it is intended should be employed when deflections at three distances are observed; and accordingly P was calculated for each combination of two distances that can be formed, viz., at 8·0 and 9·2, 8·0 and 10·5, and 9·2 and 10·5 inches; the results are as follows:—

Magnet C.1.

8·0 and 9·2 inches - - - P = −·0193
8·0 and 10·5 „ - - - P = −·0153
9·2 and 10·5 „ - - - P = −·0094

Magnet S.1.

8·0 and 9·2 inches - - - P = −·0156
8·0 and 10·5 „ - - - P = −·0173
9·2 and 10·5 „ - - - P = −·0091

The mean values were employed, viz.,—

For magnet C.1, P = −·0147· For magnet S.1, P = −·0140; and from them the following factors were obtained:—

Magnet C.1.

$\frac{1}{1 + \frac{P}{r^2}}$ Log. = 0·01459.

$\frac{1}{1 + \frac{P}{r_{\prime}^2}}$ Log. = 0·01099.

$\frac{1}{1 + \frac{P}{r_{\prime\prime}^2}}$ Log. = 0·00842.

Magnet S. 1.

$$\frac{1}{1+\frac{P}{r^2}} \quad \text{Log.} = 0{\cdot}01389.$$

$$\frac{1}{1+\frac{P}{r_{,}^2}} \quad \text{Log.} = 0{\cdot}01046.$$

$$\frac{1}{1+\frac{P}{r_{,,}^2}} \quad \text{Log.} = 0{\cdot}00801.$$

The quantity m was derived from the observations of absolute Horizontal Force, the value being calculated for the particular day on which the deflections of the dipping needle were observed.

The Total Force ϕ was found by the formula

$$\text{Log. } \phi = \text{log. } m - \text{log. } \frac{m}{\phi}$$

Log. m receiving the necessary correction to reduce it to the same temperature as that of m in the quantity $\frac{m}{\phi}$.

TABLE VI.—*Total Force.*

Date.	Magnets employed. Suspended in Circle.	Magnets employed. Deflecting Magnet.	Distance of Deflecting Magnet.	Mean Temp.	Angles of Deflection. Circle face East.	Angles of Deflection. Circle face West.	Log. Values of $\frac{1}{2}r^3 \sin u$, $\frac{1}{2}r_{,}^3 \sin u'$ &c.	Log. Values of $\frac{m}{\phi}$	Values of m at 20°.	Values of ϕ	Inclination deduced from Deflection Observations.
1848: D. H.			Inches.	°	° ′	° ′					° ′
OCTOBER 2 00	A. 1	S. 1	10·5	58·0	5 13·7	5 13·3	8·48436	8·49238	9·62889	13·513	84 46·4
01	A. 1	C. 1	10·5	67·0	6 49·6	6 45·8	8·59805	8·60647	9·71578	12·625	84 37·7
02	A. 1	C. 1	9·2	70·0	10 03·8	9 58·6	8·59896	8·60495	9·71578	12·596	84 39·2
03	A. 1	C. 1	8·0	64·0	15 22·7	15 19·1	8·59409	8·60868	9·71578	12·575	84 38·8
04	A. 1	S. 1	9·2	59·5	7 24·4	7 21·2	8·46216	8·47262	9·62889	14·136	84 38·3
05	A. 1	S. 1	8·0	56·5	11 04·8	11 02·9	8·45442	8·46881	9·62889	14·198	84 36·9
21	A. 1	S. 1	10·5	56·0	5 17·6	4 39·0	8·46283	8·47085	9·62889	14·211	84 54·0
21	A. 1	S. 1	9·2	55·0	7 28·6	7 12·4	8·45991	8·47067	9·62889	14·231	84 47·6
22	A. 1	S. 1	8·0	55·0	11 16·5	11 10·4	8·46058	8·47447	9·62889	14·097	84 48·7
23	A. 1	C. 1	10·5	57·0	6 48·7	6 28·3	8·58318	8·59680	9·71578	12·966	84 48·5
23	A. 1	C. 1	9·2	64·0	9 55·2	9 16·7	8·57554	8·58653	9·71578	12·233	84 45·4
3 00	A. 1	C. 1	8·0	65·5	15 04·3	14 58·0	8·58487	8·59946	9·71563	12·833	84 47·1
02	A. 2	S. 1	10·5	65·5	4 54·4	4 57·6	8·45948	8·46750	9·62855	14·262	84 37·2
03	A. 2	S. 1	9·2	66·0	7 13·2	7 17·3	8·45468	8·46514	9·62855	14·337	84 38·1
03	A. 2	S. 1	8·0	67·0	11 03·4	11 02·6	8·45390	8·46779	9·62855	14·245	84 38·0
04	A. 2	C. 1	10·5	66·0	6 42·0	6 44·4	8·59320	8·60167	9·71563	12·766	84 39·8
05	A. 2	C. 1	9·2	62·0	9 59·9	10 02·3	8·59389	8·60488	9·71563	12·692	84 35·7
05	A. 2	C. 1	8·0	58·0	15 12·2	15 10·3	8·58960	8·60419	9·71563	12·732	84 43·3
										13·402	84 42·0
NOVEMBER 5 21	A. 1	S. 1	10·5	47·0	5 11·4	4 27·8	8·45001	8·45808	9·61733	14·296	84 47·9
21	A. 1	S. 1	9·2	45·5	7 11·9	6 59·0	8·44484	8·45530	9·61733	14·537	84 47·7
22	A. 1	S. 1	8·0	45·5	10 51·2	10 46·6	8·44369	8·45758	9·61733	14·318	84 45·3
6 02	A. 1	C. 1	10·5	45·8	6 36·0	5 56·2	8·56316	8·57158	9·71058	12·680	84 44·1
03	A. 1	C. 1	9·2	46·0	9 25·8	9 11·8	8·56251	8·57350	9·71058	12·569	84 42·0
05	A. 1	C. 1	8·0	57·0	13 50·1	13 57·2	8·55178	8·56637	9·71058	12·734	84 21·1
22	A. 1	S. 1	10·5	46·3	4 46·2	4 41·1	8·44246	8·45048	9·61699	14·539	84 45·6
23	A. 1	S. 1	9·2	49·0	7 14·0	6 54·7	8·44373	8·45419	9·61699	14·402	84 49·0
23	A. 1	S. 1	8·0	50·5	10 52·8	10 36·6	8·44190	8·45579	9·61699	14·342	84 52·9
7 00	A. 1	C. 1	10·5	49·8	6 21·7	6 07·0	8·56169	8·56951	9·71038	12·669	84 50·2
01	A. 1	C. 1	9·2	51·2	9 28·7	8 54·5	8·55692	8·56791	9·71038	12·711	84 53·7
02	A. 1	C. 1	8·0	49·0	14 11·7	13 38·5	8·55254	8·56713	9·71038	12·748	84 36·6
21	A. 2	S. 1	10·5	51·5	4 49·0	4 49·1	8·44911	8·45713	9·61665	14·281	84 44·9
23	A. 2	S. 1	9·2	52·5	7 07·2	7 08·9	8·44748	8·45794	9·61665	14·250	84 46·6
8 00	A. 2	S. 1	8·0	53·5	10 47·0	10 46·4	8·44322	8·45711	9·61651	14·261	84 48·6
01	A. 2	C. 1	10·5	55·5	6 10·8	6 10·7	8·56081	8·57136	9·71038	[illegible]	[illegible]
02	A. 2	C. 1	9·2	53·0	9 15·1	9 18·4	8·56333	8·57187	9·71023	13·573	84 46·5
03	A. 2	C. 1	8·0	56·5	14 05·6	14 02·4	8·55706	8·57165	9·71023	13·561	84 46·6
										13·996	84 45·4

Total Force—*continued.*

Date.	Magnets employed. Suspended in Circle.	Deflecting Magnet.	Distance of Deflecting Magnet.	Mean Temp.	Angles of Deflection. Circle face East.	Circle face West.	Log. Values of $\frac{1}{2}r^3 \sin u$, $\frac{1}{2}r_1^3 \sin u'$ &c.	Log. Values of $\frac{m}{\phi}$	Values of m at 20°.	ϕ	Inclination deduced from Deflection Observations.
1848: D. H.			Inches.	°	° ′	° ′					° ′
December 13 22	A. 1	S. 1	10·5	35·0	4 59·7	4 23·8	8·43803	8·44605	9·60441	14·336	84 47·8
23	A. 1	S. 1	9·2	37·5	6 51·6	6 42·7	8·42584	8·43630	9·60441	14·650	84 47·3
14 00	A. 1	S. 1	8·0	41·5	10 32·0	10 23·8	8·43056	8·44445	9·60407	14·334	84 43·[illegible]
01	A. 2	S. 1	10·5	41·7	4 42·5	4 39·9	8·43728	8·44528	9·60407	14·305	84 43·[illegible]
02	A. 2	S. 1	9·2	42·0	6 57·3	6 53·6	8·43456	8·44502	9·60407	14·313	84 50·[illegible]
02	A. 2	S. 1	8·0	41·5	10 28·5	10 30·3	8·43158	8·44547	9·60407	14·301	84 48·[illegible]
15 01	A. 1	S. 1	10·5	40·0	4 34·4	4 27·8	8·42140	8·42942	9·60373	14·350	84 51·[illegible]
02	A. 1	S. 1	9·2	36·5	6 58·8	6 41·0	8·42880	8·43926	9·60373	14·532	84 48[illegible]
03	A. 1	S. 1	8·0	41·5	10 24·8	9 55·6	8·41828	8·43217	9·63373	14·734	84 48[illegible]
22	A. 1	C. 1	10·5	42·5	6 04·7	5 52·8	8·54267	8·55109	9·70468	14·116	84 58[illegible]
23	A. 1	C. 1	9·2	41·0	9 06·1	8 34·4	8·53987	8·55066	9·70468	14·131	84 5[illegible]
23	A. 1	C. 1	8·0	40·0	13 36·3	13 10·8	8·53611	8·55070	9·70468	14·142	84 4[illegible]
16 00	A. 2	C. 1	10·5	41·8	6 07·6	6 05·2	8·55186	8·56028	9·70453	13·819	84 4[illegible]
00	A. 2	C. 1	9·2	45·5	8 59·2	9 01·5	8·54800	8·55899	9·70453	13·840	84 5[illegible]
01	A. 2	C. 1	8·0	42·8	13 39·0	13 35·6	8·54337	8·55796	9·70453	13·888	84 5[illegible]
02	A. 1	C. 1	10·5	44·0	6 14·0	5 45·5	8·54887	8·55229	9·70453	14·063	84 5[illegible]
02	A. 1	C. 1	9·2	45·0	8 58·2	8 52·6	8·54408	8·55507	9·70453	13·968	84 4[illegible]
03	A. 1	C. 1	8·0	47·0	13 21·4	13 30·2	8·53733	8·55192	9·70453	14·059	84 4[illegible]
										14·243	[illegible]
1849: January 15 22	A. 1	S. 1	10·5	36·2	4 40·6	4 28·9	8·43416	8·44218	9·59319	14·088	84 [illegible]
23	A. 1	S. 1	9·2	41·3	6 44·2	6 40·2	8·42061	8·43107	9·59319	14·417	84 [illegible]
23	A. 1	S. 1	8·0	41·2	10 21·7	10 09·9	8·42220	8·43609	9·59319	14·252	84 [illegible]
16 22	A. 2	S. 1	10·5	34·0	4 38·2	4 35·1	8·43011	8·43813	9·59285	14·221	[illegible]
23	A. 2	S. 1	9·2	39·5	6 51·0	6 50·3	8·42954	8·44000	9·59285	14·136	[illegible]
23	A. 2	S. 1	8·0	46·8	10 18·1	10 21·2	8·42484	8·43873	9·59285	14·128	[illegible]
17 00	A. 1	S. 1	10·5	46·0	4 32·3	4 31·2	8·42236	8·43038	9·59251	14·395	[illegible]
01	A. 1	S. 1	9·2	57·7	6 54·0	6 40·1	8·42574	8·43620	9·59251	14·152	[illegible]
02	A. 1	S. 1	8·0	64·0	10 26·2	9 40·4	8·41658	8·43047	9·59251	14·303	
22	A. 1	C. 1	10·5	51·8	6 07·4	5 48·6	8·54182	8·55024	9·69973	13·932	
23	A. 1	C. 1	9·2	53·5	9 07·6	8 37·6	8·54182	8·55281	9·69973	13·840	
23	A. 1	C. 1	8·0	54·8	13 42·1	13 15·2	8·53881	8·55340	9·69973	13·814	
18 00	A. 2	C. 1	10·5	56·8	6 00·3	6 05·1	8·54747	8·55589	9·69958	13·720	
00	A. 2	C. 1	9·2	60·8	8 55·6	8 59·1	8·54560	8·55659	9·69958	13·676	
01	A. 2	C. 1	8·0	58·3	13 27·9	13 30·6	8·53912	8·55371	9·69958	13·780	
02	A. 1	C. 1	10·5	57·3	6 10·7	5 46·0	8·54219	8·55061	9·69958	13·885	
02	A. 1	C. 1	9·2	60·0	8 58·5	8 50·2	8·54327	8·55426	9·69958	13·754	
03	A. 1	C. 1	8·0	63·0	13 13·3	13 22·2	8·53302	8·54761	9·69958	13·950	
										14·02[illegible]	
February 12 22	A. 1	S. 1	10·5	36·8	4 36·6	4 16·1	8·41366	8·42168	9·58367	14·44[illegible]	
22	A. 1	S. 1	9·2	42·5	6 54·4	6 34·3	8·42286	8·43332	9·58367	14·02[illegible]	
23	A. 1	S. 1	8·0	45·5	10 11·5	9 55·3	8·41346	8·42735	9·58367	14·2[illegible]	
13 00	A. 1	S. 1	10·5	51·8	4 53·9	4 15·0	8·42665	8·43467	9·58333	13·9[illegible]	
00	A. 1	S. 1	9·2	51·5	6 51·5	6 34·1	8·42125	8·43171	9·58333	14·0[illegible]	
01	A. 1	S. 1	8·0	53·8	10 13·9	10 13·6	8·42073	8·43462	9·58333	13·9[illegible]	
02	A. 2	S. 1	10·5	55·5	4 33·6	4 34·2	8·42586	8·43388	9·58333	13·[illegible]	
03	A. 2	S. 1	9·2	55·0	6 48·6	6 46·3	8·42616	8·43662	9·58333	13·[illegible]	
03	A. 2	S. 1	8·0	57·3	10 17·9	10 17·7	8·42359	8·43748	9·58333	13·[illegible]	
22	A. 2	C. 1	10·5	45·0	6 08·4	6 05·7	8·55257	8·56099	9·69568	13·[illegible]	
22	A. 2	C. 1	9·2	49·0	9 02·6	8 59·0	8·54840	8·55939	9·69568	13·[illegible]	
23	A. 2	C. 1	8·0	56·5	13 33·7	13 33·0	8·54128	8·55587	9·69568	13[illegible]	
14 00	A. 1	C. 1	10·5	58·0	6 10·3	5 39·8	8·53855	8·54697	9·69553	13[illegible]	
01	A. 1	C. 1	9·2	61·5	9 00·7	8 52·4	8·54496	8·55595	9·69553	13[illegible]	
01	A. 1	C. 1	8·0	62·5	13 16·7	13 28·8	8·53560	8·55028	9·69553	1[illegible]	
02	A. 1	C. 1	10·5	63·0	6 10·6	5 46·0	8·54339	8·55181	9·69553	1[illegible]	
03	A. 1	C. 1	9·2	61·0	9 00·6	8 32·2	8·53677	8·54776	9·69553	1[illegible]	
03	A. 1	C. 1	8·0	59·3	13 25·6	13 02·7	8·53109	8·54568	9·69553	1[illegible]	
March 11 21	A. 1	S. 1	10·5	39·8	4 26·6	4 15·1	8·40462	8·41264	9·57449		
23	A. 1	S. 1	9·2	49·2	6 45·3	6 24·0	8·41236	8·42282	9·57449		
23	A. 1	S. 1	8·0	51·5	10 01·3	9 36·2	8·40236	8·41675	9·57449		
12 01	A. 2	S. 1	10·5	56·0	4 28·5	4 26·1	8·41529	8·42331	9·57415		
01	A. 2	S. 1	9·2	62·8	6 40·3	6 37·0	8·41672	8·42718	9·57415		
02	A. 2	S. 1	8·0	56·5	10 01·1	10 03·2	8·41253	8·42642	9·57415		
05	A. 1	S. 1	10·5	53·5	4 44·8	4 10·1	8·41545	8·42347	9·57415		
05	A. 1	S. 1	9·2	52·8	6 37·8	6 25·5	8·40906	8·41952	9·57415		
06	A. 1	S. 1	8·0	51·5	10 00·3	10 01·0	8·41146	8·42535	9·57415		
22	A. 1	C. 1	10·5	56·0	6 07·7	5 43·8	8·53903	8·54745	9·69163		
23	A. 1	C. 1	9·2	56·8	9 01·3	8 56·2	8·54672	8·55771	9·69163		
23	A. 1	C. 1	8·0	56·5	13 20·0	13 31·4	8·53728	8·55187	9·69163		
13 00	A. 2	C. 1	10·5	55·3	6 06·0	6 05·4	8·55103	8·55945	9·69148		
01	A. 2	C. 1	9·2	53·8	9 00·3	8 57·7	8·54696	8·55795	9·69148		
02	A. 2	C. 1	8·0	56·5	13 35·4	13 33·9	8·54196	8·55655	9·69148		
03	A. 1	C. 1	10·5	59·0	6 06·7	5 46·3	8·54001	8·54843	9·69148		
03	A. 1	C. 1	9·2	57·5	9 01·7	8 37·3	8·53930	8·55029	9·6914[illegible]		
05	A. 1	C. 1	8·0	56·0	13 29·3	13 07·8	8.53345	8·54804	9·6914[illegible]		

Inclination.—The Inclination was observed by direct reading of the needle, forming results independent of those deduced from the deflection experiments.

Table VII. contains the particulars of these observations. All were made with the same needle, A. 1, except in a few instances, when A. 2 was employed.

From the whole number of observations, it was found that the mean readings of "poles direct" and "poles reversed" differed by a very small amount, and less than the probable error of a single mean; so that the half determination, whether "poles direct" or "poles reversed," has been taken as an observation of the Inclination, and the monthly means found from all the observations in the month without any correction.

TABLE VII.

Date.	Needle.—No. or Mark.	Azimuth.	Poles direct.				Poles reversed.				Incli-nation.	Monthly Means.
			Face of Needle.				Face of Needle.					
			Direct.		Reversed.		Direct.		Reversed.			
			a	a'	a''	a'''	b	b'	b''	b'''		
1848: D. H. M.			° ′	° ′	° ′	° ′	° ′	° ′	° ′	° ′	° ′	
OCTOBER. 1 23 00	A. 1	–	85 14·25	84 47·25	85 03·75	84 31·50	—	—	—	—	84 54·19	84° 49′·37
2 20 30	A. 1	–	—	—	—	—	84 52·75	84 46·50	84 47·00	84 44·50	84 47·69	
3 01 30	A. 2	–	84 21·25	84 52·50	84 57·25	84 38·75	—	—	—	—	84 42·44	
6 21 15	A. 1	–	—	—	—	—	84 52·50	85 04·75	85 07·75	84 03·25	84 47·06	
7 03 00	A. 1	–	85 14·00	84 32·25	85 06·75	84 23·75	—	—	—	—	84 49·19	
10 21 00	A. 1	–	84 20·75	85 07·25	84 34·25	85 15·75	—	—	—	—	84 49·50	
11 03 00	A. 1	–	—	—	—	—	84 47·75	84 55·50	84 32·00	84 42·75	84 44·50	
13 21 00	A. 1	–	85 10·50	84 41·00	85 04·75	84 29·50	—	—	—	—	84 51·44	
14 03 00	A. 1	–	—	—	—	—	84 51·00	84 53·05	84 50·00	84 36·00	84 47·62	
14 03 30	A. 1	–	—	—	—	—	84 38·25	84 49·75	84 59·00	84 45·00	84 48·00	
20 21 00	A. 1	–	84 50·07	84 58·05	85 24·05	84 30·08	—	—	—	—	84 56·00	
21 03 00	A. 1	–	—	—	—	—	85 04·00	84 54·25	84 53·05	84 31·00	84 50·69	
24 15 00	A. 1	–	84 48·50	84 54·00	85 18·83	84 26·33	—	—	—	—	84 51·91	
24 21 00	A. 1	–	84 33·50	84 51·33	85 21·67	84 29·50	—	—	—	—	84 49·00	
25 03 30	A. 1	–	—	—	—	—	84 41·67	84 59·33	85 08·33	84 29·50	84 49·71	
25 03 00	A. 1	–	84 39·17	84 36·67	85 11·50	84 06·67	—	—	—	—	84 39·00	
26 09 00	A. 1	–	84 57·17	85 02·67	85 32·17	84 30·33	—	—	—	—	85 00·59	
31 21 00	A. 1	–	84 44·00	84 50·00	85 24·50	84 21·67	—	—	—	—	84 50·04	
NOVEMBER. 1 03 00	A. 1	–	—	—	—	—	84 13·00	84 59·67	85 16·50	84 23·17	84 50·58	84° 51′·08
1 02 30	A. 1	–	84 40·83	85 04·17	85 19·33	84 14·50	—	—	—	—	84 49·71	
3 21 00	A. 1	–	—	—	—	—	84 48·00	85 03·50	85 16·17	84 19·83	84 51·88	
4 03 00	A. 1	–	84 42·08	84 51·07	85 22·00	84 07·07	—	—	—	—	84 46·05	
5 19 50	A. 1	–	84 44·00	84 51·33	85 26·50	84 07·33	—	—	—	—	84 47·29	
6 22 00	A. 1	–	—	—	—	—	84 43·83	85 04·67	85 11·50	84 23·17	84 50·79	
7 21 00	A. 2	–	—	—	—	—	84 28·25	85 16·05	84 58·50	84 31·00	84 48·56	
10 21 00	A. 1	–	—	—	—	—	84 50·83	85 09·33	85 22·50	84 28·00	84 57·66	
11 03 00	A. 1	–	84 39·33	84 58·33	85 24·67	84 03·67	—	—	—	—	84 45·25	
14 21 00	A. 1	–	84 47·83	85 02·17	85 35·17	84 15·67	—	—	—	—	84 55·21	
15 02 40	A. 1	–	—	—	—	—	84 41·33	85 01·83	85 14·83	84 18·17	84 49·04	
17 21 00	A. 1	–	—	—	—	—	84 02·33	85 14·17	85 25·17	84 30·17	85 02·96	
18 03 06	A. 1	–	84 35·67	84 58·17	85 26·83	84 15·00	—	—	—	—	84 48·79	
21 21 00	A. 1	–	84 45·33	85 07·17	85 35·17	84 17·00	—	—	—	—	84 56·17	
22 03 05	A. 1	–	—	—	—	—	84 46 00	85 10·83	85 12·83	84 18·17	84 52·33	
28 21 00	A. 1	–	84 45·66	84 57·83	85 17·00	84 18·33	—	—	—	—	84 49·70	
29 03 05	A. 1	–	—	—	—	—	84 10·17	84 54·17	85 20·17	84 10·83	84 46·33	
DECEMBER. 13 22 15	A. 1	–	84 46·83	85 02·34	—	—	—	—	—	—	84 54·58	84° 49′·98
14 03 30	A. 2	–	—	—	—	—	23·33	85 06·05	—	—	84 44·69	
15 01 15	A. 1	–	—	—	—	—	48·00	84 57·75	—	—	84 52·87	
15 21 45	A. 1	–	—	—	—	—	54·00	85 02·25	—	—	84 58·12	
16 00 00	A. 2	–	—	—	—	—	83·25	85 13·00	—	—	84 48·12	
16 03 35	A. 1	–	84 43·00	84 49·75	—	—	—	—	—	—	84 46·37	
29 21 10	A. 1	–	84 46·83	84 52·83	85 27·00	84 10·67	—	—	—	—	84 49·33	
30 03 00	A. 1	–	84 43·50	84 50·17	85 21·83	84 07·50	—	—	—	—	84 45·75	

TABLE VII.—*continued.*

Date.	Needle.—No. or Mark.	Azimuth.	Poles direct. Face of Needle. Direct. a		Reversed. a''		Poles reversed. Face of Needle. Direct. b		Reversed. b''		Inclination.	Mo M
			a	a'	a''	a'''	b	b'	b''	b'''		
1849: D. H. M.			° ′	° ′	° ′	° ′	° ′	° ′	° ′	° ′	° ′	
JANUARY. 2 21 20	A. 1	-	—	—	—	—	84 44·00	85 08·00	85 09·83	84 33·83	84 53·91	
3 03 06	A. 1	-	—	—	—	—	84 38·00	85 05·50	84 31·67	84 29·83	84 41·25	
15 22 00	A. 1	-	84 44·67	84 46·50	—	—	—	—	—	—	84 45·58	
16 22 15	A. 2	-	—	—	—	—	84 26·50	85 12·75	—	—	84 49·62	
17 00 20	A. 1	-	—	—	—	—	84 31·50	85 06·50	—	—	84 49·00	
17 22 05	A. 1	-	—	—	—	—	84 40·00	84 56·00	—	—	84 48·00	
18 01 15	A. 2	-	—	—	—	—	84 34·00	85 06·50	—	—	84 46·25	
18 01 40	A. 1	-	84 43·05	84 47·00	—	—	—	—	—	—	84 45·25	
26 21 20	A. 1	-	84 47·83	84 52·50	85 21·00	84 19·83	—	—	—	—	84 50·29	
27 02 50	A. 1	-	84 46·00	84 51·50	85 15·50	84 23·66	—	—	—	—	84 49·16	
30 21 20	A. 1	-	—	—	—	—	84 55·67	84 59·17	85 13·00	84 24·50	84 53·08	
31 03 30	A. 1	-	—	—	—	—	84 49·00	85 12·67	85 12·67	84 24·33	84 54·67	
FEBRUARY. 2 21 05	A. 1	-	—	—	—	—	84 49·5	85 08·00	85 12·00	84 19·00	84 52·12	
3 03 00	A. 1	-	—	—	—	—	84 46·67	85 10·00	85 12·00	84 22·67	84 53·5[illegible]	
6 21 00	A. 1	-	83 59·17	85 29·50	85 16·50	84 15·17	—	—	—	—	84 45·0[illegible]	
7 03 00	A. 1	-	84 26·67	85 09·17	85 16·50	84 19·50	—	—	—	—	84 47·9[illegible]	
9 21 00	A. 1	-	—	—	—	—	85 03·67	85 15·00	85 02·67	84 44·00	85 01·[illegible]	
10 03 00	A. 1	-	—	—	—	—	84 48·00	85 05·50	84 54·00	84 38·00	84 51·[illegible]	
12 21 20	A. 1	-	—	—	—	—	84 52·00	85 13·83	—	—	84 6[illegible]	
13 01 20	A. 1	-	84 47·00	85 07·00	—	—	—	—	—	—	84 57·[illegible]	
13 01 40	A. 2	-	—	—	—	—	84 23·25	85 12·75	—	—	84 48[illegible]	
13 21 40	A. 2	-	—	—	—	—	84 22·00	85 11·00	—	—	84 46[illegible]	
14 00 05	A. 1	-	84 48·05	85 05·05	—	—	—	—	—	—	84 57[illegible]	
14 02 10	A. 1	-	—	—	—	—	84 57·25	85 11·50	—	—	84 6[illegible]	
MARCH. 6 21 05	A. 1	-	—	—	—	—	84 54·67	85 07·50	85 05·00	84 30·17	84 5[illegible]	
7 03 00	A. 1	-	—	—	—	—	84 58·83	85 09·33	85 06·00	84 27·00	84 5[illegible]	
11 21 00	A. 1	-	—	—	—	—	84 50·00	85 08·33	—	—	84 [illegible]	
12 00 10	A. 2	-	—	—	—	—	84 24·00	85 13·00	—	—	84 [illegible]	
12 03 45	A. 1	-	84 44·00	84 55·83	—	—	—	—	—	—	84 [illegible]	
12 22 20	A. 1	-	84 56·67	85 01·50	—	—	—	—	—	—	84 [illegible]	
13 00 15	A. 2	-	—	—	—	—	84 21·05	85 11·00	—	—	84 [illegible]	
13 02 20	A. 1	-	—	—	—	—	84 51·83	85 05·00	—	—	84 [illegible]	
13 21 00	A. 1	-	84 38·83	85 06·00	85 06·67	84 13·33	—	—	—	—	84 [illegible]	
14 03 05	A. 1	-	84 41·50	85 03·83	85 18·33	84 15·50	—	—	—	—	84 [illegible]	
20 21 00	A. 2	-	—	—	—	—	84 28·67	85 11·17	84 52·67	84 48·50	8[illegible]	
21 03 00	A. 2	-	—	—	—	—	84 17·00	85 09·83	84 50·17	84 23·50	8[illegible]	
27 21 00	A. 1	-	—	—	—	—	84 50·50	85 04·17	85 05·50	84 15·50	8[illegible]	
28 03 00	A. 1	-	—	—	—	—	84 41·50	84 56·67	84 54·67	84 11·00	8[illegible]	

The monthly mean results of the Inclination are—

		°	′
October	$\theta=$	84	49·4
November	$\theta=$	84	51·1
December	$\theta=$	84	50·0
January	$\theta=$	84	48·8
February	$\theta=$	84	53·9
March	$\theta=$	84	50·4

Mean by direct observation 84° 50′·6

The monthly means, by the method of deflections, regular, and somewhat less in amount, viz. 84° 45′·9.

The Horizontal Force experiments were made in s… tain periods, and not at regular intervals; there are th…

months in which no observations are made, and others that contain very few. Taking, however, monthly means of all that were observed, the results are as follows:—

October	1848	- - - - - -	X = 1·235
November	„	- - - - - -	X = 1·198
December	„	- - - - - -	X = 1·177
March	1849	- - - - - -	X = 1·210
April	„	- - - - - -	X = 1·198

The Total Force for these months, derived from the horizontal component multiplied by the secant of the Inclination, is as follows:—

October	1848	- - - - -	ϕ = 13·640
November	„	- - - - -	ϕ = 13·231
December	„	- - - - -	ϕ = 12·999
March	„	- - - - -	ϕ = 13·364
April	..	- - - - -	ϕ = 13·231

The values of ϕ, found by the direct method for the several months, are—

October	1848	- - - - -	ϕ = 13·402
November	„	- - - - -	ϕ = 13·998
December	„	- - - - -	ϕ = 14·243
January	1849	- - - - -	ϕ = 14·025
February	„	- - - - -	ϕ = 13·849
March	„	- - - - -	ϕ = 13·801

The mean Horizontal Force during the whole period is 1·205, which, multiplied by sec. 84° 50′·6, is equal to 13·407. The mean of the whole of the values of ϕ by the direct method, is 13·886. The mean of both determinations is 13·646.

Table VIII. - - - - -

Contains the hourly observations made with the Declinometer in the n
1849. These observations show correctly the *direction* of the diurnal
suspension of the Declinometer needle, as stated in page 7. In the
same object were made with the Unifilar Magnetometer, the ma
correct amount as well as direction of the diurnal variation. These
in *direction*, but indicate that the range of the diurnal variation i
are those of mean time at the station.

FORT CONFIDENCE.

Abstract of Hourly Observations during the month of October 1848.

Date.	Declinometer.										
	1h	2h	3h	4h	5h	6h	7h	8h	9h	10h	11h
	° ′	° ′	° ′	° ′	° ′	° ′	° ′	° ′	° ′	° ′	°
1	—	—	—	—	—	—	—	—	—	—	—
2	—	—	—	—	—	—	—	—	—	—	—
3	—	—	—	—	—	—	—	—	—	—	—
4	—	—	—	—	—	—	—	—	—	—	—
5	—	—	—	—	—	—	—	—	—	—	—
6	—	—	—	—	—	—	—	—	—	—	—
7	—	—	—	—	—	—	—	—	—	—	
8	—	—	—	—	—	—	—	—	—	—	
9	—	—	—	—	—	—	—	—	—	—	
10	—	—	—	—	—	—	—	—	—	—	
11	—	—	—	—	—	—	5 15	5 14	5 07	5 05	
12	—	—	—	—	—	5 30	0 30	5 21	4 55	4 51	
13	—	—	—	—	—	—	5 19	0 15	5 02	4 50	
14	—	—	—	—	—	—	5 10	—	—	—	
15	—	—	—	—	—	—	—	5 02	5 11	5 10	
16	—	—	—	—	—	—	—	—	5 10	5 00	
17	—	—	—	—	—	—	4 33	4 37	—	4 30	
18	—	—	—	—	—	—	4 50	—	4 22	—	
19	—	—	—	—	—	—	—	—	—	—	
20	—	—	—	—	—	5 18	5 10	4 25	3 55	3 52	
21	—	—	—	5 30	—	5 02	—	4 54	4 50	4 55	
22	—	—	—	—	—	5 09	5 08	5 00	5 00	5 01	
23	—	—	—	—	—	5 25	5 05	3 36	3 17	4 15	
24	—	—	—	—	4 50	4 54	4 49	4 17	4 15	4 25	
25	—	—	4 45	5 15	—	3 50	4 30	4 40	4 35	4 32	
26	—	—	—	—	—	5 24	5 15	4 50	4 47	4 32	
27	—	—	5 35	5 27	—	5 26	—	4 30	4 29	4 22	
28	—	—	—	—	—	5 05	5 01	4 10	4 14	4 2[illegible]	
29	—	—	—	—	—	4 02	4 02	3 55	4 01	—	
30	—	—	—	—	5 12	5 04	5 03	4 42	4 06	4 0[illegible]	
31	—	—	—	—	—	—	4 45	4 37	3 20	3 [illegible]	
Sums -	—	—	10 20	16 12	10 02	54 39	48 43	53 36	50 49	47 [illegible]	
Means -	—	—	5 10·0	5 24·0	5 01·0	4 58·1	4 52·3	4·28	4 14·1	418	
Diurnal Variation	—	—	0 55·9	1 09·9	0 46·9	0 44·0	0 38·2	0 13·9	0 00·0	0 0[illegible]	

Increasing numbers denot
O

- - - - - TABLE VIII.

›f October, November, and December 1848, and January and February nent, but not its *amount*, in consequence of the friction on the point of ›f December 1848, January and February 1849, observations having the vhich instrument is suspended by a silk thread, and shows, therefore, the ›ations confirm the general accuracy of the Declinometer observations ›vice the amount shown by the Declinometer. The hours in Table VIII.

FORT CONFIDENCE.

Abstract of Hourly Observations during the month of October 1848.

2^h	3^h	4^h
° '	° '	° '
—	—	—
—	—	—
—	—	—
—	—	—
—	—	—
—	—	—
—	—	—
—	—	—
—	—	—
5 27	5 25	5 30
5 11	5 15	5 15
—	—	—
—	5 00	—
—	5 12	—
5 10	5 10	4 11
5 10	0 11	—
—	5 30	—
—	—	—
—	—	—
4 06	4 38	5 16
4 46	4 55	4 59
4 30	4 40	4 50
4 50	4 10	4 29
4 32	5 29	5 30
5 00	5 06	5 10
5 15	5 15	5 23
4 48	4 58	4 58
4 55	5 00	5 30
5 00	4 59	4 54
4 31	4 20	4 34
4 30	5 07	5 01
56 43	58 35	60 33
43·6	4 52·9	
29·5	0 38·8	

›d of the needle towards the West ›ly included in the means.

FORT CONFIDENCE.

Abstract of Hourly Observations during the months of November and December 1848.

Date.	Declinometer.											
	1h	2h	3h	4h	5h	6h	7h	8h	9h	10h	11h	Noon.
	° ′	° ′	° ′	° ′	° ′	° ′	° ′	° ′	° ′	° ′	° ′	° ′
1	—	—	—	5·05	—	5 02	5 00	4 59	5 00	5 00	5 00	5 01
2	—	—	—	—	—	4 54	4 47	4 50	4 46	4 50	4 45	4 50
3	—	—	—	—	—	5 58	5 30	5 21	5 12	5 13	5 11	5 13
4	—	—	—	—	—	—	5 10	5 09	—	4 50	4 52	4 51
5	—	—	—	—	—	4 52	—	4 50	4 35	4 30	4 32	4 35
6	—	—	—	—	4 40	4 40	4 40	4 41	4 41	4 41	4 42	4 42
7	—	—	—	—	—	4 43	4 42	4 43	4 42	4 37	4 38	4 38
8	—	—	—	—	—	4 40	—	4 39	4 40	4 40	*	4 55
9	—	—	—	—	4 30	4 28	4 59	4 57	4 09	4 18	4 25	4 30
10	—	—	—	—	—	4 55	4 29	4 17	4 20	4 22	3 35	4 10
11	—	—	—	—	4 37	4 12	4 13	4 17	4 05	4 17	4 16	4 17
12	—	—	—	—	—	—	4 10	4 08	4 07	4 09	4 11	4 15
13	—	—	—	—	4 13	4 18	4 21	4 22	4 19	4 20	4 22	4 24
14	—	—	—	4·22	4 36	4 00	3 59	3 55	3 49	3 55	3 57	4 01
15	—	—	—	—	4 06	4 06	4 02	4 02	3 58	3 57	4 01	4 02
16	—	—	—	4·54	4 54	4 54	4 52	4 47	4 50	4 49	3 34	3 33
17	—	—	—	—	—	2 30	2 41	2 49	2 42	3 12	1 44	1 21†
18	—	—	—	—	4 41	—	4 47	4 30	4 22	3 55	3 30	2 30
19	—	—	—	—	—	4 08	4 10	4 06	4 07	4 11	4 20	4 29
20	—	—	—	—	—	3 16	3 18	3 19	3 30	3 50	3 49	4 05
21	—	—	—	—	4 25	4 26	4 28	4 27	4 20	3 42	3 40	3 40
22	—	—	—	—	4 30	4 32	3 20	4 27	4 10	4 21	4 19	4 25
23	—	—	—	—	—	3 38	3 35	3 48	3 46	3 51	3 57	3 52
24	—	—	6 40	6 26	6 22	6 09	5 30	5 10	5 05	4 06	4 04	4 20
25	5 01	4 55	4 49	4 50	4 47	4 46	4 44	4 42	4 39	4 40	4 42	4 43
26	—	—	—	—	—	—	—	4 45	4 44	4 45	4 46	4 46
27	—	—	—	—	—	4 29	4 25	4 26	4 28	4 27	4 25	4 08
28	—	—	—	—	—	—	4 34	4 35	4 20	4 17	4 19	4 16
29	—	—	—	4 37	4 37	4 35	4 30	4 20	4 22	4 19	4 23	4 25
30	—	—	—	—	4 23	4 22	4 20	4 26	4 17	4 10	4 15	4 10
Sums -	5 01	4 55	11 29	30 14	65 21	112 07	119 27	133 47	126 05	130 14	122 24	127 07
Means -	—	—	—	—	4 40·1	4 29·1	4 25·4	4 27·6	4 20·9	4 20·5	4 12·2	4 14·2
Diurnal Variation	—	—	—	—	0 26·9	0 15·9	0 12·2	0 14·4	0 07·7	0 07·3	0 0·09	0 01·0
1	—	—	—	—	—	—	4 29	3 42	3 42	4 02	4 11	4 20
2	—	—	—	5 00	4 59	4 57	4 30	3 56	3 52	3 56	4 34	4 31
3	—	—	—	—	—	—	5 15	5 19	5 18	5 19	5 10	5 10
4	—	—	—	—	—	5 09	5 28	4 55	4 54	4 57	4 55	4 53
5	—	—	—	—	—	4 41	4 40	4 40	4 39	4 39	4 37	4 36
6	—	—	—	—	—	3 59	3 55	3 54	3 20	3 55	4 25	4 24
7	—	—	—	—	—	4 22	4 21	4 02	3 59	4 00	4 05	4 07
8	—	—	—	—	—	—	4 28	4 20	4 22	4 15	4 10	4 12
9	—	—	—	—	—	—	4 20	4 10	4 24	4 35	4 39	4 29
10	—	—	—	—	—	4 30	3 17	1 36	2 25	2 30	2 41	3 00
11	—	—	—	—	3 37	3 35	3 32	3 22	3 19	3 17	3 22	3 25
12	—	—	—	—	—	3 40	3 44	3 46	3 54	3 54	3 55	4 00
13	—	—	—	—	—	4 10	4 08	4 07	4 02	4 05	4 06	4 0
14	—	—	—	—	—	3 45	3 50	3 40	3 52	3 32	3 30	3 3
15	—	—	—	—	—	—	3 54	3 54	3 50	3 58	3 49	3 5
16	—	—	—	—	—	—	3 58	3 48	3 37	3 37	3 39	3 4
17	—	—	—	—	—	3 36	3 31	3 35	3 36	3 32	3 30	3 3
18	—	—	—	—	5 00	4 50	4 48	4 45	4 44	4 35	4 20	4 3
19	—	—	—	—	—	4 02	4 02	4 04	4 02	3 58	3 56	3 5
20	—	—	—	—	—	—	3 38	3 42	3 40	3 37	3 37	3 4
21	3 58	3 59	2 50	3 12	3 19	3 25	3 35	3 35	3 34	3 36	3 35	3
22	—	—	—	—	—	—	4 00	3 58	3 58	3 55	3 55	3
23	—	—	—	—	3 38	3 40	3 41	3 38	3 27	3 34	3 39	3
24	—	—	—	—	—	3 46	3 42	3 41	3 45	3 40	3 42	3
25	—	—	—	—	3 32	3 36	3 46	3 48	3 26	3 25	3 27	3
26	—	—	—	2 05	2 32	2 53	2 40	2 15	2 45	2 42	3 15	3
27	—	3 33	—	—	—	—	3 30	3 29	3 19	3 25	3 27	3
28	—	—	—	3 59	3 58	3 58	3 58	3 58	3 58	3 57	3 55	3
29	—	—	—	4 03	4 03	4 03	4 02	4 02	4 02	4 00	4 00	4
30	—	—	—	—	4 46	4 46	4 35	3 29	3 35	3 37	3 46	3
31	—	—	—	—	2 58	3 02	3 04	3 08	3 10	3 13	3 12	3
Sums -	3 58	7 32	2 50	18 19	42 22	88 54	124 21	118 18	118 30	119 17	121 04	122
Means -	3 58·0	3 46·0	2 50·0	3 39·8	3 51·1	4 02·4	4 00·7	3 48·8	3 49·4	3 50·9	3 54·3	3 5
Diurnal Variation	—	—	—	—	—	0 13·6	0 11·9	0 00·0	0 00·6	0 02·1	0 05·5	0

* Declinometer moved to clean the glasses, and replaced with the foot screws in the same holes of

Increasing numbers denote a movement of the North end of the needle towards the West

FORT CONFIDENCE.

Abstract of Hourly Observations during the months of November and December 1848.

Declinometer.												
^{h}	3^h	4^h	5^h	6^h	7^h	8^h	9^h	10^h	11^h	Midn^t.	Sums.	Means.
′	° ′	° ′	° ′	° ′	° ′	° ′	° ′	° ′	° ′	° ′	° ′	° ′
47	4 55	4 55	4 55	4 55	4 55	4 55	4 55	—	—	—	84 19	4 57·6
53	4 55	4 56	5 00	5 01	5 04	5 12	5 15	—	—	—	78 48	4 55·5
13	5 11	5 12	5 11	5 11	5 12	5 12	5 13	—	—	5 12	80 30	4 40·6
51	4 52	4 53	4 54	4 55	4 52	4 52	4 54	—	—	—	68 45	4 54·6
35	4 36	4 40	4 41	4 41	4 41	4 41	4 43	—	—	—	69 45	4 39·0
48	4 48	4 42	4 48	4 44	4 45	4 45	4 44	—	—	—	79 55	4 42·1
40	4 40	4 40	4 40	4 40	4 40	4 40	4 40	—	—	—	74 37	4 39·8
58	4 58	4 59	5 00	5 00	5 00	5 00	5 00	—	—	—	68 26	4 58·3
31	4 31	4 34	4 35	4 37	4 42	4 45	4 46	—	—	—	77 46	4 34·6
19	4 27	4 30	4 30	4 30	4 35	4 30	4 33	4 35	4 34	—	79 07	4 23·7
21	4 30	4 30	4 30	4 30	4 32	5 10	4 12	4 14	—	—	79 04	4 28·6
15	4 15	4 19	4 20	4 22	4 22	4 24	4 30	—	—	—	64 02	4 16·1
28	4 29	4 27	4 30	4 30	4 30	4 30	4 31	—	—	—	74 59	4 24·6
02	4 04	4 06	4 09	4 09	4 09	4 09	4 09	—	—	—	78 31	4 05·1
04	4 04	4 06	4 14	4 14	4 15	4 15	4 20	—	—	—	69 50	4 06·5
41	3 41	3 50	3 55	4 07	4 10	4 09	4 09	—	—	—	76 29	4 14·9
15	1 15	3 10	4 15	4 00	4 47	5 05	6 00	1 50	—	—	49 56	2 56·2
32	4 52	4 42	4 48	4 47	4 58	4 48	4 46	—	—	—	68 15	4 15·9
38	4 40	4 40	4 37	4 39	4 42	4 27	4 26	—	—	—	71 06	4 26·6
20	4 20	4 19	4 19	4 20	4 20	4 20	4 25	—	—	—	64 09	4 00·6
47	3 50	3 58	4 01	3 30	4 00	4 30	4 33	—	—	—	69 04	4 03·8
12	4 23	4 28	4 29	4 40	4 40	4 43	4 48	—	—	—	75 16	4 25·6
51	3 56	4 00	4 00	4 06	4 06	4 15	3 28	—	—	—	61 55	3 52·2
23	4 40	4 39	4 44	4 45	4 41	4 45	4 45	4 46	4 46	4 46.	109 53	4 59·6
46	4 47	4 47	4 49	4 46	4 46	4 47	4 47	—	—	—	100 17	4 46·5
57	4 55	4 53	4 57	4 53	4 55	4 54	5 00	—	—	—	67 57	4 51·2
10	4 15	4 18	4 20	4 20	4 22	4 23	4 25	—	—	—	69 31	4 20·7
22	4 22	4 22	4 24	4 25	4 23	4 22	4 25	—	—	—	65 46	4 23·1
25	4 27	4 26	4 27	4 28	4 27	4 28	4 58	—	—	—	80 38	4 28·8
19	4 24	4 26	4 41	4 42	4 46	4 50	4 47	—	—	—	75 38	4 26·6
17	131 56	134 25	136 32	136 22	138 23	139 46	140 08	15 25	9 20	10 00	2,238 10	—
0·6	4 23·9	4 28·8	4 33·1	4 32·7	4 36·8	4 39·5	4 40·3	—	—	—	—	4 27·5
7·4	0 10·7	0 15·6	0 19·9	0 19·5	0 23·6	0 26·3	0 27·1	—	—	—	—	—
32	4 35	4 35	4 38	4 38	4 38	4 40	4 48	—	—	—	65 51	4 23·4
39	4 37	5 25	5 24	5 24	5 17	5 17	5 17	—	—	—	86 14	4 47·4
10	5 10	5 11	5 12	5 11	5 12	5 10	5 11	—	—	—	78 06	5 12·5
50	4 49	4 48	4 47	4 47	4 47	4 45	4 42	—	—	—	78 21	4 53·8
37	4 37	4 38	4 37	4 38	4 38	4 38	4 35	—	—	—	74 07	4 37·9
25	4 27	4 26	4 25	4 25	4 25	4 25	4 26	—	—	—	67 42	4 13·9
12	4 14	4 15	4 15	4 17	4 16	4 17	4 19	—	—	—	67 11	4 12·0
19	4 19	4 19	4 20	4 21	4 20	4 20	4 24	—	—	—	64 49	4 17·3
16	4 18	4 19	4 17	4 20	4 20	4 19	4 20	—	—	—	65 21	4 21·4
25	3 30	3 31	3 35	3 37	3 35	3 38	3 40	—	—	—	51 40	3 13·8
37	3 40	3 40	3 40	3 42	3 42	3 42	3 45	—	—	—	60 29	3 33·5
05	4 09	4 07	4 09	4 10	4 10	4 11	4 10	—	—	—	64 17	4 01·1
10	4 12	4 12	4 13	4 13	4 13	4 13	4 14	—	—	—	66 35	4 09·7
44	3 48	3 47	3 52	3 50	3 53	3 52	3 55	—	—	—	59 54	3 44·6
55	3 56	4 00	4 00	4 00	4 00	4 00	4 00	—	—	—	58 59	3 55·9
48	3 52	3 53	3 54	3 56	3 56	3 58	3 58	—	—	—	57 20	3 49·3
30	3 31	3 35	3 35	3 39	3 45	3 46	4 25	—	—	—	58 24	3 39·0
02	4 06	4 06	4 06	4 05	4 07	4 06	4 06	—	—	—	74 25	4 22·6
35	3 05	3 02	3 05	3 07	3 10	3 10	3 12	—	—	—	57 15	3 34·7
42	3 45	3 45	3 45	3 45	3 45	3 52	3 56	3 58	3 56	3 56	67 43	3 45·7
37	3 37	3 40	3 50	3 50	3 50	3 50	3 50	3 57	3 59	4 01	57 54	3 39·8
01	4 02	4 03	4 02	4 02	4 02	4 02	4 02	—	—	—	59 57	3 59·8
43	3 44	3 44	3 44	3 44	3 43	3 44	3 50	—	—	—	62 53	3 40·8
47	3 48	3 50	3 50	3 55	4 00	4 01	4 02	—	—	—	60 58	3 46·6
35	3 44	3 43	3 45	3 45	3 45	3 45	3 45	—	—	—	61 42	3 37·8
25	3 25	3 27	3 28	3 28	3 30	3 32	3 32	—	—	—	55 40	3 05·6
40	3 54	3 55	3 56	3 58	3 59	3 59	3 59	—	—	—	59 09	3 41·8
58	4 00	4 00	4 02	4 03	4 05	4 05	4 05	—	—	—	71 49	3 59·4
03	4 05	4 04	4 04	4 04	4 05	4 05	3 55	—	—	—	72 40	4 02·2
58	3 59	4 00	4 00	4 02	4 05	4 04	4 03	—	—	—	68 37	4 02·2
16	3 18	3 22	3 25	3 26	3 26	3 28	3 27	—	—	—	55 27	3 15·7
36	124 11	125 21	125 55	126 22	126 38	126 54	127 40	7 55	7 55	7 57	2,041 11	123 51
9·2	4 00·4	4 02·6	4 03·7	4 04·6	4 05·1	4 05·6	4 07·4	3 57·5	3 57·5	3 58·5	—	3 58·7
0·4	0 11·6	0 13·8	0 14·9	0 15·8	0 16·3	0 16·8	0 18·6	—	—	—	—	—

17th day noon moved soon afterwards to 0° 29′. Unifilar Magnetometer in motion from 530 to 540. Red tints of aurora in the evening. (See Notes on Aurora.)

FORT CONFIDENCE.

Abstract of Hourly Observations made during the months of January and February 1849.

Date.	Declinometer.											
	1h	2h	3h	4h	5h	6	7h	8h	9h	10h	11h	Noon.
	° ′	° ′	° ′	° ′	° ′	° ′	° ′	° ′	° ′	° ′	° ′	° ′
1	—	—	—	3 28	—	3 20	3 20	3 22	3 22	3 22	3 32	3 39
2	—	—	—	—	4 08	4 08	4 08	4 08	4 08	4 08	4 08	4 07
3	—	—	—	3 22	3 18	3 17	3 25	3 20	3 26	3 28	3 30	3 30
4	—	—	4 14	4 13	4 13	3 16	3 12	4 00	4 05	4 06	4 06	4 06
5	—	—	—	—	—	4 11	4 12	4 12	4 12	4 10	4 10	4 11
6	—	—	—	—	—	4 15	4 15	4 14	4 15	4 15	4 15	4 14
7	—	—	—	—	—	—	4 13	4 14	4 14	4 12	4 12	4 13
8	—	—	—	—	—	3 15	3 20	3 20	3 20	3 25	3 40	3 42
9	—	—	—	—	—	—	3 37	3 42	3 40	3 40	3 41	3 44
10	—	—	—	—	—	3 34	3 34	3 42	3 40	3 36	3 42	3 52
11	—	—	—	—	—	—	4 03	3 50	3 52	3 43	3 53	3 5[illegible]
12	—	—	—	—	—	—	5 03	4 50	4 48	4 48	4 45	4 45
13	—	—	—	—	3 36	3 37	3 37	3 47	3 45	3 55	3 52	3 52
14	—	—	—	—	—	—	3 47	3 49	3 48	3 42	3 28	3 30
15	—	—	—	—	4 08	3 53	3 52	3 59	4 08	3 40	3 48	3 10
16	—	—	4 06	4 07	4 07	4 08	3 59	2 05	2 08	2 30	2 55	3 10
17	—	—	—	3 59	4 00	3 58	3 50	3 55	3 55	3 55	3 56	3 58
18	—	—	4 07	4 07	4 07	4 03	4 06	4 02	3 52	3 47	3 50	3 54
19	—	—	—	3 56	3 57	3 58	3 56	3 55	3 55	3 54	3 55	3 56
20	—	—	4 01	4 00	4 00	4 00	4 00	3 57	3 58	3 55	3 44	3 45
21	—	—	—	—	—	3 50	3 54	3 56	3 55	3 54	3 46	3 50
22	4 02	4 02	4 02	4 02	4 02	4 02	4 02	4 02	4 02	4 02	4 02	4 03
23	—	—	4 28	4 28	4 29	4 30	4 28	4 27	4 13	4 22	3 58	3 51
24	—	—	—	4 16	4 14	4 14	4 18	4 12	4 07	4 07	3 57	4 1[illegible]
25	4 30	4 30	4 28	4 27	4 28	4 28	4 28	4 24	4 23	3 20	3 28	3 [illegible]
26	—	—	—	—	—	3 32	3 33	3 30	3 29	3 30	3 11	3 [illegible]
27	—	—	3 57	3 42	3 42	3 37	3 43	3 43	3 40	3 41	3 40	3 [illegible]
28	—	—	—	—	—	—	3 58	3 59	3 55	3 55	3 52	3 [illegible]
29	—	—	—	—	4 07	4 03	4 03	4 03	3 59	3 58	4 00	4 [illegible]
30	—	—	—	4 06	4 02	4 02	3 03	3 29	3 18	3 15	3 27	3 [illegible]
31	—	—	3·41	3 40	3 40	3 39	3 30	3 39	3 39	3 39	3 39	[illegible]
Sums -	8 32	8 32	37·04	59·53	72 18	96 52	120 40	119 53	119 11	117 54	118 02	11[illegible]
Means -	—	—	—	—	4 01·0	3 52·5	3 53·5	3 52·0	3 50·7	3 48·2	3 48·5	3 [illegible]
Diurnal Variation -	—	—	—	—	0 12·8	0 04·3	0 05·3	0 03·8	0 02·5	0 00·0	0 00·3	[illegible]

Date.	1h	2h	3h	4h	5h	6	7h	8h	9h	10h	11h	Noon.
1	—	—	—	4 24	4 36	4 26	4 27	4 27	3 55	4 20	4 00	
2	—	—	—	3 54	3 55	3 57	3 58	3 58	4 00	3 57	3 54	
3	—	—	—	—	4 03	4 04	4 04	4 02	4 04	4 03	4 05	
4	—	—	—	—	—	4 06	4 08	4 07	3 47	3 23	3 34	
5	—	—	—	—	3 43	3 43	3 42	3 42	3 39	3 37	3 37	
6	—	—	—	3 42	3 42	3 43	4 42	4 42	3 42	3 40	3 40	
7	—	—	—	—	3 54	3 54	4 55	4 57	3 55	3 56	3 56	
8	—	—	—	—	3 57	3 57	4 57	4 58	3 58	3·58	3 55	
9	—	—	—	—	—	4 00	4 00	3 59	4 00	3 58	3 59	
10	—	—	—	—	—	4 01	4 00	3 57	3 47	3 45	3 47	
11	—	—	—	—	—	—	3 57	3 55	3 48	3 45	3 4[illegible]	
12	—	—	—	—	4 17	4 17	4 17	4 15	4 10	3 55	3 5[illegible]	
13	—	—	—	—	—	4 02	4 01	4 00	2 58	3 07	3 1[illegible]	
14	—	—	—	—	—	3 54	3 52	3 54	3 53	3 53	3 5[illegible]	
15	—	—	—	—	—	4 34	4 34	4 33	4 29	4 25	4 2[illegible]	
16	—	—	—	—	—	4 32	4 32	4 34	4 34	3 30	4 [illegible]	
17	—	—	—	—	4 24	4 25	4 27	4 25	4 22	3 23	4 [illegible]	
18	—	—	—	—	—	—	4 31	4 30	4 20	3 12	4 [illegible]	
19	—	—	—	—	4 17	4 17	4 18	4 18	4 24	3 20	4 [illegible]	
20	—	—	—	—	—	4 35	4 33	4 31	4 28	3 50	3 [illegible]	
21	4 15	4 14	4 11	4 06	4 08	4 03	4 10	4 07	3 50	3 55	3 [illegible]	
22	—	—	—	—	—	4 19	4 10	4 00	3 32	3 28	2 [illegible]	
23	—	—	—	—	—	3 48	3 48	3 48	3 50	3 47	3 [illegible]	
24	4 24	4 24	4 23	4 21	4 21	4 18	4 16	4 12	3 50	3 53	3 [illegible]	
25	—	—	—	—	—	—	4 21	4 18	4 12	4 10	4 [illegible]	
26	—	—	—	—	—	4 16	4 13	4 11	4 06	3 32	[illegible]	
27	—	—	—	—	—	4 06	4 02	4 00	3 35	3 32		
28	—	—	—	—	—	4 42	4 47	4 42	4 29	4 28		
Sums -	8 39	8 38	8 34	20 27	49 17	104 01	116 42	116 02	111 35	109 46	1[illegible]	
Means -	—	—	—	—	—	4 09·6	4 10·1	4 08·6	3 59·1	3 55·2	3 [illegible]	
Diurnal Variation -	—	—	—	—	—	0 20·4	0 20·9	0 19·4	0 09·9	0 06·0	[illegible]	

Increasing numbers denote a mov

FORT CONFIDENCE.

Abstract of Hourly Observations made during the months of January and February 1849.

Declinometer.												
	3h	4h	5h	6h	7h	8h	9h	10h	11h	Midnt.	Sums.	Means.
′	° ′	° ′	° ′	° ′	° ′	° ′	° ′	° ′	° ′	° ′	° ′	° ′
06	4 07	4 07	4 07	4 07	4 07	4 08	4 08	4 07	4 09	4 10	77 11	3 51·5
12	4 12	4 14	4 14	4 14	4 14	4 14	4 18	—	—	—	71 08	4 10·9
33	3 37	3 38	3 38	3 33	3 38	3 38	3 40	—	—	—	68 12	3 30·7
06	4 07	4 08	4 08	4 10	4 10	4 08	4 08	—	—	—	76 42	4 02·2
11	4 12	4 13	4 14	4 15	4 16	4 17	4 18	—	—	—	67 25	4 12·8
15	4 15	4 15	4 16	4 16	4 17	4 17	4 15	—	—	—	68 04	4 15·3
12	4 15	4 10	4 15	4 15	4 16	4 10	4 16	—	—	—	63 28	4 13·9
42	3 48	3 42	3 48	3 44	3 44	3 44	3 45	—	—	—	57 31	3 35·7
50	3 50	3 50	3 50	3 50	3 50	3 51	3 52	—	—	—	56 31	3 46·1
34	3 44	3 45	3 50	3 57	3 57	3 57	3 57	—	—	—	60 25	3 46·6
59	4 08	4 04	4 05	4 07	4 09	4 10	4 50	5 15	5 07	—	71 06	4 10·9
44	3 45	4 40	3 37	4 37	4 37	4 36	4 35	—	—	—	70 55	4 48·7
02	3 04	4 05	3 06	4 07	4 08	4 08	4 09	—	—	—	66 46	3 55·6
31	3 32	3 31	3 39	3 48	3 59	4 08	4 12	—	—	—	55 55	3 43·7
25	3 30	3 36	3 50	3 40	3 55	4 05	4 11	—	—	—	64 21	3 47·1
22	3 29	3 32	3 45	3 55	3 50	4 00	4 02	—	—	—	66 39	3 30·5
59	4 00	4 08	4 05	4 05	4 07	4 08	4 09	—	—	—	72 01	4 00·1
55	3 57	3 56	3 58	3 57	3 57	3 57	3 58	—	—	—	75 27	3 58·3
57	3 59	4 00	4 00	4 00	4 00	4 01	4 02	—	—	—	71 18	3 57·7
49	3 50	3 52	3 52	3 48	3 50	3 50	3 52	3 52	—	—	77 34	3 52·7
52	3 57	3 57	3 58	3 58	3 58	4 01	4 02	4 02	4 02	4 02	74 49	3 56·2
20	4 27	4 27	4 27	4 27	4 28	4 20	4 29	4 30	4 30	4 30	101 36	4 14·0
55	3 58	4 10	4 17	4 19	4 29	4 29	4 20	4 18	—	—	85 23	4 16·2
10	4 15	4 27	4 28	4 29	4 29	4 29	4 30	4 30	4 30	4 30	90 27	4 18·4
15	3 25	3 29	3 32	3 40	3 35	3 35	3 38	—	—	—	81 42	3 53·4
23	3 30	3 47	3 48	3 48	3 55	3 57	3 57	—	—	—	57 20	3 35·0
47	3 58	3 59	4 00	4 00	4 00	4 00	4 00	—	—	—	72 40	3 49·5
56	3 57	3 58	4 01	4 02	4 02	4 03	4 04	—	—	—	59 34	3 58·3
00	4 03	4 03	4 04	4 05	4 05	4 05	4 05	—	—	—	68 43	4 02·5
31	3 31	3 32	3 37	3 40	3 42	3 42	3 41	—	—	—	64 35	3 35·3
40	3 40	3 40	3 40	3 40	3 40	3 40	3 40	—	—	—	69 35	3 39·7
13	121 52	122 50	124 04	124 42	125 33	126 08	127 03	30 34	22 18	17 12	2180 00	
·7	3 55·9	3 57·7	4 00·1	4 01·4	4 03·0	4 04·0	4 05·9	—	—	—	—	3 57·0
·5	0 07·7	0 09·5	0 11·9	0 13·2	0 14·8	0 15·8	0 17·7	—	—	—	—	—
34	3 33	3 44	3 50	3 52	3 54	3 55	3 55	3 55	—	—	75 55	3 59·7
00	4 00	4 00	4 02	4 04	4 04	4 06	4 07	4 03	—	—	76 02	4 00·1
06	4 07	4 08	4 09	4 09	4 10	4 10	4 10	4 10	—	—	73 57	4 06·5
25	3 29	3 28	3 44	3 43	3 43	3 42	3 43	3 43	—	—	62 24	3 40·2
40	3 40	3 40	3 40	3 40	3 43	3 43	4 44	4 44	—	—	66 12	3 40·7
42	3 44	3 55	3 55	3 55	3 57	3 55	4 55	4 55	—	—	71 48	3 46·7
57	3 58	4 00	4 00	3 50	4 00	4 00	4 00	3 59	—	—	71 14	3 57·4
00	4 00	4 01	4 00	4 00	4 00	4 00	4 00	4 01	—	—	71 39	3 58·8
00	4 01	4 01	4 00	4 00	4 00	4 00	4 00	4 02	—	—	68 00	4 00·0
52	3 57	3 59	4 01	4 00	4 00	4 00	4 00	4 00	—	—	66 43	3 55·5
55	4 00	4 10	4 10	4 14	4 15	4 17	4 16	4 17	—	—	64 14	4 00·9
49	3 55	4 02	4 06	4 06	4 06	4 06	4 06	4 07	—	—	72 51	4 02·8
35	3 35	3 52	3 55	3 56	3 58	3 55	3 55	3 55	—	—	62 53	3 41·9
53	3 59	4 00	4 00	4 01	4 35	4 35	4 35	4 36	—	—	69 19	4 04·6
20	4 22	4 22	4 22	4 24	3 25	3 25	4 25	4 25	—	—	75 14	4 25·5
29	4 30	4 32	4 31	4 30	3 30	3 32	4 36	4 37	—	—	76 51	4 31·2
30	4 30	4 32	4 33	4 34	3 34	3 34	4 34	4 34	—	—	80 45	4 29·2
56	3 59	4 02	4 05	4 05	3 06	3 15	4 19	4 20	—	—	66 32	4 09·5
26	3 49	3 40	3 50	3 56	3 57	4 02	3 35	3 35	—	—	73 37	4 05·4
23	3 20	3 25	3 30	3 35	3 42	3 51	4 00	4 02	4 05	4 11	73 37	3 51·9
35	3 40	3 50	3 53	3 55	4 00	4 06	4 06	4 17	4 20	4 25	96 01	4 00·0
00	3 02	3 12	3 17	3 25	3 33	3 40	3 50	3 50	—	—	58 40	3 27·1
47	3 49	3 54	4 02	4 12	4 20	4 21	4 22	4 22	4 25	4 25	75 58	3 59·9
02	4 05	4 06	4 07	4 10	4 10	4 10	4 12	4 20	—	—	91 15	4 06·9
10	4 12	4 14	4 14	4 15	4 15	4 16	4 16	4 17	—	—	67 40	4 13·8
37	3 46	3 55	3 59	4 00	4 00	4 01	4 02	4 07	—	—	66 20	3 54·1
35	3 36	3 41	3 59	4 07	4 12	4 16	4 20	4 59	5 01	—	71 21	3 57·8
25	4 25	4 25	4 23	4 24	4 23	4 22	4 23	4 22	—	—	76 04	4 28·5
43	109 02	110 50	112 19	113 10	114 32	115 15	115 25	116 40	17 51	13 01	2022 56	
·9	3 53·6	3 57·5	4 00·7	4 02·5	4 05·4	4 07·0	4 07·3	4 10·0	—	—	—	4 01·3
·7	0 04·4	0 08·3	0 11·5	0 13·3	0 16·2	0 17·8	0 18·1	0 20·8	—	—	—	—

e needle towards the West.

Fort Confidence—*continued.*

Abstract of Hourly Observations made during the months of March and April 1849.

Date. Mean Time at Station.	Declinometer. 1h	2h	3h	4h	5h	6h	7h	8h	9h	10h	11h	Noon.
	° ′	° ′	° ′	° ′	° ′	° ′	° ′	° ′	° ′	° ′	° ′	° ′
1	—	—	—	—	—	4 23	4 24	4 24	4 12	4 12	4 12	4 12
2	—	—	—	—	—	4 18	4 20	4 22	4 15	3 50	3 55	3 50
3	—	—	—	—	—	4 15	4 14	4 12	4 11	4 12	4 13	4 13
4	—	—	—	—	—	3 48	3 50	3 52	3 45	3 48	3 50	3 53
5	—	—	—	—	—	4 05	4 05	4 04	4 15	4 16	4 15	4 15
6	—	—	—	—	—	4 20	4 18	4 18	4 14	3 58	4 00	4 15
7	—	—	—	—	—	4 40	4 48	4 47	4 39	4 35	4 27	4 24
8	—	—	—	—	—	4 27	4 27	4 19	3 52	3 50	3 50	3 57
9	—	—	—	—	—	4 21	4 24	4 15	4 06	4 05	4 14	4 15
10	—	—	—	—	—	4 17	4 07	4 06	4 06	4 05	3 58	3 55
11	—	—	—	—	—	4 14	4 13	4 13	4 11	4 10	4 00	4 10
12	—	—	—	—	—	4 18	4 17	4 07	4 05	3 45	3 42	4 00
13	—	—	—	—	—	4 18	4 16	4 14	4 00	4 05	4 08	4 10
14	—	—	—	—	—	4 14	4 14	4 14	4 13	4 13	4 11	4 10
15	—	—	—	—	—	4 30	4 27	4 17	4 18	4 05	3 50	3 56
16	—	—	—	—	—	4 28	4 27	4 15	4 11	4 10	4 11	4 12
17	—	—	—	—	—	4 17	4 16	4 15	4 13	4 10	4 10	4 16
18	—	—	—	—	—	4 05	4 32	4 25	4 07	3 50	3 31	3 29
19	—	—	—	—	—	4 13	4 15	4 04	2 45	3 37	3 33	2 27
20	—	—	—	—	—	4 37	4 29	4 15	3 20	3 10	3 20	3 21
21	3 30	3 30	4 28	4 28	4 22	4 21	4 20	4 14	4 02	4 00	4 00	3 [illegible]
22	4 48	4 46	4 33	4 29	4 23	4 37	4 24	4 08	3 30	3 32	3 26	3 2[illegible]
23	—	—	—	—	—	4 15	4 01	4 03	3 40	3 37	3 40	3 4[illegible]
24	—	—	—	—	—	4 26	4 25	4 19	3 19	2 30	2 50	3 [illegible]
25	—	—	—	—	—	4 23	4 23	4 20	4 22	4 06	4 00	4 [illegible]
26	—	—	—	—	—	4 41	4 39	4 33	3 45	3 15	3 13	4 [illegible]
27	—	—	—	—	—	4 19	4 18	4 17	4 18	4 14	4 15	4 [illegible]
28	—	—	—	—	—	4 18	4 14	4 12	4 12	3 40	3 45	3 [illegible]
29	—	—	—	—	—	4 26	4 24	4 20	4 20	4 16	4 16	4 [illegible]
30	—	—	—	—	—	4 28	4 23	4 12	4 20	3 59	4 04	4 [illegible]
31	—	—	—	—	—	4 32	4 31	4 26	4 25	4 19	4 18	[illegible]
Sums -	8 18	8 16	9 01	8 57	8 45	135 08	134 25	132 04	125 11	120 48	122 41	12[illegible]
Means -	—	—	—	—	—	4 21·4	4 20·2	4 15·6	4 02·3	3 53·6	3 57·5	4 [illegible]
Diurnal Variation -	—	—	—	—	—	0 27·8	0 26·6	0 22·0	0 08·7	0 000	0 08·9	[illegible]
1	—	—	—	—	—	4 27	4 28	4 24	4 20	4 17	4 15	
2	—	—	—	—	—	4 28	4 17	2 55	2 12	3 12	3 22	
3	—	—	—	—	—	4 34	4 30	4 02	3 59	2 46	3 35	
4	—	—	—	—	—	4 13	4 14	3 35*	2 05	2 05	2 05	
5	—	—	—	—	—	4 57	4 52	4 38	4 19	4 15	4 10	
6	—	—	—	—	—	4 35	4 33	4 06	3 45	3 40	3 32	
7	—	—	—	—	—	4 30	4 15	4 10	4 05	4 06	4 10	
8	—	—	—	—	—	4 40	4 38	4 22	4 20	4 18	4 18	
9	—	—	—	—	—	4 40	4 42	4 43	1 05	3 18	3 2[illegible]	
10	—	—	—	—	—	4 30	4 30	4 30	4 45	4 02	3 1[illegible]	
11	—	—	—	—	—	4 35	4 35	4 32	4 20	4 15	4 0[illegible]	
12	—	—	—	—	4 35	4 35	4 32	4 27	4 22	4 21	4 [illegible]	
13	—	—	—	—	—	4 30	4 22	4 26	4 20	4 10	3 [illegible]	
14	—	—	—	—	—	4 20	4 33	4 20	3 22	3 17	4 [illegible]	
15	—	—	—	—	—	4 30	4 30	4 28	4 25	4 21	4 [illegible]	
16	—	—	—	—	—	4 28	4 22	4 27	2 47	2 52	3 [illegible]	
17	—	—	—	—	4 40	4 40	4 40	4 27	4 06	3 55	3 [illegible]	
18	—	—	—	—	—	4 35	4 27	4 16	3 38	4 23	4 [illegible]	
19	—	—	—	—	—	4 34	4 32	4 22	3 42	4 34	4 [illegible]	
20	—	—	—	—	4 42	4 34	4 17	3 55	4 15	4 40	4 [illegible]	
21	—	—	—	—	4 55	4 55	4 50	4 08	4 00	4 10	[illegible]	
22	—	—	—	—	—	5 10	4 45	4 58	4 58	4 25		
23	—	—	—	—	—	4 19	4 35	4 33	4 37	4 25		
24	—	—	—	—	4 44	5 00	5 15	5 07	4 30	4 31		
25	—	—	—	—	—	4 48	4 35	4 30	4 11	3 22		
26	—	—	—	—	4 20	4 25	4 20	3 25	3 17	3 03		
27	—	—	—	—	—	5 05	4 39	4 25	4 17	4 16		
28	—	—	—	—	—	4 38	4 40	4 39	4 29	4 05		
29	—	—	—	—	—	5 20	5 14	4 35	4 19	4 10		
30	—	—	—	—	—	4 36	4 34	4 35	4 29	4 22		
Sums -	—	—	—	—	27 56	139 06	137 16	130 00	116 19	114 36	1[illegible]	
Means -	—	—	—	—	4 39·3	4 38·2	4 34·5	4 200	3 52·6	3 49·2	[illegible]	
Diurnal Variation	—	—	—	—	—	0 49·0	0 45·3	0 30·8	0 03·4	0 000		

* On the 4th at 8 hours 45 minutes A.M. 1° 25′.

Increasing numb

FORT CONFIDENCE—*continued.*

Abstract of Hourly Observations made during the months of March and April 1849.

Declinometer.												
2ʰ	3ʰ	4ʰ	5ʰ	6ʰ	7ʰ	8ʰ	9ʰ	10ʰ	11ʰ	Midnᵗ	Sums.	Means.
° ′	° ′	° ′	° ′	° ′	° ′	° ′	° ′	° ′	° ′	° ′	° ′	° ′
4 18	4 17	4 12	4 19	4 20	4 20	4 20	4 20	4 20	—	—	73 05	4 17·9
4 00	4 01	4 02	4 02	4 02	4 02	4 02	4 02	4 16	—	—	69 16	4 04·5
4 15	4 15	4 15	4 16	4 17	4 17	4 17	4 10	3 55	—	—	71 42	4 13·1
4 07	4 08	4 07	4 07	4 08	4 07	4 07	4 07	4 08	—	—	67 57	3 59·8
4 18	4 30	4 26	4 21	4 22	4 24	4 22	4 23	4 23	—	—	72 58	4 17·5
4 21	4 21	4 57	4 54	4 53	4 52	4 52	4 50	4 50	—	—	76 31	4 30·1
4 22	4 22	4 24	4 24	4 25	4 27	4 26	4 26	4 28	—	—	76 34	4 30·2
4 05	4 10	4 18	4 18	4 22	4 24	4 25	4 26	4 26	—	—	71 54	4 13·8
4 25	4 21	4 22	4 20	4 25	4 25	4 25	4 25	4 25	—	—	73 36	4 19·8
3 58	4 02	4 05	4 17	4 20	4 20	4 22	4 22	4 22	—	—	74 30	4 23·5
4 19	4 10	4 10	4 10	4 14	4 15	4 20	4 21	4 22	—	—	71 42	4 12·1
4 11	4 15	4 16	4 15	4 15	4 15	4 15	4 18	4 20	—	—	70 39	4 09·4
4 10	4 11	4 11	4 12	4 14	4 14	4 15	4 15	4 15	—	—	71 18	4 11·6
4 21	4 30	4 30	4 30	4 30	4 30	4 31	4 30	4 30	—	—	74 01	4 21·2
4 07	4 12	4 18	4 25	4 25	4 26	4 27	4 27	4 25	—	—	72 40	4 16·5
4 15	4 14	4 15	4 15	4 15	4 15	4 17	4 20	4 20	—	—	72 33	4 16·1
4 11	4 15	4 15	4 15	4 16	4 22	4 30	4 42	4 42	—	—	73 09	4 18·2
3 18	3 18	3 25	3 45	3 52	4 11	4 12	4 15	4 30	—	—	65 51	3 52·4
2 45	3 35	3 37	4 15	4 25	4 25	4 40	4 35	4 35	—	—	62 20	3 40·0
3 09	4 40	3 20	3 20	3 25	3 25	3 25	3 30	3 35	3 31	3 30	67 33	3 33·3
4 20	4 24	4 35	4 38	4 28	4 30	4 42	4 41	4 49	5 02	4 49	103 49	4 19·9
4 05	4 04	4 05	4 10	4 12	4 15	4 20	4 20	4 22	—	—	91 44	4 10·2
4 10	4 20	4 25	4 30	4 31	4 30	4 30	4 30	4 30	—	—	70 54	4 10·2
3 50	4 17	4 19	4 10	4 19	4 25	4 28	4 30	4 30	—	—	67 22	3 58·3
4 53	4 40	4 47	4 47	4 45	4 45	4 45	4 45	4 42	—	—	77 56	4 35·1
4 12	4 15	4 17	4 15	4 15	4 18	4 22	4 20	4 20	—	—	71 07	4 11·0
4 12	4 18	4 10	4 15	4 15	4 19	4 20	4 20	4 20	—	—	72 40	4 15·5
4 10	4 22	4 25	4 25	4 25	4 25	4 25	4 25	4 25	—	—	70 58	4 10·5
4 15	4 15	4 17	4 20	4 25	4 24	4 25	4 35	4 26	—	—	73 42	4 20·1
4 15	4 15	4 15	4 18	4 28	4 30	4 34	4 35	4 33	—	—	73 22	4 18·9
4 21	4 22	4 23	4 25	4 25	4 25	4 25	4 25	4 27	—	—	74 42	4 22·6
127 28	130 28	131 30	132 53	134 03	134 42	135 47	136 02	136 30	8 33	8 19	2278 24	—
4 06·7	4 12·5	4 14·5	4 17·2	4 19·5	4 20·7	4 22·8	4 23·3	4 24·2	—	—	—	4 12·7
0 13·1	0 18·9	0 20·9	0 23·6	0 25·9	0 27·1	0 29·2	0 29·7	0 30·6	—	—	—	—
4 15	4 15	4 17	4 22	4 22	4 25	4 27	4 29	4 30	—	—	73 39	4 21·1
3 32	3 36	3 46	3 47	3 50	4 35	4 37	4 41	4 40	—	—	64 20	3 47·1
4 05	4 05	4 05	4 07	4 06	4 12	4 12	4 12	4 13	—	—	68 47	4 02·8
3 06	4 12	4 15	4 32	4 45	4 46	4 55	5 00	5 05	—	—	63 33	3 44·3
4 25	4 33	4 38	4 45	4 45	4 45	4 48	4 45	4 45	—	—	77·54	4 34·9
4 00	4 05	4 15	4 31	4 37	4 40	4 41	4 40	4 40	—	—	71·32	4 12·5
4 32	4 38	4 43	4 42	4 42	4 42	4 42	4 42	4 42	—	—	75 51	4 27·7
4 22	4 24	4 27	4 30	4 30	4 30	4 32	4 35	4 40	—	—	75 48	4 27·5
4 23	4 22	4 30	4 30	4 30	4 33	4 32	4 33	4 34	—	—	70 13	4 07·8
4 15	4 38	4 40	4 42	4 42	4 42	4 40	4 40	4 40	—	—	74 37	4 23·3
4 25	4 25	4 35	4 50	4 34	4 64	4 35	4 35	4 35	—	—	75 30	4 26·5
4 24	4 25	4 23	4 26	4 27	4 37	4 30	4 30	4 30	—	—	80 19	4 27·7
4 10	4 25	4 30	4 30	4 28	4 38	4 35	4 40	4 42	—	—	74 36	4 23·3
4 18	4 20	4 25	4 27	4 28	4 30	4 32	4 34	4 37	—	—	71 32	4 12·5
4 27	4 00	4 00	4 00	4 00	4 00	4 30	4 31	4 29	—	—	73 21	4 18·9
3 40	3 42	3 50	4 20	4 22	4 40	4 39	4 40	4 38	—	—	66 39	3 55·2
3 58	4 20	4 30	4 35	4 35	4 37	4 33	4 37	4 37	—	—	78 11	4 20·6
4 01	4 16	4 28	4 55	4 54	4 55	4 59	4 50	5 00	—	—	73 22	4 18·9
4 05	4 08	4 22	4 18	4 20	4 25	4 40	4 48	4 46	—	—	71 46	4 13·3
3 55	4 29	4 30	4 40	4 42	4 45	4 45	4 45	4 50	4 55	5 02	87 06	4 21·3
4 50	4 50	4 50	5 10	5 02	5 00	5 04	4 45	5 25	—	—	85 51	4 46·9
4 30	4 30	4 34	4 35	4 40	4 45	4 44	4 40	4 40	—	—	79 52	4 41·9
4 34	4 35	4 35	4 55	4 46	4 46	4 47	4 46	4 54	—	—	78 16	4 36·2
4 31	4 45	4 45	4 58	4 55	4 55	4 53	4 52	4 52	—	—	86 07	4 47·1
4 36	4 39	4 40	4 48	5 01	5 00	5 10	5 09	5 10	—	—	78 00	4 35·3
4 52	4 52	5 00	5 00	5 10	5 09	5 07	5 10	5 10	—	—	80 07	4 27·1
4 25	4 28	4 30	4 35	4 35	4 35	4 35	4 35	4 35	—	—	76 45	4 30·9
4 15	4 20	4 25	4 29	4 30	4 35	4 40	4 40	4 52	5 40	5 58	87 23	4 35·9
4 20	4 22	4 25	4 27	4 32	4 34	4 35	4 40	4 40	—	—	77 00	4 31·8
4 25	4 27	4 28	4 40	4 40	4 40	4 41	4 44	4 45	—	—	77 25	4 38·2
127 35	131 06	133 11	136 44	137 30	139 17	140 42	140 52	142 16	10 35	11 00	2275 42	4 22·6
4 15·2	4 22·2	4 26·4	4 33·5	4 35·0	4 38·6	4 41·4	4 41·7	4 44·5	5 17·5	5 30	—	4 22·6
0 26·0	0 30·0	0 37·2	0 44·8	0 45·8	0 49·4	0 52·2	0 52·5	0 55·3	—	—	—	—

he North end of the needle towards the West.

METEOROLOGICAL OBSERVATIONS

BY SIR JOHN RICHARDSON.

The following meteorological observations were made at Fort Confidence, on Great Bear Lake, in connection with the magnetic experiments. The fort (a mere log-house) stood on the banks of the lake, on limestone strata about ten feet above the level of the water, in lat. 66° 54′ N.; long. 118° 48′ 45″ W. of Greenwich or 8^h 35^m 01·5 W. of Göttingen. The observatory (a small log building, without a fire-place) was built for the reception of the Declinometer and Unifilar Magnetometer, in front of the house, or between it and the lake. The temperature of this isolated apartment was regularly recorded as often as the Declinometer was observed. On the north end of the store-house (which formed the west side of the square or yard of the house, and was parallel to the observatory), were hung a dozen spirit thermometers, constructed by Adie, for the observation of the temperature of the atmosphere in the shade. These were generally compared with each other at each observation, but one was selected for record which stood, in a mean of various trials, at −36° when plunged into freezing mercury. The temperature of a thermometer, having a bulb blackened with China ink and indigo, and enclosed in a glass bottle exposed to the sun's rays, was also noted hourly during the day. Delcros's barometer was suspended in my sleeping apartment, with the cistern about 14 feet above the surface of the lake. This barometer is constructed with a moveable brass scale, which is adjusted to the surface of the mercury in the cistern by an ivory point. The degrees were read off on the millimeter scale, and a correction made by the addition of 0·34 mill. as the mean error for capillarity and deviation from the standard barometer of Fortin.* The actual indication of the barometer was written down at the time, with the temperature shown by the attached thermometer in contact with the mercurial column; the corrections were made afterwards, and for December, January, and February were reduced to English inches for each hour, and corrected for temperature 32° Fahr. by Schumacher's table appended to the Report of the Committee of the

* The corrections for the barometer furnished by the maker were, more exactly,

Correction moyenne totale de capillarité - - -	+ 0,446
Correction du baromètre 269 donné par bar. typal. -	− 0,108
Equation des Observations brutes - - -	+ 0,338 or as applied on the tables + 0,34.

Royal Society for 1840. In that form they are presented in the tables for these three months. For October, November, March and April, the observations are printed on the millimetric scale after the correction for the mean error; and at the bottom of each column the reduction to English inches with the corrections for temperature 32° are added. Care was taken to suspend the barometer in a part of the room out of the direct radiation from the fire, and where it was sheltered as much as possible from currents of air; but it was unavoidably exposed to rapid fluctuations of temperature, since the fire when well built up heated the room rapidly, but when the door of the apartment was left open for the ten minutes which the bringing in of the daily supply of fire-wood occupied, the temperature would fall at once to the amount of 30 or 40 degrees, if at the time the external air happened to be very cold. These rapid transitions were doubtless the occasional cause of more or less error. As the surface of the mercury in the cistern tarnished rapidly, that fluid was thrice cleaned by filtering through paper in the course of the winter, the construction of the instrument permitting this to be readily done without disturbing the mercury in the tube. The wooden cistern, however, was found to shrink considerably in the extremely dry air of the apartment, and it was necessary to wind a little floss silk round it to cause it to fill its place accurately; this may perhaps have produced a little change in its capacity, but as the scale was a sliding one the error of its indications could be very trifling. An aneroid barometer was hung alongside Delcros's instrument, and a record kept of its indications; but as it was one of the earliest of its kind, and in some degree imperfect, it has not been thought necessary to print the observations made by it. In December, January and February, the aneroid stood generally between 0·020 and 0·060 inches below the Delcros's barometer when the latter was corrected for temperature 32°, no correction being made for the aneroid, but the differences were not uniform, and sometimes exceeded 0·100 inches. No correction for temperature was furnished to us with the aneroid.

The thermometers employed for ascertaining the temperatures were constructed by Mr. Adie of Edinburgh. On former expeditions I had used thermometers made by London artists of great eminence; but finding that the instruments varied greatly from each other at very low temperatures, I applied to Professor Forbes of Edinburgh, who kindly undertook to superintend the making of instruments which might be more comparable with each other in great degrees of cold. The following is an extract from a letter written on the subject by him subsequent to my return from America:—

"My Dear Sir, Edinburgh, 10th April 1851.

"My idea was in constructing the thermometers (or rather in superintending their construction) to ensure comparability, and definiteness in the principles of graduation, which you are aware does not exist in alcohol thermometers as usually made, both from uncertainty in the density of the spirit used and, especially, because only one fixed point (freezing water) is employed, the other points being taken by comparison with a mercurial thermometer. But as alcohol and mercury do not expand alike, the value of 1° of the alcohol thermometer will depend upon the point of comparison with the mercurial one.

"What I intended, and should recommend in principle, would be to use *absolute alcohol* (or as nearly so as possible), to fix the freezing point of water and of mercury, to call the latter −40° either on the centigrade or Fahrenheit's scale (which here coincide), and to divide the space uniformly, and also to graduate uniformly above 32°. There can be no possible harm in defining freezing mercury to be at 40° Fahrenheit.

"I intended to verify these fixed points myself, but I was rather seriously unwell that winter, and as your time was limited, I abandoned the freezing of mercury on a large scale, and satisfied myself with general instructions to Mr. Adie, which, I think, the results show to have been well carried out.

"The alcohol was prepared on purpose by Dr. George Wilson, chemist, and his report is enclosed. It shows that the alcohol is very nearly absolute. The tubes had round bores, and were examined in the usual way, by passing columns of mercury along them; a variation of the apparent length of the column of mercury, amounting to $\frac{1}{50}$ inch, and that, in any part of the tube, causing the rejection of the tube.

"The fixed points were 32° in ice, and 62° by comparison with a carefully corrected mercurial standard thermometer. The degrees were run up and down to the same measure.

"As you have accurately ascertained the freezing point of mercury on these thermometers, it would be easy to infer the change which my proposed method of graduation would have produced.

"Yours sincerely,

"(Signed) JAMES D. FORBES."

"'EXTRACT of a LETTER from Dr. WILSON to Professor FORBES.'"

"'Dear Sir,

"'I enclose a note* of the specific gravity of the alcohol, with such other particulars as it seemed desirable to put on record for the sake of subsequent comparison, should that be made. The uncoloured alcohol was determined with a 1,000 grain bottle. The residual coloured spirit amounted to little more than 100 grains. Its Sp. Gr. was ascertained with a bottle containing 124·18 grains of distilled water at 60°. The unused coloured alcohol, barely amounting to a quantity equal in volume to 1,000 grains of water, could not be made to fill entirely a 1,000 grain Sp. Gr. bottle when transferred from the vessel containing it; I thought it best, therefore, to determine its density in the bottle made use of for the residual alcohol. The coloured alcohols are thus directly comparable; and as the balance was delicate, and three hours were spent on the two determinations, which were repeated in each case three times, I think the results may be considered tolerably accurate.

"'It is gratifying to perceive that the difference in density between the coloured alcohols is so small, that when spread over the twenty-four thermometers (which may be supposed to contain an increasing dense spirit, in the order of their formation), it will be inappreciable.

"'Yours very sincerely,
"'(Signed) GEO. WILSON."

The mean height of the mercury in Delcros's barometer at temp. 32° Fahr. for seven months,† observed sixteen or seventeen times daily (the hours between 10 P.M. and 6 A.M. being omitted), was 29·046 inch. The lowest pressure recorded in the seven months occurred at 7 A.M. on the 25th of October, being 28·265 inches, and the highest at 8 P.M. in January, being 29·900 inch, which gives a range of 1·635 inch within little more than half a year. The last page of Table I., however, shows that the mean horary variation is very small, being only 0·006 for the same period. As during the very low winter temperatures of that locality the atmosphere

* *Note.*—Specific Gravity, at 60° Fahr., of alcohol employed in filling thermometers for Sir J. Richardson.

Uncoloured alcohol, rectified from fused carbonate of potass and unslaked quick lime	794·65
Same alcohol after being coloured with extract of cudbear (prepared by evaporating the tincture made with absolute alcohol), to dryness in a water bath, and leaving the extract over oil of vitriol *in vacuo* for two days	795·37
Residue of coloured alcohol after thermometers were filled	795·41

Feb. 28, 1848. (Signed) GEO. WILSON.

† In the first nine days of October the barometer was rarely examined, and less regularly during the remainder of that month than in the six following ones.

holds very little moisture in solution, the very small diurnal oscillation supports the opinion that it depends on the presence of vapour. The depression is greatest at night, and at noon, and in the afternoon; but the regular recurrence of two daily maxima and minima cannot be made out either in the individual months or in the aggregate of the seven months. The casual fluctuations arising from snow storms and other sudden changes in the constitution of the atmosphere appear to overlie and conceal the diurnal curves.

As there are no corresponding observations on the Arctic Sea for comparison, we can scarcely venture to assign the height of Fort Confidence from these observations. By employing Sir Edward Parry's observations at Winter Island, in latitude 66° 11′ N.* made in 1821-22, we may indeed get, as a very rough approximation, 640 feet for the altitude of Bear Lake above the sea. This is liable to the errors arising from the great distance between the places of observation, also to that from the annual fluctuations of pressure, and to the differences which most probably existed between the barometers, which were not compared with each other, nor with the same standard. From calculating the rate of descent of Bear Lake River, and of the Mackenzie below its influx, when compared with other rivers whose velocity and rate of descent were known, I had assigned 500 feet as the altitude of the lake above the sea; but this estimate is also liable to much error.

All the meteorological instruments were observed at the exact hours mean time at the station, kept by chronometers whose rate and errors were frequently ascertained by astronomical observations of the fixed stars. Göttingen time was used only on the term days, for observations on the magnets.

During the winter dense clouds or *cumuli* were never seen. The clouds generally were of the nature of thin *stratus* and *cirri*, or rarely *cirro-cumuli*, and the mean extent to which these overspread the blue sky is shown in Table VII. Very often the *stratus* was so rare that the stars shone through them previous to the rising of the moon, and their actual existence and extent became known only in the bright moonlight. It seems to be a cloud of this kind which forms the dark space near the horizon from behind which the arches of the Aurora Borealis are frequently observed to spring. On several occasions an arch of filmy cloud, of a greyish hue, was observed in the twilight crossing the magnetic meridian at or near a right angle. On watching this until daylight had wholly departed, it was seen to assume

* The mean height of the mercury in the barometer at Fort Confidence for seven months is 29·046 at 32° Fahr., the mean temperature of the air in the shade being −12·28 Fahr. At Winter Island the mean height of barometer for one year (1821-22) was 29·798 at 32° Fahr., and the mean annual temperature +9·8 Fahr.

gradually the yellow hue and brilliancy of the usual auroral arch. The clouds which accompany the most brilliant displays of the aurora are seldom so dense as to hide the larger stars, except when the moon is shining. Sometimes the stars shone through sheets of auroral light, at other times they were altogether obscured by it. I am inclined to believe that the appearance or non-appearance of the stars during displays of the aurora depends on the density of the accompanying stratus cloud. This cloud may be so rare as merely to communicate a greyish tinge to the apparently clear sky, and yet become sufficiently visible by the refraction of the moon's rays to show its true nature and extent.

Several times during the winter the auroral light was seen, both by myself and Mr. Rae, to pass in front of a mass of cloud. As we were both aware of the ease with which the eye may be deceived in such observations, we watched the displays of the phenomenon with sufficient scepticism to keep the attention on the alert, and no doubt remained on our minds of the reality of the fact. In former years I had seen seen similar occurrences more frequently, and even more manifestly.* Thirty years previously I had entertained the belief that the aurora was connected with the formation of cloud, and other changes in the constitution of the atmosphere, and the nightly observations of this winter all tended to strengthen that opinion. The great dryness of the winter atmosphere in the interior of Arctic North America may, perhaps, be the cause of the more frequent emissions of the electric light than in more southern and moister localities. Fine spiculæ of ice or minute snow were often seen falling from a clear sky, especially after a brilliant display of auroral lights. This I had also noticed many years ago.

I have written out *in extenso* the descriptions of the aurora at the hours of observation for two months. To have done so for the whole seven months would have occupied too much space. A few brief notices are substituted for the months in which the full details have been omitted. The compass bearings hereafter mentioned are true, not magnetic, unless when so expressed.

October (1848).—Aurora observed on the 1th, 13th, 17th, 18th, 19th, 20th, 21st, 23rd, 24th, 25th, 26th, 27th, 28th, and 30th. Did not occur or was not noted on the other evenings. On the 19th, after the sky had been overspread for the first half of the night by a very thin stratus, scarcely obscuring the blue vault, and from which fine icy spiculæ fell, between three and four in the morning there was a bright blue sky with flocculent clouds,

* These appearances are, however, to be understood as very rare in comparison with the common phenomenon of the auroral light issuing from behind a cloud.

which occasionally became luminous, sometimes in one quarter of the sky, sometimes in another. At 7½ P.M. in the evening of the 23rd the Declinometer was observed to move suddenly 10′, simultaneously with some quick flashes of auroral light. The aurora disappeared in a few minutes, and the needle remained stationary afterwards. On the 29th the Declinometer fluctuated upwards of 1°, the sky being wholly obscured during the whole day, without any auroral light shining through. A small snow fell in the evening. On the 31st the Declinometer ranged 2°. During the previous night there was a deposit of moisture from the atmosphere, and all the instruments in the observatory were found to be encrusted with fine crystals of ice, particularly the rough lines and lettering of the scales. The auroral arch in the evening crossed the magnetic meridian at right angles, and the light as it flashed over the stars was bright enough to dim their lustre, but not to hide the larger ones.

November 1.—Fine snow falling for seven or eight hours in the day. At 6 P.M. all the northern and part of the western horizon banked by luminous clouds, through which stars of the first magnitude shone. A few patches of light in the south also. Fine snow falling from a cloudy zenith. At 7 P.M. a curtain-like arch of the aurora, bearing north about 25° degrees high, partially in motion. A faint sheet of light spread over the rest of the sky, here and there obscured by cloud-like dark patches. Stars visible through the aurora in every part, except in the northern arch which hid them. At 8 an arch about 80° high, on the south side of the zenith. At 9 the arch in same position but fainter.

November 2.—At 6 A.M. an auroral arch, rather faint, crossing the zenith in a due east and west direction. At 9 P.M. an auroral cloud bearing south, emitting the usual yellow light. Elsewhere an uniform haze or cloud overspread the sky. At midnight a remarkably deep blue cloudless sky, with bright stars, an auroral arch rising in the N.N.W. point of the horizon, crossing the zenith, including the whole constellation of the Little Bear, and passing over Orion in the east. The arch, which often changed its form, occupied a considerable breadth of sky, and was generally made up of oval oblique bars. The moon had set before this hour.

November 3.—At 9 P.M. sky cloudless. Bright moonlight. Stars somewhat dim. An auroral arch, composed of parallel beams of light rising in the N.N.W. to the height of the Great Bear only. Auroral clouds at a greater altitude bearing north-east. At 8 P.M. several arches of light rising from one point near the N.N.E. horizon, and crossing the sky at various altitudes, so as to occupy most of the northern half of the heavens, and part also of the southern half. They became fainter after passing the zenith, and were lost in a diffused sheet of light which spread over the eastern part of the sky. The principal arch, which was brighter and more continuous than the others, passed over the tail of the Great Bear, covered the whole of the Little Bear, and as it descended in the east made a curve to the north. At 9 the aurora faint.

November 4.—At 6 P.M. aurora hanging like a curtain in the northern sky, at an altitude of about 15°, and spreading from N.N.W. to N.E. Beams of light shooting upwards from the curtain, the largest ones bearing N.N.E., but some also bearing north. The sky cloudless, and still tinged red in the west, though the sun had been set three hours. At 8 P.M. long variable streams of light rising from the N.N.W. and north to the zenith. At 9 a narrow auroral arch, extending from N.N.W. to S.S.E., its crown having an altitude of about 80°.

November 5.—At 8 P.M. oblique bars of auroral light, lying over each other, and rising from the horizon in the N.W. by N., and then dividing into several arches, one of which, crossing the Great Bear, kept to the north of the zenith. Another crossing Cassiopeia and passing to the south of the zenith, and others, which were brighter, taking a still more southerly course, but disappearing before they reached the south-west horizon. Bright moonlight in a cloudless sky. Stars not shining brightly. At 9 aurora fainter, and the arches lying for the most part to the south of the zenith.

November 6.—At 5 A.M. sky clear, with bright stars, except a dark space skirting the southern horizon, and looking like a heavy cloud, but one bright star shining through it. Along the upper border of this dark space the auroral light had the form of a series of cumulo-stratus, above which there was a light blue sky. At 11 A.M. some light cirri clouds in the north and west resembling some forms of the auroral light.

November 7.—At 1 P.M. a halo round the sun, with red rays reflected from a cloud on each side, forming small segments of an arc.

November 9.—Faint arch of the aurora in the west at 5 A.M.

November 10.—At 7 A.M. the moon, when near the horizon, had a very oval shape, the long axis being transverse. At 6 P.M. the aurora in cirrus-like streaks, 30° high, bearing south. At 6½ P.M. a falling star passed from east to west, close by Lyra. A low auroral arch terminating abruptly, as if rolled back on itself on reaching its greatest altitude in the magnetic meridian, but becoming some time afterwards more lengthened out, and at the same time fainter. The ice in the lake, which was frozen over everywhere within six or seven miles, making a rumbling noise. At 7 P.M. the aurora was in active motion. It generally formed a complete arch, along which waves of light moved rapidly, and most so about 15° above the southern horizon. The beams lay at right angles across the magnetic meridian, but their wave-like line of motion was in that meridian. The compass needles steady; sky cloudless. At 7¼ P.M. a slender auroral arch, waving to and fro, extended from the N.W. to S.E., passing across the zenith. The magnet, suspended in the Unifilar Magnetometer and loaded with the large ring (No. 5·), which was previously steady, began at this time to vibrate from 315 to 380, and the Declinometer moved from 4° 35′ to 4° 25′. At 7.50 the aurora wholly gone. The unifilar magnet was now vibrating from 270 to 340, and the Declinometer had gone back to 4° 35′. Sky clear. At 8 no aurora. Sky cloudless. Declinometer 4° 30′. At 9 P.M. a faint auroral arch lying to the south of the zenith. The Declinometer vibrating from 4° 30′ to 4° 35′. At 10^h 45^m long banks of auroral light in the south resembling stratus cloud, the uppermost of them arched, assuming at times a brighter hue, but always yellowish; the end of the arch curling back like cirro-stratus, but in a contrary direction to the light wind then blowing.

November 11.—At 5 A.M. a narrow auroral arch in the west, 30° high. At 6 P.M. the moon in the north-east quarter of the heavens grazing the upper edge of a cloud-bank, which produced a burr round her. The bank sunk below the horizon on the north point. Auroral light in detached masses and beams, the latter in form of arcs, which cross the magnetic meridian in various directions. These cloud-like masses resembled thin clouds illuminated by the moon, but were distinguishable by their variableness both in form and in the intensity of their yellowish light. At 7 P.M. the sky generally overspread by a rare stratus cloud, most visible in parts directly opposite the moon, permitting the blue sky to be seen through it elsewhere. Auroral light in arcs and streaks. The moon surrounded by an imperfect halo, 22° in semi-diameter, produced by somewhat oblique stratus clouds. The paraselenæ yielded prismatic tints. At 7¼ P.M. the aurora suddenly became active and variable, the great body of light being in the southern half of the sky. Of the prismatic tints exhibited yellow was the predominating colour, but green was occasionally seen, and the lower ends of the fringes when most vivid were crimson. In its motions the auroral light resembled the folds of a curtain made to wave to and fro; that is, the prismatic tints became visible and disappeared again in rapid succession along an arc or bar of light, first in one direction, then back in the opposite one.

This kind of motion has been denominated by some writers pulsation. After a continuance of this phenomenon in a variety of forms and places, the southern ends of the arches and of the banks of light lying to the eastward began to twist and curl on themselves, and to sway backwards and forwards before the stratus cloud, which they concealed in their passage. The cloud was strongly illuminated by the moonlight, and would have been seen had the auroral light been beyond it. In the course of the rapid evolutions of the lights, large sheets of it seemed several times to pass before the cloud, entirely concealing it, and consequently appearing to the eye to be much nearer. The needle of the Declinometer was steady at 7 P.M., but when the aurora began in the eastern part of the sky to exhibit prismatic light, it vibrated from 4° 32′ to 4° 25′, and at 7^h 45^m it had settled quietly at 4° 49′, the aurora at that hour having become comparatively inert. At this period a streak of auroral light crossed the stratus clouds under the moon, traversing the blue spaces between the clouds, and forming a continuous line in front of them, well defined on their surface. The upper bar of the lunar halo had disappeared by this time; the space within the limb of the halo, which was now three quarters of a circle, being mostly blue sky. The oval paraselenæ still gave out prismatic tints. At 8 P.M. the moon, having risen into a blue space in the heavens, the stratus cloud was less visible. The prismatic paraselenæ now emitted rays of light outwards, and a beam of auroral light stretching towards the north, and, passing near the zenith, cut off a portion of the circumference of the halo. At this time the Declinometer was vibrating slowly from 5^h 10^m to 5^h 15^m, the numbers increasing slowly. At 8½ P.M. fine snow or minute spiculæ of ice falling, occasioned a haze sufficiently dense to conceal the blue sky, but not to prevent the stars of the first magnitude from appearing. The lunar halo was this time complete, the paraselenæ distinct, and an arc showing above the halo at the distance of a quarter of its diameter. Aurora in arcs faintly seen through the mist. At 10 P.M. hazy, circle of the lunar halo very distinct, but the paraselenæ scarcely to be made out. Stars invisible, and no auroral lights.

November 12.—Hazy. At 4 P.M. the Declinometer vibrating from 4° 15′ to 4° 20′.

November 13.—Hazy.

November 14.—At 4 A.M. auroral clouds, emitting yellow light near the zenith. At 5 P.M. clear blue sky; a complete auroral arch from N.W. by N. to S.E. by S., crossing the magnetic meridian in the zenith at right angles; the arch composed of oblique yellowish beams, often moving and changing. At 6 P.M. the auroral arch occupying the same general position, but waving backwards and forwards; a few scattered masses of light in other parts of the sky. Needle steady. At 7 clear blue sky. No auroral light. At 8 a north-east wind setting in, the mercury in the barometer fell suddenly to a small extent. At 9 small, round, fleecy clouds, not dense, covering most of the sky, with blue intervals. No aurora.

November 15.—At 5 A.M. light N.E. winds, clouds coming from the N.W. At 1.40 P.M. the sun set in a halo. At 5 P.M. a broad auroral arch rising at its summit, about 18° or 20° above the southern horizon, vivid on its lower border, with quick motion backwards and forwards along the line of the arch. The upper border not defined, but fading gradually away. At 6 P.M. auroral arch in the same position, but not so bright. At 7 P.M. an auroral arch springing from the S.S.E. horizon and crossing to the N.N.W., occupying in the middle a space extending from near the zenith to within 16° of the S.S.W. horizon, but tapering towards the ends. The arch was composed of brighter streams of light, lying in the direction of its length and connected by fainter diffused lights. No auroral light was emitted from any part of the sky north of the zenith. At 8 P.M. only a few patches of auroral light remained; no clouds were visible, but the sky, generally, was greyish-blue. At 9 P.M. masses of auroral light shone dimly through the haze in the southern quarter of the sky. A very fine and slight deposition of snow, more readily felt than seen, was taking place at this hour.

November 16.—Air to-day inclined to part with moisture, evinced by the parchment windows becoming slack.

November 17.—This day the magnetic needles moved much. At 11 A.M. the magnet, suspended in the Unifilar Magnetometer, was vibrating between 520 and 540, and, subsequently, beyond the scale; and at noon the Declinometer moved suddenly from 1° 21′ to 0° 29′. At 6 P.M. clear blue sky, with some stratus cloud near the horizon, above which there was a bank of luminous clouds, having a slightly reddish tint, resembling clouds tinged with the rays of the setting sun. Shortly afterwards the red tints became more vivid, and the quick east-and-west to-and-fro movement of vertical bars was exhibited. At 7 auroral light spreading from south to west. A falling star shot from east to west past Altair, having an apparent angle of descent of 40°. At 8 the Declinometer needle was 5° 5′, but ten minutes afterwards returned to 4° 45′, the auroral light having then disappeared. At 9 P.M. a dark cloud, concealing the stars in the southern quarter of the sky, to the height of 8°. Along the arched edge of this bank a yellowish light was emitted. Unifilar magnet moved back to 95. At 10 P.M. the auroral light diverged from a point in the sky, adjoining Cassiopeia, to all parts of the horizon. The beams of light varied and moved rapidly. Soon afterwards the light had disappeared from the southern sky, and the auroral light was mostly in the north-west quarter. At the hour the Declinometer stood at 1° 50′, but moved quickly to 5°, and at $10^h \cdot 15^m$ the northern rays of the aurora had vanished, and then the needle had moved to 6° 28′.

November 18.—At 5 A.M. an auroral streak crossing the zenith from east to west.

November 19.—Snow. Sun very dimly seen at 1.

November 20.—At 6 P.M. an arch of the aurora, much like the *via lactea,* crossing the zenith from N.E. to S.W. At 7 auroral arches, having the above direction. At 8, and subsequently, no aurora.

November 21.—A burr round the crescent moon at 5 and 6 A.M. At the latter hour faint auroral light in the zenith. At 6 P.M. a bank of clouds along the southern horizon, emitting a white light. At 8 a broad low bank of yellowish light extending along the southern sky, and indented by dark clouds. Higher up several auroral arches crossed the blue sky, barred at their origin in the south-west by stratus clouds, and seated therefore beyond them. These arches did not go much beyond the zenith, but curved there in various directions. Stars pretty bright. The Declinometer moved 30′ after the aurora shone out.

November 24.—At 6 and 7 P.M. faint auroral light near the zenith. At 11 auroral clouds. At midnight a patch of an arc to S.E. by E., and another bearing S. by W., about 8° high.

November 25.—At 2 A.M. patches of auroral light in many parts of the sky. At 4 many auroral arcs. At 5 and 6 A.M. patches and beams of auroral light. At 4, 6, 7, and 8 P.M. auroral light in various forms, banks, beams, and arcs, mostly of a yellowish hue. At 9 P.M. a more than usually fine auroral display. A great curtain extended from the east to the north-west quarters of the sky, at an altitude of about 60°, appearing as if suspended from a deep blue starry sky. This luminous curtain waved up and down, narrowed and expanded, and rolled back on itself at the ends. In the south part of the sky there were clouds, from behind which flashes of light were occasionally seen to shoot.

November 26.—At 9 A.M. a mackerel sky, that is, short cirro-stratus lying across a line running north and south, and a long tract of cloud stretching from the N.W. in a S.E. direction, commencing about 12° from the horizon, and rising to about 70°; very thin and delicate, so as to be almost transparent, but appearing to the eye to lie under the mackerel sky which it crossed. Part of this cloud had a wavy and flickering motion like the ordinary auroral light, and in a few minutes it faded entirely away like the aurora. It re-appeared again more to the south somewhat altered in form, and in a minute or two vanished again. The motions of this stratus identify it with the auroral light, but had it been stationary it could not have been distinguished from a filmy cloud. I have no doubt but this variable cloud would have been luminous in the absence of daylight. At 4 P.M. an auroral arc, having a direction from N.W. to S.E., composed of detached and somewhat oblique and twisted bars. This arc occupied the site of the aurora-like cloud seen in the morning. Elsewhere a clear blue starry sky. At 5 P.M. much of the sky occupied by patches and banks of light. Several nearly contiguous arches crossed the zenith in a N.W. and S.E. direction, their ends uniting into single twisted stems as they approached the horizon. At 6 P.M. the whole sky nearly covered with auroral lights in different shapes. At 7 P.M. an arch crossing the zenith in the ordinary N.W. and S.E. directions. Patches of light elsewhere; all more or less changeable. At 9 P.M. several concentric arches, covering all the southern half of the sky from the zenith downwards. Brilliant fringes of light rising obliquely from the upper borders of the arches in continual motion; the lower edges of the arches were of more continuous light. At 10 P.M. a continuous sheet of light spread over all the southern half of the sky, but was traversed by brighter arches; a space near the horizon was the only dark part.

November 27.—At 6 A.M. auroral light in various quarters of the sky. At 7 A.M. banks of auroral light bearing south. Dawn just appearing in the east.

November 28.—At 4 P.M. an auroral arch crossed the zenith. Sky greyish; a few stars visible. At 9 a sheet of light shining faintly through clouds in the southern quarter of the sky.

November 29.—At 4 A.M. faint sheets of light in the S.W. At 5 and 6 A.M. auroral light as before, and at 7 auroral streaks still visible though the day was breaking. At 10 A.M. the suspended magnet moved in the course of two or three minutes from 330 to 350, with quick minor vibrations. At noon the magnet was vibrating in arcs of 12′, and at 4 P.M. in arcs of 10′. At 5 P.M. beams of aurora rising in the north and tending to the east, the greatest altitude 15°. At 6 P.M. a bright curtain-formed arch rising to 50° in the north, extending from N.W. to N.E. with oblique fringes of light rising from its upper edge, and inclining to the eastward. At 7 P.M. two arches rising in the N.W., and crossing the magnetic meridian; their ends on attaining the zenith curling back. At 8 P.M. an auroral arch; and at 9 P.M. an arc, not very bright, rising in the N.W., and holding a flexuouse course to Cassiopeia, where it terminated. Several rays diverging from its end there towards the south, north, east, and west.

November 30.—At 5 A.M. faint auroral beams. At 8 dawn. Suspended magnet vibrated all day in arcs varying from 5′ to 15′ in extent; and had a progressive but not uniform motion from 309′ to 254′. At 5 P.M. a broad auroral arch crossing the sky in a N.E. and S.E. direction, passing over the zenith. At 6 P.M. auroral light in patches. At 7 P.M. the arch interrupted in places, having a direction from N.W. to S.E., and touching both horizons. It was composed of detached bars and masses of light, not uniform in direction, but mostly crossing the general line of the arch obliquely. Clear blue starry sky. At 9 P.M. a faint arch having a direction from N.W. to S., and reaching both horizons; its greatest altitude about 70°. A dark cloud ranging along the southern horizon, and emitting pale light from its upper edge.

On the 7th the Declinometer fluctuated 1°, and the aurora was active in the evening. On the 8th the fluctuation was even greater. On the 17th the Declinometer ranged from 1° 20′ to 6°, its motions being unusually great. On the succeeding day it fluctuated about 3°.

December 1.—At 7 A.M. no aurora. At 9 A.M. was able to write comfortably by daylight near the window. Sun hidden at noon by Fishery Island, but visible from a gentle eminence behind the house. Aurora invisible till 8 P.M., when an arch of yellowish lights about 16° high stretched from N.W. to N.E. The arch on a north bearing was a broad sheet of light, but near the N.W. horizon it was a twisted stem. Several broad pale sheets of white light, like the Milky Way, in the northern and eastern quarters of the heavens. At 9 P.M. the sky clear and starry, several arches of light springing from the N.W. horizon, and passing through the northern half of the heavens to the E.S.E. or S.E. by E. point of the horizon. The uppermost crossed the constellation of the Great Bear, passed a little south of Cassiopeia, and faded away in the S.S.E. near the horizon. Elsewhere some streams and banks of yellowish light existed. These arches vanished, and re-appeared at short intervals, and also moved from their sites, but had little internal motion.

December 2.—At 4 A.M. faint auroral light in the north and west. At 5 an auroral arch bearing south, at an altitude of 14° and extending for 160°. At 7 A.M. considerable deposition of rime on the thermometer scales. Arch of light bearing north, 16° high; also beams of light near the zenith. Dawn of day at 7, being an hour earlier than on the 30th November. Open water in the lake producing mist. At 6 P.M. auroral light near the northern horizon. At 8 and 9 P.M. faint auroral light, ditto.

December 3.—Mercury froze solidly this day. At 5 P.M. rays of auroral light in the north 10° high. At 7 P.M. an arc of the aurora in the east, and also one in the north-west, the middle part of the arch being deficient. Deep blue sky at 8, little activity in the auroral lights.

December 4.—At 4½ A.M. bright auroral arch, with patches of light in the W. At 6 A.M. patches of yellow light near the zenith, and also in the N.W. At 7 fragments of an arch shooting up from the N.W. horizon to near the zenith, and having a direction at right angles to the magnetic meridian. [At 6½ P.M. Dr. Rae, being then in latitude 67° 12′ N., longitude 118° 16′ 24″ W., saw a falling star descending vertically on a nearly due north bearing, and passing a few degrees to the eastward of the pointers of the Great Bear.]

December 5.—[At 6¼ A.M., in latitude 67° 7½′ N., longitude 117° 58′, a falling star was observed by Mr. Rae, about 10° from the horizon, a little to the westward of north, travelling horizontally towards the east.] At 5^h 40^m A.M. an auroral arc, directed towards the east, rose from the west as high as the zenith. At 6 A.M. a bright beam of yellowish light rose from the N.E. horizon, to the height of 20°. At 7 first appearance of dawn. Mercury crystallizing in the open air in the middle of the day. At 5 P.M. faint beams of light rising from the N.W. to the height of 18°, vanishing and re-appearing rapidly. At 6 P.M. a curve of auroral light, rising abruptly and interruptedly by steps in the N.W., continued to the S.E. in a parabolic curve, but formed throughout of slender vertical rays in motion. At 7 P.M. a faint belt of light crossed the zenith at right angles to the magnetic meridian. At 8 P.M. an arch crossed the zenith from N.W. to S.E., being nearly at right angles to the magnetic meridian. A few minutes before 9 P.M. a rather brilliant aurora. The light rose in the N.E. in successive steps, like the folds of a curtain hanging obliquely, and then dividing into three streams, held on across the zenith, and making a bold convex bend to the S., the ends curved to the N.N.W., but did not approach within 30° of the zenith. The light had a yellowish colour, and a quick lateral pulsation of the fine vertical rays of which the streams were composed.

December 6.—At 6 A.M. a broad beam of light rising from the N.W. horizon to 30°, and in the S. a dull bank of light occupying the space between the horizon and 6° altitude. At 9 A.M. was able to write near the window by daylight, and at 2^h 12^m P.M. was unable to do so distinctly,—the window looking S.S.E.

December 7.—Great refraction this morning. At 7 P.M. a slight burr round the moon. Faint streaks and bands of auroral light near the zenith, the masses mostly lying across the magnetic meridian. At 8 P.M. no aurora. At 9 bright moonlight. Blue sky, with stratus cloud near the southern horizon only, an auroral arch springing from the N.W. and crossing the magnetic meridian at right angles. It frequently changed place, being sometimes in the zenith and at other times more to the southward.

December 8.—Considerable refraction; distant land much raised. At 5 P.M. a stream of auroral light rising from the N.W., crossing the zenith at right angles to the magnetic meridian, but not going onwards to the S.E. Soon afterwards this stream moved to the southward and vanished. It had an internal waving motion. At 6 P.M. a broad arch of yellowish light, extending from N.W. by W. to S. by E., and having an altitude of 20° at its crown, rose from the N.W. horizon, and without anywhere exceeding an altitude of 20°, bent round to the S. in a flexuose band, with obtuse projections to the E.S.E. It exhibited rapid changes of form, during which the suspended magnet vibrated 30′, and the Declinometer was also in motion. At 8 P.M. a broad sheet of light, including two brighter arcs, now occupied the place of the above-mentioned band, but did not extend farther to the eastward than a south bearing. At 9 only a small part of an auroral arch remained, including merely one point of the compass, and bearing S.W. and S.W. by W.

December 9.—At 7 this morning, on approaching my hand to the iron latch of my bedroom door, a spark was emitted. I was then dressed merely in my night dress, with flannel drawers. At 9.10 A.M. able to read minion type of a bible by daylight. At 4 P.M. a series of bars of light, rising obliquely in the S.E. by E., and a similar step-like succession of bars in the N.W.; there was no continuous arch across the zenith connecting these two groups of bars, but in place of it a very narrow streak of light curved boldly and convexly to the north in the zenith, and a mass of yellowish light lay more to the south. At 5, 6, 7, 8, and 9 P.M. no auroral light. Moon very clear and bright.

December 10.—At 6 A.M. bright moonlight; no aurora. At 7 A.M. some thin sheets of light distributed irregularly, several of them in the S., S.E., and S.W., having a convergence towards the zenith. At this instant the suspended magnet was observed to be moving from 390′ to 405′, and, after vibrating somewhat irregularly in a mean arc of 10°, to settle for a time at 420°. The Declinometer was then 3° 17′. After recording this observation in the bedroom, and returning to the open air, the aurora had ceased to be visible, and the suspended magnet was found at 370′. A burr at this time round the moon. At 8 A.M. faint streaks of light near the zenith, stretching to S.S.W., or nearly in the magnetic meridian; these streaks vanished and reappeared with rapidity. Burr round the moon. At 9 and 10 mist near the horizon. At 11 and noon the sun below horizon, but beams of light shooting up from it into the sky. Full moon at 3h 48m this morning, Fort Confidence time. At 6 P.M. an auroral arch from N.W. by W. stretching across the zenith, and disappearing on a S.E. bearing. A burr round the moon, but the sky elsewhere cloudless blue. 7, 8, and 9 P.M. no aurora.

December 11.—At 5 and 6 A.M. no aurora. At 7 a faint burr round the moon; and at 8 and 9 paraselenæ. At 10 great refraction. At 11 redness in the sky above the sun's place, bright and circumscribed, the sun itself hid by Fishery Island. At noon a parahelion seen to the east, where the island is lower; the sun itself invisible. At 1 the same appearance, but less distinct. (At 6.15 P.M., in latitude 67° 6′, longitude 118° 22′, Mr. Rae saw a bright falling star in the west, making in its descent an angle of 45° with the horizon. It vanished when about 14° high. At 6h 40m he observed another star falling from near the zenith towards the west, and passing to the south of Lyra. At 7h 10m he saw a falling star in the same quarter of the sky as the one he noticed at 6h 15m, and taking the same direction.) At 8 and 9 P.M., at Fort Confidence, an arch of clouds in the S.W., brightly illuminated by the moon, and not to be distinguished from some exhibitions of aurora in the absence of that luminary.

December 12.—[At 2h 32m A.M., in latitude 67° 6½′ N., longitude 118° 22′ W., Mr. Rae saw a falling star descending almost vertically, or slightly inclined northwards.] At 6 A.M., at Fort Confidence, the sky almost wholly overspread by a filmy stratus, which was rendered visible by the bright moonlight. Stars of the first magnitude visible through it. At 7 A.M. a general mistiness, with a deposit of fine snow. A dim lunar halo, with a semidiameter of 22°. Snow occasionally in the day. At 4 P.M. a broad yellowish auroral arch rising from the S.E. horizon, and passing south of the zenith in a S.W. direction, but terminating in a luminous cloud, at an altitude of 60°. At 5 P.M. two parallel arches of light rising in the S.E. and proceeding to the N.W., occupying a middle height between the southern horizon and the zenith. Considerable motion, resembling that which would be caused by a dark bar carried with extreme rapidity towards the west in front of the light. At 6 P.M. the arches of the aurora rather lower and not in motion. Their crowns are in the magnetic meridian. At 7. P.M. an arch of the aurora bearing south, and reaching from the S.E. to N.W. its crown about 20° high. It was rather broad, yellowish, and nearly motionless. At 8 P.M. two broad and fainter arches, partly blended into each other in the south, about 12° high; also some masses of light near the zenith. At 9 P.M. the southern arch now reached from N.W. only to about S., where it terminated at the height of 25°. There was no auroral light in the S.E.; but five or six arches passed from N.E. to N.W., the uppermost of them crossing the zenith, and the lowest one running near the horizon.

December 13.—At 7 A.M, early dawn. No aurora until 9 P.M., when a broad arch of yellowish light in oblique bars extended between the N.W. and S.E. horizons, passing about 30° to the north of the zenith.

December 14.—No aurora in the morning. Some fine snow deposited about noon. At 6 P.M. sky greyish, but no visible clouds; a few stars shining out. A faint but broad arch extending from N.W. to S.E., appearing and disappearing in rapid succession. At 3 P.M. a belt of pale light about 10° broad, extending from N.W. to S.E. horizons, and crossing the zenith. Sky clearer and bluer, with more stars, but a fine snow continuing to fall;—the stars shining through the auroral light. At 9 minute snow. Sky not quite so clear. Arches of light bearing south, and some masses scattered over the sky.

December 15.—About a quarter of an inch of fine snow fell in the night. At 7, 8, and 9 A.M. lunar halos. At 10 A.M. there was a light air from the W.S.W. at the height of twenty feet, and one from the N.N.E. nearer the ground, as shown by a zig-zag column of smoke from our chimney. At 6 P.M. *O Sextantis* occulted by the moon. No aurora this evening. Deep blue sky, with many stars. Fine spiculæ of snow falling thickly.

December 16th.—At 5 P.M. a sheet of pale light like the Milky Way, overspreading the southern half of the sky, with dark, narrow, oblique bars crossing it. The rapid shifting of these dark bars across the light showed it to be the aurora, otherwise it might have been thought to be twilight lingering in the sky. At 6 P.M. faint streaks of aurora rising from the N.W. horizon to past the zenith in a S.E. direction, but ending short of the Pleiades. In a few minutes this stream changed into several fainter rivulets, having the same direction, and occupying greater breadth in all. Sky dark blue, and starry. At 10 P.M. a broad luminous arch in the south.

December 17.—At 1 P.M. temperature of the atmosphere, −61°. Long prismatic crystals were formed in nitric acid, having the strength recommended in the London pharmacopeia, and at 3 P.M., when the temperature had fallen to −63·8° Fahrenheit, almost the whole of the acid in the vial (2 oz.) was frozen. Sulphuric acid had frozen solidly long before. Mercury at this time could be cut with a knife more easily and more smoothly than lead. At 7 P.M. two bright auroral arches to the southward, having a curtain-form, and a rapid to-and-fro bar-like movement; the highest was 20° from horizon. At 8 only a few patches of dull light in the south. At 9 two bright arches in the south; very changeable. They reached from W. to S.S.E., but did not in general rise above 20°. Sometimes they appeared as if twisted and bent or broken, occasionally sending shoots down towards the horizon, and exhibiting in their upper borders the quick bar-like motion, with fringes shooting upwards ometimes to the extent of 15° or 20°, or nearly half-way to the zenith. The Declinometer varied 35′ between 8 and 9. The temperature of the atmosphere was now −61° Fahrenheit; the nitric and sulphuric acids, and of course mercury, remained solidly frozen. Muriatic or hydrochloric acid was perfectly fluid.

December 18.—At 5 A.M. beams of aurora in the west. The temperature of the air at 9 A.M. was −63·9°, being the lowest observed in the winter.* Sulphuric acid had an opake white colour. At 6 P.M. a slender auroral arch from N.W. to S.E. passed across the zenith. At 7 and 8 no aurora. At 9 faint beams of light shooting towards the west from near Cassiopeia. The mean temperature for forty-eight hours was −61° Fah., or 93° below the freezing point of water. We had travelling parties out at this time.

December 19.—At 7 A.M. two beams of light rising in the west, one of them taking a course to the S.E., the other diverging from it to the E., or E. by N. They did not reach the meridian, but approached it. At noon the suspended magnet was vibrating irregularly. At 7 P.M. two broad arches crossed the zenith from N.W. to S.E. Sky generally bluish-grey. Abundance of stars overhead; none within 20° of horizon. No other sign of clouds. At 8 P.M. five broad streams of light rising in the E.S.E., and diverging in their ascent so as to spread over most of the sky. The light more dilute towards the edges of the streams, which in some points touched each other. The central streams crossed the zenith. The arches were rapid in their changes of form and extent.

* A spirit thermometer by Nosotti, constructed probably in the ordinary way alluded to by Professor Forbes at page 40, stood at −81·5°, having sunk nearly to the bulb.

December 20.—At 10 A.M. could not write by daylight at this hour. Sulphuric acid freezing partially at a temperature of −8° Fahrenheit. Nitric acid limpid. At 6 P.M. faint auroral light in the north. At 7 no aurora. At 8 P.M. faint arches rising N. by W., and extending towards the south. At 9 two faint arches of light in the south, having an altitude of 14°, and 6° at their crowns. At 10 the arches had an altitude of from 40° to 45°. At midnight patches of auroral light scattered over the sky.

December 21.—At 1 A.M. stars shining very brightly. A brilliant aurora in rapid motion, and momentarily changing its form. Its lower edge had a fine lake colour, and it was brightest on the S.W. and the W. bearings. At 2 A.M. a very bright display of auroral light. Declinometer and Dipping Needle vibrating much. At 4 A.M. faint beams of aurora in the north. At 6 A.M. curtain-shaped aurora extending north and south, with active motion. At 7 A.M. no aurora. Light variable winds. Temperature, 47·3°. At 7 P.M. a faint auroral arch rising from the N.W. to past the zenith. At 8 P.M. an auroral arch rising from the S.E. for 35°; some short beams in the N.W. At 9 P.M. a beam of light in the north.

December 22.—No aurora observed in the morning. At 5 P.M. beams of light on a S.E. bearing, and some also bearing N.W. At 6 P.M. beams in the same quarters more faint. At 7 P.M. faint auroral light bearing north. At 8 an arch of faint light crossing the zenith from the N.N.W. to the S.S.E. At 9 P.M. an arch 10° high in the S.W. extending from S.S.E. to N.N.W.

December 23.—At 5 A.M. a broad arch of light standing from W.N.W. to E.N.E. and crossing the zenith. At 6 A.M. faint rays of light in the N.W. and also to the east. At 7 A.M. an arch of light 6° high bearing S.S.W. Was able to write by daylight, when close to the window, this day for 4½ hours, viz., from 10 A.M. to 2½ P.M. At 5 P.M. auroral light rising vertically from the north horizon. At 6 P.M. a broad, irregular, and broken arch of light having a direction from N.N.W. to S.S.E., and passing south of the zenith. Its greatest altitude, 45°. At 7 P.M. a mass of light in the S.S.E. about 6° high. At 8 P.M. beams of the aurora in the south and S.W. At 9 P.M. auroral light bearing S.S.W.

December 24.—At 6 A.M. rays of light rising vertically from the eastern horizon. At 7 A.M. a faint auroral arch, its crown bearing north, and having an altitude of 12°. Another arch bearing S.W. with an altitude of 7°. At 10 A.M. stratus cloud along the E.S.E.; horizon beautifully tinged red by the sun's light. Intervals of mountain green sky in that quarter. Rest of the heavens greyish blue. At noon the sky very bright on the southern meridian for some distance above the horizon. The southern sky retained the red tints of a rising and setting sun from 10 A.M. till 2 P.M. At 6 P.M. an arch of light formed of oblique rays crossed the zenith and reached the N.W. and S.E. horizons. At 7 P.M. a similar arch in the same portion, with its tranverse bars in motion. Masses of light near the horizon all round the sky. At 8 P.M. curtain-shaped, interrupted arcs of light directed across the magnetic meridian at right angles. One of them lay a little to the south of the zenith; the others were situated a little more to the northward. The arches were separated from each other, and also interrupted in the direction of their lengths, by vertical dark spaces, which were continually changing their places and dimensions, but did not exhibit the rapid to-and-fro bar-like motion so conspicuous on other occasions. Stars moderately bright. New moon. At 9 P.M. there existed five handsome curtain-formed arches more or less twisted and uneven. One crossed the zenith; the rest were more to the southward. They occupied the whole southern half of the sky, and were directed at right angles across the magnetic meridian. The brightness of the arches varied continually, and they were occasionally connected by beams of light shooting between the contiguous arches.

December 25.—At 5 A.M. snow drift. Masses and beams of light bore E. and N.W. At 6 A.M. an auroral arch whose crown, 6° high, bore S.W. At 7 beams of aurora in the N.E. having a direction to the S.W. At 8 A.M. dawn. At 5 P.M. a bank of auroral light extending near the horizon from the N.W. by N. point of the compass round to N. and onwards to E. by N. Numerous beams shot up from it to the height of from 8° to 12° or 14°. At 6 P.M. rounded and oblong patches of auroral light near the zenith, and also in other quarters of the sky, particularly in the north. A bank of light lying along the southern horizon. At 7 P.M. faint patches of auroral light. At 8 P.M. a horizontal band of light at the height of 30° in the north. A stream of brighter light rising in the north joined the west end of the band. Faint patches of light existed elsewhere. At 9 A.M. irregular masses of aurora-like columns of mist resembling smoke in various parts of the sky; the most conspicuous are rising from the N.W. by N. points of the horizon.

December 26.—At 4 A.M. a fine auroral arch having an extent of 80° and rising in the south to an altitude of 20°; also masses of light in the east and vertical beams in the N.N.W. At 5 A.M. faint rays having a N.N.W. and S.S.E. direction. Their changes of position were rapid. At 7 A.M. aurora in masses and beams in the S.W.

quarter of the sky. Faint appearance of dawn in the east. At 7 P.M. two faint arches crossing the zenith and having a direction of from S.E. to N.W.

December 27.—At 2½ A.M. a patch of light in the N.W. at an altitude of 45°. At 7 a narrow arch crossing the zenith from the eastern to the western horizon. At 4 P.M. temperature of the air, —43·9° Fahrenheit. Nitric acid crystallized in beautiful clear crystals. At 6 P.M. a broad, yellowish, quiescent arch, extending from the N. by W. horizon to the N.E. one. Its summit not rising more than 9°. Many rounded cloud-like patches of polar light between Cassiopeia and the Great Bear. At 7 P.M. the arch near the northern horizon continued, and there was an inverted cone of light, having its base elevated 12°, and its apex touching the N.W. horizon. Also some patches of light near the zenith. At 8 A.M. the arch in the north less complete and less bright. Large masses of yellowish light lying a little to the west of the zenith. Some beams rising from the S.E. horizon and a solitary one from the N.W. At 9 P.M. two bright arches springing from the N.W. by N. point of the horizon, and spreading wider as they rose towards the zenith, where they covered 40°; thence narrowing as they advanced to the S.E.

December 28.—At 5 P.M. masses of pale light in the north forming a low, broken arch. A few patches to the north of the zenith. At 7 a dull yellowish arch from the N.W. to S.E. passing to the north of the zenith. No clouds visible, but only stars of the first magnitude shining out. At 8 P.M. the same arch, more interrupted and also connected with large cloud-like patches of light. Sky dullish, not cloudy. At 9 no aurora.

December 29.—At 4 A.M. an arch of light standing across the zenith, with patches in the S.W. and N. by W. At 5 A.M. faint patches in the east and north and near the zenith. At 6 A.M. no aurora. At 4 P.M. sky tinged yellowish in the western horizon by the sun's rays, though that luminary was considerably under the horizon. At 7 P.M. an arch of light from N.W. to S.E. passing a little to the north of the zenith. It was barred across near the zenith by layers of stratus cloud. At 8 P.M. the extremities of the arch had the same bearings, but its crown had passed to some distance south of the zenith, against the wind. At 9 P.M. five arches covering the sky from 30° north of the zenith to about 50° south of it or a zone of 80°. The ends of the arches converged in the N.W. and S.E. points of the horizon. Some internal motion existed in the arches.

December 30.—At 5 A.M. auroral rays rising vertically from the N.N.W. horizon. At 6 A.M. an arch of light crossing the zenith from east to west. At 7 A.M. an arch from W.N.W. to S.S.E., rising about 20° above the southern horizon. A cloud-like patch of light in the N.W. by W. near the horizon, and a slender curved stream passing from the same point towards the arch. A pale sheet of light diffused over the northern half of the sky, and widely spread in the east also. At 9 temperature of air $-39 \cdot 6°$. Mercury wholly fluid. Nitric acid solidly crystallized. Sulphuric acid solid, and semitranslucent. Hydrochloric acid fluid. At 10^h 40^m temperature of the air $-40 \cdot 5°$ Fahrenheit.* About the fifth part of the mercury exposed in a shallow basin frozen; the solid part lying at the bottom, and having serrated edges as usual. At 11 A.M. rays of light shooting up from the sun's place. The men who went for meat two days ago, in passing over a hill saw the sun a good way above the horizon. At 2 P.M. a bright vertical beam of light rose from the sun's place. The quantity of frozen mercury has rather increased, the temperature of the air having been for two hours $-39 \cdot 5°$ Fahrenheit. At 4 P.M. red sky in the S.W. At 5 P.M. broad vertical beams of auroral light in the north, extending from N. by W. to N. by E., separated from each other by considerable intervals of blue sky, but arranged so as to form a low interrupted arch, whose summit was 18° high. The westernmost beams were midway between the horizon and Great Bear. At 6 P.M. the interrupted arch in the northern sky had risen to the elevation of 35°. It was broadest and brightest in the N.E. At 7 P.M. two bright contiguous arches, emitting yellowish light, spanned the sky from N.W. to S.E., the uppermost of the two crossing the zenith. Their breadth varied greatly, and their limbs in approaching the horizon were much curved and twisted. At 8 P.M. a bright arch crossed from N.W. to S.E. passing above 20° south of the zenith. It was composed of oblique beams of yellowish light, and was twisted near the horizon. Two pale arches, of the intensity of the Milky Way, and covering about 30° in breadth, existed to the north of the zenith. At 9 P.M. two arches springing from the N.W. by N. part of the horizon, became fainter as they rose to near the zenith, where one disappeared; the other, passing a little north of the zenith, was prolonged to the S.E.

December 31.—At 5 A.M. rays of light bearing S. and S.W., in rapid motion. At 6 A.M. patches of light and rays in the S., S.W., and W. near the horizon. At 8 P.M. vertical beams of light in the N., and N.W. by N., and a horizontal bank bearing north. At 9 P.M. an obscure but broad arch of the aurora bearing south,

* The temperatures noted in these remarks were corrected for the error of the thermometer on the assumption that $-40°$ is the proper freezing point of mercury. The temperature actually read off at 10^h 40^m A.M. was $-36 \cdot 8°$ F.

with large inactive masses of light a little way above it. Stratus cloud running all round the horizon.

On the 1st of December the Declinometer fluctuated 1°. On the 20th the fluctuation amounted to 1½°, the sky being cloudless, but a deposition of crystals of ice on glass and rough metallic surfaces going on. On the 10th the fluctuations of the needle exceeded a degree. On the 18th it was as great. On the 26th it was 1½°. On other days it was generally below a degree, as may be observed by a reference to the table of variations of the Declinometer, the most remarkable movements only being pointed out in this summary.

On the 4th January 1849 the Declinometer fluctuated 1°. No aurora was visible, a thin haze overspreading the sky in the evening. During the low temperatures of the 6th and 7th there was little fluctuation of the needle. On the 11th the movement of the card exceeded a degree, and in the evening sheets of auroral light, with considerable changes and flashes, overspread the sky. On the 16th the needle moved 2°. Only faint appearances of aurora were observed, and the sky was perfectly cloudless all day and in the evening. On the 25th the movements of the needle again exceeded a degree; the sky was completely obscured. A fine snow fell in the evening, and no aurora was visible. This was one of a number of instances in which the needle was observed to be affected considerably when the sky was inclined to deposit a minute crystalline snow.

The aurora was comparatively seldom seen in this month. It was noticed at the hours of observation, only on the 10th, 11th, 14th, 15th, 16th, 17th, 19th, 20th, 22nd, 23rd, 24th, 27th, 28th, and 30th.

February, 1829.—The aurora was visible at one or more hours on the 1st, 2nd, 10th, 11th, 12th, 13th, 16th, 17th, 19th, 20th, 21st, 23rd, 24th, 25th, 26th, and 27th.

On the 13th the movement of the Declinometer exceeded 1°. Mackarel sky, with sheets of auroral light. On the 20th the movement was 1¼°, and there were considerable displays of auroral light at all the hours of observation in the evening. On the 21st, at 3 A.M., the temperature of the air in the shade, when corrected for the error of the thermometer for the freezing point of mercury at −40° Fahrenheit, was −56·7° Fahrenheit. At this time an ounce of nitric acid, which had been standing in a vial with a glass stopper in the open air all night, was fluid; but at 5 A.M., when the temperature of the air was −56·4°, it was solidly frozen. Sulphuric acid was also at the latter time frozen, and its upper part of an opake white colour. A bottle of creosote (4 oz.), which had been out of doors all day, began to show round opake balls at the bottom. At 4 P.M., the temperature of the air having then risen to $-35 \cdot 7°$ Fahrenheit, two or three round flat cakes existed at the sides also of the creosote bottle; the rest of the fluid was transparent, but thicker than usual. Each of the round patches was marked with concentric rings of a darker colour, like the frond of *ulva pavonia*, or a section of maple wood. At 7 P.M. these cakes had augmented in size, and had a central point. The appearances of the aurora this day are detailed in the term day observations.

On the 22d February the fluctuation of the Declinometer amounted to 1½°, the sky being overcast most of the day and no auroral light visible. On the 23rd at 3 P.M. the phials of sulphuric and nitric acids and of creosote were wholly frozen, the temperature of the air being −39° Fahrenheit, and colder than it had been previously in the day. On the 27th the fluctuation of the Declinometer was 1½° and there were some curtain-shaped, arch-like displays of the aurora.

March 1849.—The auroral light was visible on the 1st, 3rd, 6th, 8th, 11th, 12th, 14th, 15th, 17th, 18th, 19th, 20th, 21st, 22nd, 24th, 25th, 26th, and 30th.

The fluctuation of the Declinometer amounted to 1° on the 6th. The sky was wholly obscured most of the day, but in the evening there was an arc of aurora extending from the N.W. to the S.E. horizon, and passing about 50° south of the zenith; the sky then being almost cloudless and Venus shining most beautifully. On the 18th the next fluctuation of the needle to the same extent was observed, the displays of aurora being very faint, and in the southern quarter of the sky. On the 19th the Declinometer varied more than 2°. At 9 P.M. an arch of the aurora sprang from the N.W. horizon, and, passing over the zenith, descended to the S.E. On the 20th the fluctuation was 1½°. A cloudless sky with streams of auroral light and long fringes in the S.E. part of the sky. Also at 10 P.M. curtain-like expansions, rolling occasionally inwards like a scroll, and expanding again.

On the 21st the fluctuation of the Declinometer at the hours was more than 1½°. The appearances of the aurora are detailed in the term day observations. On the 22nd the Declinometer varied more than 1½°, and a few patches of aurora early in the morning were all that were seen; the sky remaining perfectly cloudless the whole 24 hours. On the 24th the Declinometer varied 2°. The sky was obscured wholly till 7 P.M., when it cleared up entirely, and as had been observed in such cases, the north end of the needle then moved more towards the east. The displays of the aurora in the evening were faint.

April 1852.—From the extent of daylight in this month the aurora was seldom seen before 10 P.M., at which hour our observations ceased for the day. It appeared, however, on the 4th, 8th, 13th, 14th, 15th, 18th, and 21st. The Declination observed by eight sets of azimuths in the fore and afternoon of March 31st, was found to be 50° 16′ 52·7″ easterly, the Declinometer being 4° 22′. On the 16th the Declination was by the mean of six sets of azimuths, 51° 58′ 52″, the north end of the Declinometer having at the time a mean direction of 3° 37′. The needle fluctuated much during these observations. On the 21st the mean variation deduced from six sets of azimuths was 49° 54′ 36″ E., the mean direction of the north end of the Declinometer being 4° 25′ The mean easterly variation by the three series was 50° 44′ 14·2″, and the mean direction of the north end of the Declinometer during the observations for azimuth, 4° 04·1′.

On the 2nd of April the fluctuation of the Declinometer was 2° 28′, the sky being obscured the whole day, and no aurora visible. On the 3rd the fluctuation was 1¾°, the sky continuing obscured, without aurora. On the 4th the sky was nearly cloudless the whole day; only a faint arch of aurora at one time in the evening; but the fluctuations of the needle were 3° 40′ in the day. On the 6th the fluctuation of the Declinometer exceeded 1°. The sky, which was cloudless all day, was obscured after sunset. On the 9th the Declinometer fluctuated 3½°. The sky was partially cloudy in the day, quite cloudless after sunset, and no aurora was seen. On the 10th the fluctuation exceeded 1°. The sky, as on the preceding day, being cloudless till 7 P.M.; after which it was more or less cloudy, but no auroral light was seen. The red tints of the setting sun had scarcely departed entirely from the sky before 10 P.M., when our observations ceased. On the 14th, the Declinometer varied more than 1°. The sky was wholly cloudless till 10 P.M., when three-tenths of the vault was cloudy, and an auroral arch crossed the zenith, having a direction from N.W. to S.E. On the 16th the fluctuation of the needle exceeded 1½°; the sky being almost wholly covered with clouds till 10 P.M., when seven-tenths of it was cloudless; no aurora was seen, On the 18th fluctuation to the extent of 1½° occurred. The sky was wholly cloudless, and a faint auroral arch was seen at 10 P.M. in the usual direction, or crossing the magnetic meridian at right angles. Daylight not wholly gone, and the red tints of the western sky extensive.

On the 19th the Declinometer varied 1°. A cloudless sky, and no aurora. On the 20th the fluctuation was also about 1°, the sky being wholly covered with clouds till 10 P.M., when it cleared up, nine-tenths becoming blue, but no aurora was seen. On the 21st the

Declinometer varied 1¼°, and very suddenly at 9 P.M., when the sky, after being quite cloudless, had become rapidly overspread. Some of the clouds resembled auroral arches; but emitted no light, the daylight not having gone. At 10 P.M. auroral arches contended with the twilight, and half an hour later their peculiar light was more apparent. After this date no aurora was recorded, the daylight being too powerful for its display at 10 P.M. The degrees of the thermometer could be read in the open air by daylight between 9 and 10 P.M. on the 23rd. The fluctuations of the Declinometer exceeded a degree on some of the subsequent days of the month, as may be perceived by a reference to the tables.

On a review of the observations made during the seven months, many instances of the simultaneous occurrence of fluctuations of the needle with movements in the auroral light were noticed; but there were also examples of fluctuations of the needle in the absence of the aurora, and very numerous ones of brilliant auroras accompanied by a stationary or sluggish needle. I cannot therefore venture to ascribe the movements of the needle in any case to those of the aurora, or to any particular direction of the beams and arches. I think, however, that the needle varied more frequently during the sudden formation of clouds than at other times; and I am also inclined to say that the formation of clouds often followed brilliant and active auroras. It is a popular belief in the fur districts that very fine displays of the aurora presage windy weather.

With respect to sounds of the aurora, the belief prevails in the arctic regions that it is occasionally audible when very bright and active, at which times it is believed by the natives to be near the earth. Having witnessed the phenomena some thousands of times without hearing it, I have become sceptical of its ever producing sounds audible on the surface of the earth. The sounds it is said to cause are likened by many to the rustling of silk; and I may observe that the curtain-like appearances and motions of the brightest auroras are likely to be associated with the remembrance of such sounds, and also that the formation of minute icy spiculæ in very cold clear nights is accompanied by a crackling in the air.

Haslar Hospital,
12 January, 1852.

DELCROS'S BAROMETER.

FORT CONFIDENCE.

Abstract of Hourly Observations in the month of October 1851.

Day. / Civil Time.	Delcros's Barometer, corrected for capillarity and mean										
	1.	2.	3.	4.	5.	6.	7.	8.	9.	10.	11.
	Millm.	Millm.	Millm.	Millm.	Millm.	Millm.	Millm.	Millm.	Millm.	Millm.	Mill
9	—	—	—	—	—	—	738·82	738·69	738·34	—	—
10	—	—	—	—	—	—	38·09	39·34	40·79	—	742·
11	—	—	—	—	—	—	35·83	35·71	35·16	734·15	—
12	—	—	—	—	—	732·19	33·69	—	—	33·79	—
13	—	—	—	—	—	—	29·90	31·14	—	—	32·
14	—	—	—	—	—	—	44·12	—	—	—	43·
15	—	—	—	—	—	—	—	35·19	34·29	33·49	33·
16	—	—	—	—	—	—	29·82	—	30·02	29·66	39·
17	—	—	—	—	—	—	32·04	32·84	—	32·32	32·
18	—	—	—	—	—	—	28·19	—	27·34	—	27·
19	—	—	—	—	—	26·84	—	27·84	—	—	—
20	—	—	—	—	—	—	33·89	33·34	33·29	33·19	32·
21	—	—	—	731·54	—	—	31·59	31·99	31·72	—	32·
22	—	—	—	—	—	32·02	33·54	33·34	31·84	32·89	32·
23	—	—	—	—	—	33·79	33·84	33·79	33·54	33·74	33·
24	—	—	—	—	728·74	28·39	27·69	26·34	25·89	24·89	23·
25	—	—	—	—	—	16·89	16·79	17·49	17·79	18·19	18·
26	—	—	—	—	—	24·07	24·19	24·04	23·54	24·59	24·
27	—	—	724·14	—	—	—	24·14	24·19	24·34	24·29	24·
28	—	—	—	—	—	27·04	28·09	27·85	28·59	29·04	29·
29	—	—	—	—	—	—	37·99	37·94	38·09	38·09	38·
30	—	—	—	—	—	42·82	43·64	43·64	43·74	43·84	44·
31	—	—	—	—	—	35·89	36·79	35·89	35·29	34·34	34·
Millimètres	—	—	724·14	731·54	728·74	729·99	732·51	732·14	731·87	731·28	732·
Inches	—	—	28·509	28·801	28·690	28·740	28·840	28·825	28·813	28·791	28·8
Corrected for 32° Fahrenheit	—	—	—	—	—	28·691	28·783	28·768	28·752	28·729	28·7
Oscillation	—	—	—	—	—	0·000	0·092	0·077	0·061	0·088	0·0

Lowest, 28·165 inches, correc

FORT CONFIDENCE.

Abstract of Hourly Observations in the month of October 1848.

1.	2.	3.	4.		
Mill^m.	Mill^m.	Mill^m.	Mill^m.	Mill^m.	Mill^m.
736·42	736·34	735·69	735·39	—	732·49
41·82	42·29	43·29	42·99	741·94	—
—	—	—	35·92	—	—
33·50	—	—	—	—	30·69
—	—	—	—	—	—
—	—	44·92	—	—	—
32·34	32·24	32·29	32·30	—	—
30·09	30·19	—	30·89	—	—
32·79	31·99	—	—	—	30·84
25·79	26·14	25·44	24·74	26·04	25·94
27·24	28·39	—	—	31·86	33·21
33·14	33·29	32·94	33·04	32·34	32·22
33·66	32·89	32·99	33·79	33·14	33·09
32·34	32·54	33·14	33·09	33·49	33·29
33·24	31·74	34·19	34·04	33·19	33·04
22·74	21·44	21·24	20·69	18·09	17·79
19·04	19·49	19·89	20·59	22·59	23·24
24·14	23·64	23·94	23·49	24·09	24·04
24·79	24·94	25·09	24·54	24·84	24·79
30·89	31·74	32·14	32·19	33·89	24·54
39·94	39·99	39·90	40·74	41·14	41·29
44·60	44·29	44·34	44·09	44·44	43·04
[illegible]	32·84	33·24	33·14	33·79	33·89
	731·49	732·49	732·05	731·61	730·44
	28·799	28·839	28·821	28·804	28·758
	28·736	28·778	28·760	28·740	28·692
	0·045	0·067	0·049	0·049	0·001

Highest, 29·278. Range, 1·102 inches.

Fort Confidence—*continued.*

Abstract of Hourly Observations in the months of November and December 1848.

	11.
	Mill^m.
1	734·69
2	44·09
3	40·94
4	47·34
5	55·34
6	54·09
7	45·19
8	35·59
9	33·04
10	38·52
11	42·12
12	27·19
13	32·46
14	38·06
15	43·18
16	24·22
17	28·87
18	25·59
19	24·96
20	29·04
21	26·09
22	27·79
23	33·77
24	42·29
25	40·79
26	41·96
27	39·72
28	32·16
29	37·07
30	36·49
	29·007
	28·960
	0·022

Lowest, 32° Fahrenheit, 28·423 inches.

	In.	In.	In.	In.
1	29·322	29·315	29·355	29·371
2	·587	·606	·625	·520
3	·684	·678	·688	·691
4	·544	·531	·526	·532
5	·312	·320	·315	·301
6	·268	·290	·307	·320
7	·463	·452	·446	·424
8	·132	·126	·120	·137
9	·198	·202	·214	·213
10	·181	·177	·167	·161
11	28·950	28·947	28·959	28·968
12	·879	·883	·861	·849
13	·680	·702	·691	·698
14	·952	·922	·933	·933
15	·735	·739	·741	·733
16	·401	·470	·459	·486
17	·840	·863	·877	·912
18	29·250	29·268	29·268	29·271
19	·173	·149	·107	·093
20	28·351	28·339	28·341	28·343
21	29·091	29·091	29·066	29·053
22	28·550	28·561	28·546	28·564
23	·841	·850	·845	·856
24	·916	·931	·920	·865
25	·801	·778	·784	·771
26	·505	·520	·501	·506
27	·517	·504	·523	·514
28	·284	·277	·258	·252
29	·620	·642	·662	·732
30	29·359	29·395	29·420	29·450
31	·333	·406	·394	·413
	28·993	28·999	28·997	28·997
	0·000	0·006	0·004	0·004
	736·41	736·59	736·51	736·51

Lowest, 32° Fahrenheit, 28·282 inches.
The Barometer in English measure is red

FORT CONFIDENCE—*continued.*

Abstract of Hourly Observations in the months of November and December 1848.

6.	7.
Millm.	Millm.
738·19	738·74
44·99	44·99
38·94	39·44
50·94	50·89
56·74	56·64
51·49	51·14
41·34	40·79
34·21	33·99
34·44	35·30
40·09	40·01
39·89	39·44
26·36	26·92
34·34	34·79
43·12	43·06
39·49	38·72
19·36	19·14
30·79	30·94
23·84	23·94
26·80	27·84
28·74	28·34
24·14	24·49
29·04	29·09
36·34	37·04
42·73	42·72
41·24	41·19
42·79	42·19
36·34	35·69
32·10	32·89
37·04	36·53
40·54	40·40
736·88	736·73
29·011	29·005
28·953	28·948
0·015	0·010

In.	In.	In.	In.	In.	In.	In.	In.	In.	In.	In.	
29·417	29·426	29·420	29·449	29·465	29·469	29·491	29·486	29·611	—	—	29
·630	·621	·632	·607	·610	·609	·610	·604	—	—	—	
·690	·679	·675	·632	·627	·611	·609	·617	—	—	—	
·530	·536	·539	·526	·506	·509	·508	·487	—	—	—	
·287	·271	·260	·220	·216	·201	·203	·191	—	—	—	
·360	·381	·387	·429	·438	·456	·543	·567	—	—	—	
·418	·307	·371	·323	·293	·270	·255	·233	—	—	—	
·138	·141	·189	·155	·102	·175	·156	·168	—	—	—	
·218	·209	·218	·232	·223	·217	·228	·226	—	—	—	
·144	·126	·106	·091	·088	·051	·048	·022	—	—	—	
28·991	28·974	·011	28·968	28·975	28·933	28·988	28·983	—	—	—	28
·829	·834	28·799	·784	·783	·773	·756	·751	—	—	—	
·707	·716	·754	·756	·772	·778	·794	·821	—	—	—	
·938	·925	·924	·910	·913	·907	·894	·879	—	—	—	
·705	·714	·720	·716	·716	·709	·691	·671	—	—	—	
·493	·504	·526	·545	·556	·576	·593	·606	28·641	—	—	
·966	·975	29·007	29·059	29·077	29·091	29·107	29·112	29·123	—	—	
29·270	29·290	·301	·298	·288	·294	·291	·288	·275	—	—	29
·048	·033	28·988	28·970	28·905	28·853	28·803	28·700	—	—	—	
28·377	28·405	·454	·578	·667	·723	·777	·824	28·870	28·928	28·959	28
·993	·968	·936	·865	·850	·800	·777	·760	·724	·702	·659	
·576	·598	·601	·670	·677	·721	·703	·721	—	—	—	
·855	·861	·868	·844	·856	·857	·853	·853	—	—	—	
·931	·942	·947	·959	·950	·942	·939	·931	—	—	—	
·724	·736	·721	·693	·679	·665	·646	·631	·622	—	—	
·483	·467	·458	·461	·455	·448	·443	·448	—	—	—	
·581	·543	·532	·533	·542	·533	·518	·491	·504	—	—	
·269	·270	·281	·304	·328	·354	·302	·387	·409	—	—	
·767	·789	·802	·872	·918	·962	·981	29·103	29·045	—	—	
29·440	29·501	29·510	29·526	29·547	29·543	29·555	·552	·541	—	—	
·361	·359	·348	·325	·315	·315	·304	·305	·260	—	—	
29·006	29·006	29·008	29·010	29·011	29·013	29·014	29·016	28·969	28·815	28·809	
0·013	0·018	0·015	0·017	0·018	0·020	0·021	0·023	—	—	—	
736·75	736·75	736·80	736·85	736·87	736·93	736·95	737·00	—	—	—	

METEOROLOGICAL OBSERVATIONS.

FORT CONFIDENCE—*continued.*

Abstract of Hourly Observations made during the months of January and February 1849.

	1.	2.	3.	4.	5.	6.	7.	8.	9.	10.	11.
	In.	In.	In.	In.	In.	In.	In.	In.	In.	In.	In.
1	—	—	—	—	—	—	29·160	29·168	29·156	29·151	29·190
2	—	—	—	—	—	—	28·731	28·798	28·696	28·663	28·651
3	—	—	—	—	—	28·829	·863	·873	·905	·969	·980
4	—	—	—	—	29·336	29·363	29·373	29·394	29·413	29·412	29·433
5	—	—	—	—	—	·708	·715	·733	·758	·758	·810
6	—	—	—	—	—	·806	·813	·821	·831	·840	·837
7	—	—	—	—	—	—	·832	·819	·817	·793	·806
8	—	—	—	—	—	·481	·436	·420	·400	·365	·364
9	—	—	29·113	—	—	—	·110	·084	·108	·176	·106
10	—	—	—	—	—	·093	·050	·076	·073	·097	·125
11	—	—	—	—	—	—	·232	·213	29·253	·247	·246
12	—	—	—	—	—	—	·018	28·963	28·923	28·858	28·800
13	—	—	—	—	—	—	28·432	·453	·493	·587	·641
14	—	—	—	—	—	—	—	·978	·983	·969	·929
15	—	—	—	—	—	28·902	·917	·937	·968	·978	29·002
16	—	—	—	—	29·458	29·486	29·514	29·520	29·534	29·570	·569
17	—	—	—	—	—	—	·648	·621	·603	·584	·581
18	—	—	—	—	—	—	·208	·201	·223	·201	·184
19	—	—	—	—	—	—	·413	·425	·433	·421	·409
20	—	—	—	—	—	—	28·795	28·800	28·859	28·901	28·933
21	—	—	—	28·633	28·606	28·557	·524	·501	·477	·439	·443
22	28·603	28·687	28·787	·841	·901	·915	·942	·977	·980	29·007	29·038
23	—	—	—	—	—	—	29·021	29·018	29·048	·045	·063
24	—	—	—	—	—	—	·169	·170	·186	·188	·183
25	29·166	29·184	29·210	29·219	29·226	29·233	·245	·263	·247	·258	·246
26	—	—	—	—	—	—	28·930	28·907	28·885	28·883	28·853
27	—	—	—	28·785	28·805	28·830	·881	·937	29·019	29·060	29·093
28	—	—	—	—	—	—	29·475	29·480	·460	·443	·424
29	—	—	—	—	28·984	·951		28·884	28·867	28·820	28·777
30	—	—	—	—	—	—		·706	·820	·837	·835
31	—	—	—	—	—	·719		·736	·764	·775	·787
	28·884	28·935	29·037	28·869	29·045	29·134		29·128	29·135	29·139	29·140
	—	—	—	—	—	—		0·000	0·007	0·011	0·012
	—	—	—	—	—	—		739·85	740·02	740·13	740·15
1	—	—	—	—	—	29·076	20·066	29·079	29·101	29·090	29·098
2	—	—	—	—	—	8·834	8·835	8·828	8·827	8·824	8·831
3	—	—	—	—	28·937	8·994	8·989	8·989	8·958	8·937	8·929
4	—	—	—	—	—	8·831	8·824	8·813	8·844	8·867	8·890
5	—	—	—	—	9·127	9·130	9·109	9·000	9·109	9·093	9·096
6	—	—	—	28·912	8·908	8·909	8·905	8·903	8·923	8·954	8·966
7	—	—	—	—	9·234	9·228	9·109	9·208	9·188	9·134	9·144
8	—	—	—	—	8·760	8·771	8·768	8·798	8·816	8·819	8·838
9	—	—	—	—	—	8·967	8·961	8·972	8·943	9·015	9·022
0	—	—	—	—	—	9·262	9·256	9·261	9·278	9·286	9·302
1	—	—	—	—	—	—	9·376	9·895	9·421	9·443	9·456
2	—	—	—	—	9·623	9·617	9·380	9·597	9·606	9·603	9·609
3	—	—	—	9·538	9·528	9·524	9·489	9·478	9·458	9·439	9·428
4	—	—	—	—	—	9·381	9·352	9·350	9·348	9·329	9·291
5	—	—	—	—	—	9·163	9·191	9·204	9·236	9·279	9·293
6	—	—	—	—	—	9·196	9·166	9·125	9·104	9·086	9·026
17	—	—	28·906	9·016	9·022	9·042	9·036	9·055	9·077	9·063	9·052
18	—	—	—	—	—	—	8·846	8·859	8·858	8·854	8·813
19	—	—	—	—	8·816	8·805	8·808	8·807	8·825	8·779	8·831
20	—	—	—	—	—	9·010	9·014	9·024	9·033	9·039	9·038
21	29·143	29·155	9·186	9·178	9·188	9·189	9·197	9·206	9·215	9·236	9·223
22	—	—	—	—	—	9·428	9·424	9·436	9·454	9·471	9·474
23	—	—	—	—	—	9·551	9·466	9·455	9·465	9·461	9·453
24	9·538	9·554	9·573	9·508	9·600	9·606	9·635	9·636	9·630	9·636	9·647
25	—	—	—	—	—	—	9·536	9·530	9·496	9·513	9·512
26	—	—	—	—	—	9·621	9·649	9·657	9·679	9·655	9·663
27	—	—	—	—	—	9·441	9·435	9·443	9·443	9·466	9·457
28	—	—	—	—	—	9·486	9·470	9·458	9·462	9·447	9·423
	29·340	29·354	29·285	29·228	29·163	29·202	29·200	29·201	29·208	29·208	29·207
	—	—	—	—	—	0·002	0·000	0·001	0·008	0·008	0·007
	—	—	—	—	—	—	—	—	—	—	—

Lowest at 32° Fahrenheit, 28·388 inches. Highest, 29·900 inches.

Lowest at 32° Fahrenheit, 28·686 inches. Highest, 29·679 inches.

Fort Confidence—*continued.*

Abstract of Hourly Observations made during the months of January and February 1849.

3.	4.	9.	10.	11.	Midnt.
In.	In.	In.	In.	In.	In.
29·109	29·093	29·015	29·000	28·950	28·952
28·645	28·634	28·657	28·649	—	—
29·059	29·100	29·221	29·243	—	—
·489	·494	·557	·559	—	—
·767	·778	·780	·786	—	—
·883	·877	·894	·896	—	—
·756	·747	·670	·665	—	—
·217	·187	·122	·104	—	—
·098	·103	·114	·113	—	—
·191	·204	·224	·225	—	—
·228	·230	·277	·263	29·257	—
28·622	28·588	28·470	28·471	—	—
·822	·832	29·014	29·038	—	—
·829	·827	28·758	28·749	—	—
29·252	29·245	29·263	29·284	—	—
·606	·620	·658	·670	—	—
·517	·402	·396	·392	—	—
·163	·182	·245	·206	—	—
·333	·319	·118	·051	—	—
·032	·033	28·971	—	—	—
28·407	28·403	·407	28·426	28·459	28·503
29·075	29·069	29·029	29·010	29·013	29·054
·076	·080	·104	·108	—	—
·176	·168	·142	·148	·154	·167
·222	·217	·138	—	—	—
28·775	28·762	28·711	28·704	—	—
29·271	29·293	29·410	29·414	—	—
·383	·380	·260	·223	—	—
28·681	28·653	28·621	28·633	—	—
·856	·852	·784	—	—	—
·878	·888	·956	·975	—	—
29·144	29·140	29·129	29·145	28·967	28·919
0·016	0·012	0·001	0·017	—	—
740·25	740·15	739·88	740·27	—	—

		3.	4.		
29·111	29·105	29·098	29·099	29·073	29·057
8·858	8·860	8·867	8·881	8·889	8·915
8·906	8·916	8·889	8·869	8·852	8·839
8·913	8·938	8·961	8·968	9·000	9·023
9·053	9·051	9·029	9·043	9·017	9·001
9·003	9·034	9·056	9·079	9·106	9·144
9·047	9·034	8·988	8·972	8·913	8·869
8·854	8·873	8·875	8·877	8·896	8·887
9·053	9·051	9·066	9·078	9·110	9·106
9·302	9·039	9·312	9·306	9·313	9·323
9·477	9·491	9·504	9·523	9·527	9·536
9·608	9·619	9·600	9·608	9·627	9·650
9·408	9·405	9·401	9·399	9·411	9·379
9·307	9·311	9·292	9·293	9·282	9·252
9·323	9·314	9·342	9·350	9·357	9·349
8·955	8·935	8·896	8·868	8·913	9·006
9·048	9·058	9·061	9·066	9·059	9·054
8·789	8·759	8·736	8·721	8·712	8·694
8·855	8·861	8·876	8·870	8·894	8·890
9·056	9·062	9·063	9·076	9·093	9·099
9·234	9·241	9·238	9·253	9·263	9·282
9·494	9·506	9·505	9·506	9·505	9·509
9·439	9·444	9·453	9·446	9·450	9·467
9·652	9·651	9·606	9·645	9·635	9·632
9·506	9·509	9·499	9·505	9·501	9·502
9·652	9·616	9·620	9·600	9·596	9·561
9·480	9·485	9·501	9·495	9·483	9·429
9·433	9·429	9·413	9·418	9·450	9·400
29·201	29·211	29·206	29·208	29·211	29·209
0·004	0·011	0·006		0·011	0·009
—	—	—			

11.	Midnt.
—	—
—	—
—	—
—	—
—	—
—	—
25·743	—
—	—
—	—
—	—
—	—
—	—
—	—
—	—
—	—
—	—
—	—
—	—
—	—
29·126	29·135
9·336	9·356
—	—
9·515	9·534
—	—
—	—
—	—
9·556	—
—	—
29·255	29·342
—	—
—	—

Range, 1·512 inches.
Range, 0·998 inches.

Fort Confidence—continued.

Abstract of Hourly Observations in the months of March and April 1849.

			11.
	Mill^m.	Mill^m.	Mill^m.
1	744·89	744·49	744·34
2	38·94	38·84	39·04
3	33·96	34·09	34·09
4	27·04	26·04	25·89
5	23·14	23·14	23·14
6	25·09	25·54	25·84
7	34·29	34·84	35·29
	43·19	43·89	44·69
	45·59	45·94	45·54
	42·79	43·14	43·76
	49·04	49·49	50·89
	54·04	54·34	54·34
	50·79	50·74	50·39
	40·49	40·34	39·84
	25·14	25·14	25·69
	25·26	25·10	25·34
	28·34	29·04	30·94
	42·84	43·50	43·09
	39·64	39·64	39·19
	30·64	30·69	30·84
	34·94	35·54	35·64
	37·84	37·34	37·84
	40·19	40·22	40·34
	39·64	40·04	39·34
	37·24	37·29	37·36
	35·44	35·84	36·04
	34·94	35·39	35·34
	—	—	—
	34·97	34·59	33·79
	37·84	38·14	38·64
	38·49	38·44	37·94
	[illegible]	[illegible]	737·46
			29·034
			28·988
			0·000

					11.
					736·79
					31·84
					34·84
					42·44
5			[illegible]	[illegible]	36·16
6			36·14	35·89	36·94
7			39·74	40·04	40·74
8			44·32	43·84	44·14
9			50·04	50·59	52·59
10			55·24	54·64	54·34
11			39·62	39·74	39·84
12			48·54	48·94	49·84
13			59·54	59·59	59·64
14			53·44	52·84	52·74
15			45·84	45·34	44·44
16			38·74	38·94	39·14
17			45·34	45·54	47·24
18			52·24	52·74	52·89
19			49·74	49·04	48·[illegible]
20			41·14	41·24	41·04
21			46·54	46·94	48·34
22			42·74	42·72	42·84
23			42·04	43·19	43·74
24			47·14	46·74	46·64
25			41·34	40·74	41·04
26			43·04	43·84	45·44
27			53·44	53·14	53·14
28			53·34	52·64	53·64
29			48·14	48·34	47·84
30			44·20	43·84	43·64
	743·86	744·66	744·77	744·74	745·07
	29·286	29·318	29·322	29·321	29·334
	29·263	29·267	29·268	29·274	29·277
		0·004	0·005	0·011	0·014

Lowest at 32° Fahrenheit, 28·396 inches; highest, 29·664 inches.

Lowest at 32° Fahrenheit, 28·710 inches; highest, 29·865 inches.

FORT CONFIDENCE—*continued.*

Abstract of Hourly Observations made during the months of March and April 1849.

from Standard Barometer, but not for temperature.

1.	2.	3.	4.	5.	6.	7.	8.	9.	10.	11.	Midn^t^.	Means.
Mill^m^.	Mill^m^.	Mill^m^.	Mill^m^.	Mill^m^.	Mill^m^.	Mill^m^.	Mill^m^.	Mill^m^.	Mill^m^.	Mill^m^.	Mill^m^.	Mill^m^.
744·44	744·59	744·64	744·54	744·14	743·82	743·94	743·54	743·19	743·24	—	—	744·26
38·64	38·74	37·74	37·84	37·94	37·59	37·64	37·04	36·24	36·26	—	—	38·27
34·14	33·79	33·74	33·94	33·44	32·89	32·49	32·69	32·44	34·94	—	—	33·63
25·84	25·34	25·32	24·84	24·34	24·09	24·24	23·54	23·14	22·64	—	—	25·39
23·24	23·34	23·44	23·74	23·64	23·74	23·94	23·89	22·89	23·54	—	—	23·36
26·68	27·34	28·14	28·54	29·14	28·94	29·59	29·34	29·44	29·94	—	—	27·23
36·44	36·14	36·14	36·44	36·34	36·74	36·89	37·34	38·04	38·64	—	—	35·79
45·10	45·44	45·54	46·04	46·44	46·49	46·84	46·89	46·89	46·89	—	—	45·02
45·36	46·09	46·09	46·04	45·74	45·44	45·70	45·64	45·01	44·94	—	—	45·61
43·18	42·94	42·94	43·14	43·54	43·34	44·09	44·00	44·74	45·04	—	—	43·52
51·49	51·84	51·94	51·84	52·44	52·59	53·14	53·44	53·49	54·09	—	—	51·14
54·89	54·36	54·24	53·84	53·74	53·74	54·09	52·84	53·49	53·39	—	—	53·90
50·39	50·04	50·09	49·64	48·84	48·49	48·09	47·74	47·84	46·94	—	—	49·59
39·04	37·64	37·24	36·34	35·99	34·64	33·56	31·94	31·11	30·28	—	—	37·25
25·14	25·04	25·24	25·64	25·64	26·24	26·54	26·44	27·29	27·24	—	—	25·66
25·29	25·39	25·44	25·49	25·19	25·29	25·19	25·24	24·64	24·44	—	—	25·19
32·79	33·34	34·89	35·84	36·34	37·54	37·94	38·34	38·84	39·34	—	—	33·39
42·94	43·24	43·14	43·44	42·84	43·19	42·94	43·29	42·94	43·49	—	—	42·88
38·74	38·44	38·49	38·14	38·19	37·09	36·94	35·74	35·44	35·84	—	—	38·19
31·54	31·34	32·14	31·34	31·39	31·74	32·34	32·14	32·24	32·39	732·54	732·84	31·55
36·04	36·54	37·09	36·74	36·84	37·14	37·06	36·85	37·14	37·34	36·77	36·91	35·70
38·26	33·94	38·34	39·04	38·79	38·64	38·64	38·64	38·59	38·34	—	—	37·83
40·34	40·74	40·79	41·54	41·34	42·29	42·14	42·34	41·84	41·64	—	—	40·84
38·49	38·74	38·59	38·54	38·04	39·14	38·34	38·24	38·34	38·84	—	—	38·87
37·52	37·34	37·04	37·54	37·94	37·74	38·34	36·94	37·74	37·94	—	—	37·50
36·04	35·44	35·64	35·44	35·19	35·04	34·74	35·54	34·59	34·44	—	—	35·43
35·94	36·19	36·49	36·74	38·14	38·24	37·84	37·89	38·14	38·44	—	—	36·37
—	—	—	—	—	—	—	—	—	—	—	—	} 35·02
33·44	33·54	33·24	33·44	33·04	34·09	33·99	34·79	34·74	34·64	—	—	
38·89	39·04	39·09	39·24	39·44	41·32	40·29	39·94	39·94	40·33	—	—	38·91
38·14	37·54	37·64	37·59	36·99	37·39	37·29	36·49	36·64	36·33	—	—	37·73
737·61	737·61	737·68	737·75	737·73	737·82	737·83	737·63	737·57	737·66	734·65	734·87	737·46
29·040	29·040	29·043	29·046	29·045	29·049	29·049	29·041	29·069	29·048	28·924	28·988	29·034
28·993	28·988	28·900	28·988	28·990	28·993	28·992	28·993	28·990	28·996	—	—	28·991
0·005	0·000	0·002	0·000	0·002	0·005	0·004	0·005	0·002	0·008	—	—	0·005
736·64	737·24	737·84	737·99	738·24	738·79	738·94	738·94	738·94	738·94	—	—	737·16
30·64	30·69	30·84	30·64	31·04	31·29	31·14	31·24	31·24	31·44	—	—	31·84
36·72	37·64	37·54	38·74	38·64	39·04	39·26	39·54	39·64	39·69	—	—	37·03
43·16	43·44	43·34	43·49	42·84	43·74	43·14	42·14	42·24	41·94	—	—	42·31
35·89	35·74	35·94	35·84	36·14	36·34	37·54	36·24	36·49	36·74	—	—	36·18
36·74	36·54	36·64	36·64	36·64	36·54	36·44	36·54	36·19	36·84	—	—	36·40
42·04	42·06	43·14	42·94	42·84	44·64	43·19	43·79	43·14	43·24	—	—	41·83
44·69	45·29	45·74	46·64	47·04	47·19	47·20	47·69	47·64	46·12	—	—	45·67
53·54	53·74	54·14	54·64	54·44	54·44	54·74	54·94	54·79	55·19	—	—	52·99
52·74	52·79	51·64	50·94	49·34	48·34	47·54	47·04	45·14	44·29	—	—	51·35
40·04	41·24	41·04	41·04	41·24	41·94	41·94	42·34	43·14	44·24	—	—	40·90
52·14	52·94	53·14	53·54	55·04	55·54	55·94	57·04	56·74	56·84	—	—	52·30
59·84	59·64	59·04	59·04	59·34	59·04	58·84	58·94	57·84	57·34	—	—	59·05
51·44	50·84	50·24	49·84	49·44	49·54	49·54	48·94	48·04	47·94	—	—	51·10
43·94	43·34	42·84	42·24	41·69	41·24	41·34	41·29	40·64	40·44	—	—	43·08
39·39	39·94	40·29	39·99	40·32	40·94	40·74	40·84	40·94	41·20	—	—	39·79
47·94	48·34	48·84	49·64	50·34	50·84	51·54	51·74	50·84	51·34	—	—	48·03
53·39	53·24	53·04	53·19	53·24	53·49	52·84	52·99	52·64	52·24	—	—	52·83
48·24	47·14	47·04	45·84	45·84	45·14	44·24	43·74	43·34	43·09	—	—	46·81
42·04	41·24	41·34	41·29	41·26	40·74	41·04	41·89	41·34	41·54	741·74	741·94	45·58
49·04	49·24	49·74	49·14	48·09	48·54	48·14	47·84	47·64	47·34	—	—	47·74
42·94	42·84	43·14	43·19	43·04	43·74	42·88	42·54	41·49	40·19	—	—	42·58
44·99	45·04	44·94	44·89	45·09	45·44	45·74	45·84	46·24	46·34	—	—	44·41
45·94	43·64	44·64	44·54	44·04	43·74	43·74	42·69	42·14	42·14	—	—	45·21
41·89	42·04	42·24	42·34	42·49	42·54	42·74	43·24	42·74	43·34	—	—	41·85
47·34	48·44	48·84	49·14	50·14	50·64	50·74	50·84	51·24	51·44	—	—	47·24
58·31	53·14	53·14	53·14	52·04	53·14	52·94	52·94	52·34	53·14	—	—	53·03
52·74	52·64	52·14	52·14	51·74	51·34	50·74	50·54	50·64	50·14	49·74	49·84	51·80
46·81	46·74	46·64	46·24	46·64	45·84	45·74	45·84	45·94	45·84	—	—	46·01
44·04	43·74	43·74	43·84	44·24	43·74	43·84	43·34	43·54	43·64	—	—	43·53
745·35	745·42	745·43	745·42	745·48	745·58	745·48	745·47	745·16	745·19	745·74	745·89	745·12
29·345	29·348	29·348	29·348	29·350	29·354	29·350	29·350	29·338	29·339	29·360	29·366	29·336
29·280	29·284	29·290	29·278	29·280	29·282	29·280	29·286	29·280	29·278	29·282	29·297	29·277
0·017	0·021	0·017	0·015	0·017	0·019	0·017	0·023	0·017	0·015	—	—	0·019

Range, 1·268 inches.

Range, 1·155 inches.

FORT CONFIDENCE—*continued.*

Months.	1.	2.	3.			[illegible]
1848-9.	In.	In.	In.	In.		In.
October -	—	—	—	—		28·783
November -	—	—	—	—		8·940
December -	—	—	—	—		9·661
January -	—	—	—	—	9·130	9·129
February -	—	—	—	—		9·200
March -	—	—	—	—		9·600
April -	—	—	—	—		9·267
Means -	—	—	—			29·046
Oscillation -	—	—				

FORT CONFIDENCE—*continued.*

2.	3.	4.	5.	6.	7.	8.		
In. 28·736	In. 28·778	In. 28·760	In. 28·737	In. 28·748	In. 28·705	In. 28·740	In. —	In. 28·746
8·961	8·959	8·960	8·963	8·953	8·948	8·973	-	8·961
9·006	9·008	9·007	9·010	9·011	9·013	9·014	-	9·009
9·139	9·144	9·140	9·138	9·134	9·133	9·131	-	9·130
9·211	9·206	9·208	9·211	9·209	9·210	9·213	-	9·210
8·988	8·990	8·988	8·990	8·993	8·992	8·993	-	8·991
9·284	9·280	9·278	9·280	9·282	9·280	9·286	-	9·277
29·046	29·052	29·049	29·047	29·047	29·040	29·050		29·046
0·006	0·012	0·009	0·007	0·007	0·000	0·010		0·006

FORT CONFIDENCE.

Abstract of Hourly Observations in the month of October 1848.

Civil Time.	1.	2.	3.	4.	5.	6.	7.	8.	9.	10.	11.
	°	°	°	°	°	°	°	°	°	°	°
1	—	—	—	—	—	—	—	—	—	—	—
2	—	—	—	—	—	—	—	—	—	—	—
3	—	—	—	—	—	—	—	—	—	—	—
4	—	—	—	—	—	—	—	—	—	—	—
5	—	—	—	—	—	—	—	—	—	—	—
6	—	—	—	—	—	—	—	—	—	—	—
7	—	—	—	—	—	—	—	—	—	—	—
8	—	—	—	—	—	—	—	—	—	—	—
9	—	—	—	—	—	—	10·4	10·3	13·4	—	—
10	—	—	—	—	—	—	4·1	10·1	11·6	—	16·4
11	—	—	—	—	—	4·6	—	11·0	17·4	10·0	—
12	—	—	—	—	—	9·4	15·8	—	—	14·4	—
13	—	—	—	—	—	—	3·6	8·6	—	—	8·4
14	—	—	—	—	—	—	7·6	—	—	—	11·2
15	—	—	—	—	—	—	—	9·4	9·9	11·7	13·6
16	—	—	—	—	—	—	10·4	—	11·6	9·1	8·8
17	—	—	—	—	—	—	9·6	9·6	—	12·1	13·4
18	—	—	—	—	—	—	10·8	—	11·1	—	11·6
19	—	—	—	—	—	12·1	—	14·6	—	—	—
20	—	—	—	—	—	—	12·1	12·6	14·6	13·5	13·4
21	—	—	—	9·4	—	—	6·6	6·6	7·6	—	6·4
22	—	—	—	—	—	6·4	14·1	11·1	9·7	10·4	12·0
23	—	—	—	—	—	13·6	18·8	9·6	10·1	10·6	11·1
24	—	—	—	—	9·1	11·6	11·6	13·2	10·6	11·6	5·1
25	—	—	—	—	—	9·6	11·1	13·6	10·6	11·6	12·6
26	—	—	—	—	—	10·6	12·1	10·1	9·1	14·1	11·6
27	—	—	8·6	—	—	—	7·6	8·6	8·6	9·2	11·0
28	—	—	—	—	—	9·6	13·1	—	9·6	11·1	10·8
29	—	—	—	—	—	—	9·1	10·1	11·1	11·8	7·6
30	—	—	—	—	—	4·6	10·1	10·1	8·6	14·8	10·6
31	—	—	—	—	—	2·2	13·6	8·6	12·8	6·1	11·1
Hourly Means	—	—	8·60	9·40	9·10	8·57	10·36	10·43	11·00	11·38	
Fahr. Scale	—	—	47°·48	48°·92	48°·38	47°·43	50°·66	50°·77	51°·80	52°·48	

FORT CONFIDENCE.

Abstract of Hourly Observations in the month of October 1848.

orrected for Mean deviation from Paris Standard Thermometer, −0·88.

1.	2.	3.	4.	5.	6.	7.	8.	9.	10.	11.	Midnt.	Means.
°	°	°	°	°	°	°	°	°	°	°	°	°
—	—	—	—	—	—	—	—	—	—	—	—	—
—	—	—	—	—	—	—	—	—	—	—	—	—
—	—	—	—	—	—	—	—	—	—	—	—	—
—	—	—	—	—	—	—	—	—	—	—	—	—
—	—	—	—	—	—	—	—	—	—	—	—	—
—	—	—	—	—	—	—	—	—	—	—	—	—
—	—	—	—	—	—	—	—	—	—	—	—	—
—	—	—	—	—	—	—	—	—	—	—	—	—
11·1	14·0	15·6	11·6	12·4	—	—	—	9·1	—	—	—	11·90
8·8	7·6	14·1	10·4	9·4	4·6	—	8·6	—	—	—	—	8·82
—	10·4	—	16·2	12·6	—	—	—	—	—	—	—	11·83
11·5	—	—	—	—	—	—	—	12·9	—	—	—	12·92
—	—	—	—	—	—	—	—	—	—	—	—	7·05
—	—	9·6	—	—	—	—	—	—	—	—	—	9·95
12·4	12·8	13·1	8·5	—	—	12·4	—	—	—	—	—	11·79
—	7·6	7·6	—	—	—	11·6	—	—	—	—	—	9·39
10·1	—	7·6	—	11·6	—	—	—	9·6	—	—	—	10·45
12·1	10·6	5·4	4·1	4·6	—	8·6	8·8	15·0	14·6	14·2	13·8	10·39
11·6	—	—	—	13·4	16·1	14·6	14·1	14·6	—	—	—	13·17
13·5	11·3	10·6	11·7	12·1	12·1	11·4	12·6	13·6	—	—	—	12·39
15·8	10·1	7·6	11·6	11·4	9·6	11·1	10·8	10·8	—	—	—	9·77
6·8	11·4	14·6	10·6	13·1	13·1	13·8	14·6	15·1	—	—	—	11·71
13·1	17·1	10·4	11·4	12·1	13·1	8·6	10·6	10·2	—	—	—	11·70
5·4	6·0	8·6	8·9	12·3	14·2	9·4	12·6	13·8	—	—	—	10·00
14·4	9·6	12·1	8·6	13·8	15·1	12·1	16·1	16·1	—	—	—	12·54
11·9	10·6	9·4	9·1	10·1	10·6	13·4	13·8	12·1	—	—	—	11·39
14·1	14·6	11·6	10·6	10·1	15·6	10·8	12·8	10·6	—	—	—	11·03
9·6	13·6	13·8	13·6	13·0	10·6	6·9	10·8	11·6	—	—	—	11·39
14·0	12·6	9·2	12·1	11·6	9·2	12·6	9·4	10·6	—	—	—	11·01
16·8	13·8	14·6	15·6	17·1	13·1	13·6	10·1	8·6	—	—	—	12·32
12·6	10·6	14·6	13·6	9·6	13·6	14·4	14·6	14·6	—	—	—	11·48
1·61	11·35	11·06	11·07	11·68	12·19	11·58	12·02	12·29	14·60	14·20	13·80	11·20
°2·90	52°·43	51°·91	51°·93	53°·02	53°·94	52°·84	53°·64	54°·12	58°·28	57°·56	56°·84	52°·16

Fort Confidence—*continued.*

Abstract of Hourly Observations in the months of November and December 1848.

							9.	10.	11.
	—	—	8·1	12·1	9·1		9·1	13·6	9·1
	—	—	—	—	9·1		8·4	9·1	10·4
	—	—	—	—	2·8		10·1	7·6	10·6
	—	—	—	—	0·1		2·1	5·6	−4·1
	—	—	—	—	−2·4		2·6	5·1	6·4
	—	—	7·6	8·1	8·2		5·4	3·7	7·0
	—	—	—	—	2·1		9·2	9·6	11·1
	—	—	—	—	—		12·4	12·4	12·3
	—	—	—	—	0·6		8·4	3·4	5·2
	—	—	—	2·1	10·3		8·8	9·4	7·5
	—	—	—	−6·4	1·8		3·6	5·1	2·7
	—	—	—	—	2·1		6·8	8·6	7·4
	—	—	2·1	6·1	5·4		8·4	8·8	9·8
	—	—	—	1·1	6·3		5·3	5·4	2·6
	—	—	—	—	1·3		5·1	4·6	7·6
	—	—	4·6	7·3	11·2		12·1	13·8	11·1
	—	—	7·8	—	10·1		9·3	10·5	10·9
	—	—	6·6	12·6	4·1		14·1	14·2	4·6
	—	—	—	—	5·1		9·5	9·3	9·1
	—	—	—	—	4·2		11·1	11·1	9·3
	—	—	—	7·6	7·2		10·6	10·1	6·1
	—	—	—	—	11·3		10·6	13·1	13·2
	—	—	—	—	11·8		14·1	7·4	8·0
	—	—	—	4·1	9·8		9·1	9·6	9·4
	7·6	3·2	1·8	−0·8	9·5		4·6	5·1	3·6
	—	—	—	—	—		10·3	3·5	3·4
	—	—	—	—	5·0		4·4	11·3	12·4
	—	—	—	—	−1·7		4·0	7·6	3·4
	—	—	—	—	−3·9		6·1	4·6	7·2
	—	—	—	5·8	3·9		7·6	2·6	3·1
	7·6	3·2	5·51	4·97	5·16		8·11	8·19	7·02
	—	—	—	—	41·29		46·60	46·74	45·72
—	—		—	—	14·12	33·44	39·92	36·50	33·44
—	—		—	20·84	14·18	28·40	46·04	39·38	33·98
—	—		—	—	11·48	31·28	31·28	32·18	37·58
—	—		—	24·98	41·72	34·16	29·48	30·74	30·38
—	—		—	16·88	39·38	33·26	34·88	30·38	31·46
—	—		—	7·88	25·52	27·50	31·28	36·32	31·82
—	—		—	—	35·78	39·38	38·12	31·28	41·72
—	—		—	—	15·96	27·68	31·64	43·88	40·28
—	—		—	—	4·28	23·72	42·44	33·48	40·22
—	—		—	20·12	36·32	33·08	36·14	39·56	39·20
—	—		23·18	26·96	34·88	29·12	31·28	31·82	39·02
—	—		—	12·92	41·18	47·48	29·66	41·18	45·32
—	—		—	26·78	42·44	37·58	25·88	32·36	35·06
—	—		—	20·48	32·90	36·14	37·58	30·56	32·36
—	—		—	—	11·48	36·50	40·64	33·44	39·56
—	—		—	—	16·88	42·98	30·02	44·78	43·52
—	—		—	13·28	26 42	42·44	36·32	36·32	29·48
—	—		4·28	15·08	22·46	31·28	23·18	18·50	23·18
—	—		—	13·28	33·08	42·98	33·26	41·00	41·00
—	—		—	—	25·70	44·78	50·18	44·78	42·26
41·00	47·12		36·50	30·38	35·42	38·48	38·12	40·64	33·08
—	—		—	—	28·04	38·48	25·88	38·48	39·38
—	—		—	—	18·68	36·32	27·32	30·92	31·28
—	—		—	—	38·48	42·08	24·08	40·28	34·88
—	—		—	—	33·80	51·98	39·56	36·96	41·72
—	—		—	—	23·00	42·62	35·24	39·56	42·26
—	—		—	—	22·28	24·08	46·04	42·08	41·72
—	—		—	—	29·12	47·12	34·88	45·68	46·04
—	—		—	—	37·58	45·14	40·64	44·78	37·04
—	—		—	—	23·00	35·78	34·16	36·32	43·52
—	—		—	—	18·68	40·28	38·66	39·74	41·36
41·00	47·12		21·32	19·22	26·92	36·95	34·96	37·06	37·52

The temperature observed in December is reduced to 32° Fahre

Fort Confidence—*continued.*

Abstract of Hourly Observations in the months of November and December 1848.

1.	2.	3.		5.	6.	7.	8.	9.	10.	
° 10·6	° 14·6	° 7·1		° 9·1	° 17·6	° 14·6	° 16·6	° 15·1	° —	
14·1	13·2	15·1		11·4	12·4	15·1	12·1	13·6	—	
11·6	10·1	10·6		8·1	11·2	13·3	14·1	9·4	—	
7·1	5·8	11·0		4·1	11·4	8·1	10·1	12·6	—	
10·6	8·4	8·1		9·6	12·6	11·8	12·1	11·2	—	
3·6	7·4	8·6		7·1	3·6	13·8	0·8	12·1	—	
16·2	10·1	10·0		7·1	11·4	12·1	10·4	10·6	—	
9·5	9·2	12·2		13·1	8·0	7·3	4·1	9·4	—	
10·1	8·8	10·8		—	16·6	14·2	10·3	12·8	15·1	
8·8	3·4	7·4		14·6	15·6	15·8	5·4	5·0	2·1	
5·4	3·2	5·4		6·4	1·6	1·4	4·4	5·1	5·2	
8·8	12·4	9·1		9·1	5·2	12·1	7·1	8·7	—	
10·6	12·4	14·1		14·1	18·1	19·8	17·1	10·2	—	
3·8	4·7	5·6		7·1	9·9	10·7	5·1	9·8	8·8	
—	19·1	12·4		9·1	9·2	7·7	9·7	7·1	—	
10·3	14·5	14·6		12·6	11·8	14·8	13·9	13·2	—	
12·8	16·7	16·1		12·7	17·6	15·1	16·4	19·1	10·6	
14·1	11·1	8·6		11·2	12·7	11·8	10·1	10·1	—	
10·2	10·2	11·1		13·4	5·6	6·1	3·4	12·5	—	
9·4	4·8	8·2		3·1	9·0	5·4	8·4	10·6	—	
8·0	16·4	9·2		9·1	8·4	9·8	11·4	12·4	—	
11·7	16·1	14·1		11·3	10·6	7·8	9·0	7·6	—	
7·6	8·6	9·9		10·5	11·4	12·6	11·1	10·0	—	
8·2	2·1	5·3		7·6	3·6	6·0	5·3	7·1	9·8	
4·2	5·8	7·4		6·8	7·4	8·6	7·6	2·6	—	
11·1	4·6	6·4		12·9	10·6	9·6	11·3	10·4	14·6	
12·5	9·4	11·9		7·1	10·1	9·6	9·6	4·6	—	
4·1	7·0	7·2		9·6	11·4	11·6	11·6	0·1	—	
−0·3	2·0	4·4		10·6	6·8	−0·4	10·4	9·6	—	
2·6	11·1	2·8		9·6	7·6	7·6	1·6	4·8	—	
8·87	9·44	9·49		9·59	10·63	10·46	9·35	9·58	9·46	
° 47·97	° 48·99	° 49·08		° 49·26	° 51·12	° 50·83	° 48·83	° 49·24	—	
° 37·40	° 24·98	° 30·92	° 43·88	° 44·78	° 43·34	° 43·88			° 46·94	° —
25·88	33·44	30·56	29·84	38·66	46·04	46·94			—	—
37·94	42·08	44·24	46·76	37·22	42·98	37·94			—	—
40·28	36·86	39·38	41·18	40·46	46·94	47·48			—	—
31·28	30·92	26·78	33·98	34·88	28·04	32·90			—	—
33·80	35·78	41·18	41·18	37·76	42·26	45·68			—	—
34·16	36·68	39·38	45·68	37·58	42·08	42·80			—	—
40·28	49·10	45·32	44·24	39·38	39·38	15·08	[illegible]	[illegible]	—	—
36·32	42·08	35·78	53·78	42·98	42·98	51·98	44·78	38·48	—	—
34·52	33·98	43·88	44·24	36·68	39·56	36·86	14·72	35·24	—	—
11·12	27·68	35·60	38·48	39·38	38·60	43·88	20·84	39·74	—	—
40·46	40·28	44·78	44·96	52·16	40·28	44·06	49·28	52·88	—	—
44·06	42·98	43·88	51·08	41·18	40·28	32·36	25·88	35·78	—	—
36·50	38·48	40·28	43·52	38·48	38·66	35·06	41·34	50·00	—	—
42·08	46·22	42·98	50·18	44·06	41·18	42·80	40·28	38·12	—	—
32·00	36·32	43·52	45·32	39·56	44·60	42·44	41·72	37·40	40·28	—
36·68	33·26	33·44	45·68	31·64	37·40	34·88	26·78	30·38	27·68	—
36·86	36·86	33·08	36·68	36·68	41·18	40·28	38·48	25·88	34·16	—
42·26	39·38	41·54	42·08	47·66	50·90	53·78	43·16	47·48	—	—
47·48	45·68	48·92	51·62	48·38	52·52	40·64	45·08	40·46	39·38	44·24
38·30	42·98	40·28	42·98	40·28	38·48	41·18	42·08	47·48	38·48	33·68
39·38	30·38	51·08	45·68	36·68	35·78	33·98	43·88	36·68	—	—
47·48	40·28	37·58	37·94	45·68	42·98	38·12	41·00	31·28	—	—
40·82	30·38	37·76	42·08	34·52	38·84	38·12	40·64	39·38	—	—
38·84	42·28	47·48	43·52	47·84	46·58	35·78	49·28	47·30	46·04	—
38·48	44·60	47·30	49·46	43·70	46·04	45·68	46·94	46·76	—	—
42·20	40·28	44·24	48·56	45·68	45·32	43·52	48·56	47·30	47·30	—
40·82	40·82	47·48	47·30	48·20	60·98	53·78	51·08	52·52	50·72	—
51·08	49·28	52·52	51·98	50·54	52·32	50·90	52·34	15·80	49·64	—
40·46	51·44	51·26	52·34	50·18	44·96	48·20	43·52	37·76	43·52	—
42·62	41·18	43·88	44·42	53·24	46·94	45·50	47·48	48·38	55·94	—
38·42	38·97	41·49	44·54	42·13	43·18	41·79	39·23	39·73		

nd corrected for deviation from Paris Standard Thermometer.

Fort Confidence—*continued.*

Abstract of Hourly Observations in the months of January and February 1849.

	C[illegible] e Thermometer, attached to Delcr[illegible]s's Barometer corre[illegible]										
	[illegible]	2.	3.	4.	[illegible]	6.	7.	8.	9.	10.	11.
	°	°	°	°	°	°	°	°	°	°	°
1	—	—	—	—	—	—	45·7	49·5	48·0	39·4	35·1
2	—	—	—	—	—	—	39·4	49·6	47·8	50·2	44·1
3	—	—	—	—	—	41·7	57·2	58·3	57·4	54·7	52·2
4	—	—	—	—	26·2	54·3	50·7	49·5	48·0	51·1	41·2
5	—	—	—	—	—	37·4	45·7	43·2	37·6	43·0	43·0
6	—	—	—	—	—	13·6	31·3	39·9	34·9	40·5	44·8
7	—	—	—	—	—	—	18·7	31·3	33·6	34·9	32·2
8	—	—	—	—	—	17·2	44·6	47·1	38·8	47·5	43·3
9	—	—	30·4	—	—	—	34·0	48·4	35·1	47·8	44·8
10	—	—	—	—	—	24·1	37·6	39·9	39·6	43·5	40·3
11	—	—	—	—	—	—	29·5	48·9	36·7	42·3	37·6
12	—	—	—	—	—	—	15·1	34·5	33·1	37·0	32·2
13	—	—	—	—	—	—	46·4	58·6	50·5	46·8	42·1
14	—	—	—	—	—	—	—	19·0	40·3	41·2	45·0
15	—	—	—	—	—	17·1	29·5	34·7	29·5	31·3	36·1
16	—	—	—	—	9·3	24·1	24·1	26·4	24·1	25·0	27·0
17	—	—	—	—	—	—	2·5	23·2	30·4	37·0	36·7
18	—	—	—	—	—	—	12·6	37·0	34·5	42·8	49·3
19	—	—	—	—	—	—	16·9	40·6	23·2	37·0	37·0
20	—	—	—	—	—	—	27·7	48·4	52·9	44·8	47·3
21	—	—	—	37·4	45·7	47·1	48·4	55·0	57·4	46·6	54·7
22	55·0	51·6	45·0	37·6	33·8	28·0	45·7	45·9	41·7	42·8	47·3
23	—	—	—	—	—	—	28·9	44·8	44·1	47·5	43·2
24	—	—	—	—	—	—	25·9	40·3	32·2	45·3	46·8
25	36·7	37·4	45·5	39·6	36·0	34·0	34·0	34·5	35·2	34·0	43·3
26	—	—	—	—	—	—	28·8	48·6	50·9	45·7	52·2
27	—	—	—	30·2	31·3	44·8	44·6	48·9	43·2	46·4	46·6
28	—	—	—	—	—	—	13·3	34·5	28·8	31·6	35·6
29	—	—	—	—	14·2	18·3	32·7	41·9	42·4	43·5	34·0
30	—	—	—	—	—	—	28·1	49·6	34·9	46·0	42·8
31	—	—	—	—	—	40·3	54·0	45·3	36·7	43·9	42·1
	45·85	44·50	40·3	36·20	28·07	32·21	33·12	42·49	39·31	42·36	41·95
1	—	—		—	—	15·1	28·8	32·4	27·0	34·2	36·3
2	—	—		—	—	12·4	16·2	34·9	31·6	39·9	30·7
3	—	—		—	18·7	35·4	39·2	39·7	32·5	39·2	39·6
4	—	—		—	—	21·9	22·6	42·4	37·4	37·0	42·1
5	—	—		—	29·8	28·2	33·8	51·6	33·3	41·0	27·0
6	—	—		32·5	30·2	26·8	27·0	43·9	38·1	38·5	41·7
7	—	—		—	18·7	16·5	37·4	43·5	31·6	34·0	36·5
8	—	—		—	26·2	21·0	34·0	47·5	34·5	42·8	44·6
9	—	—		—	—	25·7	34·3	44·8	37·0	39·4	37·8
10	—	—		—	—	13·6	21·6	30·4	31·6	36·3	39·2
11	—	—		—	—	—	10·4	28·8	29·5	37·4	37·8
12	—	—		—	12·4	11·8	29·1	34·0	25·9	28·2	31·6
13	—	—		22·3	23·4	22·3	20·5	33·8	26·4	33·1	35·8
14	—	—		—	—	22·8	35·6	39·9	46·4	43·5	48·9
15	—	—		—	—	39·2	47·5	45·0	48·7	55·4	59·4
16	—	—		—	—	46·2	55·6	58·3	50·2	56·1	54·3
17	—	—		39·9	40·3	35·8	44·8	44·6	36·0	36·7	37·9
18	—	—		—	—	—	23·4	38·8	38·5	37·0	33·3
19	—	—		—	14·0	19·8	34·3	36·0	33·8	34·7	35·6
20	—	—		—	—	4·3	28·6	26·6	23·4	23·7	24·1
21	36·3	32·4		25·9	22·6	19·6	18·7	20·7	27·3	27·1	32·7
22	—	—		—	—	10·0	23·9	25·9	25·2	27·0	30·6
23	—	—		—	—	10·0	25·2	34·3	28·6	25·5	30·2
24	35·6	34·2		32·5	30·2	27·7	22·3	24·6	30·6	29·5	30·7
25	—	—		—	—	—	−1·8	21·6	16·0	17·8	18·0
26	—	—		—	—	−0·9	6·1	20·1	8·8	18·3	24·4
27	—	—		—	—	7·0	22·6	26·4	27·0	32·2	31·3
28	—	—		—	—	14·0	16·0	39·2	32·7	34·9	37·8
				30·62	24·23		27·06	36·06	31·77	35·01	36·07

Fort Confidence—*continued.*

Abstract of Hourly Observations in the months of January and February 1849.

1.	2.	3.	4.	5.
48°·6	39°·4	48°·2	49°·6	54°·0
47·3	49·6	53·4	50·9	55·0
55·9	52·0	54·9	56·5	52·3
41·2	45·7	48·4	47·5	57·0
36·9	45·1	44·1	43·0	47·5
47·8	34·9	34·0	39·9	48·5
36·7	41·2	34·9	35·6	40·3
36·7	42·1	45·3	51·1	51·1
44·6	45·7	45·7	45·9	51·3
40·3	40·3	40·6	41·7	34·5
28·4	28·0	33·8	37·6	37·6
42·8	37·9	36·3	44·2	43·9
39·0	44·1	42·1	43·9	44·4
32·7	37·6	39·0	40·1	47·1
36·7	38·5	41·4	36·7	34·9
30·4	28·9	30·6	28·6	28·8
31·3	31·8	34·2	40·6	35·8
46·8	52·9	48·6	48·4	53·2
38·5	37·6	35·8	38·5	33·1
43·5	41·0	45·9	42·8	46·8
44·1	44·4	41·2	50·7	54·0
45·3	47·1	46·0	45·7	42·1
43·3	39·4	51·1	43·9	38·7
45·7	41·7	31·3	35·6	35·6
32·7	42·1	43·0	45·0	43·9
45·7	46·8	46·8	47·5	44·8
37·2	35·8	41·0	41·5	42·8
39·2	33·4	28·8	27·0	39·2
40·6	43·9	48·9	43·0	48·4
47·5	46·7	44·2	42·6	43·9
36·9	41·7	41·2	41·9	45·7
40·79	41·27	41·96	42·82	44·23

30·1	36·3	34·9	34·3	38·5	41·7	39·2	32·7	—	—	32·
30·7	38·5	37·4	31·8	36·5	38·3	38·8	38·8	—	—	33·
35·8	28·8	33·8	45·1	41·4	53·4	33·6	44·6	—	—	38·
39·0	45·0	46·4	45·7	45·0	51·8	46·4	53·6	—	—	42·
44·8	42·4	54·0	50·7	47·7	53·1	37·4	49·3	—	—	42·
43·9	47·5	41·9	44·8	42·8	43·2	42·8	50·0	—	—	40·
41·2	44·2	40·5	39·4	43·0	49·3	43·0	39·0	48·0	—	38·
46·6	46·8	48·2	48·9	53·6	47·1	46·9	48·7	—	—	43·
38·8	43·2	43·0	43·2	43·0	45·7	43·7	28·6	—	—	39·
39·7	43·9	41·2	49·6	50·0	41·5	44·1	41·0	—	—	37·
41·0	34·2	38·3	41·9	40·3	44·2	36·5	38·7	—	—	35·
33·1	36·7	39·6	43·9	46·0	43·3	42·1	44·2	—	—	34·
27·3	31·3	40·6	40·6	41·0	47·1	34·3	43·9	—	—	34·
48·0	48·6	51·8	52·9	54·0	53·6	53·8	56·8	—	—	48·
58·3	55·1	56·8	58·8	56·5	51·4	56·5	52·9	—	—	53·
50·7	50·2	45·0	52·7	54·1	52·3	54·7	55·4	—	—	53·
39·6	40·6	42·4	42·3	43·2	42·1	37·8	41·4	—	—	40·
31·6	34·2	41·7	37·9	42·3	40·6	42·8	43·5	—	—	37·
35·4	39·7	40·6	36·9	35·4	36·1	38·1	39·0	—	—	34·
36·3	30·9	30·6	36·3	34·5	26·6	31·3	34·3	33·8	36·3	29·
33·1	29·8	32·4	34·9	34·9	37·8	36·7	34·7	37·6	32·2	30·
28·6	36·9	32·7	25·5	28·9	37·0	33·8	32·9	—	—	29·
31·3	33·1	33·4	33·1	38·7	35·4	34·5	29·0	29·8	36·8	31·
38·1	34·5	35·4	32·7	33·3	35·8	30·0	39·6	—	—	32·
23·4	27·9	30·9	30·2	29·1	27·9	23·9	23·8	—	—	23·
27·7	27·1	32·4	30·6	31·3	35·6	32·2	30·2	—	—	24·
39·9	38·1	34·9	43·9	46·6	39·6	34·5	37·6	42·4	—	33·
50·5	47·5	51·4	43·9	47·5	33·4	30·9	45·0	—	—	39·
			41·16	42·18	42·33	39·30		38·32	34·98	

FORT CONFIDENCE—*continued.*

Abstract of Hourly Observations in the months of March and April 1849.

	4.	5.	6.	7.	8.	9.	10.	11.
	°	°	°	°	°	°	°	°
1	—	—	−3·9	3·4	4·6	4·1	4·2	2·1
2	—	—	6·8	8·4	7·6	4·1	10·4	13·6
3	—	—	−0·9	3·6	3·1	7·4	6·6	10·0
4	—	—	0·6	6·4	14·8	6·2	3·2	5·1
5	—	—	−0·4	3·1	10·1	6·7	8·2	8·4
6	—	—	1·6	5·5	12·0	8·0	11·0	13·1
7	—	—	−5·8	2·4	5·6	2·7	4·0	6·4
8	—	—	4·6	4·6	2·8	5·8	5·6	9·2
9	—	—	−8·4	−2·4	3·1	−0·6	2·8	2·8
10	—	—	−10·9	4·4	−0·4	−3·6	1·0	5·1
11	—	—	−8·4	−8·0	1·6	−0·4	−0·1	6·4
12	—	—	−8·8	0·7	1·4	2·1	5·1	8·1
13	—	—	−6·8	8·6	6·2	3·5	9·7	7·8
14	—	—	−1·4	4·7	6·0	1·2	2·9	4·1
15	—	—	−3·2	3·3	8·7	9·5	12·2	14·1
16	—	—	1·1	3·0	9·1	7·6	7·6	10·6
17	—	—	0·2	6·2	10·6	8·6	8·1	9·1
18	—	—	−6·4	7·2	7·1	4·1	7·4	6·1
19	—	—	−7·4	1·4	1·9	−0·9	3·4	4·0
20	—	—	−5·9	6·8	5·8	3·0	3·4	8·6
21	2·1	−0·4	1·9	2·3	9·6	4·6	8·6	7·8
22	0·8	−0·6	3·2	0·4	3·1	8·5	5·6	8·1
23	—	—	−4·4	8·1	9·6	6·6	6·1	7·4
24	—	—	−10·0	−1·2	4·3	2·8	12·1	8·8
25	—	—	−8·2	7·0	6·6	1·0	3·3	6·3
26	—	—	−9·9	−3·2	1·7	−1·4	2·6	6·6
27	—	—	−3·2	2·3	3·1	4·0	4·7	4·8
28	—	—	−10·4	4·8	0·7	—	—	—
29	—	—	—	—	—	14·4	9·6	8·1
30	—	—	0·2	8·3	11·5	6·9	9·1	9·4
31	—	—	−8·1	5·4	6·9	1·8	4·4	6·7
	1·45	−0·50	−3·75	3·58	5·96	4·28	6·09	7·62
	34°·61	31°·10	25°·23	38°·44	42°·73	39°·70	42°·96	45°·72

1	—	—	—	−4·9	4·8	4·3	5·6	3·1	9·2
2	—	—	—	7·4	9·3	7·3	5·1	10·1	10·3
3	—	—	—	9·9	10·7	12·4	10·9	13·2	14·4
4	—	—	—	−2·3	8·1	9·4	4·8	7·4	11·0
5	—	—	—	0·3	10·1	11·9	10·3	13·5	13·3
6	—	—	—	0·1	8·8	13·1	7·9	14·6	11·9
7	—	—	—	−1·2	10·1	8·0	8·4	9·4	6·6
8	—	—	—	2·5	13·0	13·5	7·4	8·8	8·1
9	—	—	—	−4·4	4·9	2·6	4·1	7·6	11·4
10	—	—	—	7·2	4·1	6·8	3·6	9·8	9·6
11	—	—	—	2·6	14·0	14·1	15·4	13·1	15·3
12	—	—	11·0	8·6	10·5	7·8	7·3	6·6	6·1
13	—	—	—	−3·9	3·1	8·6	7·0	7·5	7·0
14	—	—	—	−1·7	11·1	4·6	5·1	9·6	11·0
15	—	—	—	−0·4	−0·5	12·8	9·4	10·6	10·6
16	—	—	—	4·3	12·0	14·4	10·8	12·0	11·6
17	—	—	3·0	12·1	11·6	9·6	6·0	7·1	11·0
18	—	—	—	−2·4	11·0	7·5	7·1	8·2	9·0
19	—	—	—	0·7	8·2	9·1	6·2	7·4	10·0
20	—	—	2·3	11·1	10·6	9·8	9·2	12·8	10·6
21	—	—	6·6	11·0	13·4	14·2	11·6	11·4	10·8
22	—	—	—	2·8	13·0	12·4	9·4	9·1	12·4
23	—	—	—	6·1	12·2	12·6	12·1	10·6	12·1
24	—	—	2·6	8·9	14·7	12·1	10·8	13·6	13·8
25	—	—	—	5·9	14·1	12·1	12·2	14·6	13·4
26	—	—	7·6	15·7	12·8	11·6	11·3	14·0	13·6
27	—	—	—	−2·4	2·4	7·6	6·6	4·0	6·6
28	—	—	—	−5·9	−1·3	7·6	1·2	5·2	1·8
29	—	—	—	0·4	−0·1	−7·6	5·8	10·6	9·8
30	—	—	—	1·1	9·1	6·6	5·0	3·8	6·2
	—	—	5·52	2·97	8·86	9·73	7·92	9·71	10·30
	—	—	41°·94	37°·35	47°·95	49°·51	46°·26	49°·46	50°·54

Fort Confidence—*continued.*

Abstract of Hourly Observations in the months of March and April 1849.

aris Standard Thermometer.

1.	2.		4.	5.	6.	7.	8.				
°7·0	°5·6	°7·8	°9·8	°8·6	°5·4	°7·1	°7·4	°5·8		°—	°5·62
11·9	15·5	11·6	13·6	14·1	11·3	12·4	10·0	8·1		—	10·74
10·1	11·0	9·8	12·4	12·0	10·6	10·8	12·1	12·6		—	9·16
8·0	7·2	10·4	9·6	7·3	9·8	10·4	9·4	11·6		—	8·12
10·4	12·6	14·5	14·5	10·6	11·0	13·1	10·6	3·1		—	9·23
11·6	14·4	16·4	16·4	15·3	11·6	14·0	9·8	7·4		—	11·31
10·4	8·2	9·0	11·0	9·5	10·6	11·1	11·0	10·8		—	7·43
8·4	6·8	6·4	9·1	6·6	6·1	7·4	8·4	5·4		—	6·56
2·1	7·6	11·0	11·6	10·4	9·1	10·1	8·4	5·2		—	4·58
0·6	0·2	1·5	4·1	5·6	2·2	6·6	4·6	6·1		—	2·24
7·7	7·4	6·3	5·1	6·8	5·4	6·1	7·6	6·1		—	3·70
10·6	10·6	10·8	8·6	11·5	9·6	13·4	4·1	10·2		—	7·02
12·4	11·3	14·6	13·6	12·8	13·7	9·6	10·2	13·0		—	9·52
8·8	5·1	8·4	8·4	9·6	11·0	9·3	4·6	4·4		—	5·62
8·6	8·6	9·5	11·6	10·8	11·4	11·1	6·9	11·0		—	8·89
1·1	14·6	15·0	14·4	11·2	13·6	14·8	14·8	9·1		—	10·48
5·0	5·8	6·8	7·1	5·8	6·2	6·2	6·1	4·8		—	6·26
3·3	5·9	4·4	3·8	2·4	6·1	5·7	7·6	7·6		—	4·97
7·1	7·1	9·6	9·6	9·3	10·5	9·9	4·6	7·5		—	5·61
2·3	11·1	9·5	12·6	11·6	10·8	12·4	9·5	7·6		12·8	8·29
9·6	11·8	13·9	11·1	12·0	12·5	8·8	5·9	11·0		4·6	7·71
1·6	15·1	11·8	16·1	15·0	14·6	14·6	12·9	12·6		—	8·06
3·6	6·4	6·1	11·6	8·8	14·6	12·6	12·8	9·3		—	7·60
8·5	13·5	11·6	13·8	10·6	12·4	11·4	10·2	9·8		—	8·01
2·6	6·6	3·6	3·6	6·2	7·4	11·2	−4·1	5·7		—	4·04
9·6	6·9	10·7	9·6	8·6	6·6	5·6	8·4	7·4		—	4·97
7·1	6·1	8·6	10·6	9·8	7·2	8·0	7·3	7·4		—	5·87
—	—	—	—	—	—	—	—	—		—	—
7·1	10·1	7·8	9·1	12·1	12·3	7·2	14·4	13·3		—	8·36
8·8	9·9	7·3	8·8	9·1	14·0	10·0	7·3	9·3		—	8·78
6·2	8·3	11·0	12·8	10·1	14·0	14·5	10·4	12·2		—	7·82
·07	9·04	9·54	10·47	9·80	10·03	10·18	8·44	8·51		8·70	Cent.
·53	48°·27	49°·17	50°·85	49°·64	50°·09	50°·32	47°·19	47°·32			
·0	10·1	11·0	11·1	11·1	12·6	12·3	12·0	11·0	8·7		7·92
·8	12·5	11·6	9·6	12·6	14·4	13·1	11·3	9·3	10·6		10·59
·4	13·1	10·8	14·0	10·0	11·3	11·4	10·3	10·8	10·2		12·18
·1	9·9	10·8	13·8	14·2	19·1	15·8	10·6	11·4	11·7		10·36
·8	12·2	14·8	14·4	14·6	15·0	17·6	7·6	13·6	14·4		12·21
·6	12·2	13·0	14·4	14·0	15·3	13·1	13·6	12·1	13·6		11·96
·6	8·1	12·1	10·4	9·4	13·0	10·8	11·1	7·9	7·6		8·87
·6	8·2	8·7	12·2	11·4	11·4	11·3	10·6	10·2	8·6		9·47
·8	13·8	16·6	18·6	16·2	13·1	12·6	14·1	8·0	11·6		10·54
·2	12·4	13·6	18·4	15·0	14·3	15·1	15·2	12·1	11·1		10·96
·1	13·8	18·4	16·0	15·6	18·2	16·1	13·1	14·1	16·4		14·68
·4	9·6	9·8	8·1	13·6	11·6	11·6	11·0	11·2	10·0		9·61
·8	9·1	8·0	9·1	10·8	14·6	13·7	14·8	9·1	7·6		8·41
·6	13·4	13·2	12·6	11·6	14·0	15·4	14·6	11·6	12·5		10·87
·1	10·1	11·1	12·4	13·4	12·0	15·6	14·3	13·8	13·8		10·57
·8	13·1	15·2	14·4	13·6	17·6	12·6	12·4	13·6	14·1		12·62
·1	10·4	11·6	13·6	16·2	15·6	16·8	14·2	9·8	9·1		11·11
·6	9·2	8·0	10·9	11·8	11·4	10·4	12·8	9·6	7·2		9·06
8	15·1	18·2	15·1	18·2	15·9	15·8	12·0	11·6	12·8		12·45
0	13·6	14·9	16·5	14·8	14·0	16·6	16·0	16·0	18·6		13·60
5	18·5	21·4	15·6	18·0	17·8	16·4	14·6	15·4	15·4		14·98
6	11·6	11·9	13·6	14·6	16·0	16·0	14·1	9·9	8·6		11·72
6	16·8	13·9	15·8	14·6	13·6	12·6	9·6	12·6	9·8		12·82
1	11·1	11·1	12·6	11·3	13·1	14·6	13·1	10·1	12·6		11·82
4	16·4	18·0	16·5	15·6	14·1	14·8	16·2	11·8	14·0		14·25
4	14·4	13.4	12·0	10·0	9·2	7·9	8·2	2·6	3·2		10·82
4	9·1	8·0	9·6	8·1	9·0	6·0	6·2	−0·2	7·2		5·90
6	7·6	8·3	10·8	11·2	10·1	8·6	9·4	12·2	14·0		6·85
0	7·6	9·6	9·1	12·6	9·1	9·8	10·1	10·0	8·1		7·85
0	6·1	6·8	8·4	10·1	7·2	6·7	5·4	4·2	6·6		6·81
8	11·83	12·46	13·01	13·14	13·45	13·04	11·78				
5	52°·93	54°·43	55°·42	55°·65	56°·21	55°·47					

FORT CONFIDENCE.

Abstract of Hourly Observations made during the month of October 1848.

	4.	5.	6.	7.	8.	9.	10.	11.	No
	°	°	°	°	°	°	°	°	
1	—	—	—	—	—	—	—	—	
2	—	—	—	—	—	—	—	—	
3	—	—	—	—	—	—	—	—	
4	—	—	—	—	—	—	—	—	
5	—	—	—	—	—	—	—	—	
6	—	—	—	—	—	—	—	—	
7	—	—	—	—	—	—	—	—	
8	—	—	—	—	—	—	—	—	
9	—	—	29·0	29·0	29·0	29·0	29·0	29·0	
10	—	—	—	18·0	18·0	18·0	18·0	18·0	
11	—	—	22·0	22·0	22·5	23·0	—	—	
12	—	—	16·0	19·0	19·0	—	20·0	19·0	
13	—	—	—	12·0	11·2	12·0	—	14·0	
14	—	—	—	6·0	—	—	—	18·0	
15	—	—	—	—	22·0	24·0	23·2	23·0	
16	—	—	—	23·0	23·0	24·0	25·0	27·0	
17	—	—	—	16·0	16·0	—	18·0	19·5	
18	—	—	—	25·0	—	27·0	—	30·0	
19	—	—	23·0	23·0	22·0	—	—	—	
20	—	—	15·0	15·0	15·2	16·0	16·0	16·0	
21	10·0	—	—	9·0	9·5	10·5	13·0	13·0	
22	—	—	12·0	11·0	10·0	12·0	11·8	14·5	
23	—	25·0	24·6	24·0	24·0	25·3	26·4	27·6	
24	—	23·0	23·0	23·0	24·0	26·3	27·4	27·5	
25	28·0	—	27·0	27·0	26·5	26·7	27·3	28·0	
26	—	—	25·0	25·0	26·0	21·4	25·7	27·3	
27	19·4	—	20·6	21·0	21·0	24·0	25·1	25·6	
28	—	—	21·5	21·5	21·7	22·5	23·6	25·0	
29	—	—	20·9	21·0	21·2	22·0	21·8	21·0	
30	—	—	4·3	8·8	9·7	6·5	6·0	9·0	
31	—	—	—	9·0	5·3	10·4	16·0	20·0	
	19·13	24·00	20·28	18·56	18·90	20·03	20·74	21·52	
	—	—	3·72	2·00	2·34	3·47	4·18	4·96	

NOTE.—All the numbers are above zero this mo[illegible]
No corrections made this month.
Thermometer suspended in the shade, five

FORT CONFIDENCE.

Abstract of Hourly Observations made during the month of October 1848.

1.	2.	3.	4.	5.	6.	7.	8.	9.	10.
°	°	°	°	°	°	°	°	°	°
—	—	—	—	—	—	—	—	—	—
—	—	—	—	—	—	—	—	—	—
—	—	—	—	—	—	—	—	—	—
—	—	—	—	—	—	—	—	—	—
—	—	—	—	—	—	—	—	—	—
—	—	—	—	—	—	—	—	—	—
—	—	—	—	—	—	—	—	—	—
—	—	—	—	—	—	—	—	—	—
31·0	32·0	32·0	29·0	28·0	—	—	—	—	—
18·2	19·5	19·0	17·0	15·5	—	—	—	—	—
—	25·0	25·0	25·0	24·8	—	—	—	—	—
17·0	—	—	—	—	—	12·0	—	9·4	—
—	—	—	—	—	5·5	—	—	—	—
—	—	21·0	—	—	—	—	—	—	—
23·0	22·8	22·9	23·0	—	—	—	—	—	—
—	26·0	—	—	—	—	—	—	—	—
24·0	—	23·5	—	23·0	—	—	—	—	—
30·0	—	—	—	—	—	21·0	22·0	22·0	23·0
26·0	—	—	—	24·0	22·5	22·0	20·8	20·8	—
17·2	17·0	15·8	14·2	13·2	13·0	12·7	12·0	12·0	—
11·0	11·8	11·0	6·8	3·5	7·6	7·0	8·5	11·0	—
19·0	20·0	21·5	22·5	23·0	22·0	22·5	23·2	23·0	—
28·0	27·7	26·2	26·0	25·0	21·6	17·0	21·0	22·2	—
26·7	27·8	28·1	28·3	28·2	28·0	28·0	28·5	27·5	—
28·8	27·9	28·1	26·0	25·2	24·6	24·3	22·0	23·0	—
25·7	23·1	20·4	19·3	17·0	15·2	16·0	16·0	16·0	—
26·8	26·3	26·2	25·5	25·0	24·6	24·0	23·2	24·0	—
23·6	23·5	22·9	21·8	21·4	21·3	20·6	20·0	20·0	—
19·5	16·8	16·0	16·3	16·7	17·0	17·2	18·8	19·2	—
10·5	9·0	7·0	3·0	−5·0	−6·0	−6·5	−3·0	−1·0	—
20·0	18·0	16·6	16·5	16·0	15·0	14·2	18·0	19·0	—
23·42	22·01	21·29	19·96	19·32	16·56	16·80	17·93	17·82	23·00
5·86	5·45	4·73	3·39	2·76	0·00	0·24	1·37	1·26	—

for five hours in the afternoon of the 30th.

No corrections made this month.

the ground, facing the north. Fahrenheit's scale.

Fort Confidence—*continued.*

Abstract of Hourly Observations made during the months of November and December 1848.

	2.	3.			8.	9.		11.	Noon
	°	°			°	°	°	°	°
1	—	—			16·0	16·0	16·1	17·0	16·
2	—	—			−7·0	−5·0	−2·0	1·1	2·
3	—	—			−5·1	−2·5	−0·6	0·1	0·
4	—	—			−21·5	−17·5	−13·8	−10·4	−7·
5	—	—			−14·0	−12·0	−13·0	−9·0	−6·
6	—	—			−17·0	−9·0	−5·5	−4·3	−0·
7	—	—			3·0	4·3	5·0	5·5	3·
8	—	—			8·8	8·2	8·5	8·8	8·
9	—	—			3·2	4·2	6·5	7·8	7·
10	—	—			−25·0	−23·0	−21·0	−17·5	−17·
11	—	—			−30·0	−28·4	−22·5	−19·5	−15·
12	—	—			−4·0	−3·7	−0·0	−0·0	0·
13	—	—			−17·0	−20·5	−19·8	−13·8	−12·
14	—	—			−10·0	−14·0	−13·5	−13·4	−12·
15	—	—			−11·0	−9·5	−8·0	−6·8	−3·
16	—	—			10·0	11·2	14·0	14·5	15·
17	—	—			+5·5	5·0	4·8	5·9	6·
18	—	—			+8·8	13·2	14·0	17·0	18·
19	—	—			11·8	11·5	10·0	9·0	8·
20	—	—			11·0	11·3	12·2	13·5	14·
21	—	—			13·0	15·0	12·5	14·0	15·
22	—	—			10·7	9·0	6·8	7·0	11·
23	—	—			12·8	12·0	12·3	12·7	12·
24	—	9·0			8·0	2·0	−0·0	−3·0	−3·
25	−8·6	−15·0			−19·7	−20·6	−17·5	−14·3	−11·
26	—	—			−11·8	−10·7	−11·5	−12·0	−12·
27	—	—			−6·7	−6·8	−6·8	−7·0	−6·
28	—	—			−3·5	−1·5	−1·2	−2·0	−1·
29	—	—			−13·8	−15·2	−15·1	−14·0	−12·
30	—	—			−15·0	−14·0	−15·0	−16·0	−15·
	−8·60	−3·00			−3·05	−3·03	−2·14	−0·97	0·
	—	—			−4·23	−3·33	−2·34	−1·07	0·
	—	—			0·05	0·95	1·94	3·21	4
1			—	−30·8	−31·0	−29·0	−28·0	−27·8	−2
2			−36·6	36·5	35·0	36·5	36·8	33·8	3
3			—	37·3	39·0	39·5	39·5	39·7	3
4			24·0	24·0	23·0	22·5	21·9	22·2	2
5			36·8	37·6	39·5	37·5	36·0	34·0	3
6			38·0	39·8	38·8	38·4	37·0	35·0	3
7			37·0	36·0	36·8	37·8	37·0	32·0	2
8			—	27·8	28·0	31·0	32·0	29·9	2
9			—	30·0	40·5	40·8	40·4	39·6	4
10			41·0	42·5	42·4	43·0	42·3	42·5	4
11			35·5	39·0	39·0	37·0	40·0	39·2	3
12			31·0	27·4	23·5	23·3	22·5	22·2	2
13			44·0	43·0	42·5	43·5	43·5	44·2	4
14			46·0	48·8	45·5	46·3	44·7	42·5	4
15			—	37·2	36·1	35·0	34·8	35·0	3
16			—	31·3	31·0	34·2	34·3	36 5	3
17			57·0	55·0	55·0	56·1	57·0	53·0	5
18			56·5	57·0	55·0	58·1	54·0	55·0	5
19			42·0	40·0	40·0	40·5	42·0	38·0	3
20			—	9·5	9·5	8·8	8·8	8·0	
21			43·0	42·2	40·7	41·5	42·0	41·6	4
22			—	39·6	36·6	36·3	35·5	37·7	3
23			45·5	42·5	41·8	43·2	44·0	42·2	4
24			47·6	45·5	45·9	46·0	44·0	44·0	4
25			25·2	22·5	22·0	21·0	19·0	18·5	1
26			26·5	27·5	23·0	23·0	20·5	19·0	1
27			—	34·0	31·8	32·7	34·2	35·5	3
28			15·8	16·0	16·9	15·2	14·0	14·2	1
29			18·8	18·0	16·0	14·7	14·0	14·0	1
30			37·5	35·5	35·2	36·0	37·0	37·0	3
31			17·1	16·0	16·0	11·0	11·2	9·9	
			−36·56	−34·80	−34·06	−34·15	−33·82	−33·15	−3
			−40·22	−38·28	−37·47	−37·56	−37·20	−36·46	−3
			0·00	1·94	2·75	2·66	3·08	3·76	

—Thermometer suspended in the shade fa

Freezing point of mercury −36·0 in thermometer u

The whole of the reading

Fort Confidence—*continued.*

Abstract of Hourly Observations made during the months of November and December 1848.

1.	2.	3.	4.	5.		7.	8.	9.	10.	11.
16·3	16·3	14·6	13·8	12·0		9·0	7·0	5·6	—	—
4·0	4·7	5·0	5 0	5·1		2·5	2·0	1·0	—	—
0·0	−2·8	−4·4	−8·0	−9·0		−10·5	−11·5	−16·4	—	—
−7·0	−6·5	−8·5	−13·5	−15·5		−17·0	−18·5	−15·3	—	—
−6·3	−7·2	−10·0	−15·0	−17·0		−17·6	−12·5	−14·2	—	—
1·0	−0·0	−1·5	−2·5	−1·5		−2·5	−2·3	−2·0	—	—
6·0	6·0	6·0	7·0	6·0		6·5	7·8	9·5	—	—
8·6	8·0	7·6	8·5	7·5		6·0	5·0	4·2	—	—
7·5	7·2	7·0	6·4	—		2·0	2·0	2·0	—	—
−16·6	−18·0	−25·0	−27·0	−28·0		−29·8	−29·0	−30·8	−31·2	−30·0
−16·0	−17·4	−20·0	−20·8	−24·0		−23·0	−20·5	−15·3	−14·3	—
1·0	1·8	1·5	1·0	0·8		−0·0	−0·1	−2·0	—	—
−12·8	−13·4	−20·0	−16·8	−13·5		−11·0	−10·5	−10·0	—	—
−12·0	−18·0	−21·8	−22·0	−23·0		−28·0	−27·2	−23·0	—	—
−4·0	−3·0	−3·0	−3·5	−11·5		−12·0	−7·8	−6·0	—	—
15·8	16·0	17·5	18·0	19·0		16·0	15·8	14·0	—	—
7·0	7·5	7·0	6·0	5·5		−0·6	−4·0	−2·0	−2·0	—
17·5	18·0	18·5	18·0	18·7		17·0	16·0	14·0	—	—
8·0	7·2	6·0	6·0	6·0		5·7	6·1	6·9	—	—
14·0	13·8	14·0	14·4	15·0		7·5	3·5	1·2	—	—
14·0	15·0	15·2	16·0	15·2		14·0	12·5	11·9	—	—
11·0	11·5	12·0	12·0	12·0		11·5	11·6	11·0	—	—
12·0	12·9	13·2	13·0	13·0		12·0	12·0	11·5	—	—
−3·1	−3 0	−1·4	0·9	1·0		0·2	−2·0	−7·0	−6·0	−10·9
−5·6	−7·5	−5·5	−8·5	−9·8		−9·0	−9·0	−9·6	—	—
-11·5	−12·5	−12·4	−14·5	−15·0		−11·8	−11·6	−12·0	—	—
5·8	−6·0	−5·5	−5·4	−5·3		−5·2	−3·5	−3·2	—	—
−1·1	−2·0	−2·0	−2·2	−2·0		−2·5	−2·8	−3·0	—	—
-12·0	−12·0	−13·2	−14·4	−15·0		−15·6	−14·9	−13·8	—	—
-17·7	−19·0	−19·8	−21·0	−21·2		−22·0	−23·0	−24·0	—	—
0·41	−0·08	−0·96	−1·64	−2·57		−3·61	−3·51	−3·89	−13·37	−20·45
0·41	−0·09	−1·06	−1·80			−3·97	−3·86	−4·28	—	—
4·69	4·19	3·22	2·48			0·31	0·42	0·00	—	—
	−28·5	−27·0	−27·8	−27·5			−28·2	−28·5	—	
	34·5	34·3	33·5	33·5			33·8	33·4	—	
	38·5	38·7	40·0	38·5			35·5	38·5	—	
	23·0	27·0	29·5	30·5			32·0	31·8	—	
	37·8	34·9	34·8	33·8			37·5	37·7	—	
	39·0	39·0	39·4	40·0			35·0	32·5	—	
	31·5	31·0	29·0	29·2			27·7	32·0	—	
	33·0	35·0	36·5	38·8			38·0	38·5	—	
	42·0	42·2	42·0	42·5			40·5	40·0	—	
	43·0	42·5	39·0	42·4			42·0	35·5	—	
	40·8	40·5	40·5	41·0			38·4	38·0	—	
	23·0	26·5	28·0	32·2			35·8	38·0	—	
	42·0	42·5	45·0	45·1			46·0	46·0	—	
	41·0	42·0	39·5	39·5			35·2	34·5	—	
	39·5	42·0	41·4	39·0			41·1	36·0	—	
	37·5	36·5	40·2	41·0			47·2	47·0	—	
	56·8	58·0	55·0	57·2			57·5	55·5	—	
	56·5	52·5	54·5	53·8			51·0	48·0	—	
	28·0	26·5	25·0	21·0			19·5	18·0	—	
	8·6	11·1	12·4	14·6			23·8	27·6	−29·5	
	42·4	43·5	41·0	43·8			41·0	44·9	43·9	
	40·4	38·9	39·0	36·1			33·6	35·4	—	
	44·1	43·8	44·0	45·0			46·6	47·6	—	
	42·0	40·8	36·0	39·1			36·0	37·0	—	
	26·5	29·0	29·8	30·5			27·8	29·0	—	
	16·2	16·5	17·0	19·8			24·1	22·0	—	
	37·0	39·0	39·9	36·3			35·5	35·0	—	
	20·1	22·0	22·6	20·0			15·0	14·0	—	
	8·8	8·5	8·5	10·0			12·0	18·0	—	
	35·5	36·5	36·2	37·5			38·5	37·1	—	
	7·2	7·5	6·1	6·3			7·0	6·0	—	
	−33·70	−34·05	−33·97	−34·37	−34·22		−34·38	−34·29	−36·70	
	−37·07	−37 45	−37·37	−37·81	−37·64		37·82	−37·72	—	
	3·15	2·77	2·85	2·41	2·58		2·40	2·50	—	

north, five feet above the ground.
renheit's scale −. Observations recorded without correctio.
ember are −quantities.

Fort Confidence—*continued.*

Abstract of Hourly Observations made during the months of January and February 1849.

Spi

Day	3.	4.	7.	8.	9.	11.	Noon.
	°	°	°	°	°	°	°
1	—	0·5	−0·2	−1·4	−2·0	−1·0	−2·[illegible]
2	—	—	−4·5	−6·0	−4·0	−3·8	−4·[illegible]
3	—	1·0	−0·1	−1·1		−5·0	−6[illegible]
4	−14·2	−14·6	−17·0	−17·6		−16·5	−16·[illegible]
5	—	—	−27·0	−31·5		−36·5	−37·[illegible]
6	—	—	−43·0	−43·0		−40·5	−41·[illegible]
7	—	—	−40·0	−40·0		−38·5	−37·[illegible]
8	—	—	−17·5	−16·8		−14·0	−14·[illegible]
9	—	—	−18·5	−14·0		−10·8	−9[illegible]
10	—	—	−34·0	−33·2		−25·0	−25[illegible]
11	—	—	−25·0	−26·5		−34·0	−29[illegible]
12	—	—	−23·0	−20·0		−12·0	−11[illegible]
13	—	—	10·8	9·5		1·8	−3[illegible]
14	—	—	−14·5	−14·2		−9·5	[illegible]
15	—	—	−36·0	−36·0		−35·0	−3[illegible]
16	−49·5	−47·6	−51·0	−49·0		−46·0	−4[illegible]
17	—	−28·2	−28·8	−29·8		−25·2	−2[illegible]
18	−14·2	−13·6	−14·0	−13·0		−12·5	−1[illegible]
19	—	−33·6	−38·0	−39·5		−35·5	−3[illegible]
20	−6·5	−5·7	−3·6	−14·0		−17·0	−1[illegible]
21	—	5·0	17·4	19·6		20·5	[illegible]
22	6·6	2·7	−3·5	−3·5		−8·6	[illegible]
23	−5·3	−3·5	−0·5	−0·5		0·2	
24	—	−16·5	−25·8	−24·5		−19·8	[illegible]
25	−33·4	−36·3	−30·0	−28·1		−21·8	[illegible]
26	—	—	−2·3	−3·0		−6·2	[illegible]
27	−8·9	−10·4	−15·5	−16·8		−20·0	[illegible]
28	—	—	−39·6	−42·5		−36·0	[illegible]
29	—	—	−18·3	−18·2		−17·5	[illegible]
30	—	−30·2	−31·6	−32·5		−27·0	[illegible]
31	−12·2	−12·5	−20·5	−23·7		−17·6	[illegible]
			−19·20	−19·70		−18·46	[illegible]
	—	—	−21·12	−21·67		−20·31	[illegible]
	—	—	0·74	0·19		1·55	

Day								
1	—	—	−38·5	−40·5	−43·0	−43·5	−42·0	−36·0
2	—	—	−32·0	−33·9	−31·0	−28·8	−30·0	−25·2
3	—	—	—	−16·0	−14·0	−11·0	−9·0	−8·0
4	—	—	—	—	−19·5	−18·2	−15·2	−12·2
5	—	—	—	−7·2	−6·3	−5·0	−4·8	−2·5
6	—	—	−20·0	−25·2	−28·8	−26·0	−27·0	−22·0
7	—	—	—	−40·5	−33·6	−37·0	−25·0	−23·5
8	—	—	—	−10·6	−8·7	−8·8	−8·0	−8·0
9	—	—	—	—	−15·8	−13·5	−11·0	−10·0
10	—	—	—	—	−37·3	−35·7	−30·2	−29·2
11	—	—	—	—	−45·3	−42·0	−37·0	−35·0
12	—	—	—	−32·1	−26·3	−22·5	−19·0	−18·0
13	—	—	—	—	−29·3	−28·0	−25·0	−22·2
14	—	—	—	—	−2·0	0·0	2·6	7·5
15	—	—	—	—	25·0	22·5	23·5	22·2
16	—	—	—	—	21·0	22·8	22·8	22·5
17	—	—	—	+3·0	−9·0	−8·4	−8·0	−7·5
18	—	—	—	—	−6·0	−6·0	−5·5	−3·0
19	—	—	—	−21·5	−22·4	−22·5	−21·8	−21·0
20	—	—	—	—	−29·5	−30·0	−30·1	−29·5
21	−46·5	−51·5	−51·0	−51·3	−50·2	−45·0	−40·0	−38·0
22	—	—	—	—	−32·8	−31·5	−30·4	−30·5
23	—	—	—	—	−31·4	−29·8	−27·0	−26·[illegible]
24	−47·0	−46·3	−46·5	−49·0	−49·3	−43·5	−39·0	−30·[illegible]
25	—	—	—	—	−52·6	−48·5	−46·0	−40·[illegible]
26	—	—	—	—	−54·5	−50·5	−47·0	−41·[illegible]
27	—	—	—	—	−42·8	−40·0	34·8	−32·[illegible]
28	—	—	—	—	−27·5	−20·0	−16·0	−17·[illegible]
	−46·75	−48·90	−37·60	−27·07	−25·10	−23·23	−20·71	−18·[illegible]
		—	—	—	−27·61	−25·55	−22·78	−20·[illegible]
		—	—	—	0·77	2·83	5·60	7·[illegible]

Suspended in the shade.

Fort Confidence—*continued.*

Abstract of Hourly Observations made during the months of January and February 1849.

2.	7.	8.	9.	10.	
° −1·0	° −3·0	° −4·9	° −3·5	° −3·2	° −1·86
−5·5	−2·5	−2·0	−5·5	—	−4·59
−8·9	−14·0	−11·6	−11·5	—	−6·49
−18·5	−19·5	−19·8	−16·2	—	−16·77
−39·0	−41·5	−42·0	−43·0	—	−36·62
−43·0	−40·8	−43·5	−42·0	—	−42·09
−36·5	−35·2	−35·0	−33·8	—	−37·13
−15·0	−13·5	−15·0	−14·0	—	−14·73
−8·0	−12·5	−14·0	−15·5	—	−11·79
−26·8	−22·5	−22·5	−23·8	—	−25·89
−33·5	−35·0	−35·0	−35·0	−36·5	−32·25
−8·0	2·5	2·1	0·5	—	−9·39
−3·1	−11·5	−13·5	−14·0	—	−1·26
−9·9	−16·0	−22·0	−26·0	—	−14·82
−38·0	−39·5	−41·0	−45·2	—	−37·55
−44·0	−46·9	−47·5	−42·8	—	−46·35
−25·8	−17·0	−16·0	−16·0	—	−24·01
−12·0	−18·0	−20·0	−22·0	—	−14·49
−32·5	−20·8	−18·9	−17·5	−15·0	−30·67
−15·4	−17·0	−20·0	−13·5	−12·0	−12·56
25·2	22·5	22·6	23·0	23·0	20·28
−7·9	−8·0	−7·0	−3·8	−1·5	−3·50
0·4	−3·0	−2·5	−3·0	−3·5	−1·74
−21·6	−26·6	−29·7	−30·0	−30·5	−25·40
−14·0	−9·0	−6·0	−5·0	—	−21·45
−12·3	−9·8	−8·0	−7·9	—	−8·06
−23·8	−31·5	−33·0	−34·8	—	−22·25
−30·0	−28·0	−25·0	−22·5	—	−31·86
−16·8	−19·0	−18·0	−18·0	—	−17·73
−22·0	−22·0	−21·5	−18·5	—	−26·83
−21·2	−32·5	−33·0	−35·0	—	−22·52
−18·98	−19·13	−19·46	−19·22	−9·90	
−20·88	−21·04	−21·41	−21·14	—	
0·98	0·82	0·45	0·72	—	

−33·0	−33·5	−38·0	−42·0	−38·5	−38·8	−39·23
−18·0	−21·2	−26·0	−26·0	−25·0	−27·2	−27·30
−7·3	−8·0	−10·4	−13·5	−16·0	−19·5	−13·13
−9·0	−9·9	−10·6	−11·2	−11·0	−10·8	−12·96
−3·0	−7·8	−8·0	−9·4	−14·2	−19·5	−9·78
−20·5	−24·2	−29.5	−34·4	−36·5	−39·5	−29·09
−18·0	−16·5	−15·8	−12·0	−10·8	−11·6	−21·80
−6·2	−8·6	−7·8	−7·2	−7·5	−8·8	−8·57
−12·0	−16·0	−21·4	−26·5	−24·0	−26·5	−19·15
−26·0	−26·2	−29·8	−33·0	−36·0	−37·2	−32·42
−31·0	−33·0	−34·2	−37·5	−35·0	−39·0	−37·00
−17·0	−19·5	−21·0	−23·6	−29·0	−27·0	−24·37
−15·5	−13·0	−14·0	−20·5	−18·5	−16·2	−19·86
9·0	10·0	10 2	11·0	13·5	14·8	7·94
27·0	25·0	24·0	23·6	21·0	21·4	23·18
20·8	19·5	19·5	20·8	15·8	13·0	18·28
−7·8	−7·5	−7·5	−6·0	−5·5	−4·2	−5·62
−3·0	−2·5	−3·0	−3·0	−3·0	−3·0	−4·24
−20·0	−21·8	−23·8	−24·6	−26·0	−27·5	−23·44
−30·0	−31·8	−33·0	−34·5	−36·0	−37·5	−33·48
−31·0	−32·0	−32·5	−34·1	−32·0	−30·5	−38·95
−31·0	−31·2	−31·3	−33·5	−34·5	−33·5	−32·03
−25·5	−27·0	−29·3	−33·0	−36·0	−36·4	−32·96
−35·0	−35·0	−36·2	−42·5	−47·8	−50·5	−44·09
−35·7	−36·5	−38·3	−44·0	−49·0	−47·5	−44·22
−35·6	−34·6	−35·4	−37·5	−35·4	−35·0	−41·26
−31·0	−31·0	−31·8	−36·0	−38·0	−38·0	−36·42
−11·0	−10·8	−11·5	−10·0	−11·0	−22·0	−17·58
−16·30	−17·30	−18·80	−20·72	−21·64	−22·79	
−17·93	−19·03	−20·68	−22·79	−23·80	−25·59	
10·45	9·35	7·70	5·59	4·58	2·79	

Fahrenheit's scale. Observations recorded without correction.

Fort Confidence—*continued.*

Abstract of Hourly Observations made during the months of March and April 1849.

Spiri Thermometer constructed by Adie.

		2.	3.	4,		6.	7.	8.	9.	10.	11.	Noon.
	°	°	°	°		°	°	°	°	°	°	°
1	—	—	—	—		−18·0	−24·3	−20·9	−11·7	−10·9	−6·0	−4·
2	—	—	—	—		−4·2	−7·0	−8·0	−7·5	−2·9	−1·0	3·
3	—	—	—	—		−15·4	−13·3	−9·4	−5·0	−2·2	0·0	1·
4	—	—	—	—		−17·2	−17·7	−9·0	−4·0	0·2	4·0	6·
5	—	—	—	—		−1·6	−1·3	−1·0	1·0	2·5	5·0	6·
6	—	—	—	—		3·0	2·5	2·6	3·5	3·8	4·5	6·
7	—	—	—	—		−30·5	−31·0	−25·0	−20·5	−16·5	−13·8	−9·
8	—	—	—	—		−21·9	−22·5	−21·7	−20·1	−19·0	−17·0	−16·
9	—	—	—	—		−35·6	−34·0	−32·4	−28·0	−24·5	−22·8	−19·
10	—	—	—	—		−44·0	−43·4	−41·0	−31·8	−25·0	−20·0	−19·
11	—	—	—	—		−26·3	−26·3	−25·2	−23·0	−21·0	−18·3	−17·
12	—	—	—	—		−30·3	−28·4	−24·0	−21·0	−18·0	−15·2	−13·
13	—	—	—	—		−21·2	−22·0	−19·8	−17·0	−12·4	−10·5	−10·
14	—	—	—	—		−17·5	−17·2	−16·0	−13·0	−12·0	−11·5	−9·
15	—	—	—	—		−9·7	−8·2	−7·0	−3·2	−1·2	3·0	1·
16	—	—	—	—		−1·3	−1·0	1·8	4·2	8·0	7·5	10·
17	—	—	—	—		−16·5	−18·2	−16·8	−16·0	−16·0	−16·0	−15·
18	—	—	—	—		−32·5	−29·4	−26·0	−24·0	−22·2	−20·8	−18·
19	—	—	—	—		−41·5	−37·2	−32·6	−27·0	−22·8	−20·0	−18·
20	—	—	—	—		−24·8	−21·8	−17·4	−12·4	−11·2	−9·8	−9·
21	−26·0	−26·5	−29·5	−32·3		−35·3	−28·6	−27·0	−21·0	−15·8	−14·8	−12·
22	−35·2	−37·8	−37·2	−40·2		−39·8	−37·2	−32·2	−25·4	−21·5	−20·6	−15·
23	—	—	—	—		−29·0	−25·5	−24·7	−20·0	−15·8	−12·5	11·
24	—	—	—	—		−31·8	−27·0	−21·7	−15·0	−12·5	−9·8	−8·
25	—	—	—	—		−31·3	−29·4	−25·2	−21·0	−20·0	−18·3	−16·
26	—	—	—	—		−29·0	−28·5	−26·0	−20·0	−16·5	−15·2	−13·
27	—	—	—	—		−39·0	−33·2	−27·0	−20·8	−16·0	−12·0	−13·
28	—	—	—	—		−42·5	−35·3	−28·5	−21·4	−15·5	−12·5	−9·
29	—	—	—	—		−1·0	0·0	1·5	2·0	4·0	6·3	9·
30	—	—	—	—		−12·0	−4·8	−1·3	2·0	3·0	0·6	2·
31	—	—	—	—		−34·7	−30·4	−24·5	−18·2	−17·5	−12·5	−10·
·	−30·60	−32·15	−33·35	−36·25		−23·63	−21·99	−18·88	−14·69	−11·82	−9·68	−7·
tions	—	—	—	—		−25·99	−24·19	−20·77	−16·16	−13·00	−10·65	−8·7
		—	—	—		0·00	1·80	5·22	9·83	12·99	15·34	17·
1	—	—	—		—	−26·5	−19·5	−13·7	−6·5	−8·0	−5·4	−3·
2	—	—	—		—	−17·2	−9·5	−3·0	3·0	5·0	8·0	10·
3	—	—	—		—	—	5·2	6·0	16·3	17·8	19·8	14·
4	—	—	—		—	−20·2	−13·8	−11·2	−7·5	−5·0	−4·0	−2·
5	—	—	—		—	0·7	2·2	5·4	8·0	9·6	12·0	11·
6	—	—	—		—	−15·0	−10·6	−7·5	−3·6	−0·3	0·0	1·
7	—	—	—		—	−11·3	−8·0	1·0	2·0	7·5	6·5	7·
8	—	—	—		—	3·8	6·8	8·0	15·0	10·0	11·1	10·
9	—	—	—		—	−21·5	−18·0	−16·0	−10·0	−8·0	−6·9	−5·
10	—	—	—		—	−34·0	−28·2	−24·0	−16·0	−12·0	−10·0	−7·
11	—	—	—		—	−5·5	−3·5	−2·0	2·0	5·5	10·5	+6·
12	—	—	—		−9·8	−12·0	−12·5	−12·0	−11·0	−9·8	−8·0	−6·
13	—	—	—		—	−24·0	−18·0	−14·0	−10·0	−9·9	−7·5	−5·
14	—	—	—		−21·0	−18·5	−15·0	−5·5	−1·5	1·4	4·0	4·
15	—	—	—		—	−2·0	−1·8	1·2	3·0	7·0	8·3	10·
16	—	—	—		—	4·8	8·0	6·0	9·0	14·0	13·0	11·
17	—	—	—		−11·5	−11·0	−5·0	−5·0	−1·2	−1·0	−0·5	1·
18	—	—	—		—	−22·5	−17·0	−17·0	−15·0	−13·0	−11·0	−10
19	—	—	—		—	−18·9	−9·5	−7·0	−2·0	1·0	3·0	3
20	—	—	—		−6·0	−8·0	−0·0	3·5	4·8	8·0	10·0	10
21	—	—	—		−2·5	−2·0	−1·0	−0·0	−0·0	2·0	3·5	4
22	—	—	—		—	3·5	6·0	11·0	11·2	14·4	15·8	17
23	—	—	—		—	17·0	13·0	13·7	14·0	16·8	18·0	18
24	—	—	—		1·0	2·0	3·2	6·0	7·6	10·0	12·5	11
25	—	—	—		—	15·0	17·5	17·5	10·5	18·5	18·3	21
26	—	—	—		6·0	6·0	7·5	10·8	9·0	8·8	3·0	[illegible]
27	—	—	—		—	−17·5	−9·0	−4·6	1·0	3·0	3·4	5
28	—	—	—		—	−9·0	−7·8	−4·3	−0·8	3·0	4·0	5
29	—	—	—		—	−5·0	−2·5	−1·0	0·5	0·5	1·3	3
30	—	—	—		—	−7·0	0·5	2·5	10·0	9·0	7·9	11
	—	—	—		−6·26	−8·82	−4·68	−1·84	1·69	3·53	4·71	5·
	—	—	—		−6·89	−9·70	−5·15	−2·02	1·69	3·53	4·71	5·
	—	—	—		2·81	0·00	4·55	7·68	11·39	13·23	14·41	15·

Suspended in the shade.

FORT CONFIDENCE—*continued.*

Abstract of Hourly Observations made during the months of March and April 1849.

	2.	3.	4.	5.	6.	7.	8.	9.	10.	11.	Midnt.	
	-3·0	-3·0	-5·5	-9·5	-15·0	-17·2	-17·8	-15·0	-12·5	—	—	-11·59
	4·2	4·0	2·0	-0·9	-6·0	-13·5	-13·8	-13·2	-11·2	—	—	-4·18
	-0·2	-1·0	-4·4	-5·0	-9·5	-11·5	-11·0	-7·5	-5·0	—	—	-5·75
	6·5	7·5	7·0	5·8	5·0	4·8	2·0	1·0	0·4	—	—	0·48
	7·0	7·5	6·2	5·1	4·1	3·5	3·5	3·8	3·5	—	—	3·63
	2·4	1·0	1·8	-3·0	-6·5	-14·5	-17·5	-20·5	-22·2	—	—	-2·73
	-5·5	-7·0	-6·0	-7·0	-10·2	-9·8	-9·0	-12·5	-10·0	—	—	-13·73
	-14·8	-14·5	-17·0	-20·0	-24·0	-25·0	-25·0	-26·0	-25·5	—	—	-20·29
	-21·2	-20·8	-21·2	-23·5	-28·0	-30·2	-32·2	-38·0	-39·0	—	—	-27·73
	-18·0	-18·0	-21·0	-22·5	-23·5	-23·5	-23·5	-22·5	-22·5	—	—	-25·80
	-20·0	-20·0	-21·2	-23·5	-25·0	-25·0	-24·4	-24·4	-26·0	—	—	-23·72
	-13·5	-15·0	-16·0	-18·6	-20·5	-22·6	-26·5	-32·6	-31·5	—	—	-21·28
	-8·0	-8·5	-9·5	-10·5	-12·5	-14·5	-14·8	-15·0	-15·8	—	—	-13·58
	-10·0	-9·0	-10·0	-10·8	-15·6	-17·0	-18·2	-16·8	-16·8	—	—	-13·60
	4·0	-0·5	0·0	-1·2	-2·5	-2·4	-2·5	-8·0	-8·0	—	—	-2·44
	9·0	7·2	3·8	2·5	1·0	-1·5	-4·2	-4·0	-4·0	—	—	2·82
;	-16·0	-15·4	-16·0	-16·2	-20·0	-21·0	-22·5	-28·0	-30·5	—	—	-18·61
)	-17·0	-16·9	-17·5	-19·0	-21·5	-24·5	-25·2	-26·0	-30·5	—	—	-22·85
)	-16·4	-14·6	-15·8	-16·8	-21·5	-21·8	-26·0	-24·0	-24·0	—	—	-23·44
3	-6·5	-7·0	-8·0	-11·2	-16·2	-20·4	-23·0	-24·0	-24·0	-25·2	-28·0	-16·14
b	-14·0	-13·8	-13·6	-17·0	-20·0	-28·0	-31·0	-32·5	-32·5	-37·0	-35·3	-24·30
)	-16·5	-16·0	-16·0	-16·5	-20·0	-23·0	-27·2	-29·0	-30·0	—	—	[illegible]
0	-12·5	-13·0	-14·0	-17·0	-19·3	-29·2	-30·8	-34·0	-30·0	—	—	-20·66
5	-4·2	-6·5	-6·0	-7·8	-12·9	-18·8	-23·0	-25·5	-25·5	—	—	-15·52
6	-17·0	-16·6	-17·5	-18·0	-21·5	-28·2	-28·8	-30·8	-31·4	—	—	-22·71
0	-14·0	-15·5	-15·2	-17·5	-21·5	-26·0	-30·5	-31·5	-32·5	—	—	-21·44
0	-10·0	-10·5	-11·0	-13·8	-17·5	-20·0	-26·5	-31·0	-39·0	—	—	-20·66
5	-6·5	-6·5	-6·8	-8·3	-12·8	-20·5	-20·2	-25·3	-22·4	—	—	-17·65
6	10·5	11·8	12·8	9·0	7·5	6·6	6·5	4·5	4·8	—	—	6·20
8	-0·5	0·0	-2·0	-4·8	-6·5	-11·0	-16·5	-20·4	-22·0	—	—	-5·44
·8	-8·0	-7·8	-6·8	-8·0	-14·0	-21·5	-21·5	-27·0	-30·0	—	—	-17·79
45	-7·41	-7·69	-8·53	-10·50	-13·75	-16·94	-18·74	-20·51	-20·83	-31.10	-31·65	
20	-8·15	-8·46	-9·38	-11·55	-15·12	-18·63	-20·61	-22·56	-22·91	—	—	
20	17·84	17·53	16·61	14·44	10·87	7·36	5·38	3·43	3·08	—	—	
·5	1·0	0·0	0·0	-2·0	-5·0		-15·5	-19·5	-21·0			-9·09
·0	12·0	8·8	9·8	8·0	7·0		5·0	4·0	4·0			4·46
·0	7·8	8·0	4·9	3·0	1·1		-3·0	-5·0	-6·2			6·31
·2	-0·2	1·0	1·8	1·0	-3·0		-16·0	-14·0	-15·0			-7·16
2·2	12·0	11·5	7·4	7·5	5·0		-6·0	-8·0	-11·0			4·61
3·5	4·5	4·5	4·3	2·0	-1·0		-6·5	-6·5	-6·5			-2·48
6·0	6·0	6·2	7·5	5·0	3·0		3·2	3·0	3·0			3·02
1·0	4·0	2·2	1·0	-2·0	-4·8		-6·2	-9·0	-14·5			2·42
6·2	-4·5	-6·0	-7·5	-9·8	-12·0		-26·0	-22·5	-26·3			-13·08
8·0	-8·0	-7·6	-3·6	-5·8	-10·5		-14·5	-12·0	-10·0			-13·31
5·2	4·8	2·5	2·0	-2·0	1·0		0·5	1·0	-3·0			1·78
5·2	-5·0	-5·0	-5·5	-6·5	-8·0		-11·0	-12·5	—			-8·84
5·0	-4·0	-4·0	-3·5	3·5	-5·5		-18·0	-22·0	-21·8			-10·84
7·0	8·0	8·0	7·0	6·0	4·0		-3·0	-2·5	-2·5			-1·07
9·5	11·5	13·0	14·0	12·5	11·0		7·0	5·0	5·0			7·32
12·0	11·0	11·0	10·0	10·0	6·0		0·0	-4·0	-3·5			7·12
2·2	3·0	2·2	1·5	0·2	-2·5		-9·0	-15·0	-20·0			-4·31
·8·8	-7·5	-6·5	-5·5	-4·8	-6·0		-13·0	-20·0	-21·5			-12·18
4·0	5·0	5·8	6·4	7·0	6·0		0·0	1·0	-1·2			0·68
16·2	16·0	17·0	16·2	13·0	13·0		13·4	14·5	-14·0			8·24
52·0	5·5	6·2	6·0	5·0	0·0		-6·0	-8·0	-10·0			0·31
4·0	22·0	19·0	20·0	18·5	17·0		13·0	13·2	13·0			14·94
18·0	16·0	14·0	12·0	8·5	5·0		-1·0	-3·0	-2·5			10·68
12·5	12·5	11·5	11·0	13·0	11·8		9·0	9·0	9·0			9·12
24·5	23·5	24·0	20·3	17·2	14·5		9·0	5·0	4·5			10·80
6·0	4·8	6·0	5·5	3·5	1·0		-3·0	-10·0	-19·0			2·91
7·0	5·8	5·0	6·0	6·0	1·0		-2·0	-13·5	-16·5			-1·14
5·0	10·2	13·0	13·8	11·5	6·0		2·0	-4·0	-6·5			1·42
4·0	4·0	6·0	7·0	5·5	3·0		0·0	-4·0	-7·0			1·09
12·0	12·5	11·2	12·0	10·0	8·5		5·0	3·5	3·0			7·03
6·70	6·47	6·28	6·06	4·72	2·22		-3·09	-5·10	-7·19			
6·70	6·47	6·23	6·06	4·72	2·22		-3·40	-5·71	-7·91			
16·40	16·17	15·98	15·76	14·48	11·08		[illegible]	[illegible]	1·79			

Fahrenheit's scale. Observations recorded without correction.

Fort Confidence—*continued.*

peratures in the Shade for the Months at the Hours of Observation, and for the entire Period.

	2.	3.	4.	5.	6.	7.	8.	9.	10.	11.	Noon.
—	—	—	—	—	° —	° 18·56	° 18·90	° 20·03	° 20·74	° 21·52	° 22·[illegible]
—	—	—	—	—	−2·83	−3·85	−4·23	−3·33	−2·34	−1·07	0·11
—	—	—	—	—	−40·22	−38·28	−37·47	−37·56	−37·20	−36·46	−35·79
—	—	—	—	—	—	−21·12	−21·67	−21·15	−20·73	−20·31	−19·54
—	—	—	—	—	−27·41	−28·38	−27·61	−25·55	−22·78	−20·79	−19·10
—	—	—	—	—	−25·99	−24·19	−20·77	−16·16	−13·00	−10·63	−8·77
—	—	—	—	—	−9·70	−5·15	−2·02	1·69	3·53	4·71	5·71
—	—	—	—	—	—	−19·93	−18·95	−11·72	−10·25	−9·01	−7·85
—	—	—	—	—	—	−30·27	−29·88	−29·02	−27·80	−26·70	−25·62

Thermometer used stood at 36° in
Observations in this table corrected for the error

METEOROLOGICAL OBSERVATIONS.

Fort Confidence—*continued.*

Iean Temperatures in the Shade for the Months at the Hours of Observation, and for the entire Pe

1.	2.	3.	4.	5.	6.	7.	8.		10.	11.	Midn^t.	
23°·42	22°·01	21°·29	19°·95	19°·32	16°·56	16°·80	17°·93	17°·83	°—	—	—	19°·62
0·41	−0·09	−1·06	−1·80	−2·83	−3·32	−3·97	−3·86	−4·28	—	—	—	−2·40
35 78	−37·07	−37·45	−37·37	−37·81	−37·64	−37·78	−37·82	−37·72	—	—	—	−37·46
19·61	−21·88	−20·34	−21·01	−21·86	−21·09	−21·04	−21·41	−21·14	—	—	—	−20·86
18·14	−17·93	−19·03	−20·68	−22·79	−23·80	−25·59	−25·31	−25·92	−25·25	—	—	−23·29
-8·20	−8·15	−8·46	−9·38	−11·55	−15·12	−18·63	−20·61	−22·56	−23·91	—	—	−15·59
6·70	6·47	6·28	6·06	4·72	2·22	−0·27	−3·40	−5·71	−7·91	—	—	−0·82
-7·46	−7·95	−8·40	−9·18	−10·40	−11·74	−12·92	−13·47	−14·65	—	—	—	
25·33	−26·14	−26·46	−27·23	−28·29	−28·42	−29·07	−29·12	−29·20	—	—	—	

eezing mercury. Zero point correct.
' −4° between the zero point and −40° Fahrenheit.

FORT CONFIDENCE.

Abstract of Hourly Observations in the month of October 1848.

Spirit Thermometer by Adie, Fahrenheit's scale. Kept within the Observatory. Stands

	1.	2.	3.	4.	5.	6.	7.	8.	9.	10.	11.	Noon.
			°	°	°	°	°	°	°	°	°	°
1	—	—	—	—	—	—	—	—	—	—	—	—
2	—	—	—	—	—	—	—	—	—	—	—	—
3	—	—	—	—	—	—	—	—	—	—	—	—
4	—	—	—	—	—	—	—	—	—	—	—	—
5	—	—	—	—	—	—	—	—	—	—	—	—
6	—	—	—	—	—	—	—	—	—	—	—	—
7	—	—	—	—	—	—	—	—	—	—	—	—
8	—	—	—	—	—	—	—	—	—	—	—	—
9	—	—	—	—	—	—	—	—	—	—	—	—
	—	—	—	—	—	—	—	—	—	—	—	—
	—	—	—	—	—	—	—	—	—	—	—	—
	—	—	—	—	—	—	—	—	—	—	—	—
	—	—	—	—	—	—	—	—	—	—	—	—
	—	—	—	—	—	—	—	—	—	—	16·0	19·0
	—	—	—	—	—	—	—	21·5	21·5	22·0	22·0	22·3
	—	—	—	—	—	—	—	24·0	24·0	25·0	23·0	27·0
	—	—	—	—	—	—	25·0	25·0	—	18·0	28·0	—
	—	—	—	—	—	—	—	—	—	—	37·0	32·0
	—	—	—	—	—	—	—	—	—	—	—	—
	—	—	—	—	—	26·0	26·0	25·0	26·0	26·0	25·0	27·5
	—	—	—	23·0	—	—	22·0	22·0	21·5	23·0	24·0	25·0
	—	—	—	—	—	19·0	20·0	20·0	20·0	19·8	18·5	18·5
	—	—	—	—	—	24·0	24·0	24·0	25·0	27·8	28·0	28·0
	—	—	—	—	25·0	25·0	25·0	25·0	25·0	—	—	—
	—	—	28·0	28·0	—	28·0	28·0	—	—	—	29·0	29·0
	—	—	—	—	—	28·0	28·0	29·0	30·0	30·0	31·0	32·0
	—	—	28·0	27·0	—	—	26·5	27·0	28·0	29·0	30·0	31·0
	—	—	—	—	—	27·5	28·0	29·0	29·0	30·0	30·0	30·0
	—	—	—	—	—	26·5	26·5	26·8	—	—	26·0	26·0
	—	—	—	—	21·4	22·0	21·0	20·5	21·0	22·0	22·0	20·0
	—	—	—	—	—	—	13·5	14·0	14·6	15·0	16·0	17·0
	—	—	28·00	26·00	23·20	25·11	24·12	23·77	23·80	23·97	25·53	25·61
	—	—	—	—	—	1·34	0·35	0·00	0·03	0·20	1·76	1·84

All the Temperatures above zero.

FORT CONFIDENCE.

Abstract of Hourly Observations in the month of October 1848.

agnets.

3.	4.	5.	6.	7.	8.	9.	10.	11.	Midn[t].	Means.
°	°	°	°	°	°	°				°
—	—	—	—	—	—	—	—	—	—	—
—	—	—	—	—	—	—	—	—	—	—
—	—	—	—	—	—	—	—	—	—	—
—	—	—	—	—	—	—	—	—	—	—
—	—	—	—	—	—	—	—	—	—	—
—	—	—	—	—	—	—	—	—	—	—
—	—	—	—	—	—	—	—	—	—	—
—	—	—	—	—	—	—	—	—	—	—
—	—	—	—	—	—	—	—	—	—	—
—	—	—	—	—	—	—	—	—	—	—
—	—	—	—	—	—	—	—	—	—	—
—	—	—	—	—	—	—	—	—	—	—
—	—	—	—	—	—	—	—	—	—	—
21·0	—	—	—	—	—	—	—	—	—	18·67
22·0	23·0	—	—	—	—	—	—	—	—	22·03
—	—	—	—	—	—	—	—	—	—	25·67
26·0	—	23·0	—	—	—	23·0	—	—	—	24·00
—	—	—	—	—	—	32·0	—	—	—	32·50
—	—	—	29·0	29·0	29·0	29·0	—	—	—	29·00
29·0	26·5	28·0	27·0	28·0	27·2	27·2	—	—	—	26·68
22·0	21·6	21·3	21·0	21·0	20·4	20·0	—	—	—	22·05
20·0	21·0	21·0	21·0	21·0	21·2	22·0	—	—	—	20·15
26·0	27·0	26·0	27·0	27·0	27·2	26·0	—	—	—	26·56
—	—	28·0	28·0	28·0	28·0	28·0	—	—	—	25·50
28·8	28·6	28·7	28·6	27·5	28·5	28·5	—	—	—	28·46
33·0	32·5	29·0	29·0	29·0	29·0	29·0	—	—	—	30·28
31·0	31·0	30·5	31·0	31·0	30·0	29·0	—	—	—	29·24
28·0	—	—	—	—	—	—	—	—	—	29·15
25·5	25·5	25·3	25·2	25·2	25·0	25·0	—	—	—	25·72
19·2	19·0	—	17·0	13·5	—	13·0	—	—	—	19·39
18·0	17·7	17·8	17·5	17·0	18·0	18·0	—	—	—	16·65
25·11	24·85	25·33	25·11	24·77	25·77	24·98	—	—	—	24·90
1·34	1·08	1·50	1·34	1·00	2·00	1·21	—	—	—	—

All the Temperatures above zero.

Fort Confidence—*continued.*

Abstract of Hourly Observations in the months of November and December 1848.

Spirit … neter by Adie, ke … ervatory. Stands at 36°

			3.	4.		6.	7.			10.	11.	Noon.
			°	°		°	°			°	°	°
1		—	—	—		19·5	20·5			19·5	20·0	19·1
2		—	—	—		14·5	14·0			13·5	15·4	14·5
3		—	—	—		9·2	8·3			7·1	7·2	7·2
4		—	—	—		—	0·0			0·0	−0·2	1·0
5		—	—	—		−2·5	—			−3·0	−2·3	0·0
6		—	—	—		−4·0	−3·0			−3·0	−1·6	−1·0
7		—	—	—		2·3	3·0			4·0	4·0	4·0
8		—	—	—		9·5	—			10·0	—	10·0
9		—	—	—		9·0	9·0			—	—	15·0
		—	—	—		−9·0	−9·0			2·0	6·0	0·0
		—	—	—		−10·0	−10·0			−10·0	−10·0	−9·0
		—	—	—		−8·0	−8·5			−4·0	−4·0	−2·8
		—	—	—		1·0	0·0			2·0	3·0	3·0
		—	—	−0·5		−0·8	−0·8			1·8	1·8	0·0
		—	—	—		−7·5	−7·0			−4·0	−4·0	−5·0
		—	—	−2·3		−2·0	−1·0				4·7	5·2
		—	—	—		9·8	9·5				10·1	9·8
		—	—	—		7·4	7·5				12·0	9·3
		—	—	—		13·5	13·8				13·5	13·6
		—	—	—		11·3	11·0				14·0	14·0
		—	—	—		10·9	11·0				13·0	14·0
		—	—	—		13·5	13·0				15·2	15·5
		—	—	—		13·6	13·0				16·8	16·5
		—	13·8	13·6		13·5	13·0				12·0	12·1
		—	—	—		—	—				—	—
		—	—	—		—	—				0·0	0·0
		—	—	—		−2·5	−3·0				−1·2	−2·0
		—	—	—		—	0·5				1·0	1·0
		—	—	1·0		0·2	0·0				−0·8	−1·0
		—	—	—		−4·0	−4·2				−4·5	−4·2
		—	13·80			4·17	3·87				5·00	5·56
						1·54	0·24				1·37	1·93
1	—	—		—			−10·0	−10·3	−10·8	−11·1	−11·0	−11·8
2	—	—		−14·0	−14·4	−15·0	15·2	15·0	16·0	16·0	16·0	16·0
3	—	—		—	—	—	18·0	19·0	19·0	19·2	19·0	19·3
4	—	—		—	—	20·0	20·5	20·0	19·8	19·0	19·0	17·5
5	—	—		—	—	19·0	19·0	19·0	19·0	19·0	20·0	20·0
6	—	—		—	—	22·0	21·5	21·5	21·5	21·6	21·5	22·0
7	—	—		—	—	22·0	22·0	22·0	22·0	22·0	22·0	22·0
8	—	—		—	—	—	20·0	20·0	20·0	20·0	19·8	19·8
9	—	—		—	—	—	23·0	22·9	22·0	23·0	23·0	23·0
	—	—		—	—	25·0	24·4	24·6	25·0	25·0	25·0	25·2
	—	—		—	25·5	25·0	25·0	25·0	25·0	25·0	25·0	25·0
	—	—		—	—	25·5	25·5	25·0	24·8	24·0	23·5	23·0
	—	—		—	—	24·4	24·0	24·2	24·6	25·0	25·3	25·3
	—	—		—	—	29·0	29·0	29·2	29·2	29·4	29·2	29·0
	—	—		—	—	—	26·0	26·0	26·0	25·9	25·9	25·5
	—	—		—	—	—	26·8	26·5	26·5	25·4	25·5	25·4
	—	—		—	—	31·0	31·0	31·5	32·0	32·2	32·2	32·6
	—	—		—	36·6	37·0	37·2	38·0	38·0	37·0	35·0	36·0
	—	—		—	—	34·0	33·8	34·0	34·0	34·0	32·0	32·0
	—	—		—	—	—	18·6	18·0	17·8	17·8	15·5	15·[illegible]
	−9·0	−9·9		−10·6	−10·8	11·8	12·1	11·8	11·1	12·0	13·0	14·5
	—	—		—	—	—	24·0	24·3	24·3	24·5	24·4	24·4
	—	—		—	25·6	26·0	26·5	26·9	27·0	27·3	27·5	27·[illegible]
	—	—		—	—	30·2	30·0	30·8	31·0	31·0	31·0	31·0
	—	—		—	27·3	27·0	26·0	25·7	25·1	25·1	24·6	24·9
	—	—		22·0	22·0	22·0	22·0	22·0	22·0	22·0	21·5	21·0
	—	—		—	—	—	19·3	19·6	19·7	20·4	20·5	20·[illegible]
	—	—		21·8	21·2	21·0	20·0	20·1	19·9	19·0	19·0	19·[illegible]
	—	—		14·6	14·5	14·5	15·0	15·0	14·9	14·9	14·0	14·[illegible]
	—	—		—	14·2	15·0	15·0	15·0	17·0	17·0	13·8	19·[illegible]
	—	—		—	21·7	21·0	21·0	21·0	19·5	19·0	18·6	17·[illegible]
	−9·00	−9·90		−6·76	−16·67	−22·15	−21·00	−22·04	−22·03	−21·99	−21·65	−21·[illegible]
				—	−18·34	−25·36	−23·10	−24·24	−24·23	−24·19	−23·81	−23·[illegible]
				—	—	—	1·09	0·00	0·01	0·05	0·38	

Temperatures below zero mark

All the O

Fort Confidence—*continued.*

Abstract of Hourly Observations in the months of November and December 1848.

when mercury freezes. 8

1.	2.	3.	4.	5.	6.	7.	8.					
° 19·4	° 19·4	° 19·4	° 20·0	° 19·0	° 19·5	° 19·0	° 18·7	° 18·6	° —	° —	° —	° 19
14·0	14·3	17·0	14·3	14·8	15·0	16·0	15·0	14·0	—	—	—	14
7·0	7·7	8·0	7·6	7·5	7·0	7·0	7·0	6·0	—	—	—	7
1·5	1·0	1·5	1·0	2·0	2·0	—	—	1·0	—	—	—	0
0·0	−0·2	−0·2	0·0	0·0	0·0	0·0	0·0	0·0	—	—	—	−0
0·0	0·0	1·0	1·5	1·2	1·2	1·5	1·2	1·2	—	—	—	−0
5·0	5·0	6·0	6·0	6·0	6·0	6·5	6·8	7·0	—	—	—	4
10·5	10·0	10·4	10·1	10·5	10·5	10·5	10·8	10·2	—	—	—	10
15·0	15·0	16·0	15·0	—	13·0	12·0	12·0	12·0	—	—	—	12
0·0	0·0	−1·0	−1·0	−1·0	—	—	−3·0	−3·0	−5·0	−6·0	—	−2
−9·0	−8·0	−8·5	−8·5	−8·5	−8·2	—	−7·5	−8·5	−8·0	—	—	−9
−3·0	−2·0	−2·0	−2·0	−1·0	−1·0	0·0	0·0	0·0	—	—	—	−3
4·0	3·0	4·0	1·5	3·0	3·0	4·0	1·5	0·0	—	—	—	1
0·5	1·2	2·0	0·0	−1·5	−2·0	−2·4	−3·0	−4·0	—	—	—	−0
−2·5	−1·3	−2·0	−2·0	−3·0	−4·0	−3·0	−3·0	−3·0	—	—	—	−4
6·0	6·5	5·2	5·8	6·5	7·0	7·0	8·0	9·0	—	—	—	3
12·0	10·0	10·5	11·0	10·0	10·0	9·5	9·5	8·5	9·0	—	—	10
9·0	9·2	11·0	11·0	11·5	11·8	12·0	12·0	12·0	—	—	—	9
13·5	13·5	13·0	13·2	13·0	13·0	12·7	12·4	12·2	—	—	—	13
14·0	14·5	14·5	14·0	14·0	13·7	13·5	13·5	13·2	—	—	—	13
14·5	14·5	15·0	16·0	14·8	15·0	14·0	15·0	14·4	—	—	—	13
15·5	14·0	14·0	14·0	14·0	13·8	13·8	13·8	13·9	—	—	—	14
16·5	15·0	15·8	14·9	17·5	16·5	15·0	14·8	14·5	—	—	—	15
12·0	—	—	—	—	—	—	—	—	—	—	—	12
—	9·0	6·5	5·3	5·0	5·0	4·5	4·0	4·5	—	—	—	5
0·0	0·0	−0·3	−0·5	−1·0	−1·0	−1·2	−1·5	−2·1	—	—	—	−0
−1·0	−2·0	−1·2	−1·3	0·0	1·0	2·0	1·5	−0·5	—	—	—	−1
1·5	1·5	1·5	1·8	2·0	2·0	2·0	2·0	2·0	—	—	—	1
−1·5	−2·0	−2·0	−1·5	−2·0	−2·8	−2·6	−3·0	−3·0	—	—	—	−1
−4·5	−5·0	−5·1	−6·0	−6·0	−6·0	−6·5	−7·0	−7·0	—	—	—	−5
5·86	5·65	5·86	5·56	5·30	5·75	6·42	5·41	4·93	−1·33	−6·00	—	
2·23	2·02	2·23	1·93	1·67	2·12	2·79	1·78			—	—	
−12·0	−12·0	−12·0	−12·0	−12·6	−13·0	−12·6	−13·0	−13·5	—	—		−11
16·0	16·5	16·2	17·0	17·0	17·0	17·3	17·3	17·5	—	—		9
19·5	19·6	20·0	19·6	20·0	20·8	20·8	20·2	20·5	—	—		19
17·8	17·9	17·5	17·2	17·0	17·0	17·5	17·5	17·0	—	—		18
20·0	20·0	20·0	20·0	20·0	20·0	20·2	20·5	20·5	—	—		19
22·0	22·2	22·0	22·0	22·0	22·0	22·0	22·2	22·0	—	—		21
22·0	22·0	22·0	21·0	21·0	21·0	21·0	21·0	21·0	—	—		21
19·0	18·5	19·0	19·5	19·5	19·6	20·0	20·0	20·4	—	—		19
23·0	23·0	24·0	23·4	23·6	23·8	24·0	24·2	24·5	—	—		23
25·5	25·2	25·5	25·5	25·4	25·4	25·5	25·5	26·0	—	—		25
25·0	25·0	25·0	25·0	25·0	25·3	25·6	25·5	25·0	—	—		25
22·6	22·5	21·5	21·6	21·5	21·4	21·0	21·1	21·0	—	—		22
25·5	25·2	25·6	26·0	26·0	26·0	26·5	26·5	26·5	—	—		25
29·2	29·1	29·2	29·0	29·0	28·9	28·5	28·5	28·5	—	—		28
25·0	25·5	25·5	25·5	25·5	25·4	25·6	26·0	26·1	—	—		25
23·5	25·0	25·4	25·6	26·0	25·8	26·0	27·0	27·0	—	—		25
33·2	33·5	34·0	34·8	35·0	35·0	35·0	35·2	35·2	—	—		
36·0	35·0	36·0	36·5	35·0	36·0	36·0	36·0	36·2	—	—		
30·0	29·0	30·5	30·0	29·0	29·0	28·0	27·0	22·0	—	—		
15·2	11·5	9·8	9·0	8·0	8·2	7·1	7·5	7·3	−7·0	−7·5		
14·5	17·0	19·0	20·0	20·0	20·0	22·0	22·1	22·1	23·0	23·5		11
21·0	19·5	22·6	23·5	21·0	21·4	23·2	23·8	24·0	—	—		23
27·6	28·0	28·0	28·2	28·3	28·5	28·6	26·7	29·0	—	—		27
31·0	31·0	31·0	31·0	31·0	30·8	30·8	30·5	30·1	—	—		30
23·5	23·5	23·0	22·3	22·5	22·5	22·5	22·5	22·5	—	—		24
21·0	20·5	20·0	20·0	20·0	19·5	19·0	19·0	18·9	—	—		20
20·5	22·0	21·9	22·0	22·2	21·8	23·0	23·0	23·0	—	—		21
17·6	18·0	18·5	17·0	17·0	17·0	17·0	17·0	17·0	—	—		18
14·0	14·0	13·0	13·0	13·0	13·0	12·1	12·0	12·0	—	—		13
18·6	19·0	19·5	19·6	20·0	20·4	21·0	21·0	21·0	—	—		18
16·8	16·0	16·0	15·0	15·0	14·5	13·8	13·0	11·5	—	—		17
−21·42	−21·38	−21·59	−21·54	−21·38	−21·42	−21·55	−21·56	−21·35	−8·00			
−23·56	−23·52	−23·75	−23·69	−23·52	−23·56	−23.70	23·72	23·46	—			
0·63	0·67	0·44	0·50	0·67	0·53	0·49	0·37	0·71	—			

above, without a prefixed sign.
below zero in December.

FORT CONFIDENCE—*continued.*

Abstract of Hourly Observations in the months of January and February 1849.

	1.	2.	3.	4.	5.	6.	7.	8.	9.		11.	Noon.
	°	°	°	°	°	°	°	°	°	°	°	°
1	—	—	—	−9·0	—	−8·0	−7·0	−7·8	−7·5	−7·0	−7·2	−6·0
2	—	—	—	—	−4·1	−4·2	−4·0	−4·5	−4·5		−4·9	−4·0
3	—	—	—	−3·2	−3·0	−3·0	−3·0	−2·5	−2·5	−2·5	−2·5	−2·3
4	—	—	−5·6	−6·0	−6·0	−6·4	−6·8	−6·8	−9·5	−7·9	−7·8	−7·8
5	—	—	—	—	—	−11·0	−11·0	−11·5	−12·0	−11·8	−12·0	−14·0
6	—	—	—	—	—	−22·0	−23·0	−22·5	−22·5		−23·5	−24·1
7	—	—	—	—	—	—	−27·0	−27·0	−27·4	−27·5	−27·4	−27·4
8	—	—	—	—	—	−23·5	−23·0	−22·5	−21·0	−19·0	−17·5	−19·0
9	—	—	—	—	—	—	−15·5	−15·0	−15·2	−15·0	−15·0	−14·5
0	—	—	—	—	—	−15·5	−16·0	−10·0	−17·0	−17·0	−17·0	−17·0
1	—	—	—	—	—	—	−13·0	−13·5	−13·5		−18·5	−18·5
2	—	—	—	—	—	—	−23·0	−23·0	−23·5	−23·0	−22·0	−20·5
	—	—	—	—	−4·0	−3·8	−3·5	−3·0	−2·0	−1·8	−2·0	−2·0
4	—	—	—	—	—	—	−9·0	−9·0	−9·0	−9·1	−9·5	−9·0
5	—	—	—	—	−14·2	−15·0	−15·2	−16·5	−17·0	−17·0	−17·5	−17·5
6	—	—	−24·0	−25·0	−25·2	−25·8	−26·8	−27·0	−27·5	−27·5	−27·5	−27·5
7	—	—	—	−27·5	−27·0	−26·6	−26·4	−27·0	−26·0		−25·0	−25·0
8	—	—	−16·7	−16·5	−16·0	−15·5	−15·0	−15·0	−14·0	−14·0	−14·0	−14·0
9	—	—	—	−15·8	−16·2	−16·7	−17·0	−17·0	−17·0	−18·8	−19·0	−19·4
	—	—	−16·3	−16·0	−15·2	−14·0	−13·4	−13·8	−13·6	−17·5	−12·5	−12·5
	—	—	—	−9·6	−9·0	−7·8	−6·7	−3·1	−3·0	−3·0	−2·0	−1·2
	8·0	8·0	8·0	7·8	7·6	7·3	7·0	6·4	5·5		4·0	3·0
	—	—	−1·3	−2·0	−2·0	−1·8	−1·6	−2·0	−1·8	−1·5	−1·5	−1·5
	—	—	—	−2·5	−3·0	−3·3	−4·0	−4·5	−4·8	−5·0	−6·2	−5·6
	−4·9	−5·0	−6·1	−6·9	−6·5	−6·6	−6·7	−6·2	−7·0	−6·3	−5·6	−5·8
	—	—	—	—	—	−5·5	−5·5	−5·3	−5·2	−5·0	−5·0	−5·0
	—	—	−5·8	−5·7	−6·0	−6·0	−6·0	−6·0	−6·8	−7·0	−7·0	−7·0
	—	—	—	—	—	—	−16·5	−17·5	−17·8	−18·0	−17·6	−17·5
	—	—	—	—	−17·6	−17·5	−17·0	−17·0	−17·0	−17·0	−17·0	−17·0
	—	—	—	−14·5	−15·0	−15·0	−15·2	−15·6	−15·6		−16·0	−16·0
	—	—	−14·4	−14·0	−15·2	−13·6	−13·6	−13·7	−12·5	−12·5	−12·5	−12·5
	15·5	1·50	−9·16	−10·40	−10·40	−11·23	−12·40	−12·43	−12·52		−12·54	−12·52
	—	—	−10·08	−11·44	−11·44	−12·35	−13·64	−13·67	−13·77	−14·04	−14·79	−13·77
	—	—	—	—	2·35	2·44	1·15	1·12	1·02		0·00	1·02
1	—	—	—	−17·7	−18·0	−18·6	−19·2	−19·7	−20·0	−20·0	−20·0	−20·0
2	—	—	—	−22·0	−22·0	−22·0	−22·2	−22·0	−22·0	−22·0	−22·0	−21·2
3	—	—	—	—	−19·2	−18·8	−18·5	−18·4	−18·0	−17·8	−17·0	−16·0
4	—	—	—	—	—	−14·0	−14·0	−14·2	−14·0	−14·0	−14·0	−14·0
5	—	—	—	—	−10·0	−9·6	−9·5	−9·4	−10·9	−9·0	−8·5	−8·5
6	—	—	—	−10·1	−10·4	−10·8	−11·3	−11·5	−12·0	−12·0	−12·0	−11·2
7	—	—	—	—	−18·2	−18·7	−19·1	−19·3	−20·0	−19·5	−19·0	−18·0
8	—	—	—	—	−10·4	−10·4	−10·4	−10·3	−10·0	−10·0	−10·0	−9·0
9	—	—	—	—	—	−8·2	−8·3	−8·5	−8·5	−8·8	−8·8	−9·8
	—	—	—	—	—	−14·3	−14·3	−14·8	−15·2	−15·0	−15·0	−15·0
	—	—	—	—	—	—	−20·5	−20·8	−21·0	−21·0	−21·0	−20·3
	—	—	—	—	−21·4	−21·7	−21·6	−21·5	−21·8	−20·8	−20·0	−19·8
	—	—	—	—	—	−19·5	−19·5	−19·7	−20·0	−20·0	−20·0	−18·5
	—	—	—	—	—	−12·0	−11·6	−11·0	−7·0	−9·5	−9·0	−8·5
	—	—	—	—	—	7·3	7·5	7·6	9·0	8·5	10·0	10·0
	—	—	—	—	—	13 0	13·0	13·0	13·8	14·0	13·8	13·8
	—	—	—	—	9·8	9·4	1·5	6·5	6·0	4·5	3·5	2·0
	—	—	—	—	—	—	−0·5	−0·5	−1·0	−1·0	−1·2	−1·2
	—	—	—	—	−4·3	−5·7	−5·8	−6·6	−7·0	−8·0	−8·0	−8·0
	—	—	—	—	—	−17·0	−17·2	−17·6	−17·6	−18·0	−18·0	−18·0
	−21·5	−22·0	−22·5	−22·5	−23·2	−23·7	−24·3	−24·6	−25·5	−25·6	−25·4	−25·0
	—	—	—	—	—	−23·7	−24·0	−24·2	−23·5	−23·0	−23·0	−23·0
	—	—	—	—	—	−23·3	−23·3	−23·4	−23·4	−22·8	−23.0	−24·0
	−15·5	−15·0	−15·6	−16·3	−16·2	−17·3	−17·1	−17·5	−17·1	−17·4	−17·9	−17·4
	—	—	—	—	—	—	−28·0	−28·5	−28·5	−28·2	−28·0	−27·9
	—	—	—	—	—	−29·6	−30·0	−30·3	−30·5	−30·0	−29·8	−29·8
	—	—	—	—	—	−27·0	−27·0	−27·0	−27·0	−26·5	−26·0	−26·0
	—	—	—	—	—	−23·3	−23·2	−23·0	−22·2	−21·8	−21·0	−19·8
	−18·50	−18·50	−19·05	−17·72	−13·63	−14·38	−14·94	−14·90	−14·82	−14·81	−14·65	−14·40
	—	—	—	—	—	−15·82	−16·43	−16·39	−16·30	−16·29	−16·11	−15·84
	—	—	—	—	—	0·61	0·00	0·04	0·18	0·14	0·32	0·30

Temperatures below zero marked — ;

FORT CONFIDENCE—*continued.*

Abstract of Hourly Observations in the months of January and February 1849.

				6.	7.	8.	9.	10.	11.	Midnᵗ.	
				° -5·7	° -5·0	° -5·0	° -5·0	° -5·0	° -5·0	° -5·0	° -6·15
				-5·0	-5·0	-5·0	-5·0	—	—	—	-4·58
				-2·8	-2·6	-4·0	-4·4	—	—	—	-2·85
				-8·8	-9·0	-9·0	-9·0	—	—	—	-7·91
				-17·0	-18·7	-19·0	-19·2	—	—	—	-14·86
				-25·5	-25·6	-25·6	-26·0	—	—	—	-24·32
				-27·5	-27·0	-26·5	-27·0	—	—	—	-27·11
				-17·5	-16·4	-16·0	-16·0	—	—	—	-18·70
				-12·0	-12·0	-12·0	-12·0	—	—	—	-13·57
				-18·0	-18·5	-18·5	-18·8	—	—	—	-17·32
				-20·0	-20·1	-20·5	-20·8	-21·0	-21·0	—	-18·58
				-15·0	-15·0	-14·0	-14·0	—	—	—	-19·20
				-3·0	-3·5	-3·5	-4·1	—	—	—	-2·79
				-9·0	-9·5	-9·5	-10·0	—	—	—	-9·17
				-20·0	-20·0	-20·5	-21·2	—	—	—	-17·93
				-23·0	-23·4	-29·0	-23·6	—	—	—	-27·21
				-22·0	-21·5	-21·0	-20·0	—	—	—	-24·41
				-13·0	-13·0	-13·5	-13·0	—	—	—	-14·25
				-20·0	-20·0	-20·0	-19·5	-19·0	—	—	-18·71
				-12·5	-12·6	-12·4	-13·0	-13·0	—	—	-13·53
				4·3	4·9	5·0	5·0	6·5	7·5	7·9	0·33
				-1·0	-1·0	-1·9	-2·0	-1·5	-1·4	-1·4	2·94
				-1·2	-1·7	-1·2	-1·2	-1·5	—	—	-1·49
				-3·2	-3·2	-3·2	-3·4	-3·6	-4·2	-4·6	-3·97
				-9·5	-9·2	-9·2	-9·0	—	—	—	-7·25
				-5·5	-5·5	-5·0	-5·0	—	—	—	-5·14
				-9·0	-9·5	-10·0	-11·0	—	—	—	-7·37
				-17·5	-17·5	-17·8	-18·0	—	—	—	-17·45
				-14·0	-14·0	-14·0	-14·0	—	—	—	-15·83
				-15·8	-15·5	-15·5	-15·6	—	—	—	-15·52
				-13·0	-13·5	-14·0	-14·0	—	—	—	-13·29
				-12·51	-12·55	-12·62	-12·74	-7·26			
				-13·76	-13·80	-13·88	-14·01	—			
				1·03	0·99	0·91	0·78	—			
		-20·0	-20·0	-20·0	-20·0	-20·8	-20·8	-21·0	—	—	-19·74
		-19·9	-19·2	-19·0	-19·8	-20·0	-19·0	-19·2	—	—	-20·76
		-14·0	-13·0	-14·0	-14·0	-13·0	-13·2	-13·5	—	—	-15·76
		-13·0	-12·5	-11·0	-12·0	-12·0	-11·5	-11·0	—	—	-13·01
		-7·5	-7·5	-7·4	-7·4	-7·5	-8·0	-8·0	—	—	-8·48
		-11·0	-10·9	-11·5	-12·0	-12·5	-13·0	-13·8	—	—	-11·53
		-16·0	-15·0	-14·7	-14·2	-14·0	-13·9	-13·0	—	—	-16·87
		-7·5	-8·0	-7·8	-7·8	-7·5	-7·5	-7·0	—	—	-8·78
		-8·0	-8·0	-8·0	-9·0	-9·0	-9·4	-10·0	—	—	-8·55
		-15·8	-16·0	-16·0	-16·0	-16·8	-17·0	-17·0	—	—	-15·61
		-19·0	-19·5	-19·2	-19·8	-19·8	-20·0	-20·5	—	—	-20·26
		-17·0	-17·0	-17·0	-17·0	-17·0	-17·0	-17·0	—	—	-19·02
		-17·0	-16·0	-16·0	-16·2	-16·0	-15·0	-15·0	—	—	-17·79
		-4·8	-4·0	-3·0	-2·0	2·0	1·0	1·0	—	—	-5·61
		11·2	11·5	12·0	12·0	12·2	12·2	12·5	—	—	10·29
		14·2	14·0	14·7	14·5	14·0	13·8	13·0	—	—	13·80
		0·5	0·5	-0·2	-0·2	-0·2	0·0	0·0	—	—	2·70
		-1·1	-1·1	-1·1	-1·1	-1·1	-1·3	-1·4	—	—	-1·06
		-10·0	-10·4	-10·5	-11·5	-12·0	-12·0	-12·0	—	—	-8·88
		-18·0	-18·5	-19·1	-19·5	-20·0	-20·5	-20·4	-20·4	-20·4	-18·64
		-24·0	-23·0	-23·0	-23·0	-24·0	-23·0	-23·0	-23·5	-23·0	-23·68
		-23·5	-23·5	-23·5	-23·5	-24·0	-24·0	-24·0	—	—	-23·52
		-16·6	-15·1	-13·2	-14·0	-13·8	-13·3	-13·3	-14·4	-15·9	-18·58
		-18·2	-19·0	-19·0	-20·0	-20·0	-21·0	-22·0	—	—	-17·82
		-25·2	-25·4	-25·6	-25·6	-26·0	-26·0	-26·4	—	—	-26·71
		-27·2	-27·0	-27·0	-27·0	-27·0	-27·0	-26·7	—	—	-28·44
		-23·0	-22·0	-22·0	-22·0	-22·0	-22·0	-22·0	-22·0	—	-24·18
		-15·0	-14·0	-14·0	-13·6	-13·9	-14·0	-14·0	—	—	-17·95
		-13·05	-12·84	-12·72	-12·92	-12·92	-12·94	-13·02	-20·08	19·77	
		-14·35	-14·12	-13·99	-14·21	-14·21	-14·23	-14·32	—	—	
1·00	1·53	2·08	2·31	2·44	2·22	2·22	2·20	2·11	—	—	—

those above, without a prefixed sign.

Fort Confidence—*continued.*

Abstract of Hourly Observations in the months of March and April 1849.

Spirit Thermometer by Adie, kept within the Observatory. Stands at 36° when

	1.	2.	3.	4.	5.	6.	7.	8.	9.	10.	11.	Noon.
	°	°	°	°	°	°	°	°	°	°	°	°
1	—	—	—	—	—	−14·0	−15·0	−15·0	−14·0	−13·0	−10·5	−10·0
2	—	—	—	—	—	−8·5	−8·5	−8·4	−7·0	−5·0	−3·5	−2·5
3	—	—	—	—	—	−4·5	−5·0	−5·0	−3·0	−1·0	0·0	2·2
4	—	—	—	—	—	−2·3	−2·6	−3·0	−3·2	−3·1	−2·5	−1·2
5	—	—	—	—	—	−0·2	−0·2	0·0	3·4	3·4	4·0	4·2
6	—	—	—	—	—	3·5	3·0	3·0	4·8	5·0	6·0	6·2
7	—	—	—	—	—	−4·7	−5·6	−6·3	−6·8	−3·5	−2·0	−2·0
8	—	—	—	—	—	−8·0	−8·8	−8·6	−9·5	−8·5	−6·5	−5·4
9	—	—	—	—	—	−13·7	−14·4	−15·0	−11·0	−11·2	−9·2	−8·0
10	—	—	—	—	—	−17·6	−18·3	−18·5	−18·3	−17·0	−16·5	−16·0
11	—	—	—	—	—	−15·8	−16·2	−16·0	−16·2	−16·4	−15·3	−15·0
12	—	—	—	—	—	−16·6	−16·8	−17·4	−13·0	−11·5	−10·2	−10·0
13	—	—	—	—	—	−15·4	−15·4	−15·0	−11·5	−9·4	−8·2	−7·0
14	—	—	—	—	—	—	−11·5	−11·6	−10·0	−10·0	−8·0	−6·0
15	—	—	—	—	—	−8·8	−8·7	−8·0	−8·0	−7·2	−6·2	−5·1
16	—	—	—	—	—	−3·5	−3·3	−3·2	−3·0	−2·0	−2·0	−1·0
17	—	—	—	—	—	−0·7	−1·5	−1·9	−1·8	−2·0	−2·0	−1·5
18	—	—	—	—	—	−2·5	−14·0	−14·2	−15·0	−14·0	−13·8	−13·0
19	—	—	—	—	—	−17·7	−18·2	−17·8	−16·5	−15·8	−15·0	−13·5
20	—	—	—	—	—	−16·2	−16·7	−15·2	−12·0	−10·0	−10·0	−6·8
21	−9·0	−9·8	−10·1	−10·8	−11·4	−12·2	−12·7	−12·8	−10·8	−9·0	−8·8	−6·0
22	—	—	—	—	—	—	—	—	—	—	—	—
23	—	—	—	—	—	−13·5	−13·5	−13·5	−11·2	−11·2	−11·2	−11·0
24	—	—	—	—	—	−13·2	−13·5	−13·6	−13·0	−13·0	−10·2	−9·1
25	—	—	—	—	—	−13·2	−13·5	−13·2	−12·8	−12·4	−11·5	−10·3
26	—	—	—	—	—	−13·0	−13·4	−13·3	−12·8	−11·5	−10·2	−10·0
27	—	—	—	—	—	−14·5	−15·2	−14·9	−13·0	−13·0	−12·0	−10·0
28	—	—	—	—	—	−15·0	−15·5	−15·2	−14·5	−13·0	−12·5	−10·5
29	—	—	—	—	—	−8·5	−6·8	−5·2	−4·6	−3·0	−1·5	−0·2
30	—	—	—	—	—	0·8	0·3	0·2	0·1	0·5	0·6	0·1
31	—	—	—	—	—	−9·0	−9·6	−9·5	−7·6	−8·0	−7·5	−6·6
	−9·00	−9·80	−10·10	−10·80	−11·40	−9·60	−10·37	−10·27	−9·06	−8·19	−7·21	−6·18
	—	—	—	—	—	−10·56	−11·41	−11·30	−9·97	−9·01	−7·93	−6·80
	—	—	—	—	—	0·85	0·00	0·11	1·44	2·40	3·48	4·61

	1.	2.	3.	4.	5.	6.	7.	8.	9.	10.	11.	Noon.
1		—	—	—	—	−10·0	−10·0			−9·2	−7·2	−6·1
2		—	—	—	—	−6·0	−6·7			−5·0	−4·0	−3·0
3		—	—	—	—	—	3·0			5·5	6·5	7·3
4		—	—	—	—	−1·0	−1·5			−1·0	−1·0	2·0
5		—	—	—	—	0·8	1·0			5·6	7·0	8·2
6		—	—	—	—	3·0	2·3			—	8·0	10·0
7		—	—	—	—	3·0	2·2			11·0	13·5	7·9
8		—	—	—	—	5·0	5·0			7·0	8·0	8·5
9		—	—	—	—	−1·0	−1·5			2·0	3·5	4·0
		—	—	—	—	−8·0	−9·0			−3·0	−2·5	−2·0
		—	—	—	—	−5·0	−5·0			0·0	1·0	2·0
		—	—	—	−0·2	−0·2	−4·8			−4·0	−3·0	−2·0
		—	—	—	—	−8·0	−8·0			−8·0	−5·0	−4·0
		—	—	—	—	−4·8	−6·0			−0·8	0·1	3·0
		—	—	—	—	0·0	0·0			4·0	6·0	7·[illegible]
		—	—	—	—	6·0	6·0			9·0	9·0	12·0
		—	—	—	5·0	4·5	4·0			6·1	7·0	7·0
		—	—	—	—	−1·6	−1·0			−2·0	−2·0	−2·0
		—	—	—	—	−7·0	−7·0			−1·0	2·2	—
		—	—	—	3·0	5·0	5·5			8·0	7·0	10·0
		—	—	—	8·8	8·0	10·0			9·6	13·0	15·0
		—	—	—	—	7·5	7·0			9·0	9·0	10·2
		—	—	—	—	14·2	15·0			18·6	20·0	20·0
		—	—	—	7·9	7·2	10·0			9·0	9·2	10·2
		—	—	—	—	12·0	14·0			15·0	17·5	[illegible]
		—	—	—	16·0	15·0	17·0			19·0	17·0	19·0
		—	—	—	—	5·0	4·5			7·0	8·0	9·0
		—	—	—	—	1·0	1·0			3·0	4·5	6·[illegible]
		—	—	—	—	1·0	1·0			5·2	6·5	[illegible]
		—	—	—	—	2·0	2·0			5·0	6·0	7·0
		—	—	—	6·75	1·64	1·07			4·30	5·40	
		—	—	—	—	0·00	0·03			2·66	3·85	

Temperatures above zero without a prefixed sign.

FORT CONFIDENCE—*continued.*

Abstract of Hourly Observations in the months of March and April 1849.

1.	2.	3.	4.	5.	6.	7.	8.	9.	10.	11.	Midn^t.	
°	°	°	°	°	°	°	°	°	°	°	°	
−5·0	−4·5	−3·0	−0·0	−5·5	−6·0	−6·0	−6·0	−6·5	−6·5	—	—	−8
−0·8	−1·0	0·8	0·5	−0·0	−0·0	−0·0	−0·2	−0·5	−0·5	—	—	−2
3·0	4·8	2·0	2·0	2·0	1·5	1·1	1·0	1·0	1·0	—	—	0
−1·0	−0·2	0·0	0·5	1·2	2·0	2·2	2·0	1·8	−1·6	—	—	−0
5·0	6·0	6·0	6·0	4·0	3·9	3·5	3·5	3·5	3·5	—	—	3
8·0	8·2	8·0	5·4	5·0	5·0	4·4	4·0	3·5	3·0	—	—	5
−1·0	0·0	−1·0	−2·0	−2·0	−1·8	−1·8	−2·0	−2·5	−3·0	—	—	−2
−4·0	−6·0	−6·0	−6·0	−6·0	−6·0	−8·0	−8·0	−9·0	−10·0	—	—	−7
−7·0	−6·0	−3·5	−8·0	−8·0	−8·8	−9·5	−9·8	−10·5	−11·0	—	—	−9
−15·5	−15·0	−15·0	−15·0	−14·5	−14·0	−14·0	−14·0	−14·5	−15·0	—	—	−15
−13·5	−13·4	−13·0	−12·5	−12·5	−12·0	−13·0	−13·2	−14·0	−14·0	—	—	−14
−10·5	−10·0	−10·0	−10·0	−10·0	−10·8	−11·0	−11·0	−11·5	−12·0	—	—	−11
−7·0	−6·5	−6·5	−7·0	−7·0	−7·5	−7·5	−8·0	−8·2	−8·5	—	—	−9
−5·0	−4·5	−1·0	−4·0	−4·6	−5·0	−6·0	−7·0	−7·4	−8·0	—	—	−6
−4·8	−3·0	−3·0	−3·0	−3·0	−2·5	−2·0	−2·0	−2·2	−3·0	—	—	—
0·8	1·8	2·5	3·0	3·5	3·5	4·0	4·0	3·0	3·0	—	—	
−1·0	−1·0	0·0	−0·4	−1·0	−1·0	−1·5	−2·0	−3·0	−3·5	—	—	—
−13·0	−13·0	−12·5	−12·5	−12·5	−11·5	−12·0	−12·0	−12·5	−13·0	—	—	−1
−13·0	−12·0	−12·0	−12·0	−10·0	−10·0	−11·0	−10·0	−10·0	−11·0	—	—	−1
−6·5	−4·8	−3·0	−1·0	−4·0	−5·0	−5·0	−5·0	−6·0	−7·0	−7·8	−8·0	—
−6·0	—	—	−2·0	−1·0	—	—	0·5	—	—	—	—	—
—	−6·5	−6·5	−6·3	−6·0	−7·0	−7·0	−7·0	−7·6	−8·5	—	—	—
−8·5	−7·0	−5·8	−4·0	−3·8	−3·4	−4·8	−6·0	−7·0	−7·8	—	—	—
−7·5	−7·6	−7·0	−7·0	−7·0	−7·0	−7·0	−7·0	−7·5	−7·8	—	—	—
−9·5	−8·0	−7·5	−7·5	−6·0	−6·0	−6·2	−7·0	−7·4	−8·0	—	—	—
−8·2	−8·0	−7·0	−6·7	−6·5	−6·0	−6·0	−7·0	−7·5	−8·2	—	—	—
−9·0	−8·0	−6·5	−6·5	−5·5	−5·2	−5·5	−6·0	−7·0	−7·0	—	—	—
−7·0	−7·0	−6·0	−5·5	−5·0	−5·0	−5 0	−5·0	−5·2	−5·4	—	—	—
1·0	3·0	3·0	4·0	4·0	3·6	4·0	4·2	4·0	3·5	—	—	
2·0	2·0	2·0	2·5	2·0	1·5	1·0	0·0	−0·2	−0·8	—	—	
−5·0	−4·0	−2·0	−1·5	−1·0	−0·8	−0·8	−1·0	−1·8	−2·0	—	—	—
−4·98	−4·37	−3·78	−3·76	−3·89	−4·04	−4·35	−4·42	−5·09	−5·53	−7·80	−8·00	
−5·48	−4·81	−4·16	−4·14	−4·28	−4·44	−4·78	−4·86	−5·60	−6·08	—	—	
6·93	5·60	7·25	7·27	7·13	6·97	6·63	6·55	5·81	5·33	—	—	
								°				
−4·5	−3·0							−0·5	−0·8	—	—	
−2·0	−1·0							3·0	2·6	—	—	
9·2	9·0							5·0	5·0	—	—	
3·0	4·5							6·0	5·0	—	—	
14·0	12·5							12·0	11·5	—	—	
10·0	12·0							9·5	9·2	—	—	
8·0	8·0							7·0	7·0	—	—	
9·0	9·9							8·5	7·0	—	—	
4·0	6·0							3·5	2·0	—	—	
0·0	−0·5							−1·0	−2·0	—	—	
3·1	4·0							4·2	4·0	—	—	
−0·8	0·5							0·0	—	—	—	
−3·6	−1·0							2·0	1·2	—	—	
7·5	7·5							5·0	4·0	—	—	
9·9	10·0							9·0	8·5	—	—	
12·0	13·0							10·0	9·5	—	—	
7·5	9·0							8·0	7·5	—	—	
0·0	1·0							1·5	1·0	—	—	
—	—							7·0	6·0	—	—	
9·0	9·0							12·0	12·0	12·0	12·0	
14·0	14·0							14·0	12·0	—	—	
10·5	11·0							14·0	14·0	—	—	
19·5	18·0							13·0	11·5	—	—	
10·0	12·0							12·0	12·0	—	—	
19·5	20·0							20·0	20·0	—	—	
18·0	18·0							16·0	14·0	—	—	
10·0	10·0							11·5	10·2	—	—	
7·0	8·0							9·5	8·5	7·0	7·0	
8·0	8·0							7·3	7·5	—	—	
7·5	8·0							10·0	10·0	—	—	
7·56	8·19							7·97	7·58	9·50	9·50	
5·92	6·55							6·33	5·94	—	—	

Temperatures below zero marked −.

Table I.

Directions of the Winds at Fort Confidence within 30 feet of the Ground.

Direction.	October.	November.	December.	January.	February.	March.	April.	Seven Months.
	Hours.	Hours.	Hours.	Hours.	Hours.	Hours.	Hours.	Hours.
North.	3	12	8	11	11	10	2	57
N. by E.	8	9	6	4	2	1	5	35
N.N.E.	5	15	6	24	7	8	7	72
N.E. by N.	21	11	9	9	1	7	3	61
N.E.	38	90	41	29	42	42	21	303
N.E. by E.	34	28	12	3	—	6	4	87
E.N.E.	23	41	45	28	14	16	14	181
E. by N.	16	28	75	20	7	21	6	173
East.	29	57	114	62	102	97	86	547
E. by S.	15	28	48	23	7	45	8	174
E.S.E.	36	84	18	39	13	30	78	298
S.E. by E.	9	17	2	5	5	16	8	62
S.E.	27	24	12	13	13	32	34	155
S.E. by S.	3	—	3	4	—	1	1	12
S.S.E.	—	2	4	—	2	4	2	14
S. by E.	—	—	1	1	—	1	—	3
South.	—	1	2	4	6	1	—	14
S. by W.	—	—	3	—	2	1	1	7
S.S.W.	—	—	1	9	2	5	1	18
S.W. by S.	—	—	4	5	5	—	—	14
S.W.	—	2	15	27	16	7	6	73
S.W. by W.	—	—	1	7	2	—	3	13
W.S.W.	—	—	8	28	12	10	14	72
W. by S.	—	—	—	7	6	22	16	51
West.	4	13	4	35	54	61	125	286
W. by N.	—	15	8	26	16	17	15	97
W.N.W.	3	3	3	24	39	6	13	91
N.W. by W.	—	—	—	3	2	—	—	5
N.W.	2	3	2	33	34	—	25	99
N.W. by N.	—	—	1	4	3	—	2	10
N.N.W.	4	2	2	5	1	3	5	22
N. by W.	1	4	3	7	1	1	3	20
Calm	27	15	49	58	75	63	7	294
Hours of Observation	308	504	510	557	502	534	515	3,430 3,136
Mean Direction	N. 70° E. or E. by N. ¾ N.	S. 84½° E. or E. ½ S.	N. 80° E. or E. by N.	S. 16° E. or S. by E. ½ E.	S. 13° E. or S. by E. ¼ E	S. 45° E. or S.E.	S. 7° E. or S. ¾ E.	S. 53° E. or S.E. ½ E.

Of 3,430 hours of observation 294 were calm, and in 3,136 there was wind of various strength, from a storm down to an air just sufficient to move a light vane. For the mean strength of the winds, see the following Table (III.)

Table II.

Table of the Mean Force of the Winds at Fort Confidence.

Direction.		November.	December.	January.	February.	March.		
North.		1·17	0·75	0·91	1·00	1·80	2·00	1·23
N. by E.		1·78	2·17	1·12	1·00	2·00	1·20	1·45
N.N.E.	1·20	2·13	0·67	0·88	1·00	1·88	1·14	1·27
N.E. by N.		2·36	2·89	1·11	1·00	2·29	1·34	1·90
N.E.	3·68	2·45	1·18	1·24	1·01	1·98	1·67	1·89
N.E. by E.	2·09	3·57	1·37	3·67	—	1·16	2·25	2·35
E.N.E.		1·56	1·53	0·55	1·64	2·31	2·79	1·77
E. by N.	3·56	2·11	1·41	1·97	1·14	1·21	2·17	1·94
East.	3·90	2·05	1·02	1·44	1·41	1·64	1·78	1·89
E. by S.	3·07	2·64	2·13	2·09	1·00	1·96	1·38	2·04
E.S.E.	3·72	3·13	1·61	2·56	2·85	2·45	2·33	2·66
S.E. by E.	8·22	4·24	3·50	3·20	4·40	5·69	1·38	4·38
S.E.	4·48	2·60	2·50	2·46	1·30	2·30	3·09	2·65
S.E. by S.	2·14	—	1·67	3·75	—	3·00	5·00	3·11
S.S.E.		3·00	1·25	—	0·50	3·25	9·50	3·50
S. by E.		—	2·00	3·00	—	3·00	—	2·67
South.		1·00	3·50	3·75	0·83	3·00	—	2·42
S. by W.		—	1·33	—	1·00	3·00	5·00	2·58
S.S.W.		—	0·50	2·55	2·00	2·60	1·00	1·73
S.W. by S.		—	1·50	3·20	2·60	—	—	2·43
S.W.		0·75	1·80	3·00	3·69	0·86	2·50	2·10
S.W. by W.		—	4·25	4·00	1·00	—	4·33	3·39
W.S.W.		—	1·06	4·75	2·00	1·40	3·00	2·44
W. by S.		—	—	3·86	2·58	1·78	3·12	2·84
West.	1·50	3·15	1·37	4·62	2·26	2·44	3·09	2·63
W. by N.		2·87	1·38	4·19	3·68	2·76	1·80	2·73
W.N.W.	3·67	2·33	1·33	3·50	4·68	1·67	3·38	2·94
N.W. by W.		—	—	6·33	8·50	—	—	7·41
N.W.		0·67	1·00	4·85	7·44	—	3·12	3·03
N.W. by N.		—	0·50	6·75	8·67	—	2·50	4·60
N.N.W.	2·00	1·25	0·50	3·10	1·00	2·00	2·00	1·69
N. by W.	1·00	2·00	1·67	1·29	1·00	3·00	5·33	2·18
Mean Force	2·99	2·47	1·33	2·46	2·01	1·91	2·48	
Calm Hours -		15	49	58	75	63	7	

The force of the wind is denoted by figures, as recommended by Rear-Admiral Sir Francis
Thus, 12 denotes a hurricane, 11 a storm, 10 a whole gale, and 1 a light breeze, j
will be observed, by looking at Table, that though the N.E., East, and E.S.E. wi
they were comparatively light, and that the N.W. winds were stronger.

fort, K.C.B.
perceptible. It
were most frequent,

TABLE III.

Table of the Mean Extent of Cloudy Sky at Fort Confidence for each Month, and for Seven Months, with the Number of Hourly Observations.

Periods.	October.	November.	December.	January.	February.	March.	April.	Seven Months.
Proportions of cloudy sky	8·47	6·25	2·34	4·87	4·65	3·74	4·36	4·95
No. of observations	241	492	510	557	502	534	515	3,351

NOTE.—A sky totally covered with clouds, whether rare or dense, or obscured by mist or snow, so that the blue sky is wholly hidden, is denoted by 10·00.

LONDON:
Printed by George E. Eyre and William Spottiswoode,
Printers to the Queen's most Excellent Majesty.
For Her Majesty's Stationery Office.

Lightning Source UK Ltd.
Milton Keynes UK
UKHW021247080119
335202UK00013B/636/P